WORDS INTO TYPE

WORDS INTO TYPE

Based on Studies by
Marjorie E. Skillin, Robert M. Gay
    and other authorities

THIRD EDITION,
Completely Revised

PRENTICE-HALL, INC.
Englewood Cliffs, New Jersey

62574218

OPTOMETRY

Library of Congress Cataloging in Publication Data
Skillin, Marjorie E.
  Words into type.
Includes bibliographies.
  1. Authorship—Handbooks, manuals, etc. 2. Printing,
Practical—Style manuals. I. Gay, Robert Malcolm, 1879–
1961, joint author. II. Title.
PN160.S52 1974   808'.02   73-21726

PN160
S521
1974

© 1974 by Prentice-Hall, Inc., Englewood Cliffs, New Jersey

Printed in the United States of America

10   9   8   7   6   5   4   3   2

ISBN Number: 0-13-964262-5

PRENTICE-HALL INTERNATIONAL, INC., LONDON
PRENTICE-HALL OF AUSTRALIA, PTY. LTD., SYDNEY
PRENTICE-HALL OF CANADA, LTD., TORONTO
PRENTICE-HALL OF INDIA PRIVATE LIMITED, NEW DELHI
PRENTICE-HALL OF JAPAN, INC., TOKYO

CONTENTS

xii

xiv

PREFACE

In the years since its first publication, *Words Into Type* has become a classic among style manuals, an invaluable reference source for many of the fine points of grammar, usage, style, and production methods. Writers, editors, copy editors, proofreaders, compositors, and printers have turned to *Words Into Type* with the confidence that whatever the problem, they would find help there. With the passing of years, however, there has been a gradual change in the editorial practices of publishing and in the way in which the language is used. Grammar has become less prescriptive; rules of capitalization and punctuation have become less rigid; a greater flexibility of style is allowed by book publishers as well as by periodicals and newspapers. Developments in photolithography and the introduction of phototypography and cathode ray tube (CRT) equipment, computerized typesetting, electrostatic printing, and cold type methods have increased the technical and design possibilities of printing. Design techniques formulated by such innovators as Jan Tschichold and Hermann Zapf have had a profound influence on typographical design.

The third edition of *Words Into Type* provides an overview of the many processes of transforming the written word into type—from the preparation of the copy to the market-ready work. For the writer, an understanding of the technical problems of copy-editing and typographical style may facilitate his preparation of the manuscript, while a knowledge of the various stages in production may explain what is happening during the seemingly interminable wait for the final printed and bound product. For the editor, copy editor, compositor, proofreader, production worker, or printer, an understanding of the publishing processes outside his particular field may prove equally beneficial, while for the student this survey offers a practical introduction to the world of publishing.

The book is organized into seven parts, each dealing with a different aspect of production but interrelated with the others in the information provided. Part I concerns the preparation of the manuscript—the physical form of the manuscript, the use of editorial marks, the manner of handling topical heads, block quotations, footnotes, bibliographies, tables, and illustrations—and also provides a brief discussion of the legal aspects of authorship—copyright, libel, and the right of privacy. Part II discusses the handling of copy, galleys, and page proofs, with a separate section on the procedures for indexing. Part III is a comprehensive guide for the copy editor, and for the writer and editor as well, to the complexities of copy-editing style — abbreviations, symbols, numbers, italics, capitalization, punctuation, and compound words. Part IV covers the specifics of page makeup and typographical style, including the problems of typography of various forms of writing and of foreign languages. Parts V and VI complement parts III and IV with specific discussions of grammar and the use of words. (In this connection, the publisher wishes to express deep appreciation to the publishers of *The American Heritage Dictionary of the English Language* for permission to cite and quote from the dictionary.) Part VII deals with the methods, materials, and equipment involved in the composition of type and the reproduction of illustrations. It is accompanied by a "Glossary of Printing and Allied Terms." (We are grateful to the American Association of Advertising Agencies and the Advertising Typographers Association of America for permission to quote definitions from their *Phototypography*, Report No. 1, March 1972.) A comprehensive index makes all information in the edition readily accessible.

The third edition of *Words Into Type* has been prepared under the general editorship of Catherine B. Avery with the assistance of Linda Pelstring, both of Appleton-Century-Crofts. The publisher wishes to express appreciation to the following contributors for material they prepared and for their editing and revision of the original edition: James Boylan, adjunct associate professor of journalism, Columbia University, and contributing editor, *Columbia Journalism Review;* Philip James, free-lance writer, copy editor, and indexer based in New York; Margaret Nicholson, author of *A Dictionary of American-English Usage, A Manual of Copyright Practice for Writers, Publishers, and Agents,* and *A Practical Style Guide for Authors and Editors;* and Betsy Wade, chief of the Foreign Copy Desk, *The New York Times.* In addition, the publisher wishes to thank the following consultants for material that they prepared on their specialties and for their advice and suggestions: Curtner B. Akin, Jr., President, Typo/Graphics; Albert L. Arduino, Vice-President, Black Dot; Michael Mahony, President, Mahony and Roese and Company; Leonard B. Schlosser, President, Lindenmeyr Paper Corporation; and Herbert E. Youngkins, ComCom, Division of Intext.

WORDS INTO TYPE

PART I. MANUSCRIPT

TECHNICALITIES OF FORM

Certain conventions of form and arrangement have become established by usage as the clearest way to present various details in a manuscript. These standards need not be slavishly followed, but any radical departure from them is usually undesirable. Forms so unusual that they attract attention are likely to take the mind of the reader away from the thought of the text and lessen the effect of the author's words. Observance of the practices recommended in the following pages will lead to a satisfactory manuscript with a saving of time and expense.

PHYSICAL FORM

Typewritten manuscript. Regardless of the subject matter, the manuscript copy that is to be sent to the printer should be in the best possible form. Since no handwriting can be read as rapidly and accurately as typewriting, most publishers refuse to accept handwritten copy. When the final manuscript is typed, at least two copies should be made: the original "ribbon copy," showing a distinct black impression of every letter, to be sent to the publisher, and a duplicate to be retained by the writer. Many publishers, however, require a second copy for house use (the original for the editor, a duplicate for the designer or production department). If the author has already made publishing arrangements, he should consult his editor—or his contract—before having the final manuscript typed. If not, and an extra copy is needed, a photocopy will usually be satisfactory.

The typing paper should be uniform in size and of good quality, preferably white or cream-colored; thin manifold or highly glazed or onionskin paper should never be used for printer's copy. Using cheap paper or paper wider than standard is being penny wise and pound foolish. Typesetters are well-paid skilled workers, and any condition that slows up their production or makes it less accurate costs more in the end than any saving in cost of paper or typing. Paper too wide to go into the copy holder on the typesetting machine and copy that is hard to read because of thin paper or gray or smudged impression are unsatisfactory. The same typewriter should be used throughout, or at least the same size type—pica or elite. Pica is usually preferred as easier to read.

Margins. A liberal margin of at least 1 inch should be left on both sides to provide room for editorial instructions or corrections. For paper 8½ inches wide a satisfactory result can be obtained by setting the typewriter for a 6-inch line centered on the page; in other words, 72 spaces in elite typing, 60 spaces in pica. The typist should try to have the lines as nearly as possible the same length, but should not divide words arbitrarily in order to gain this evenness. The instruction sometimes given not to divide words at all is ill-advised; it inevitably results in lines varying so much in

-142-

Cather established that this mission would not have
been so overwhelming to one of a different character;
to illustrate this she offers the example of Laertes,
who was brought to essentially the same position as
Hamlet but whose purpose did not require a spiritual
grappling with his soul--it was simply a task upon
which he would act whenever the opportunity presented
itself. For Hamlet and for Myra Henshawe this mission
becomes so much a part of the essential self that it
serves to delineate their tragic stature above the
other characters. Each is so attuned to the inner life
that he, or she, has become removed from the rest of
mankind. The Myra of Part II has become like Hamlet,
who "had to contend with the realities not only of
this world, but of a world of his own."[2] Each has
found realities in life so different from those of
other men that these realities have created another
world, a world like the distilled essence of all that
man is able to realize in life--but at the same time
all that mankind does not realize.

    The singleness of purpose thus displayed is not
without its toll:

> The real tragedy of the play is [the] breaking of
> Hamlet's heart fiber by fiber, muscle by muscle.
> The final snap of the last quivering cord merely
> closes the tragedy. Hamlet died at the very close
> of the play, but he had been dying ever since the
> first act.[3]

Figure 1
Page of a Typewritten Manuscript

Footnotes for Chapter 4

1. Willa Cather, <u>My</u> <u>Mortal</u> <u>Enemy</u>, (Vintage Edition; New York: Alfred A. Knopf, 1961), p. 48.

2. Willa Cather, <u>The</u> <u>Kingdom</u> <u>of</u> <u>Art</u>, ed. Bernice Slote (Lincoln: University of Nebraska Press, 1966), p. 432.

3. Ibid. p. 431.

---

from the rest of mankind. The Myra of Part II has become like Hamlet, who "had to contend with realities not only of this world, but of a world of his own."[2] Each

---

2. Willa Cather, <u>The</u> <u>Kingdom</u> <u>of</u> <u>Art,</u> ed. Bernice Slote (Lincoln: University of Nebraska Press, 1966), p. 432.

---

has found realities in life so different from those of other men that these realities have created another world, a world like the distilled essence of all that man is able to realize in life—but at the same time all that mankind does not realize.

The singleness of purpose thus displayed is not without its toll:

The real tragedy of the play is [the] breaking of Hamlet's heart fiber by fiber, muscle by muscle. The final snap of the last quivering cord merely closes the tragedy. Hamlet died at the very close of the play, but he had been dying ever since the first act.[3]

---

3. Ibid. p. 431.

Figure 2
Manuscript Copy for Two Ways of Typing Footnotes:
Top, on a Separate Page; *Bottom,* on the Text Page

length that an accurate estimate of the number of words in the manuscript is made doubly hard. It is desirable also to have pages approximately uniform in length, with at least a 1-inch margin at top and bottom.

Spacing. Double or triple spacing should be used throughout the manuscript, including the bibliography and footnotes. Long excerpts that are to be set in type smaller than the rest of the text should also be double-spaced but with extra indention on one or both sides and extra space above and below. Poetry should be centered on the page and aligned as in the original.

Footnote reference numbers or symbols should be typed above the line — no slash, parenthesis, or other mark is necessary (see Fig. 1).

Most publishers prefer to have the footnotes themselves numbered consecutively for each chapter and typed, double-spaced, on a separate page, even though each will appear on the same page as its reference in the printed book. Another way of handling footnotes, preferred by some publishers, is to type them immediately following the line in which the reference is made, separated from the text by a rule above and below (see Fig. 2).

Although double-spaced typing is more easily read and more rapidly set by the typesetter, there are times when single spacing is preferred. If the page shown in Figure 1 were for a thesis to be microfilmed or presented in typescript form, or a monograph being prepared for photographic or similar reproduction, it should be typed in a slightly different way. Footnotes would be typed single-spaced at the foot of the page; long excerpts would also be single-spaced and indented on both sides. In short, the manuscript page should approximate the appearance of a printed page as closely as possible (see Fig. 3). The publisher usually gives detailed instructions for the preparation of material to be reproduced in this way and should be consulted before the final manuscript is typed.

Reprint copy. Clippings from printed matter — called reprint copy by the printer — should be pasted on sheets of paper the size of the rest of the copy. If any part of the manuscript is in the form of leaves printed on both sides, the best method is to use two copies of such matter and paste each page on a sheet of the copy paper. Sometimes, however, only one copy of printed matter is available. Then the most satisfactory method is to make photocopies of alternate pages.

Additions. Nothing should ever be typed or written on the back of a page of manuscript copy, and insertions should never be written or typed on narrow strips and attached to the edge of the copy. These slips, called *flyers* or riders, are likely to be torn off and lost. Short additions or corrections can be made between the lines or in the margin — horizontally, not vertically. If there are many insertions on a single page, or if material amounting to a page and a half or more must be inserted, the page or pages affected should be retyped, even though the result is a manuscript with pages of uneven length. The only other safe way to make long insertions is to type or paste them on a full-size sheet of paper, mark them "Insert A," "Insert B," and so on, and indicate clearly in the margin of the manuscript

-142-

Cather established that this mission would not have

been so overwhelming to one of a different character;

to illustrate this she offers the example of Laertes,

who was brought to essentially the same position as

Hamlet but whose purpose did not require a spiritual

grappling with his soul—it was simply a task upon

which he would act whenever the opportunity presented

itself. For Hamlet and for Myra Henshawe this mission

becomes so much a part of the essential self that it

serves to delineate their tragic stature above the

other characters. Each is so attuned to the inner life

that he, or she, has become removed from the rest of

mankind. The Myra of Part II has become like Hamlet,

who "had to contend with the realities not only of this

world, but of a world of his own."[2] Each has found

realities in life so different from those of other men

that these realities have created another world, a

world like the distilled essence of all that man

is able to realize in life—but at the same time all

that mankind does not realize.

> The real tragedy of the play is [the] breaking of
> Hamlet's heart fiber by fiber, muscle by muscle.
> The final snap of the last quivering cord merely
> closes the tragedy. Hamlet died at the very close
> of the play, but he had been dying ever since the
> first act.[3]

---

2. Willa Cather, The Kingdom of Art, ed. Bernice
Slote (Lincoln: University of Nebraska Press, 1966),
p. 432.
   3. Ibid. p. 431.

Figure 3
Page of Typewritten Manuscript Prepared for
Photographic Reproduction

the place of insertion. These pages should be inserted in the manuscript after, not preceding, the page on which they are to be inserted, and numbered as described in the following paragraph.

Numbering. Manuscript should be numbered consecutively throughout the text in the upper right-hand corner of the page. If additions are made later, after page 5, for instance, the new matter can be numbered 5a, 5b, and a note made on the bottom of page 5 that pages 5a, 5b, follow. If pages 8 and 9 are taken out after the manuscript has been numbered, the seventh page should be marked 7–9. If there have been many or substantial additions or deletions, the manuscript should be renumbered throughout, preferably with a hand-numbering machine.

Shipping. Pages of manuscript should be loose — not fastened together in any way. They should be sent to the publisher flat — not rolled. A typing-paper box makes a good container. All the text should be sent at one time, together with the title page, the table of contents, and other preliminary pages, and all end matter except the index.

If any illustrative matter to be used as final art is sent at this time, it should be in a separate envelope or mailing tube and protected from being bent or otherwise damaged.

FINAL READING OF THE MANUSCRIPT

The author who wishes to secure for his work in print the best possible appearance for the least expenditure of time and money will read the final draft of his manuscript with as much care as if he were reading the proof. By so doing he will avoid the experience of finding in the proof flaws that he would like very much to change but cannot, because of the expense involved; he will reduce materially the charge for author's alterations (see below); and he will make his work on the proofs much easier and less time-consuming. These results are well worth the effort required to make the manuscript as nearly letter-perfect as possible. This does not mean that the typewritten sheets should be as flawless in appearance as the final copy for a typescript book to be photographed, with no changes whatever indicated. Additions and deletions can be made as described earlier. Words can be corrected or added by printing neatly above the line or in the margin. But when the final manuscript is set in type in accordance with the editorial marks and corrections, the proof should be so nearly what the author wants that almost no changes will be needed.

Primarily this final reading is a check-up on details of form rather than of content, upon which the writer's attention has been centered in all previous readings. He should not expect that the typist will have observed consistency in all details, nor should he expect that the editor or printer will correct "obvious errors." Therefore in this reading he should check the typist's accuracy and his own success in following the style he planned — in spelling, capitalization, abbreviation, and other details to which his attention will be called in the following pages.

The spelling and punctuation in quotations should be carefully verified, for such excerpts are set by the printer strictly according to copy (see pp. 17–18).

All formulas and equations should also be carefully checked for accuracy. The distinction between capital letters ("caps") and small letters should be clear; any letters or symbols not on the standard typewriter keyboard should be identified in the margin. There is often confusion between the small letter l and the numeral 1, also between zero and the letter o. The author should be sure that the distinction is clear in the manuscript and also that all subindexes ("subscripts" or "inferiors") and exponents ("superscripts" or "superiors") are plainly marked: B_1, A_2; C^3, C^4. If equations or formulas are to be displayed—centered on a line by themselves—either type the line exactly as it is to appear or else pencil a marginal note of instructions to the printer. Mixed fractions should be clear, so that 1 1/2, for instance, could not be interpreted 11/2. (In typewriting, the only difference between these is a space between the 1's.) The author should indicate whether fractions are acceptable with a slant or should be set with a horizontal separating line: a/b or $\frac{a}{b}$; 12/7 or $\frac{12}{7}$.

The spelling of all foreign words should be checked carefully and all necessary accents marked in.

All lists supposed to be in numerical or alphabetical order should be checked.

In short, all necessary checking and verification should be done at this time in order that none will be required when the proofs are being read later.

AUTHOR'S ALTERATIONS

Any change from copy made by an author after the manuscript has been set in type is an author's alteration (AA) and is charged for on a time basis. Most of the money spent for author's alterations in proof is money thrown away, as far as the author is concerned, because the changes could have been avoided by his reading the manuscript as carefully as if it were proof.

The printer is responsible only for following copy and reproducing it accurately in the type format specified. He is not responsible for consistency in spelling, punctuation, and other details, and if he makes any changes from copy he does so at his own risk. He cannot be held responsible for errors resulting from illegible copy.

A certain amount of changing is unavoidable in making a book, and the publisher makes provision for it in his contract with the author. A percentage (usually 10 percent) of the original cost of composition is commonly allowed the author for alterations; but the cost of changing 10 percent of the text type is far in excess of 10 percent of the original cost of composition. The publisher ordinarily assumes the cost of arranging type beside narrow cuts, of adding or deleting to make pages the right length, and of other special adjustments.

In Linotype, the insertion of a comma or one letter in galley or page proof necessitates the resetting of the whole line. The insertion of two or three words in one line might involve resetting several lines. Notice the following paragraph, for example. In order to insert five letters in the first line, that line and all following lines in the paragraph must be reset, as is shown:

will furnish future generations∧with a supply of lumber. There is much land that is unsuited for farming because of the type of soil or the slope of the land. Usually land of this type is suitable for reforesting. Trees should be planted that will do well in each particular type of soil. The trees should be set in rows to aid future lumbering. At frequent intervals, a firebreak should be made. This is an open roadway that is kept plowed to prevent the growth of underbrush and weeds.

/of men

will furnish future generations of men with a supply of lumber. There is much land that is unsuited for farming because of the type of soil or the slope of the land. Usually land of this type is suitable for reforesting. Trees should be planted that will do well in each particular type of soil. The trees should be set in rows to aid future lumbering. At frequent intervals, a firebreak should be made. This is an open roadway that is kept plowed to prevent the growth of underbrush and weeds.

A correction like this made in galleys is expensive. If such a correction is made after the work has already been made up in pages, the expense is much greater, especially if more than one page has to be handled. The remedy in the case above is to rewrite and reword the first two lines, as is indicated:

will furnish future generations of men with a supply of lumber. There is much land unsuited for farming, because

Corrections in plates are very expensive. Even a slight change requires resetting part or all of the line, reading, casting, cutting the incorrect part out of the plate and inserting the new piece. A correction that causes a change of more than a line is even more expensive and necessitates having the whole plate remade.

EDITORIAL MARKS

Editors use certain marks and abbreviations that are universally understood by printers (see Fig. 4). Everyone who prepares manuscript or who marks proofs should conform to these established practices.

Capitals and small letters. To indicate that a small letter—called "lowercase" by printers—is to be capitalized, draw three lines under it. If a capital is to be set lowercase, draw an oblique line through it.

Small capitals are the shape of capitals but only the height of lowercase letters. They are indicated by two-line underscoring.

The abbreviations for the forms noted above are *caps* (capitals), *lc* (lowercase), *sc* (small capitals), *c* & *lc* or *c/lc* (capitals and lowercase, "cap-and-lower"), *c* & *sm* or *c* & *sc* (capitals and small capitals, "cap-and-small").

Italic and boldface. Italic is indicated by single underscoring, boldface by wavy underscoring.

Paragraphs. Paragraphs are indicated by a paragraph mark (¶). (The newspaper copyreader uses a downstroke with a right-angle turn.)

If a paragraph is made where none is necessary, a connecting run-in line should be used.

Insertions. If a letter or word is to be inserted in a line, make a caret (∧) at the point of insertion and write the word above it if there is room, otherwise in the margin. If anything is written in the margin, write it horizontally across the margin as nearly opposite the caret as possible, with a slash next to the line in which it is to be inserted.

Avoid writing on copy lengthwise in the margin if possible, but if it must be done, always write upward from the line in which the correction is made.

Circle instructions written in the margin if there is any possibility of their being mistaken for part of the copy.

Deletions. If anything in the copy is to be omitted, erase it or mark through it unmistakably; do not enclose it in brackets or parentheses.

If a hyphen is to be deleted, use close-up marks (⌒) or a space mark (#) to indicate whether the word is to be one word or two.

Restorations. If words have been crossed out by mistake, place dots under the words that should remain and write *stet* in the margin opposite. (*Stet* is a Latin imperative meaning "Let it stand.")

Indentions. Mark matter that is to be indented by using the sign for an em space (□) before the line to be indented.

Alignment. Vertical alignment is indicated by a line drawn downward beside the characters to be set in line. Horizontal alignment is indicated by parallel lines drawn horizontally above and below the characters to be set in line.

Spaces. A straight line down between letters indicates a space; and curves (close-up marks) above and below a space mean to close up spaces incorrectly placed.

Transpositions. To transpose adjacent letters, words, or phrases draw a line curving over one and under the other.

Abbreviations. A circle around an abbreviation or number means that it should be spelled out. This use of a circle, however, is not always safe, especially with abbreviations; the change can often be made more clear by crossing out the incorrect form and writing the correct form above.

Superiors and inferiors. Figures, letters, or symbols that are to be set above or below the line—called superiors or superscripts, inferiors or subscripts—often require special indication in the manuscript. An inverted caret may be placed under a superior, or superior letters and figures may be indicated by typing a "shelf": the superior is underscored and followed by a stroke (2/). This need not be carried to the excess of marking asterisks, daggers, the symbols for degrees, minutes, seconds, and the like, which are always above the line. An inferior or subscript may be indicated by placing a caret over it.

Hyphens and dashes. When it seems advisable to mark a hyphen or a dash so that it will not be taken for some other mark, indicate a hyphen by two short lines—a double hyphen—and a dash by drawing a line down each side of the dash. A hyphen occurring at the end of a line in the manuscript should be marked as a double hyphen if it is to be retained or be marked through and closed up to indicate a single, unhyphenated word.

A long dash, called a 2-em dash, is marked by a 2 above the dash and *m* under it.

Peculiar spellings. Spellings so unusual or peculiar that the editor or proofreader might question their accuracy should be verified and stet marks (a series of dots) placed under them.

STYLE

When publishers and printers speak of style, they do not mean literary style, but are referring to spelling, punctuation, capitalization, the use of abbreviations and italics, the expression of numbers, and many details of typographical form and practice. Printed matter should conform to recognized standards; inconsistencies annoy the reader or arouse the suspicion that an author inaccurate in details may be inaccurate also in his thought. (See Copy-Editing Style, pp. 99ff.)

Spelling. An author should make his spelling consistent, with itself at least. It should conform to some one dictionary—Webster's International, American Heritage, Random House, or others (see pp. 469, 487). Most publishers like to have the privilege of following their house style. Magazines, technical publications issued periodically, and some textbook houses usually have an office ruling about the dictionary spellings to be followed, and writers should conform to these rulings. For example, the writer on chemical subjects might well acquaint himself with the reports and rulings on nomenclature of the American Chemical Society. The author or the typist should verify the spelling of technical and unfamiliar words, proper names, and foreign words and check all accents and hyphens.

Punctuation. A writer should give careful attention to punctuation, for the clearness and force of his writing may be spoiled by misuse of punctuation or a failure to observe established conventions. Even if punctuation is used sparingly, good usage cannot be disregarded without inviting criticism. Rules for the use of punctuation marks are given on pages 172ff.

Cather established that this would not have been over/
whelming to one of a different character; to illustrate,
she offers the example of Laertes, who was brought es/
sentially to the same position as Hamlet, but *whose* purpose
did not require a spiritual grappling with his soul—;
it was simply a task upon which he would act whenever
the opportunity pre~s~ented itself. For Hamlet, and for
Myra Henshawe, this mission ~became~ *becomes* so much a part of
the essential self that it serve~s~ to delineate their
tragic stature above the other characters. Each ~has~
~become~ *is* so attuned to ~his~ *the* inner life that he *s or she s* has become
removed from the rest of mankind. In Part II Myra
has become ~much~ like Hamlet, who "had to contend with
the realities not only of this world, but of a world
of his own." *Each has* ~They have~ found realities in life so
different from those of other men that these realities
have created another world ~for them~, a world ~that is~
like the distilled essence of all that man~kind~ is
able to realize in life—;but at the same time all
that mankind does not realize.

This singleness of purpose is not without its self=;
imposed toll, for Myra as much as for Hamlet:

> The real tragedy of the play is [the] breaking of
> Hamlet's heart fiber by fibre, muscle by muscle.
> The final snap of the last quivering cord merely
> closes the tragedy. Hamlet died at the very
> close of the play, but he had been dying eversince #
> the first act.

Figure 4
Typewritten Manuscript, Copy-Edited for Compositor

Capitalization. Having adopted one style, a writer should follow it throughout. When he uses capitals for special purposes, he should be strictly consistent in their use. A free use of capitals may be appropriate in some kinds of writing, but an indiscriminate use of them without logical purpose is a serious defect. See the rules for capitalization on pages 142ff.

Abbreviations. Abbreviations should not be used with the expectation that the editor or printer will spell them out. Whenever a form of abbreviation has been established by usage or authority, this form should be used in preference to any other. Each branch of writing has its list of commonly used abbreviations, and sometimes the same letters may stand for different words, according to the text. Rules and lists of abbreviations are given on pages 100ff.

TOPICAL HEADS

Topical headings, called *heads*, mark definite and well-thought-out divisions of the subject under consideration. They are the concern and the responsibility of the writer himself. If his work is intended for study or reference, a writer should give careful attention to this detail of form, for a logical and well-planned use of headings enhances the usefulness of any writing intended to instruct.

Construction. Topical heads should indicate concisely the subject discussed in the text that follows. They are usually nouns, or phrases built about nouns:

Early Marriages   The Use of Fireplaces
Civilizations as Regressions   among the Indians

Phrases like *Description of, Discussion of,* and *Statement of* are superfluous.

Subheads. Subheads and sub-subheads may be differentiated by different sizes or kinds of type. If further subordination is necessary it is ordinarily represented by side heads, or by successive indentions in the form of an outline, sometimes using numerals or letters, either in parentheses or followed by a period but not both together. Indication of the type in which heads are to be set is the responsibility of the publisher (see pp. 247–250), but the author should see that his copy shows clearly the comparative value of each head. Three typescript forms are suggested in Figure 5. The first style leaves the editor free to indicate placement, type, and capitalization. The second would be used in a manuscript to be reproduced photographically. The third style, with all lines flush on the left, is most economical of the typist's time.

Side heads. Depending on the design of the book, side heads may be set flush on the left, with space above and below; or they may have regular paragraph indention with the text running in after a period and a space. Three styles are shown below:

Chapter V

Isaiah

Judah During the Eighth Century B.C.

The Political Situation. The political, social,
moral, and religious conditions in Judah during the
activity of Isaiah and his younger contemporary Micah

Chapter V

I S A I A H

JUDAH DURING THE EIGHTH CENTURY B.C.

The Political Situation. The political, social,
moral, and religious conditions in Judah during the
activity of Isaiah and his younger contemporary Micah

Chapter V

I S A I A H

JUDAH DURING THE EIGHTH CENTURY B.C.

The Political Situation

The political, social, moral, and religious
conditions in Judah during the activity of Isaiah and
his younger contemporary Micah were essentially

Figure 5
Three Typewritten Forms of Headings

Side Heads

Side heads may be set flush at the left with space above and below—a *shoulder head.*

Side heads. Side heads may be regular paragraph indention with the text running on after a period and a space; this is sometimes called a paragraph heading, or a *run-in side head.*

Side heads: The capitalization in side heads may be the same as if it were a centered head, or only the first word and proper nouns may be capitalized ("sentence-style"). Sometimes a colon is used after a side head, but a period usually is better.

Each side head should contain a noun, and the words immediately following the head should not be mere repetition. Appearance, appropriateness, usefulness, and expense are considerations entering into the choice of form for side heads. Sometimes the decision is made by the author, sometimes by the editor or the designer. In any case the manuscript should

```
                         OUTLINES

Topic outline

     A topic outline consists of headings that indicate

the important ideas in a composition, and their

relation to each other. . . .


Sentence outline

     A sentence outline is expressed in complete

sentences punctuated as in ordinary discourse. . . .


Indention

     Indent and number all headings properly. . . .


Parallel form

     Express in parallel form all ideas that are

parallel in thought. . . . .
```

Figure 6
Typewritten Copy to Be Edited as a Simple Side Head, a
Shoulder Head, a Cut-in Head, or a Marginal Note

be typed in a consistent manner. The style of capitalization should be made clear: *headline-style,* in which all important words are capitalized, and *sentence-style,* in which only the first word and proper nouns or adjectives are capitalized.

In some books *cut-in side heads* are effective. They are set in type smaller than the text and are inserted in a space left for them when the **Cut-in** text type is set. They are used sparingly, however, because they **side** necessitate so much extra work compared with the setting of **head** ordinary side heads and therefore are much more expensive. Also, since they must be short to give the best appearance, the style is limited in adaptability.

Also effective in some texts is a "running gloss," words or phrases placed in the margin, sometimes only at the beginning of a paragraph, but often opposite the line of text to which the word or phrase applies. Such notes are sometimes used to form an outline of the subject matter without inter- **Marginal** rupting the flow of the text. The printer calls them *marginal* **notes** *notes.* Usually sentence-style capitalization is used. Like cut-in heads they must be concise. They too are expensive by comparison with simple side heads.

The typescript form in Figure 6 could be edited as any of the forms described above.

EXCERPTS

When a writer quotes directly from the writing of another, he should indicate clearly what is quoted. He should also keep a record of his sources, for checking in manuscript and proof and for use in compiling his bibliography, if there is to be one.

Unless the reader is advised that the borrowing author has made changes, all quotations should be exactly as they were in the original. The only changes without notice that are generally considered permissible are (1) to capitalize the first word of a quotation that has been formally introduced and is itself structurally a sentence, even though preliminary words have been omitted; (2) to change a capital to a small letter in the first word of a quotation if the quoted matter is a grammatical part of the writer's sentence;[1] and (3) to change punctuation at the end of the quoted matter if it is not appropriate in the writer's text.

He says: "Any actual learning is always expressed as a change in attitude of the individual . . ."

---

[1] In some scholarly works in which the exact text is significant, even these two changes are not allowed.

When he says that "any actual learning is always expressed as a change in attitude," we may suppose that he means . . .

"Beauty is truth, truth beauty,—that is all
Ye know on earth, and all ye need to know"

well illustrates how Keats . . .

[The original has a period after *know.*]

However, when the quoted matter is followed by a new sentence or a new paragraph, a period should be inserted at the end in place of other punctuation—comma, semicolon, or colon—that may appear in the original, except when the original ends with a question mark.

Emily Dickinson thought often of death, and not with fear:

Because I could not stop for death,
He kindly stopped for me.

[The original has a semicolon after *me.*]

Any other changes should be brought to the reader's attention ("italics ours"; "adapted from . . .") and, if the material is used by permission, should be made only with the consent of the copyright proprietor (see pp. 49–50).

Excerpts denoted by quotation marks. The most common method of denoting short excerpts is by enclosing them in quotation marks. For example:

According to the *MLA Style Sheet*, "Unless special emphasis is required, prose quotations up to 100 words in any language should always be run on, in quotation marks . . . as part of your text." The University of Chicago Press *Manual of Style* advocates running on anything less than eight or ten typed lines, except in special circumstances.

A prose excerpt of more than one paragraph should have opening quotation marks before each paragraph and closing quotation marks at the end of the last paragraph. If the excerpt is a letter, quotation marks may be placed before the first line and after the last:

"2671 Fifth Ave.
Sacramento, Cal.
Nov. 23, 1971

My Dear Mr. Keene:
You have heard, no doubt, that the proposition advanced by our good friend Mr. Smith has . . .

Yours sincerely,
Charles M. Butler"

A less simple style is to set opening quotes before each line of the heading, each line of the salutation, each paragraph of the letter, and each line of the complimentary close and signature, with closing quotes at the end. This style is preferred if the letter is not complete or if it is given as being spoken aloud by a character in a book.

> "My Dear Mr. Keene:
> "You have heard, no doubt, that the proposition advanced by our good friend Mr. Smith has . . .
>
> > "Very sincerely yours,
> > "Charles M. Butler"

An excerpt of two or more stanzas of poetry, if not set in reduced type (see below), should have opening quotation marks before each stanza and closing quotation marks at the end of the last stanza (see p. 214). Double quotation marks in the original should be changed to single ones and arranged as follows:

> "'Why are you cross, Sammy Squirrel?'
> Questioned his mother one day.
> 'You quite forgot your manners
> When Squeaky Mouse came to play.'"

If more than one line of poetry is quoted in the running text, the lines should be separated by a solidus (slash or slant):

> There she sat, "singing alone,/Combing her hair," like Tennyson's Mermaid.

Reduced type. Instead of using quotation marks to indicate excerpts, the publisher often sets excerpts in a smaller size type than that used for the rest of the text. Letters, poetry, extracts from plays, and the like are usually so set and are said to be reduced. The printing term for such reduced quotations is *extracts*. The appearance of brief quotations in reduced type is unpleasing; the method is therefore inappropriate for prose quotations of less than five lines. Shorter ones may be reduced, however, if they occur close to longer extracts.

In the manuscript, extra indention indicates that the excerpt is to be reduced, and extra space should always be left above and below an extract (see Fig. 1). Since reducing the type indicates that the matter is a quotation, it is unnecessary to use quotation marks to enclose extracts. Therefore quotation marks within the excerpt should remain as in the original.

Even short excerpts from poetry are generally reduced (see p. 254). Quotation marks are not necessary when type is reduced, but if type the size of the extracts is used to any extent for matter other than quotations, quotation marks may well be used to avoid any possibility of misunderstanding. In the example from Keats on page 18, the quotation marks are in the original poem.

If a poetical extract begins with part of a metrical line, this line should be set to indicate the fact clearly.

                              long and level lawn,
                    On which a dark hill, steep and high,
                    Holds and charms the wandering eye.

Points of ellipsis. Three points, properly called points of ellipsis, are used to indicate an omission. They are therefore often used in excerpts. The use of points of ellipsis within an extract is usually no problem, but there is often some confusion about whether they should be used at the beginning of an excerpt.

If a prose quotation begins in the middle of a sentence, the fact is sufficiently indicated by writing the first word with a small, or lowercase, letter.

As Mrs. Ashleigh explains matters to Tom, the "young, wild spirit
of youth . . . tramples rudely on the grave-mount of the Past . . ."

If the last portion of an excerpt is further reduced it should start with a small letter and be flush with the following lines of the excerpt.

Mary began to think of all the other guests at the party who

    had taken advantage of the Queen's absence, and were resting in the shade;
    however, the moment they saw her, they hurried back to the game. . . .

It should be noted that terminal punctuation is retained before points of ellipsis. Thus in the example above, the first point is the period, indicating the end of the sentence quoted, and the three points of ellipsis follow.

Points of ellipsis at the beginning of an excerpt starting in mid-sentence merely emphasize a fact already marked by the small letter for the first word, but in some scholarly and technical works, this emphasis is desirable. If an excerpt is formally introduced, as in the example below, and the first letter is not capitalized, the ellipsis points should be used to avoid the appearance of violating the rule that a quotation so introduced should begin with a capital.

According to Sahlins and Service:

    . . . evolution moves simultaneously in two directions. On the one side, it creates diversity
    through adaptive modification . . . On the other side, evolution generates progress . . .

Using points of ellipsis before an excerpt beginning with a complete sentence in the original, or at the end of an excerpt when the last sentence is complete, is equivalent to saying, "This is not all the writer of these words had to say on the subject." That fact would in most cases be understood by the reader without the ellipsis, but if it should be emphasized, the phrase "in part" in the sentence introducing the excerpt is often preferable to using the ellipsis points.

If an extract of poetry begins in the middle of a metrical line, it is sometimes good to use points of ellipsis at the beginning, even though deep indention and a small letter would make it clear that words at the beginning of the line have been omitted.

It might have been the original of Poe's

> . . . dank tarn of Auber
> In the ghoul-haunted woodland of Weir.

Style within excerpts. Even in quotations from works in the public domain—the Bible, public records, manuscripts, archaic works, and standard editions of classics of all kinds—the original should be followed exactly. Editor and printer would not venture to depart from copy in such quotations, and the author may depend upon his copy being followed. Since responsibility for accuracy in copying is entirely his, the author should compare his copy of such excerpts with the original to catch any errors in typing.

It is often desirable, however, to modernize archaic spelling and punctuation. Even modern British spelling like *honour, connexion, waggon, kerb,* are sometimes Americanized, as, for instance, in a school text for elementary grades. The author should use judgment and discretion in these details and be sure that he is consistent.

If a quotation contains spelling or punctuation that is so contrary to ordinary usage that it might be thought a typist's error made in copying, the author should indicate that the copy is correct by marking it *stet* or *sic* or *O.K.* in the margin. Otherwise the editor might be tempted to change it or at least to query it.

If the author italicizes for the sake of emphasis any words not italicized in the original, he should indicate the fact in a footnote or in brackets immediately following the italicized words. Any insertions of his own within the quoted matter should also be enclosed in brackets; parentheses are not sufficient for this purpose.

> "It may be fairly said that this necessary act *fixed the future frame and form of their government* [Webster's italics]. The character of their political institutions was *determined* [our italics] by the fundamental laws respecting property."

Acknowledgments. When a writer uses excerpts from the works of another, he should acknowledge his indebtedness either by a footnote or by a credit line. An excerpt used to supplement the writer's own composition, like that shown in Figure 1, should be accompanied by a bibliographical footnote, explicitly noting the source. The content and arrangement of such footnotes are noted in detail in the section immediately following. A short excerpt, such as an epigraphical quotation at the head of a chapter or article, should be followed by a credit line, which may be only the author's name, or include also the title of the work.

> It is a strange desire, to seek power and lose liberty, or to seek power over others and to lose power over a man's self.
> —Francis Bacon, *Of Great Place*

The quotation of more than a few lines of a poem, story, play, or article, still in copyright, requires a footnote stating that the original publisher or copyright owner has permitted its use (see pp. 24–25).

In an anthology or book of readings the name of the author may be given under the heading or in a credit line at the end. The word *Adapted* should appear in the credit whenever the material is simplified or condensed in any way, even if it is in the public domain. Such might be the case of a French story reproduced in English in a form more suited for young readers than a literal translation would be.

> —Alphonse Daudet (*Adapted*)

FOOTNOTES

Footnotes are used (1) for the acknowledgment of borrowed material, (2) for notifying the reader of the source of statements or quotations, (3) for the presentation of explanatory or supplementary material not appropriate to the text, and (4) for cross references to other parts of the work.

References to footnotes. Superior figures—small figures set above the line—are generally used as references to footnotes. The footnote itself is identified in most cases by a corresponding superior figure.

> Each formed a single type, or rather *logotype*.[1] The symbols . . .
>
> [1] It must be recalled here that the Chinese language is not an alphabetic but an ideographic language.

Some designers, however, place the identifying numeral, followed by a period, on the line with the footnote and in the same size type.

> Each formed a single type, or rather *logotype*.[1] The symbols . . .
>
> 1. It must be recalled here that the Chinese language is not an alphabetic but an ideographic language.

Unless footnotes are few and do not occur close together, it is best to number them consecutively throughout a chapter or article. If there are many notes and they are numbered by pages of manuscript, with the intention that when the type is made up into pages the first note on each page will begin a new series of numbers, a great deal of expensive correcting will be required; the first note on a manuscript page is seldom the first note on a page of type. Sometimes every line in which a superior figure is changed will have to be reset. For these reasons the system of numbering footnotes by articles or chapters is much to be preferred.

Reference marks may be used instead of superior figures. If several marks are used on a page the proper sequence is:

* asterisk or star
† dagger
‡ double dagger
§ section mark
‖ parallels
¶ paragraph mark

If more are needed, the same marks are used double: **, ††, and so on.

Because their number is so limited, reference marks are unsatisfactory when there are many notes. A new series of marks should begin on each new page; therefore all references have to be verified when the type is made up in pages. These marks are useful, however, when superior figures might be confusing, as in algebraic matter: $A = b^2 + c^2$.† They are also used for footnotes within quoted matter, to avoid confusion with the author's numbered footnotes. Notes to tables may be referred to by the reference marks or by superior letters. The notes are then placed directly under the table. (See also Tables, pp. 32ff.)

Both superior figures and reference marks are used sometimes — superior figures referring to citations, which are all given in a group at the end of the article or chapter of a book; reference marks for explanatory footnotes placed at the bottom of each page.

Position of references. Reference figures or marks should be set after any mark of punctuation except the dash or a closing parenthesis if the reference relates to matter within the parentheses. They should be placed after, not before, a word or paragraph that is explained or amplified.

Each formed a single type, or rather *logotype*.* The symbols . . .

*It must be recalled here that the Chinese language is not an alphabetic but an ideographic language.

Reference to the footnote citing the source of an excerpt should stand at the end of the excerpt, not in the text that precedes it. Thus placed, the reference does not distract the attention of the reader as it would elsewhere.

A scholarly critic writes:

I do not say that "etc." is not to be used, but its use should be rare, and chiefly for the omission of parts of quotations and the like. When used by the author to eke out his own matter or to save himself trouble the reader is disposed to exclaim, "If you have anything more to say pray say it; if not, finish your sentence properly; 'etc.' conveys no meaning at all."[1]

[1]Clifford Allbutt, *Notes on the Composition of Scientific Papers*, 2nd ed., pp. 158–59.

A possible result of this placement of the reference is, however, that a note may not be on the same page as the beginning of the extract, a separation considered undesirable by some editors. If so, the reference can be placed after the name of the author if the wording of the text makes this possible.

The following extracts from an article written by G.O. Smith,[1] a former Director of the Survey, are suggestive and highly significant:

[1]G.O. Smith, "Plain Writing," Science, new ser., Vol. 42 (Nov. 5, 1915), pp. 630–32.

A reference should not be placed after a name in the possessive form: "In Dr. Mann's analysis,[3]" not "In Dr. Mann's[3] analysis."

Many designers dislike having references placed after display type of any kind. This can often be avoided by shifting it to a suitable place in the text. In the following example it could well be placed at the end of the first sentence:

Excerpts[1]

The author is responsible for careful indication of all matter quoted from the work of another. For that reason detailed instructions . . .

[1]See also Legal Provision Affecting the Author.

When a selection is given with its title and is the only one starting on a page, the credit line may be given in an unnumbered footnote.

KUNG BUSHMAN BANDS
Lorna Marshall
[Selection begins]

Reprinted from Africa, 30, No. 4 (1960), by permission of the International African Institute.

If there are other footnotes on the page and a reference mark must be used, many designers use a symbol after display type rather than a number.

Explanatory and supplementary footnotes. Although annotation is essential in most scholarly works, a succession of long footnotes will inevitably create problems in page makeup. Each note must start on the page on which the reference mark occurs and sometimes has to be run over to the following page; if there are many reference marks on that page too, the result may be a page of more annotation than text—displeasing both to the reader and the designer. The author should therefore consider carefully (1) whether his notes are worded as briefly as possible; (2) whether there could be more intervening text between references to long supplementary notes; (3) if there are many long notes throughout the work, whether some or all of the material might better be given in the text or in a special section in the appendix.

Permission footnotes. When a copyright owner gives permission to quote from his publication he may specify the form in which he wishes the acknowledgment of indebtedness. Such specifications should of course be scrupulously followed. They always include the title of the book, the name of the author, and usually the name of the publisher and the copyright owner. The phrase "by permission" is often used, and bibliographic data may also be required.

[1]From Bernard Malamud, *The Tenants,* p. 119. Copyright © 1971 by Bernard Malamud. Reprinted by permission of Farrar, Straus and Giroux, Inc.

Bibliographical footnotes. Footnotes notifying the reader of the source of statements or quotations that he might wish to look up should be in a form sufficiently detailed to enable the reader to find the reference as easily as possible. References to books that have been printed in innumerable editions, like the Bible, Shakespeare, Milton, or Blackstone, should specify the book, chapter, and verse; the act, scene, and, if possible, the line; the canto, stanza, and line; or the section and paragraph.

I Cor. 13:4-7.
1 Henry VI, iii. 2. 14.
The Lady of the Lake, canto V, stanza xxi.
Summa Theologica, II, ii, q. 9, ad. 1, English translation (London, 1917), pp. 114–15.

References to manuscripts should give the author, title, location, and date.

Eugene O'Neill, "The Revelation of John the Divine," unpublished. The Houghton Library, Harvard University, Cambridge, Mass. [1924]

When authorities are cited in books of law, the reference is made as definite and concise as possible (see also pp. 282–283).

[1]See M. E. Cohen, *Censorship of Radical Materials by the Post Office,* 17 St. Louis Law Review 95 (1932).

[4]62nd Cong., 8th sess., H. Doc. 341.

The first citation of a work not in the categories noted, especially if there is no full bibliography in the book, should, if possible, contain the following information, given in the order here noted.
  (1) The author's name, not inverted, as given on the title page
  (2) The title of the book, including the subtitle if there is one
      (Underscore this in the manuscript. Titles of books and serial publications are more often in italics than in any other form. If the writing is being done for a publication known to follow a different practice, conform to its style.)
  (3) The edition, if other than the first
      (The title page or the copyright page shows the number of the edition. Do not confuse it with the number of printings or impressions.)
  (4) The title of the series, if significant
      (Do not underscore or enclose in quotation marks. The series

title may often be omitted in footnotes, especially if there is a full bibliography later.)
(5) The place of publication
(Give the home office when several place names appear on the title page. A list of home offices is given in *Books in Print* and in *LMP*, both published annually by R. R. Bowker.)
(6) The publisher
(If the name is shortened, use the form given in *Books in Print*.)
(7) The date of publication
(Use the date given on the copyright page. The date on the title page, if any, may be changed with every printing.)
(8) Volume and page numbers

When articles in periodicals are cited, the following information should be given:
(1) The name of the author
(2) The title of the article (enclosed in quotation marks)
(3) The name of the periodical (underscored in manuscript)
(4) The volume or number
(5) The date of the issue
(6) The page numbers

Citations of unpublished material should include
(1) The name of the author
(2) The title (in quotation marks)
(3) The type of work (dissertation, thesis, work in preparation)
(4) The collection, if any, and location
(5) The date, if known

MODELS FOR BIBLIOGRAPHICAL FOOTNOTES

Note: These models follow a generally accepted basic form:

Author, *Title: Subtitle* (Place: Publisher, date), vol. no., p. no.

It should be understood, however, that variant forms of punctuation, capitalization, and abbreviation may be preferred by some publishers as a matter of house style (see pp. 59 and 99).

Book—one author:

[1]Norman Cohn, *The Pursuit of the Millennium* (New York: Essential Books, 1957), p. 96.

Book—two authors:

[1]W.W. Tarn and G.T. Griffith, *Hellenistic Civilization*, 3rd ed. (London: E. Arnold, 1952), ch. 10.

Book—more than three authors:

[1]R.C. Majunder et al., *Advanced History of India*, 2nd ed. (London: Macmillan, 1950). [Some publishers prefer "and others" to *et al.*]

Book—association the author:

[1]American Council of Learned Societies, *Studies in the History of Culture* (Menasha, Wis.: George Banta Publishing Co., 1942). [The state is included only when the place is relatively unfamiliar or to avoid confusion.]

Book—no author named:

[1]*A Manual of Style*, 12th ed. rev. (Chicago: The University of Chicago Press, 1969), p. 137. [Do not use *Anon.*]

Book—author and translator:

[1]Max Weber, *General Economic History*, trans. Frank H. Knight (Glencoe, Ill.: The Free Press, 1927), pp. 244–45.

Book—author, editor, and translator:

[1]Hermann Kees, *Ancient Egypt: A Cultural Topography*, ed. T.G.H. James and trans. Ian F.D. Morrow (Chicago: University of Chicago Press, 1961), pp. 47ff.

Book—one volume of a work of several volumes:

[1]Charles W. Eliot, ed., *The Harvard Classics*, Vol. II, *The Meditations of Marcus Aurelius*, trans. G. Long (New York: P.F. Collier & Son, 1909–10).

Book—one of a series:

[1]Allen Chester Johnson, *Egypt and the Roman Empire*, The Jerome Lectures, 2nd Series (Ann Arbor: University of Michigan Press, 1951), p. 20.

Technical bulletin, government report, or similar publication:

[1]*Subway Riders and Manhattan Autos*, Tri-State Regional Planning Commission, Regional Profile, Vol. 1, No. 14 (New York: 1971), p. 6.
[2]Joseph W. Masselli, Nicholas W. Masselli, and M. Gilbert Burford, *Controlling the Effects of Industrial Wastes on Sewage Treatments*, New England Interstate Water Pollution Control Commission, Technical Report 15 (Boston: 1970), p. 49.

Part of a book:

[1]Bertrand Russell, "Modern Homogeneity," *In Praise of Idleness and other Essays* (London: George Allen and Unwin, 1960), pp. 117–18.

Part of a yearbook:

[1]Wesley Towner, "The Rise of Bibliomania," *The Old and the New*, 1971
AB Bookman's Yearbook, Part II (Newark, N.J.: Bookman's Weekly, 1971), p. 14.

Part of a report:

[1]Phyllis Baxendale, "Content Analysis, Specification and Control," *Annual
Review of Information Science and Technology*, American Documentation Institute, Vol. I (New York: John Wiley and Sons, 1966), p. 86.

Article in a periodical:

[1]V. Gordon Childe, "War in Prehistoric Societies," *Sociological Review*,
Vol. XXXIII (1941), p. 130.
[2]I.J. Gelb, "The Ancient Sumerian Ration System," *JNES*, XXIV (1965), 242.

Unpublished work:

[1]Joseph W. Elder, "Industrialism in Hindu Society," Ph.D. Thesis, Harvard
University, 1957.
[2]Siegfried von Raht, "Corneal Transplants in Rats: A Method Developed in
the University of Erewhon by the Author," (Translation of a paper delivered at
the Ophthalmological Society of Erewhon, March 18, 1973).

Shortened forms. When reference has been given once in full, many
publishers prefer to use a shortened form when it is given later in the same
chapter. This is an alternative to the use of *ibid., op. cit.,* and other Latin
abbreviations, as described later.

If the author's name is in the text, it is not necessary in the footnote:

[1]*The Century Collegiate Handbook*, p. 186.

The authors' names may be shortened:

[2]Greever and Jones, *The Century Collegiate Handbook*, pp. 186–90.

A long title may be shortened after it has been given once in full within
the chapter:

[3]Greever and Jones, *Collegiate Handbook*, pp. 169–72.

Repeated references to the same book — unless more than one book
by the same author is being used — may be shortened to the author's surname and the page reference:

[4]Greever and Jones, p. 60.

If the author and the book are well known, the possessive may be used:

[5]Darwin's *Origin of Species.*   [6]Adam Smith's *Wealth of Nations.*

Use of "op.cit.," "ibid." If the shortened form of references is not used, much unnecessary repetition in footnotes can be avoided by using *op. cit., loc. cit., ibid.,* and *idem* (see pp. 109ff.). These abbreviations are now usually printed in roman, though some publishers prefer to use italic. Unless the author knows his publisher's preference, it is better in the manuscript not to underscore them, leaving the decision to the editor. When a page reference follows one of the abbreviations, it may be set off by a comma or not—"Ibid. pp. 46–52" or "Ibid., pp. 46–52"—depending on the publisher's preference.

The first two abbreviations—*op. cit.,* for *opere citato,* "in the work cited," and *loc. cit.,* for *loco citato,* "in the place cited"—may be used only when accompanied by an author's name, which may be either in the text or at the beginning of the footnote. When a book or a magazine article is cited, the full title should be given the first time it is referred to and, for the convenience of readers, the first time it is referred to in each chapter. In later references *op. cit.* or *loc. cit.,* with or without a page reference, may replace the title.

First reference:

[1]Garland Greever and Easley S. Jones, *The Century Handbook of Writing,* 4th ed. (New York: D. Appleton-Century Company, 1942), pp. 148–59.

Later reference:

[2]Greever and Jones, op. cit. pp. 151–58.

First reference:

[1]H.V. Kaltenborn, "Wanted: Economic Peace," *The Commentator,* Vol. 3, No. 4 (May 1938), pp. 70–74.

Later reference:

[2]Kaltenborn, loc. cit.

More specific reference:

[3]Kaltenborn, loc. cit, p. 74.

Author's name in the text preceding the reference index:

[4]Loc. cit.

Note that *op. cit.* cannot be used if more than one book by the same author has been cited.

The abbreviation *ibid.*, for *ibidem*, "in the same place," may be used when reference is made to the same work referred to in the preceding footnote if it appears within the same two-page spread. (Some editors modify the rule to the extent of using *ibid.* when referring to the preceding note in the chapter, but this is not convenient for the reader.)

First reference:

[1]Richard Corson, *Stage Makeup*, 4th ed. (New York: Appleton-Century-Crofts, 1967), pp. 119–62.

Any later reference:

[2]Corson, op. cit. p. 4.

Immediately following either of the above:

[3]Ibid. p. 119.

Identical with the preceding note:

[4]Ibid.

Ibid. is understood to replace all words that are identical in consecutive notes:

[1]*British Documents,* Vol. IV, No. 257, p. 279.
[2]Ibid. No. 261, p. 283.
[3]Ibid. p. 287.

In the second note *Ibid.* stands for "*British Documents,* Vol. IV." In the third note *Ibid.* stands for "*British Documents,* Vol. IV, No. 261."

The Latin word *idem*, occasionally used in footnotes to refer to the same person or reference as the preceding one, is falling into disuse, perhaps because its use is not a matter of common agreement.

Abbreviations. Abbreviations may be used freely in footnotes. *Volume, Part, Book, chapter, section, page, line,* and other words used like them are regularly abbreviated. (See the list on p. 116.) Besides these, used in locating the reference, the titles of books and periodicals may be abbreviated. Classical references listed once in complete bibliographical form may subsequently be given as briefly as possible.

[7]Jornandes, de. Reb. Get. c. 30, p. 654 (p. 87, ed. Lugd. B. 1597).

Abbreviated forms of titles of periodicals are commonly used when citations of them are frequent and the abbreviations are unquestionably intelligible to the reader; as *Pol.Sci.Rev.* for *Political Science Review; Acad.Sci.* for *Académie des Sciences, Belles-lettres et Arts.* Well-known works or publications may even be referred to by symbols, provided their meaning is explained somewhere in the publication. For instance: *HDB* for *Hastings' Dictionary of the Bible; OED* for *Oxford English Dictionary; PMLA* for *Publication of the Modern Language Association; JAMA* for *Journal of the American Medical Association.*

When many of the same abbreviations are used throughout the text, a note in the front matter is often helpful:

ABBREVIATIONS

The following abbreviations are used in the footnotes:
CAH Cambridge Ancient History
JNES Journal of Near East Studies
NH Pliny, *Natural History*

If only one or two titles are abbreviated, the form used may be given after the first citation:

¹See *The Literary Market Place,* most recent edition (New York: R.R. Bowker Co., published annually); hereafter cited as *LMP.*

Uniformity in abbreviating is imperative, and writers should follow approved forms. Since there are alternative forms of many words, the writer in a particular field should obtain a list of the publications in that field:

For medicine: *Style Book and Editorial Manual* (Chicago: American Medical Assoc., 1965).

For law: *A Uniform System of Citation* (Cambridge: Harvard Law Review Assoc., 1967).

For chemistry: *Chemical Abstracts List of Periodicals* (Washington, D.C.: American Chemical Society, 1961.)

Capitalization. Footnotes ordinarily begin with a capital and end with a period, but occasionally, in a book in which capitals are used sparingly, footnotes may begin with a lowercase letter.

¹p. 63   ²op. cit.   ³ch. 9

The nature of the notes may make lowercase more logical and appropriate, as in the following from *Nineteenth-Century Spanish Plays:*

¹*por la posta,* "in all haste"
²*ab intestato,* "intestate," "without having made a will"

The library style of capitalizing in book titles only the first word and proper nouns is not often used in trade book footnotes, although it has been adopted by publishers of some scientific books and technical journals.

[1]A.S. Hill, *The principles of rhetoric,* pp. 18–22.

The abbreviations for *page* (p.) and *line* (l.) are never capitalized, except at the beginning of a note. (The abbreviation for *line* should be used with caution, for it can easily be mistaken for the numeral 1.) Other abbreviations of parts of a book, like *vol., chap.,* or *ch.,* and *sec.,* are now more often lowercase than capitalized, though both forms are common. The form of the numerals used after these abbreviations usually varies accordingly: Vol. X., Chap. V, *or* vol. 10, ch. 5.

Page references. As a rule, numerals referring to front-matter pages should be in lowercase roman type, the rest arabic. Numbers in page references may be elided: 189–90; 171–72 or 171–2; 170–72; but not the teens: 14–15, 118–19. If elisions are used at all, they should be used consistently. (See also pp. 77, 131.)

TABLES

Although the author has full responsibility for the construction of any tables that form part of his manuscript, no final work on long or complicated tables should be undertaken until he has conferred with his publisher. The following paragraphs point out general principles he should keep in mind. The publisher will be responsible for typographical details (see pp. 255ff.).

Copy. Tables in typescript should approximate the appearance of printed tables (see Fig. 7). Copy for long or complicated tables should be on separate sheets, with the place of insertion carefully marked in the margin of the text manuscript and of the table. Short tables with no rules, or with horizontal rules only, can be typed in the text where they are to be placed.

Numbering. When tables are referred to in the text they should be numbered consecutively throughout the work, not beginning a new series of numbers with each new chapter. (Sometimes, as in a symposium or collection of articles by various persons, it is best to use a system of double numeration: "Table 7–9" would indicate the ninth table in chapter, or article, seven.) Reference in the text should be to table number, not to a specific page. The table numbers may be either roman numerals or arabic and may either be set on a separate line or run in with the caption.

TABLE II
Marsh Herbs

Table 2: Immigrant Aliens Admitted to
the United States

Captions. Table numbers and captions are usually set above the table itself (figure numbers and legends are usually below the figure). The caption should be concise and not divided into two or more clauses or sentences. Unnecessary words should be omitted. Facts shown in the column headings or pointed out in the text need not be duplicated in the caption. For example, suppose that the description in the text for the table shown in Figure 7 were

Table I shows the relative responsibility of author, editor, and printer for errors that are found in the average printed book.

A caption such as the following would then be unnecessarily wordy and repetitious:

Table I
Relative Responsibility of Author, Editor, and Printer
for Errors in Printed Books.

Credit. If a table is not original, credit should be given to the source, either below the table as an unnumbered note (as in Fig. 7) or under the caption.

Table 2. Population Growth Since 1900
Source: Bureau of Census, Department of Commerce

If it is taken from a copyrighted work and the copyright notice must also be given, it should usually appear below the table in small type (see Fig. 7).

Footnotes. Uncommon abbreviations and qualifications or limitations of data should be explained in footnotes below tables. Reference marks (star, dagger, double dagger, and so on) or superior letters are generally used as references to footnotes so that there may be no confusion with the data of the table or with text footnotes. Reference marks should follow the item explained or amplified and be arranged in order reading across the columns. A note applying to the table as a whole may be given as an unnumbered footnote with no reference mark (see Fig. 7).

Construction. Short tables are clearer and more forceful than long ones. A large, unwieldy table, therefore, should be broken up into separate smaller tables if the data will allow. All data should be arranged compactly so as to occupy a minimum of space and at the same time be clear.

Numbers in a column are more easily compared than numbers in a line. Only as many significant digits should be used as the precision of the data justifies. In a column of numbers all denoting even dollars, decimal points and ciphers to the right of the decimal points should be omitted.

A table in which every number ends in three ciphers can be made more

```
                              Table I

              Responsibility for Errors in Printed Matter

                       (in estimated percentages)

                              Author      Editor      Printer

      Dates                     70*         25*         10†

      Misspellings              40*         40*         20†

      Factual                   90          10‡          0

      Typographical             10§         10§         80†
```

Source: Personal observation of editors. ©Meredith
   Publishing Co. 1971

Note. This table is fictitious, created for
   illustration only.

*Should have verified in ms and proof.

†Should have caught typos in galleys, even if author
   and editor did not.

‡Should have called author's attention to any
   obvious errors.

§Might have been caught in proof.

Figure 7
Typewritten Copy for a Table

compact and easy to grasp by omitting the three ciphers and noting the omission in the caption.

Wheat—Receipts at Primary Markets
(All figures in thousands of bushels)

The note may be placed in the column head instead of in the caption.

Product	Quantity (*thousands of unit specified*)	Value (*thousands of dollars*)

If all the numbers in a column denote dollars, the dollar sign is used with the first number in the column and after every break in the column.

Assets

Cash	$ 80,000
Real estate	17,000
Furniture and fixtures	3,000
	$100,000

If leaders are used to indicate that there is no data available for a particular line (see p. 257), the dollar sign is placed before the first number in the column, not at the top of the column.

OUTPUT	PROFIT
12345	. . . . .
3456	. . . . .
4567	$1,456
5645	2,776

If the numbers in a column do not all denote dollars, the dollar sign should be used with every one that does.

Colleges and normal schools	15
Students in public schools	100,000
Teachers in all schools	600
Property for higher education	$ 60,000
Annual expenditure for education	$700,000

Columns of whole numbers are aligned on the right; decimals are aligned by the decimal points. Dollar signs, plus and minus signs, and the like are aligned. (See p. 257.)

Column heads. Classifications of the data presented are placed at the top and left of the data. Column headings indicate the classification of the figures or facts beneath; they should not be omitted unless the text immediately preceding the table clearly states what the table shows. The first column of a table may present the classification of figures or facts along the same horizontal line. Left-hand columns of this nature are called *stubs*. If a heading above the stub is not needed for clarity, it may be omitted.

Column heads should be concise, with no unnecessary words. For example

Births per 100,000 population	1952	1972

is better than

Number of Births per 100,000 Population	In 1952	In 1972

Capitalization in the stub should be consistent; the better style is to capitalize only the first word and proper nouns and proper adjectives ("sentence-style"). No punctuation is needed before leaders, but a colon should be used after lines that are in the nature of side headings.

MALE GRADUATES

Number graduated   76
    Number employed:
        In business for self 4
        In business with father 2
        Working for others . . . . <u>56</u> 62

    Number in college   1
    Number sick   1
    Number out of work   2
    Number who did not reply . . . .   <u>10</u> 76

The word *Total*—always used in the singular—may be omitted when the footing is obviously a total.

Leaders should be used only when they are essential to guide the eye.

	Commercial	Scholarly
Manufacturing	$ .36	$ .66
Editorial costs	.02	.03

is better than

Commercial		Scholarly
Manufacturing . . . . $ .36	. . . .	$ .66
Editorial 02	. . . .	.03

The author should try to visualize his tables in type and foresee possible difficulties in setting. Much handwork would be required to set a table like the one below in Linotype or Monotype. (It would be routine in photocomposition, see p. 516.)

CAT-TAIL FAMILY (Perennial herbs with stemless leaves)		BUR REED FAMILY (Flowers in separate spherical heads)		WATER PLANTAIN FAMILY (Herbs with long-stemmed leaves)	
Typha latifolia	*Typha angus-tifolia*	*Sparganium eurycarpum*	*Sparganium simplex*	*Alisma plantago*	*Sagittaria variabilis*
Yellow-Brown, June, July	Yellow-brown, June, July	Brown-white, May-Aug.	Brown-white, June-Aug.	White or pale pink, July-Sept.	White, July-Sept.

Rearranged as follows, it loses neither force nor clearness, and adjusts itself readily to any size page.

Cat-Tail Family (Perennial herbs with stemless leaves)	*Typha latifolia:* Yellow-brown, June, July *Typha angustifolia:* Yellow-brown, June, July
Bur Reed Family (flowers in separate spherical heads)	*Sparganium eurycarpum:* Brown-white, May-Aug. *Sparganium simplex:* Brown-white, June-Aug.
Water Plantain Family (Herbs with long-stemmed leaves)	*Alisma plantago:* White or pale pink, July-Sept. *Sagittaria variabilis:* White, July-Sept.

BIBLIOGRAPHIES

If bibliographical lists form a part of an article or book, the writer should give careful attention to their construction and maintain consistency in their form (see also pp. 266–269).

Content. Bibliographies range from the simple list of references (which is sufficiently complete when the only data given are the author's name, the title of the book, and the name of the publisher in a shortened form) to the detailed annotated bibliography (which informs the reader of all the facts of publication and gives a critical appraisal of the book). Therefore the first thing to consider is the purpose of the list. The reader who will use it must be kept in mind and the bibliography made as valuable to him as possible. For instance, a list for young readers to use in finding books at the library needs only a simple form:

The Church Mouse	by Graham Oakley
The Serpent's Teeth	by Penelope Farmer

Reading lists, which are often used at the ends of chapters in text-books, are frequently condensed to three items: surname of author, title of book, and page or chapter reference; the publisher may also be included. All books so referred to are usually given a full bibliographical listing elsewhere in the book. (In lists like the second below, the period is often omitted at the end of each entry.)

READINGS

Breasted, *Ancient Times,* Chaps. 4–18. Davis, *Readings, I,* Chap. 7; *A Day in Old Athens.* Seignobos, *History of Ancient Civilizations.* Van Loon, *Story of Mankind,* Chap. 16. Wells, *The Outline of History,* Chap. 15.

Brewster, Dorothy, *A Book of Contemporary Short Stories,* The Macmillan Company.

Harte, Bret, *Selections from Poems and Stories,* ed. C.S. Thomas, Houghton Mifflin Company.

A bibliography listing the sources used in writing the text or one listing books that would supplement the information given in the text should give all the details a mature reader would need in order to find a reference in a library or bookstore.

Full bibliographies like this differ from bibliographical footnotes (see pp. 25–28) only in minor details of form. The author's names are inverted for greater ease in alphabetical arrangement and are usually followed by a period. The publication data are not enclosed in parentheses but are separated from the other items by a period. A well-made bibliography of this kind presents the following information:

For a book:
    (1) The name of the author in the form appearing on the title page, with the surname first
    (2) The title of the book, and the subtitle, if any (underscored); the series, if any, and volume number
    (3) The edition, if other than the first (shown on the title page or the copyright page)
    (4) The place of publication (the home office)
    (5) The publisher
    (6) The date of publication (the date on the copyright page, not the one on the title page)
    (7) The number of volumes, if more than one

For an article:
    (1) The name of the writer, surname first
    (2) The title of the article (enclosed in quotation marks)
    (3) The title of the periodical in which the article was published (underscored)
    (4) The volume number
    (5) The date of issue of the periodical (month, day, and year)
    (6) The page number

For unpublished material:
    (1) The name of the writer (if known), surname first
    (2) The title (in quotation marks)
    (3) The collection, if any, and location
    (4) The date, if known

MODELS FOR BIBLIOGRAPHIES

Note. Variant forms of punctuation, capitalization, and abbreviation may be preferred by some publishers as a matter of house style. Some of these are discussed later.

Book—one author:

Cohn, Norman. *The Pursuit of the Millennium.* New York: Essential Books, 1957.

Book—two authors:

Tarn, W.W., and G.T. Griffith. *Hellenistic Civilization,* 3rd ed. London: E. Arnold, 1952.

Book—more than three authors:

Majunder, R.C., et al. *Advanced History of India,* 2nd ed. London: Macmillan, 1950. [Some publishers prefer "and others."]

Book—association the author:

American Council of Learned Societies. *Studies in the History of Culture.* Menasha, Wis.: George Banta Publishing Co., 1942.

Book—no author named:

Manual of Style, A. 12th ed. rev. Chicago: University of Chicago Press, 1969.

Book—author and translator:

Weber, Max. *General Economic History,* trans. Frank H. Knight. Glencoe, Ill.: The Free Press, 1927.

Book—author, editor, and translator:

Kees, Herman. *Ancient Egypt: A Cultural Topography,* ed. T.G.H. James and trans. Ian F.D. Morrow. Chicago: University of Chicago Press, 1961.

Book—one volume of a work of several volumes:

Eliot, Charles W., ed. *The Harvard Classics,* Vol. II, *The Meditations of Marcus Aurelius,* trans. G. Long. New York: P.F. Collier & Son, 1909–10.

Book—one of a series:

Johnson, Allen Chester. *Egypt and the Roman Empire,* The Jerome Lectures, 2nd Series. Ann Arbor: University of Michigan Press, 1951.

Technical bulletin, government report, or similar publication:

Subway Riders and Manhattan Autos. Tri-State Regional Planning Commission, Regional Profile, Vol. 1, No. 14. New York: 1971.
Masselli, Joseph W., Nicholas W. Masselli, and M. Gilbert Burford. *Controlling*

the *Effects of Industrial Wastes on Sewage Treatment.* New England Interstate
Water Pollution Control Commission, Technical Report 15. Boston: 1970.

Roché, Ben F. *Chemical Weed Control in Pasture and Range.* Washington State
University, Cooperative Extension Service Bulletin 3452. Pullman, Wash.:
1971.

Part of a book:

Russell, Bertrand. "Modern Homogeneity," *In Praise of Idleness and other
Essays.* London: George Allen and Unwin, 1960.

Part of a yearbook:

Towner, Wesley. "The Rise of Bibliomania," *The Old and the New.* 1971 AB
Bookman's Yearbook, Part II, pp. 5–19. Newark, N.J.: Bookman's Weekly, 1971.

Part of a report:

Baxendale, Phyllis. "Content Analysis, Cost Specification and Control," *Annual
Review of Information Science and Technology.* American Documentation
Institute, Vol. I, pp. 71–106. New York: John Wiley and Sons, 1966.

Article in a periodical:

Childe, V. Gordon. "War in Prehistoric Societies." *Sociological Review,* Vol.
XXXIII (1941), pp. 126–38.

Gelb, I.J. "The Ancient Sumerian Ration System." *JNES,* XXIV (1965) 230–43.

Unpublished work:

Elder, Joseph W. "Industrialism in Hindu Society." Ph.D. Thesis, Harvard
University, 1957.

VARIATIONS IN BIBLIOGRAPHICAL STYLE

Arrangement. Bibliographies are usually alphabetical, but chronology
or value may determine the order. If so, the author should make his
intention clear:

Biblical Commentaries, Listed in Order of Publication.

Although the arrangement used in the models—Author. *Title.* Place:
Publisher, date—is by far the most common, publishers in some scientific
fields, notably anthropology, have adopted a different form:

Beatty, J.H.M.
1955 "Contemporary Trends in British Social Anthropology," *Sociologus,* 5:1–
14.
1964a *Other Cultures: Aims, Methods and Achievements in Social Anthropology,* New York: Free Press.

1964b "Kinship and Social Anthropology," *Man,* 64:101–3.

In the text the credits are given in a brief bracketed note at the end of the extracts: [Beatty, 1964a, p. 32]

Names of authors. In the bibliography, the form of the names of authors, whether initials or spelled out, should correspond with that on the title page or article, unless common practice requires that it be spelled out. For bibliographies of scientific works, a form often used is the author's last name and his initials. Unless the author has definite instructions from his publisher, he should use cap-and-lower in the manuscript, which will be easier for the copy editor to mark in the style preferred.

If there are two or more authors, most readers find it less confusing if the name of only the first is inverted. Some house styles, however, call for inversion of the names of all authors.

Hart, F.W., Peterson, L.H., and Grimes, T.L.

If two or more books by the same author are listed, a long (3-em) dash is usually substituted for the name in the second and following entries:

Kael, Pauline. *I Lost It at the Movies.* Boston: Little, Brown, 1965.
———. "Raising Kane," *The Citizen Kane Book.* Boston: Little, Brown, 1971.

The dash in such cases is understood to replace the names, one or more, given in the preceding entry.

Groves, E.M.
———, and P. Blanchard.
———. [i.e., Groves and Blanchard, because all books by Groves alone would precede those by Groves and Blanchard.]

In some bibliographical styles, the names are repeated:

Boak, Arthur E.R.
Boak, Arthur E.R., and William G. Sinnigen.

A less usual practice is to use the Latin *idem,* "the same person," instead of repeating the name. Whatever method is adopted, it should be followed consistently.

Titles. Book titles and titles of periodicals are usually italicized and titles of chapters and articles are usually in roman, quoted. In some fields, however, neither italics or quotation marks are used:

33. LEWIS, T., The pathological changes in the arteries supplying the fingers in warm-handed people and in cases of so-called Raynard's disease. Clin. Sc. 3:287–319, 1938.

Titles of unpublished works are usually in roman, quoted, but it should be remembered that a work is not "unpublished" if copies of it are available to the public, regardless of the means of production:

> Kaiser, Christine R. "Calvinism and German Political Life." Ph.D. Thesis, Radcliffe College, 1961.
> Krishna, Gopal. "The Development of the Organization of the Indian National Congress." Mimeographed; in preparation.
> Braendel, Felix Werner, Jr. *Fable and History in "Beowulf": A Literary Study.* Ph.D. Thesis, Cornell University, 1970 (Ann Arbor, Mich.: University Microfilms, 1971).

Titles of bulletins, reports, and similar works may be roman cap-and-lower without quotation marks. Occasionally book titles are also so listed.

> Curriculum Laboratory Publications, No. 35
> Educational Monographs, No. 4
> University of Iowa Studies in Education, Vol. IX, No. 3
> Department of Education Bulletin No. 11

If book titles are set in roman, a foreign book title should not be italicized.

> Geddes, A.E.M., "Meteorology: An Elementary Treatise."
> Suring, Reinhard, "Leitfaden der Meteorologie."

Library lists and some specialized fields (notably medical publications) frequently use sentence-style capitalization (the first word and proper nouns and adjectives), but this is seldom followed in trade book or periodical lists of references.

> Huey, E.B., *The psychology and pedagogy of reading.*

Publication data. The place of publication, the name of the publisher, and the date are usually grouped together, but see the example under Arrangement, above. Enclosing these items in parentheses seems unnecessary unless a page reference follows.

Annotations. For annotated bibliographies, annotations should be brief and may be run in after the entry or set as a separate paragraph.

> Carroll, Lewis. *The Annotated Alice: Alice's Adventures in Wonderland and Through the Looking Glass.* Edited, with introduction and notes, by Martin Gardner. Illustrated by John Tenniel. New York: Clarkson N. Potter, 1960. Marginal annotations explicate text in terms of chess, mathematics, nursery rhymes, historical events. Introduction to Looking Glass section details structure of action as it relates to chess moves.
> ———
> Udall, Stewart L. "'. . . and miles to go before I sleep': Robert Frost's Last Adventure." *New York Times Magazine* (June 11, 1972), pp. 18-19ff. Account of Robert Frost's 1962 trip to the Soviet Union, his Black Sea meeting with Nikita Khrushchev, and repercussions during the Cuban missile crisis (1962).

Book lists. In addition to the usual bibliographical information — author, title, publisher, and place and date of publication — book lists may include other information, such as details of contents, illustrations, number of pages, binding, and price.

Bellow, Saul, *Mr. Sammler's Planet*. New York: The Viking Press, 1970. pp. vi, 313. Casebound. $6.95

Faulkner, William
   Go Down Moses   Modern Library $2.95
      Includes "The Fire and the Hearth," "The Old People," and "Go Down, Moses" as well as the full text of "The Bear."
   Absalom, Absalom!   Modern Library $2.95
   Light in August   Modern Library $2.95

ILLUSTRATIONS

In any work in which illustrations will play a major role, the author, artist, editor, and designer must work in close association. Before he does anything about illustrations, therefore, the author should consult his editor.

The discussion that follows offers only general suggestions for the writer of a manuscript in which illustrations — including diagrams, maps, plates, as well as drawings in black and white or color — are only incidental. Technical details are discussed on pages 520–524.

Line drawings. The author may make his own drawings for line cuts if he is sufficiently skilled, or he may submit sketches to the publisher, who will have finished drawings made by an expert. Maps should always be drawn by a professional. They are then usually submitted to the author for approval before they are sent to the engraver for reproduction.

Blueprints are not suitable for reproduction, and photostats should not be used if any better form of copy is procurable.

Copy for line cuts should be made with black ink on white paper. Waterproof ink is preferable for this work and should be used on smooth or medium-rough white paper that can be presented flat. Art work offered for reproduction should never be folded.

If graph paper is used for drawings, a blue-lined paper must be chosen if the cross-lines are not to show, because blue does not photograph in this process. A graph or similar diagram in which the coordinate lines are to be used in the cut should be drawn on brown-lined or black-lined paper.

Lettering. Lettering on drawings should be large enough to reproduce well when the drawing is reduced. Coordinate lettering should usually be incorporated in the drawing and not left to be set in type. If large enough and if made in black ink on light areas and in white on dark areas, lettering will show up well on line cuts.

Halftones. The halftone is used for reproducing a photograph, wash-drawing, crayon drawing, pencil sketch, etching, mezzotint, or any other picture that is made up of shading and tone. A halftone may also be made by photographing an object directly. The best halftone copy is a glossy photograph in black and white, with clearness and considerable contrast. A halftone made from a print of a halftone is unsatisfactory.

Photographs that are to be used by an engraver as copy for halftones should always be kept flat, amply protected by corrugated board or beaverboard, so that they will not become cracked or creased or torn at the edges. They should not be pasted on sheets or mounted except by an expert. Nothing should ever be fastened to a photograph by a metal clip, unless protected by a paper pad, for this will leave a permanent impression of the clip on the surface of the photograph. It is best not to write on the back of an unmounted photograph, for even the slightest pressure may leave an impression that will show in the reproduction. A sheet of thin paper bearing necessary instructions can be lightly pasted on the edge of the back and folded over the front.

Regardless of size, the copy for each halftone illustration to be used in a printed work should be on a separate sheet of paper, and each sheet should bear a notation of the number and title the illustration will have when printed. A complete list of these titles should be furnished separately.

Numbering. Text illustrations, both photographs and line drawings, should be numbered consecutively throughout the text in a single series, using arabic numerals. The preferred location for each illustration should be indicated by the author on the manuscript, using the same numbers. Even if illustrations are to be unnumbered in the text, they must be numbered for identification. "Map 1," "Map 2," in the margin of the manuscript, the map copy, and the legend copy, is more easily identifiable for the editor, the designer, and the compositor than "Map of Australia" or "Map of Taiwan."

Captions and legends. Copy for captions and legends should be typed on separate pages, not on the illustrations themselves, and numbered as indicated above.

Only a few conventions apply to the placement and wording of legends. Beyond these, the author and editor are free to follow their own ideas.

It is customary to use a period after a complete sentence used as a legend, but when the legend is topical no period is necessary. When parts of the illustrations are identified in the legend, only as much identification (set in italic) should be used as is really needed.

> Our wedding (*top left*); my father; my mother-in-law; my father-in-law ["Top left" is not really needed.]
> _____
>
> James Joyce *(center)* with Joel *(on his left)* and a neighbor
> _____
>
> *(Left to right)* henbane; monkshood; deadly nightshade

Sometimes the credit line is run in with the legends.

> My cousin, Dorothy Reed (Mendenhall). *(Portrait by N. Vollet)*

If a quotation from the text is used as a legend, it may or may not be enclosed in quotation marks.

"William stepped forward to greet the captain."

They all walked across the wide plain to the sea.

Credits. Acknowledgment must always be made for illustrations reproduced by permission. The author should keep a file of the written authorizations for their use. The simplest credit is a line, usually in italic, close under the cut, flush left or right. For example:

Courtesy U.S. Geological Survey   *W.H. Jackson*

Fremont Peak From Seneca Lake

© *Field Museum of Natural History 1973*

Certain picture agencies and copyright owners may require the inclusion of specified credit lines in the legend.

Figure 5
The Development of the Roman Alphabet
By permission of the
Norman T. A. Munder Company

Even if an illustration is taken from a book no longer in copyright, for which no permission is necessary, acknowledgment of the source is usual.

An Elizabethan ship, from an early seventeenth century engraving
(courtesy of the National Maritime Museum)

Credits for illustrations are sometimes given in the list of illustrations, in a special grouping in the acknowledgment section, or on the copyright page. The author must be sure, however, that there are no restrictions about placement of credit lines by the owners of the material.

SPECIAL RESPONSIBILITIES OF THE BOOK WRITER

The author of a book also has certain other duties. In addition to the body of the book, his manuscript should include front matter and end matter, if there is to be any.

The makeup of a book depends upon the nature of the subject matter and the simplicity or elaborateness of treatment. A simple book may contain only body matter preceded by a title page. Most books have also a copyright page, table of contents, and preface, and often an introduction and list of illustrations. These parts constitute the *front matter,* or "preliminaries," because they all precede the text. A serious work—historical, educational, scientific—usually contains *back matter* ("end matter" or "subsidiaries"). In any such work the back matter should include an index and in addition may include an appendix—notes, practice material, a glossary, or other supplementary information regarding the text (see pp. 68–69).

FRONT MATTER

The author should supply copy for any of the following parts of the front matter he wishes to use in his book: imprimatur, frontispiece, title page, dedication, table of contents, preface, list of illustrations, introduction, and epigraph.

Imprimatur. The wording of the imprimatur—the approval of the Roman Catholic Church for publication—is usually prescribed and is usually not obtainable until after the work is in proof. If the author is assured that he will receive it, however, he should so advise his publisher so that space may be left for it.

Title page. For the title page the author should supply copy showing the title of the book and the form in which he wishes his name to appear. His degrees and associations and other books he has written should not be on the title page copy but should be given to the publisher on a separate page.

Dedication. If the author wishes to have a dedication page, he should phrase the dedication as simply as possible.

Table of Contents. The writer should include a complete table of contents with his manuscript, listing front matter, chapter titles (with subdivisions, if desirable, in a textbook or other nonfiction), and back matter. (See Fig. 8.) Some publishers prefer to have the page numbers indicated by 000 (to be filled in after the work is paged), with the manuscript page number given in the margin, in pencil and circled.

Preface. Any remarks the author wishes to make to the reader should be placed in the preface. Acknowledgments of indebtedness may also be included, or they may be under a separate heading. At the end the author should sign his initials and note the place and date of completion. If a preface is written by someone other than the author, it should be signed by the full name of the writer of it. If the preface has not been written when the author delivers his manuscript, he should put in a page on which he writes "Preface to come."

The terms *preface, foreword,* and *introduction* are often used interchangeably. It is desirable, however, to differentiate them. A *preface* or *foreword* deals with the genesis, purpose, limitations, and scope of the book and may include acknowledgments of indebtedness; an *introduction* deals with the subject of the book, supplementing and introducing the

Contents

Contents

Figure 8
Typewritten Copy for Simple Tables of Contents

text and indicating a point of view to be adopted by the reader. The introduction usually forms a part of the text; the preface does not.

List of illustrations. Copy for the list of illustrations is easy to prepare. Brief titles of all the illustrations are listed. If there are both plates and text figures, they may or may not be listed separately.

Epigraph. If the book is to have an epigraph (a short quotation appropriate to the content), it should be included in the manuscript on a separate page. Quotation marks are not necessary, but the source should be given — the author's name, the title of the book, or both. Other bibliographical details are unnecessary unless the quotation is used by permission and the granter of the permission insists on an acknowledgment (see p. 50).

BACK MATTER

Each section to appear in the appendix should start on a new page in the manuscript and should be typed double-spaced and numbered consecutively with the text. Notes, glossary, bibliography, and any other material should be arranged in the order in which they are to appear. A general index cannot be made, of course, until the work has been paged, but if there is to be an alphabetical index of authors or an index of first lines in an anthology, the copy should be submitted with the manuscript. If there is to be a general index, a final page saying "Index to come" should be included. Whether the author plans to prepare the index himself or to have it prepared by a professional indexer, he should be familiar with the general principles of compiling indexes (see pp. 76ff.).

LEGAL PROVISIONS AFFECTING THE AUTHOR

Everyone who writes for publication should be aware of the provisions of three laws that concern authors particularly — the Copyright Law, laws against libel, and laws protecting the right of privacy. A writer may be held responsible for violation of these laws; he should therefore be informed of their provisions.

THE COPYRIGHT LAW

The common law protects the rights of an author in an unpublished manuscript. After publication this protection ceases. Unless the published work carries a legal notice of copyright, it becomes public property, "falls into the public domain."

Copyright can be secured on books, pamphlets, periodicals, newspapers, gazetteers, addresses, lectures, sermons, dramatic or musical compositions, charts, maps, engravings, prints, photographs, paintings, drawings, and plastic works. The protection lasts twenty-eight years and may be renewed for a second term. During the terms of copyright, reprinting, copying, translation, adaptation, and abridgments are prohibited in the absence of specific permission by the copyright proprietor.

Infringement of copyright. There is no copyright on the title of a book, but if there occurs a clear case of dishonest imitation, legal action may be taken.

Quotation is not always infringement. Long custom has recognized the right of reviewers to quote passages of sufficient length to illustrate their criticism and indicate the quality of the book under review (though not to the extent of materially robbing it of its value as literary property), and a writer of a technical work, such as a medical or legal treatise, usually intends it to be cited and quoted for professional purposes in subsequent works on the same subject.

Such uses all come under the concept of *fair use.* Many publishers who are members of the Association of American University Presses have subscribed to a resolution that allows quotation without permission from books published under their imprints, for "accurate citation . . . for criticism, review, or evaluation," provided that appropriate credit is given. This does not apply, of course, to "quotations that are complete units in themselves (as poems, letters, short stories . . . maps, charts, graphs, tables, drawings, or other illustrative materials)," or to a quotation "presented as primary material for its own sake (as in anthologies or books of readings)." These exceptions are never considered fair use.

In interpreting the law, the courts look to the nature, quantity, and value of the material used and determine whether its use was unfair to the quoted author. A quotation so large or significant as to injure the worth of the original and in a measure satisfy the market for it is beyond question an infringement. The number of words quoted is immaterial. Three hundred words quoted from a short story, or two lines from a lyric poem might constitute an infringement. On the other hand, limited extracts in textbooks and reasonable extracts for illustration would not generally be called unfair.

There is no copyright on ideas or information, only on the form in which they are expressed, but a paraphrase or a condensation, though it may be more freely used than a literal quotation, should be acknowledged.

The author should be sure he can secure all necessary permissions for borrowed material before he sends his final copy to the publisher. For short excerpts there is usually no charge, but copyright proprietors usually require a credit line to protect their rights. If permission should be refused or the charge should be exorbitant, the author might want to substitute other material.

Securing permission to quote. Written permission must be secured from the copyright proprietor to use copyrighted material, including charts, tables, and illustrations. This usually means from the publisher, who grants permission on behalf of the author. Some publishers who control many copyrights have a form that they send to writers who make requests for permission to use excerpts from books on their lists. One such form requires the following information:

(1) Title of the book to be quoted from
(2) Nature of the work for which the selection is requested
(3) Name of the author or compiler
(4) By whom the book is to be published
(5) The intended time of publication
(6) Selections desired (These are to be specifically designated by giving the first word or phrase of each and the page on which it appears, and the last word or phrase of each and the page on which it appears.)
(7) Total pages or total lines
(8) The market for which rights are requested (for example, U.S. only; U.S. and Canada; or world in English language)

Acknowledgments. Sometimes the copyright proprietor permits the borrower to phrase the credit lines. The borrower can then approximate a uniform style, placing the acknowledgments on the verso of the title page or in a footnote thus:

> [1]Alistair MacLean, *Force 10 from Navarone* (London: Heron Books, 1968), p. 81. Copyright © 1968 by Cymbeline Productions Ltd.

A chart or table often has a credit line in small type directly below it on the left, or the credit may be set within parentheses directly under the caption. Occasionally it follows the legend. A photograph sometimes has an acknowledgment in small type directly under it at the left, and beside it, at the right, the photographer's name.

When a publisher specifies the placement and the phrasing to be used in the credit line, his instructions must be followed precisely. For example, a condition in granting the permission might be that a notice such as the following be carried as a footnote on the page on which the excerpt appears.

> [2]From "John Brown's Body" in *Selected Works of Stephen Vincent Bénet.* Copyright 1927, 1928 by Stephen Vincent Bénet. Copyright 1955 by Rosemary Carr Bénet. Reprinted by permission of the publisher, Holt, Rinehart and Winston, Inc.

Securing a copyright. Copyright for a book, short story, article, or other similar work is secured by the act of publishing the work with the prescribed copyright notice on the copies. Copyright cannot be secured for a manuscript of such material, which is protected under the common law prior to publication. (Unpublished lectures, dramatic works, music, and certain other classes of works may, however, be copyrighted in manuscript form.)

Whether the author retains the copyright or transfers it to the publisher is a matter for agreement between them. In most cases the publisher prepares the copyright page and attends to the registration of the claim.

Promptly after publication two copies of the work, a completed application form, and a $6 fee is sent to the Register of Copyrights, Copyright Office, Library of Congress, Washington, D.C., 20559. "Publication" is defined as the placing on sale, sale, or public distribution of copies. It does not refer to the method of reproduction, which may be by printing, microfilming, or some other method.

There are a number of different application forms for the various classes of works. Form A is for books printed and first published in the United States. There are other forms for books first published outside the United States. Form B is for periodicals, and Form BB for contributions to periodicals. Application forms may be obtained from the Copyright Office without charge, as well as information and special circulars on subjects such as renewal, the copyright notice, recording assignments, new versions, and the like.

A new copyright cannot be secured for a reprint, but a registration may be made for each new version of a work containing substantial new copyrightable material, provided the requirements for the form of the copyright notice are met. Copyright on a new version does not extend the term of the original work.

Under the Copyright Law now in effect (1973), the term of copyright is twenty-eight years, with the privilege of renewing for a second term of twenty-eight years. (The application for renewal must be made in the last year of the first term.) However, the copyright on many books whose second term would have expired in 1962 or succeeding years has been extended on a year-to-year basis in the expectation that a revision of the law will be enacted. Hence it is unsafe to assume that any work with a copyright date of 1906 or later is in the public domain.

Works of United States citizens may be protected in many foreign countries, but the methods for securing protection differ. Copyright is secured in the countries currently parties to the Universal Copyright Convention, which became effective in 1955, if all published copies of a work bear the notice prescribed by the Convention, that is, the symbol ©, the name of the copyright owner, and the year date of first publication. No registration in these countries is necessary to secure copyright, but the claim to United States copyright on a book first published here must of course be registered in the Copyright Office in the usual way.

Great Britain and most continental countries are also members of the International Union for the Protection of Literary and Artistic Works, better known as the Berne Union. Citizens of a country that is not a member of the Union, such as the United States, can secure the rights granted by the Berne countries only if they publish their works first or simultaneously in a country of the Union.

Works in the public domain. The copyright on all works published in the United States before September 1906 has now expired. The copyright on any work published over 56 years ago that was not renewed at the proper time has expired. These works are now in the public domain. Note, however, that "new matter" contained in any revised editions that were properly copyrighted after those dates is probably still protected.

Works by citizens of Berne Union countries and of many South American countries, and works first published in those countries are protected in those countries until (usually) fifty years after the author's death, regardless of their status in this country.

Works published or generally distributed and sold in the United States with no legal notice of copyright are in the public domain in the United States.

Publications of the United States Government and official state publications are in the public domain. This is not always so of writings or statements by public officials or employees, however, nor of material quoted in official publications, even if there is no specific notice of copyright.

News—that is, the facts of current events—is public property, but not the editorials, comments, or opinions expressed about news.

If in doubt about the copyright status of any borrowed material, the author should always consult his publisher. His editor can usually advise him about obtaining copyright information, such as by checking in the *Catalog of Copyright Entries,* which lists the facts not only of original copyright registrations, but also of renewals; or by having a search made by the Copyright Office.

Bibliography. Books on copyright written for the layman rather than the specialist are necessarily over-simplified. Among those that the writer may find useful are

A Copyright Guide, 4th ed., by Harriet F. Pilpel and Morton David Goldberg. New York: R. R. Bowker, 1969.
A Manual of Copyright Practice, 2nd ed. with a new Preface, by Margaret Nicholson. New York: Oxford University Press, 1970.
Protection of Literary Property, by Philip Wittenberg. Boston: Writer, Inc., 1968.

Not for the layman but a reference on domestic and international copyright law is

Nimmer on Copyright, by M.B. Nimmer. New York: Matthew Bender and Company; looseleaf, brought up to date by annual supplements.

LIBEL

Any writer whose work, in whole or in part, discusses or comments on real living people, should bear in mind the words of Sir Frederick Pollock: "A man defames his neighbor at his peril."

Libel is "any written or pictorial statement that damages a person by defaming his character or exposing him to ridicule" *(American Heritage Dictionary).* There is no one legal definition of libel, for laws against libel differ from state to state. But again and again the courts have used the same words: libel, as opposed to slander, is defamation in "permanent form"; it is printed, published, expressed in words, signs, symbols, or other permanent methods; it exposes the subject to "public hatred, shame, ridicule, or contempt," "disgraces him as a member of his community," "degrades him in his business, profession, or office."

Publishers of newspapers, periodicals, and books dealing with current affairs are well aware of the perils of libel (for the publisher is deemed as culpable as the writer) and usually consult their own legal advisers about any material that might be considered libelous. But libel may be disguised or inadvertent. A character in a novel may be libelous, even if only partly modeled on a real person. The name used in the novel and the reprehensible actions of the character may be wholly fictitious, but if the neighbors of the real person can recognize him from his description in the book, or by the milieu in which he is pictured as moving, the writer may face a suit for libel.

Or a writer may describe a purely fictitious editor of a literary journal as being essentially self-promoting and malicious, and choose a name for the magazine that he believes never has existed. If unhappily there is such a magazine and the editor imagines that he is being maligned, he may threaten suit. The note often included in novels to the effect that all characters are fictitious and any resemblance to living people is accidental is useless if defamation, malicious or otherwise, can be proved, but might be effective against a person, unknown to the author, who claimed imaginary resemblance.

A word or group of words may be considered libelous in one community or at one time, but not so in different circumstances. For example, to say a man is a homosexual may be libel or may be a statement of acknowledged fact. Slang meanings, understood by large segments of the community but not found in any dictionary, may libel a man's character. Some of the "vulgar" uses, although given in many recent dictionaries, might not alert the copy editor's attention. Or a writer might use a word without realizing its implications. Usually "deliberately false or malicious" occurs in definitions of libel, but neither is necessarily essential.

Political and editorial comment is not considered libelous if it does not contain false statements of fact. (But merely saying "alleged"—by some unknown source—is no protection.) A commentator may express his opinions, however unfavorable they may be, if they are clearly given as his *opinions* and nothing more. A literary or dramatic critic may demolish the work he is reviewing, but he may not demolish the character of the creator of the work. For example, he may say a work is imitative and has no originality, but if he says or implies that the author is a plagiarist he invites a libel suit.

Biographers of living persons should be especially careful about the possibilities of libel. Documented facts about the life of public figures are not libelous; usually the danger lies in statements about their friends and relations, who are not public figures. As to persons who are dead, the serious biographer or historian usually has nothing to fear; there is danger only if it can be proved that there has been a malicious intention to injure the family or descendants of a deceased person and expose them to defamation.

It is impossible here to give an account of libel adequate for the writer. A comprehensive treatment of the subject is given in *Libel* by Robert Phelps and Douglas Hamilton (New York: Macmillan, 1966). If there is any question in the mind of author or editor about the possibility of libel, he should seek legal advice.

THE RIGHT OF PRIVACY

This is another field with possible legal dangers to the writer or photographer of contemporary events. Again, each state has its own definitions, and certain restrictions are recognized under common law.

In general, "public figures"—actors, writers, politicians, artists, athletes, musicians, or anyone who voluntarily offers his person or his product to the public—cannot object to comment and discussion by the public (including writers) about what he has offered. His name in a way has become public property. He can object, however, to having his name or his likeness used for commercial purposes without his permission. The public figure may be regarded as being in the public domain as a subject for factual biography, news, discussion, public information. But not for fiction, advertising, or purposes of trade.

So far as "private figures" are concerned, the only safe thing for the writer to do who wants to use their names or likenesses is to get their consent in writing. Oral consent might be binding, depending on the circumstances, but publishers know that even persons whom the writer has described as personal friends may change their minds after they see their names or pictures in print.

PART II.   COPY AND PROOF

WORKERS ON COPY AND PROOF

After the writer of a book has sent his manuscript to the publisher and all contractual matters are in order, he will still be obliged to wait some time for the processes of manufacture to get under way. During this period the editor will read the manuscript critically and attend to the numerous details that are his responsibility. Another reading will then be given the manuscript by a copy editor (sometimes called a manuscript editor), with attention to typographical details and preparation for the printer.

The duplicate copy of the manuscript is turned over to the production department, where it is designed and all details of format and manufacture are worked out, usually in consultation with the editorial, sales, and publicity departments.

But the allocation of these responsibilities differs in various publishing companies. The editor may also do much of the copy-editing, and he may also decide upon many matters of design and format, or there may be a managing editor who takes care of these details (see p. 491). In the following pages, therefore, the word "editor" should be interpreted loosely. It may mean *editor, copy editor,* or *production man,* depending on the type of organization in the publisher's office, as well as on the size of the company.

THE EDITOR

In general, it may be said that the work of the editor is to see that the manuscript is editorially and factually correct.

The editor's first reading of a manuscript is critical of content, appraising its value, detecting its weaknesses and its strong points, judging its literary style and the skill, or lack of it, shown in the assembling of the subject matter. He will focus his attention not only on the larger aspects of content and arrangement, but also on details of what the author has written and how he has expressed it. He will see to it that the sensibilities of the readers of the book have been respected and not unnecessarily offended. Local prejudices will be kept in mind and respected whenever possible.

He will also look at the book from the standpoint of marketability. His acquaintance with the books of other publishers on the same or similar subjects may lead him to suggest changes in scope or emphasis. Expansion of some parts and simplification of others may seem to him desirable. He may, indeed, advise the rewriting of large portions.

EDITOR OR COPY EDITOR

After the editor finds that the manuscript is substantially satisfactory, a second reading is necessary; sometimes this reading is by the editor, sometimes by the copy editor. In this critical reading the whole matter of expression is carefully considered. More quickly than the author himself, the editor will notice a habit of using clichés, or of repeating pet expressions. He will be quickly aware of an overabundance of elliptical phrases, of too frequent use of *but* at the beginning of sentences, of shifts from third person to second where parallel construction would be preferable. In short, the diction, grammar, and rhetoric of the writer are subject to critical inspection. The editor has learned from experience that the possibilities of error are limitless, and he does not expect the author to have detected his minor blunders.

Especially in nonfiction works, the editor must be alert at all times to inaccuracies and conflicting statements. Geographical features must be carefully checked, an exacting task in some books. Classical, historical, and literary references may be inexact and must be scrutinized. Dates and all references or statements in which there is a time element must be examined with care. At the same time the editor must keep printing conventions in mind. His experience with the work of many writers has shown him in general the sort of inaccuracies he may find. He knows that tables are often not constructed to the best advantage; that footnotes and bibliographical references are often incomplete; that cross references are not always in the most desirable form. It is better, for instance, to write, "in Figure 1" than "in the illustration above," since the exact placement of figures and illustrations cannot be determined until the work is in page proof.

The editing of a work of fiction often proves trying. Unless the writing has been carefully done by an experienced author, the editor is likely to find puzzling capitalization, faulty punctuation, and misused words. The story may be full of dialect or of localized words not to be found in the dictionary. It may contain slang that the author could not decide how to spell. Briticisms may be used inaccurately, or foreign words and phrases may be used to excess. Misquotations and the misspelling or misuse of common expressions are not uncommon; such expressions, for instance, as *lay* (not *lie*) *of the land, whited sepulcher, to the manner born, combat à outrance* are often misspelled. Anachronisms may appear, and inconsistent statements: sometimes the girl who had blue eyes in the beginning of the story changes them for brown later on. In short, only by thoughtful concentration on details can fiction be satisfactorily edited.

The application of familiar expressions must be checked for correctness, since the exact meaning of such expressions is sometimes not clearly understood by authors. For example, "the devil's advocate," signifying someone who depreciates for a good cause, is often inaccurately used. A similar fault is often observed in the use of foreign phrases. *Vice versa* is a common term that needs watching; it is often used carelessly. The propriety of using non-English words is often questionable. Whenever an equally expressive English word can serve the same purpose, it should be substituted.

If illustrations are part of the manuscript, the editor must examine them closely, since unusual and unexpected inconsistencies are often discovered in them. They must show exactly what text and legends say they show. If a children's book describes the snow "drifting slowly down, like petals of apple blossoms," the illustration should not look as if an arctic blizzard were in progress. Drawings sometimes show surprising inaccuracies, such as a violinist playing with the bow in his right hand although the text has made a point of his being left-handed.

Finally, the editor must make sure that no libelous remarks have been made and that there are no infringements of copyright or trademarks.

No single book would bring all these problems to the editor, but in the course of his work he would meet them all more or less often. Many of them are specifically considered in the following pages.

THE COPY EDITOR

To the person inexperienced with publishing, it would seem that after all the reading and preparation described in foregoing pages, a manuscript should certainly be ready for the printer. The copy editor knows otherwise. Enough remains to require his concentrated attention and painstaking care to complete the work satisfactorily. He is probably the last one who will see the book as a whole until it is completed. From him it will go to the printer, who may give it to several different compositors to set in type. Half a dozen typesetters and proofreaders may be assigned to work on it; therefore it will not be satisfactory to write "Capitalize this throughout," "Italicize when so used"; the style must be carefully marked from beginning to end.

At the same time the copy editor is expected to read the text with care. He will make sure that proper names are spelled correctly, that dates are accurate, that quotations from Scripture and from classical works have been copied correctly. Realizing the multiplicity of details the editor had on his mind, the copy editor does not expect that the editor will have noticed all errors, and he will be alert himself to detect what may have been overlooked.

The copy editor must be familiar not only with the techniques for preparing manuscripts, described in Part I, and with the conventions of book makeup (pp. 61–69), but also with the details of copy-editing style, typographical style, grammar, and the use of words, as given in Parts III, IV, V, and VI. Some authors, in concentrating on content, overlook errors and inconsistencies in the manuscript that they would see immediately—and deplore—on the printed page.

Style. The extent to which the copy editor corrects punctuation, capitalization, and other details of style depends on the kind of work and the proficiency of the author. Publishers of textbooks, scientific books, and the like usually adhere strictly to house style and the copy editor is authorized to make all necessary changes to achieve uniformity. In editing literary work, he is more cautious. When the author is an experienced writer and the manuscript gives evidence of having been meticulously prepared, the copy editor will usually not quibble over punctuation. If the use or omission of a comma can be justified by any recognized rule of punctuation, no change should be made without consulting the author; particular care should be taken not to change the meaning of a sentence, as can so easily be done by changing the punctuation.

He should, however, correct or query improper capitalization, irregular compoundings, wrong paragraphing, and inconsistencies of spelling. It is generally understood that if spelling is inconsistent and the company follows a particular dictionary, he must change all spellings not conforming; otherwise all the copy editor can do is to call the inconsistencies to the attention of the author.

Grammar and rhetoric. The copy editor should catch inadvertent errors in grammar: disagreement of verb with subject, failure of a pronoun to refer clearly to its antecedent, the fragmentary sentence, and so on. He must remember, however, that in fiction the author sometimes has stylistic reasons for disregarding conventional rules. (Imagine Molly Bloom's soliloquy in *Ulysses* neatly punctuated and paragraphed!) Whenever major points of style are in doubt they should be called to the attention of the author—in other words, *queried.* The copy editor may correct or query such common errors as the use of two *that's* where only one should be used; and redundant expressions like *funeral obsequies, present incumbent, first beginning* (see also pp. 407–410).

Queries to the author. It is usually necessary for the copy editor to refer various questions to the author. If there are relatively few, he may incorporate them in a letter, citing manuscript page and line. If there are many queries, however, he often attaches *flyers* to the manuscript pages opposite the queries and sends the whole manuscript to the author. In a covering letter the author is asked to answer each question, but not to detach the flyers or make changes in the manuscript itself.

Marking the manuscript for the printer. When all questions have been settled to the satisfaction of the author and the publisher, it is the duty of the copy editor to mark the manuscript for the printer. He will have received from the editor or the production department a schedule of the types to be used for text, extracts, footnotes, tables, and headings, and instructions about any special forms called for. He will be held responsible for marking the copy throughout in accordance with this schedule. He will see that chapters, sections, numbered tables and illustrations, and other numerical sequences are in order; that footnotes and bibliographies follow a style consistently. He will note throughout the use of italics, abbreviations, figures, capitals, and other details, and correct any inconsistencies.

Even though the printer will also receive a copy of specifications drawn up by the production department, the copy editor enters on the first page of each chapter the opening indention and style of the initial letter, as well as type size, typeface, and leading. He must indicate all matter to be set as extracts and how they are to be set, especially if they include poetry. Subheads, figures, and tables in nonfiction works must be marked clearly for placement, capitalization, and punctuation. Footnotes must be clearly indicated and marked for placement and style. The more meticulously the manuscript is marked, the sooner it can be set in type and the fewer changes will be necessary in proof.

THE PRODUCTION DEPARTMENT

While a general understanding of the methods and the materials concerned in book manufacture is often helpful to the editor, the actual design and manufacture is something that is usually supervised by a trained production man. The production department decides upon the details of format and manufacture: typeface and type size, leading, size and bulk of the book, paper, binding, dimensions of the printed page and similar matters (see pp. 491ff.). If the book has footnotes, extracts, figures, tables, or illustrations, all questions relating to them must be answered.

In close cooperation with the editorial department, the production department prepares a *layout,* a detailed design or plan of the completed book.

When all decisions have been made regarding the physical appearance of the book, the editor will usually ask the production department to get sample pages from the compositor that has been selected to set the book. Copy for diagrams, graphs, equations, examples of text matter with heads and folios, symbols, non-roman alphabets or other unusual characters are sent with the text to be set. Sample pages are for the purpose of illustrating page design and typography only; they are not, as the authors who see them sometimes think, to be judged for correctness of spelling or typographical errors.

LAYOUT

With the exception of the placement of the copyright notice, the arrangement of the various parts of a book has been established by custom rather than by rule, and the editor may vary it in any way that his copy seems to make desirable. Generally speaking, however, it is wise to follow an order such as the one given below, although few books would have all these parts.

Lowercase roman numerals are usually used for the front matter, and the text starts on page 1 (see pp. 269–270). Right-hand pages (*rectos*) are odd-numbered and left-hand pages (*versos*) are even-numbered.

ARRANGEMENT OF THE PARTS OF A BOOK

Front Matter

— Certificate of limited edition (usually not counted in numbering)
  i Bastard title (also called first half-title or false title)
 ii Book card, imprimatur, series title, or blank
— Frontispiece (usually unnumbered)
iii Title page
 iv Copyright page
  v Dedication
 vi Blank
vii Table of contents
viii Table of contents (continued), or List of illustrations
 ix List of illustrations, or Acknowledgments
  x List of illustrations (continued), or Acknowledgments (continued), or blank
 xi Editor's preface
xii Editor's preface (continued), or blank
xiii Author's preface
xiv Author's preface (continued), or blank
 xv Foreword or Introduction (if not part of text)
xvi Foreword (continued), or blank

Text

  1 Second half-title, or part title
  2 Epigraph, or blank
  3 First page of text

Back Matter

Appendix (supplementary material)
Notes
Glossary
Bibliography
Index
Colophon

FRONT MATTER

The book card, frontispiece, and copyright page are always left-hand pages. Other sections traditionally appear, or at least begin, on right-hand pages. When space is limited, however, short units, such as the imprimatur, the dedication, and the epigraph, are placed wherever they can best be accommodated.

Certificate of limited edition. This page, called limit page or limit notice, is an announcement of the number of copies printed, worded somewhat like this: "This edition is limited to 1000 copies of which this is No.   ." The number is written in, and sometimes, also, the signature of the author or the publisher. Copy for this page must, of course, be supplied by the editor.

Bastard title. The bastard title, also called false title or first half-title, is ordinarily exactly like the half-title that precedes the text. If, however, the volume is divided into "Books," "Parts," or "Units," each preceded by a half-title, the bastard title should bear the title of the book, not the part title used on the second half-title. A series title may precede the bastard title, follow it, or appear on the title page.

Card page. The book card, also called card page or face title, is a list of books by the same author or a list of books in the same series.

Imprimatur. If it is expected that the book will receive an imprimatur ("let it be printed," granted by the Roman Catholic Church), its placement should be planned when the book is designed. It usually appears on page ii, but may be placed on any other unused or unfilled page in the front matter.

Frontispiece. The frontispiece is usually an insert tipped in to face the title page and is unnumbered. It may, however, be printed on the same kind of paper as the text and be included in the pagination.

Title page. The title (and the subtitle if there is one) should be the most prominent information on the title page, followed by the name of the author. At the foot of the page should appear the publisher's imprint and the place of publication (the publisher's home office). If a translator, editor, compiler, or illustrator has made a major contribution to the work, his name also appears, usually after the author's. A nonfiction book that has been revised often carries the number of the edition on the title page: "Third edition, revised and enlarged."

Other items may appear if the editor so chooses, but before adding anything he should consider carefully whether the title page is the proper place for the addition. For instance, if the book is one of a series, that information can be given on the title page but often would better head a list on the preceding page or be placed on the bastard title. Lines under the author's name, "*Author of . . .*" are also more appropriately presented on the card page. Some publishers carry the year date of the edition or reprinting, but this is now more often given only on the copyright page.

In modern trade books, the designer is often given a free hand. Sometimes the title page is a two-page spread, extending over pages ii and iii. In illustrated children's books the artist may contribute to the design.

Copyright page. The editor usually supplies the copy for the copyright page, since he is familiar with the legal requirements and the forms adopted by the publisher.

The law specifies that the copyright notice shall be "on the title page or the page immediately following," and shall consist of the word *copyright* or the copyright symbol © accompanied by the name of the copyright proprietor and the year of publication. The order is not important so long as the three elements appear in juxtaposition.

© Jane Carew 1970 *or* © 1970 by Jane Carew

Each revision must be copyrighted and the new copyright information

added to all former notices. If the copyright owner is the same, it is necessary only to add the new date.

© Jane Carew 1970, 1973

If the copyright owner of the new matter in the revision is not the
same, however—the original author has died, for example, and the revisions were made at the expense of her estate or of the publisher—a
second notice must be added.

© Jane Carew 1970   © Jane Carew 1970
© by the Estate of Jane Carew 1973   © 1973 by Tufts Publishing Company

The publisher of a translation may be required by contract to carry the
copyright notice of the original work:

Originally published in France under the title *L'Histoire de l'Empire romain*
© J. Laurens 1969
English translation © 1972 by Tufts Publishing Company

Some publishers prefer to use the word copyright as well as the copyright symbol ©. This is not legally necessary, however, especially for
books published after 1955, when the Universal Copyright Convention
became effective.

Since one requirement for securing copyright on a book first published
in the United States is that it shall be manufactured here, there should
also appear on the copyright page the words: "Printed in the United
States of America."

If a book is published under a different title from that under which
it has been previously published (in a periodical or newspaper, or under
a different imprint, for example) the previously used title must appear
on the copyright page. (The former title must also appear on the front
flap of the jacket and be mentioned in catalogs and circulars.) This is to
comply with a ruling of the Federal Trade Commission.

The foregoing items of information are required. In addition, copyright pages may carry other information:

"All rights reserved," which gives the work copyright protection
in most South American countries.

"Published simultaneously in Canada by———*"* followed by the
name of the Canadian publisher, if arrangements have been made for
Canadian publication in order to protect the copyright under the
Berne Convention. (Simultaneous publication in any other Berne
Convention country would be equally satisfactory.)

The Standard Book Number (SBN), assigned according to an
arrangement with the R.R. Bowker Company.

The Library of Congress Catalog Number (L.C.No.), assigned on
application to the Library of Congress Descriptive Cataloging Division.

The printing history of the book: "First printing 1960; Seventh printing 1972," for example. Note that "printing" refers to "impression," that is, reprintings with no changes, or only minor changes, in the text. If substantial changes have been made, the printing is a new edition.

In any new printing it is advisable (though not required) to include the copyright renewal date if the copyright has in fact been renewed:

Copyright 1940, 1968 by Morris Hamson

Or, if the author is dead and his widow renewed the copyright:

Copyright 1940 by Morris Hamson, copyright renewed 1968 by Phyllis Jean Hamson

(The copyright symbol is not appropriate for books first published before 1955 or for copyright renewals.)

The copyright page in dramatic works often has a special warning against infringement, similar to the following:

Caution. Professionals and amateurs are hereby warned that this play, being fully protected under the Copyright Law of the United States of America, the British Empire, including the Dominion of Canada, and the other countries of the copyright union, is subject to a royalty; and anyone presenting the play without the consent of the author, or his authorized agent, will be liable to the penalties provided by law. All applications for the right of amateur production must be made to [name and address of agent]. All applications for the professional rights must be made to the author.

If not given on the title page, credit is sometimes given on the copyright page to special contributors to the book: "Designed by————";
"Edited by————"; "Translated by————".

Dedication. If the author wishes a dedication he should be urged to word it as briefly and simply as possible. Ideally it should be on the first right-hand page after the title page, followed by a blank page.

Table of contents. Formerly it was the convention to place the table of contents (now usually headed *Contents*) and list of illustrations after all other preliminary matter except the introduction, second half-title, and epigraph. This practice is still followed by many publishers, on the theory that *Contents* refers to the contents of the text only. When a short preface is the only other front matter, the placement is not important, but in books that have substantial preliminary material it is a convenience to the reader to have the contents page as early as possible and to have all sections listed in it — Preface, Foreword, Introduction, and even Acknowledgments. A growing number of publishers, therefore, follow this practice, even in works of fiction.

Although the writer should supply copy for the contents page, it should be checked carefully by the editor against the text itself. Page numbers, to be filled in after the book is in page proof, are indicated by 000.

The contents page should be in harmony with the character of the book. The designer no longer follows a stereotype form but works in close cooperation with the editor. Notice how different the following arrangements look:

CONTENTS

Introduction, *by George Soule*	vii
The Runaway Boy	3
The Circus-Day Parade	15

Contents
Acknowledgments, x
Prologue, xi
Part One: The Island, 1
Part Two: The Flight, 29

CONTENTS

THE PLAN OF THE WHOLE WORK	v
PREFACE TO VOLUME I	ix
ABBREVIATIONS	xiv
AN INTRODUCTION TO ECONOMIC HISTORY	1
The Nature of Economic History	1
The Sources	4
I. THE NEOLITHIC PERIOD	8
The Neolithic Economy	8
What is a Peasant?	12

A long table of contents may be set in two columns, or the subheads may be run in. Leaders are used only when necessary for clarity.

CONTENTS

Preface	v	Author	61
I		Books	69
Background of Copyright Law	3	Choreography	73
The Copyright Office	9	Classification	74
The Subject Matter of Copyright	13	Common Law Rights	79

CONTENTS

CHAPTER I. ECONOMIC CAUSES OF WAR 3
    Economic factors neither exclusive nor final, 3. War as an end in itself, 4. Economic Objectives of War, 5
        The quest for "living-space," 5. The quest for raw materials, 11. The quest for market, 16. The quest for investment outlets, 19.
    Economic Conditions Making for War, or, the Economics of Peace, 26.

A contents with page references within a paragraph, as above, should not be set in type until the book is in pages and correct page numbers can be filled in.

List of illustrations. Like the table of contents, the list of illustrations should also be submitted by the author but must be carefully checked by the editor against the illustrations themselves. Long captions need not be given in full, but may be a shortened identification.

When there are both text illustrations and plates, they should be listed under separate heads.

ILLUSTRATIONS

Positive Health	*frontispiece*
Early Diagnosis of Pregnancy	5
Thorough Obstetrical Supervision	6
Protection of the Expectant Mother	16
Safe Delivery in an Adequate Hospital	17
Relative Proportions of the Body at Various Ages	22
Intrauterine and Extrauterine Environments	23
The Newborn Personality	*facing* 24
The Relation between Heart Shape and Body-Build	*facing* 45
The Technique of Nursing	51
Equipment for the Preparation of the Formula	56

ILLUSTRATIONS
Plates
Between pages 80 and 81
1. The Young Shakespeare
2. Sir Francis Drake
3. Elizabeth I
4. Sir John Hawkins

Facsimiles
A page from *The Merchant of Venice*   38

ILLUSTRATIONS IN THE TEXT
1. Two Types of Magnet   5

Errata. An errata list, if required, is often placed after the contents, but sometimes at other points in the front matter or even at the end of the book. If the book is already bound when the errata are discovered, a separate leaf may be inserted.

When constructing copy for this page, the place the error occurs must first be given and then the words *for* and *read* used in this manner:

> Page 177, line 8: *For* Charleston represented *read* Charleston, favoring manu-
> factures as a relief to poor whites, represented
> Page 202, footnote 100: *For* Sherill *read* Sherrill
> Page 214, line 25: *For* adjured *read* abjured

No punctuation should be used at the end of corrections unless it is part of the correction.

Preface. The preface may precede the contents page or follow it (see p. 65). It always starts on a right-hand page.

An editor's preface (whatever it is called) usually precedes the author's preface; that to a second edition precedes the one to the first.

If there is no separate acknowledgment page, acknowledgments may be incorporated in the preface, sometimes under a separate centered subhead.

Epigraph. The epigraph sometimes appears on the title page, sometimes on the back of the dedication; it may replace the second half-title or be on the back of it, facing the first page of text.

The information on the source should be only enough to identify it—the writer's full name or his last name only, if he is well known; the title of the work; or both. It is not necessary to give page or line numbers or any other bibliographical data. (If it is used by permission, acknowledgment can be made elsewhere, preferably on the copyright page.) The type of the credit, whether cap-and-small, cap-and-lower, or italic, the punctuation before and after, and the position in relation to the quotation or to the right-hand margin are all matters of style requiring the editor's attention.

> "If your will want not, time and place will
> be fruitfully added."   *King Lear*
> _____
>
> He that hath never done foolish things never
> will be wise. — Confucius

If there is not room enough for the credit on the last line of the quotation, the credit should be dropped to the next line and aligned at the right, not divided.

BACK MATTER

Appendix. The appendix should begin on a right-hand page and be preceded by a half-title if there is room. If there are several sections, they may be headed *Appendix A, Appendix B,* and so on, each followed by its own title and each starting on a new, preferably right-hand page. Discursive sections should come before lists such as Notes, Glossary, or Bibliography.

Glossary. A glossary usually precedes the bibliography. If many of the entries are short, as in a textbook vocabulary, a two-column arrangement may save space. If economy of cost needs to be considered, the typography should be kept as simple as possible. For instance, lightface roman, italic, and cap-and-small are more economical to set than these three faces with boldface in addition.

> *Legend.* The descriptive line or lines accompanying an engraving . . .
> Legend. The lines of descriptive matter accompanying an illustration . . .
> **legend** The title or short description accompanying an illustration . . .
> LEGEND An explanatory caption accompanying a map, chart, or illustration . . .

The editor should check the alphabetization and also check for uniformity of style in the definitions.

Bibliography. The bibliography should be checked carefully by the editor, and all rules for content and order presented on pages 37–38 and shown in the models on pages 38–43 should be observed.

Index. The index or indexes (in books of poetry there may be an index of first lines as well as of authors, for example) should usually be a separate section, not part of the appendix, and may be arranged in various ways. When the typographical form of the index has been decided upon, the editor should check the work of the indexer. (See p. 96.)

Colophon. In some books that are specially designed, and perhaps published in limited editions, the publisher may elect to include a colophon. This is a brief bibliographical note giving details of design, typography, and general makeup. The designer or production department usually supplies the facts; the editor is usually responsible for writing the copy.

The following appears on the last printed page of *For the Love of Books*, by Paul Jordan-Smith, published by Oxford University Press, 1945.

> This edition of *For the Love of Books*, abridged
> from the original version, consists of 600
> copies specially printed and bound
> for private distribution. It was
> designed by John Begg and
> set in Times New Roman.

Note: In the early days of bookmaking, the colophon ("finishing touch") appeared on the last page of the book and gave most of the details now shown on the title page. Hence the use of the term for publisher's device, trademark, or symbol, which used to appear at the end of the book, and now often appears on the spine and the title page.

WORK ON PROOF

As soon as is practicable, the manuscript, now spoken of as "copy," will be sent to the printer, who will have it set in type and read before sending proofs to the publisher.

Typographical errors. It is next to impossible for the average worker to set type without making errors that need correction. The compositor therefore employs trained proofreaders to detect such typographical errors and to mark them for correction.

Of first importance is the detection of poor spacing and alignment, imperfect letters, crooked lines, protruding spaces, letters of wrong style or size, "outs" (a word or words omitted), "doubles" (a word or words set twice), and wrong indentions. Next come the misspelled words, of which there are two classes: first, such spellings as *ecstacy, sieze, her's, childrens'*; second, such errors as *stationery* instead of *stationary*, *therefore* for *therefor*. Errors such as these are marked for correction before proofs are sent to the author, and are not considered "author's alterations."

Proofreader's marks. The proofreader uses many of the editorial marks mentioned on pages 10–12 and several others applicable only to proofs. Every correction in a proof must be indicated in the margin; therefore each one requires two marks, one within the text and one in the margin. The one within the text is usually a caret or a line indicating where the correction is to be made. (See Fig. 9.)

Author's reading of proof. The author is responsible for the factual proofreading and for returning the proof promptly to the publisher. If he takes more time than has been allotted for this purpose, he may delay the whole publishing schedule for his book.

GALLEY PROOFS

The first proofs to come to the author will ordinarily be in strips called galley proofs or galleys. (A simple text that has been well prepared may be made up into pages before proofs are sent to the author, an economical and time-saving procedure that could be adopted more generally if authors could be persuaded to prepare their copy with care.) He will usually receive two sets, a marked set—sometimes called the *master proof*—and a duplicate set. (Some publishers do not send the master proof to the author, but only a duplicate. When it is returned to the publisher the copy editor then transfers the author's corrections to the master set.) The author will also receive the manuscript, but only for use in proofreading. No marks of any kind should be made on it now.

The author must read galley proofs with care, checking them against copy (the manuscript). If the text is complicated—a scientific work with many formulas, for example, or an anthology of previously published material—he will find his reading more accurate if he adopts the method used by the printer, using a copyholder to read with him. If he does not have anyone to act as a copyholder, however, he will find it helpful to use a card or a ruler as he reads, sliding down the galley from line to line, so that he concentrates on only one line at a time.

The copyholder. The most important requirement of a copyholder is that he be accurate. He reads aloud from the copy as the reader follows on the proof, and it is essential that he read exactly what is in the copy. This requires care. It is easy indeed, for instance, to read *judicial* for *juridical*, or *alteration* for *alternation*.

Reading practices. Italics, boldface, and capitals must be clearly denoted, and most of the punctuation read. In order to read accurately, the copyholder must be able to interpret correctly the marks and abbreviations used in editing and proofreading (see pp. 11–12, 71).

Reading type and punctuation. Boldface and italic words should be indicated, but not roman unless they appear in the midst of boldface or italic type. Paragraphing should be mentioned. Punctuation marks should be named, with the exception of obvious uses of the period and the comma.

Proofreader's Marks

delete or take out	set in lowercase letters	lc / lc
delete and close up	set in capital letters	cap
close up; no space	set in small caps	sc
insert space	set in italics	ital
insert letter, punctuation, word	set in roman type	rom
[move left	set in boldface type	bf
move right]	set in lightface type	lf
move 'up' or down	insert period	⊙
align	insert comma	
align horizontally	insert semicolon	
broken type	insert colon	
make new paragraph	insert apostrophe or single quotes	
no new paragraph	insert quotation marks	
transpose	insert reference	
words transpose	insert hyphen	=
lines transpose	insert dash	
run on	do not make correction indicated	stet
break run over line	spell out numeral: ①	sp
turn over	set as numeral: one	fig
	query, verify: Mosel ?	

An early colophon of **D. Appleton and Company** was the punning device of the apple tree, with an open book at its foot, bearing the letters "d. a. & Co." Around it was a scroll with the words *Inter Folia Fructus* and the statement "Established in 1825."

The colophon of the Century Company, designed about 1875 by Stanford WHITE—was a blazing sun with an open book at its center. The original D. Appleton-Century Company colophon bore the apple tree with the sun against its trunk and in the middle of the sun an open book bearing the dates 1825, 1870, 1933. The surrounding scroll bore the old Appleton motto with the name of the old company at its foot; in 1945 a new colophon, designed by W. A. Dwiggins, displayed the tree, the suns rays, the dates 1825 and 1870, and the initials AC, the whole surround with a wreath. The old motto was omitted. The colophon of Appleton-Century-Crofts, Inc., was an adaptation of this, with the initials A C C and the single date 1825. The colophon of Appleton Century Crofts, Division of meredith Publishing Company, show only the tree and the initials A C C.

A 2nd meaning of the term colophon is, "a statement placed at the end of a book, giving information about the book's production." Included would be such information as one. name of the designer and printer 2. kind and size type used 3. type of paper used be included in the colophon. Specific details concerning the particular edition may also

---

Figure 9
Example of Marked Proof

For instance, it is not necessary to read the punctuation in *A.M., P.M., New York, N.Y.,* or in dates, unless conventional punctuation is omitted. NYC, for example, would be read "NYC no points." The hyphen should not be read in words that are always spelled with a hyphen, such as *twenty-five, sister-in-law.*

Reading names. Any variation from the common form of spelling a name should be told the reader. For instance, *Laurence* should be read "Laurence with a u," because otherwise the reader may suppose it is spelled *Lawrence.* Similarly, "Elisabeth with an s."

Reading numbers. Numbers should be read as clearly and concisely as possible. "One sixty-five" is the proper way to read 165, not "one hundred and sixty-five"; "four o eight" for 408. A four-figure number in which no comma is used should be read in twos: "twenty-six forty-five" for 2645, but "two, six forty-five" for 2,645. A number of five or more figures should be broken at the comma: "three eighty-nine, one fifty-six" (389,156). If the figure preceding the comma is a cipher—as in 30,156—the number is better read "three o, one fifty-six," so that it cannot be mistaken for "thirty-one fifty-six (3156). Fractions are read with the word *over* or *slant*: $\frac{4}{5}$ is "four over five," ⁴/₅ is "four slant five." Roman numerals are read by the letters of which they are composed; that is III is read "three i's," LXXIV is "l, two x's, i, v." Since the capitalized form is the common one, the lowercase form is the one specifically designated.

Reading accents and signs. Accents and signs must be recognized and accurately named.

Accent or Sign	Copyholder's Term
acute (´)	acute
grave (`)	grave
circumflex (^)	circum *or* flex
tilde (~)	tilde *or* Spanish
cedilla (¸)	soft
macron (-)	long
breve (˘)	short
dieresis (¨)	di
(⊥)	mod
(°)	circle
asterisk	star
superior figure	sup (2/ is sūp 2)
inferior figure	sub (₂ is sŭb 2)
foot or minute (')	prime
inch or second (")	double prime
multiplication sign	mul
division sign	div
percent sign	per sign
cents (¢)	cut c
per, shilling (/)	slant
thousand (M)	canceled M
dollar sign ($)	dollar
pound sign (£)	pound

Application of methods. The foregoing instructions may be more clearly understood by comparing the paragraph below with the paragraph as read by a copyholder.

> **B.C. and A.D.** (for *before Christ* and *anno Domini*). Set in small capitals, with no space between the letters. Place date before B.C. and after A.D.: 14 B.C., A.D. 28.

As the copyholder would read them: "B.C. and A.D., all bold, paren, for before Christ, two ital, and anno cap Domini, two ital, close paren, point. Set in small capitals, com, with no space between the letters. Place date before B.C., small caps, and after A.D., small caps, colon, fourteen B.C., small caps, com, A.D., small caps, twenty-eight."

Marking proofs. The marked set of proof should be used by the author when he reads. At the top of the galley may be words and numbers not a part of the text. These may be circled, but they should never be obliterated. They have significance for the compositor, and he will discard them when he makes up the pages.

The author should use ink of a different color from that of any markings already on the proofs so that the author's alterations will be clearly distinguishable from the printer's errors. Corrections should always be marked in the margin, because a correction written within the page with no mark in the margin is more than likely not to be noticed by the printer. If more than one error occurs in a certain line, the corrections in the margin should always read from left to right. That is, if two or more corrections are written in the left margin, the first one should be written far enough to the left so that the next one can be written to the right of it. The conventional proofreading marks shown on page 71 should be used.

New material and insertions of more than one line should be typed, and the place of insertion clearly indicated, so that no time will be lost in deciphering handwriting. Small slips (*flyers*) bearing insertions should never be pinned to proofs; they may be pasted *on* the margin but not *to* it.

Correction. Proofs are pulled on cheap paper and they are never so clear-cut as the final printing will be. It is therefore a waste of time for the author to bother about evenness of printing. Rows of crosses opposite a line in which the type looks broken are useless and really defeat their purpose.

The author's attention in reading should be directed to accuracy of the content, rather than to literals—that is, typographical errors—although any of these missed by the printer should of course be corrected too. The author should verify carefully all names, all tables and numerical calculations and formulas, and the spelling and division of foreign words and phrases. He should not make any changes in style, form of abbreviations, or similar details without consulting the publisher. He should make only changes that are necessary or that are of vital importance. If he makes an addition, he should try to make a compensating deletion, thus lessening the expense of the change (see Author's Alterations, p. 9).

General directions, such as to spell a word a specified way throughout, or "All instances like this should be arranged as here marked," are not effective. The only way the author can be sure of obtaining the form desired is to mark each instance himself.

Queries. *Qy., Qu.,* or a circled question mark is the proofreader's way of calling the author's attention to a particular detail that it seems to him might be a slip. The author should carefully read the passage questioned and answer the queries. If the change suggested meets with his approval and he wishes it made, he should cross off the question mark or *Qy.*; if the change is not to be made, he should cross out the entire query so that there will be no doubt about what he wants. It is not enough simply to write *O.K.,* because it is not always clear *what* is O.K. For instance, suppose the query is *?/tr,* and the author simply writes *O.K.* The compositor will not know whether he means "The phrase is O.K. as it is," or "All right, make the transposition."

The author should never erase any marks made in the margin by the proofreader or publisher. If they indicate a change that the author does not want made, they should be crossed through and the words "O.K. as set," or similar phrasing, written in the margin.

Under no circumstances should any corrections or changes be marked on the manuscript.

Illustrations. Illustrations that are numbered will be inserted by the compositor as near the point of reference as the limitations of makeup will allow. If an illustration *must* be inserted near a given place, however, the fact should be clearly indicated on the galley proof.

Some publishers submit to the author a *cut dummy*—proofs of all illustrations, pasted on sheets of uniform size, with the figure numbers in the lower corner. The author should transfer from the original drawings to the cut dummy the figure numbers and titles for the cuts; he should mark there also all corrections to be made in the illustrations. These cut dummies should be returned with the corresponding corrected galleys.

Return of proofs. The marked set of proof, together with the corresponding copy, should be returned as promptly as possible to the publisher. The duplicate set should be kept by the author.

PAGE PROOFS

Page proofs are sent to the author principally for the verification of cuts and captions, for the adjustment of long or short pages, for filling in of page numbers in cross references, table of contents, and list of illustrations, and for the making of an index (see p. 76). These proofs also are on cheap paper, and the author should not be disturbed because the pages do not line up, are not square on the paper, or vary in heaviness of impression.

Revision. In printing terminology, revising is the process of comparing a proof with a previous proof of the same matter to see if all corrections marked on the earlier proof have been made accurately. Page proofs are always revised in the compositor's office, and the compositor is responsible for accuracy in correcting, but the author will do well to read his page proofs carefully, at the same time checking the headings at the top of each page ("running heads") and the folios (page numbers). (See below.) It should be remembered that often a whole line or more must be reset for a change of a single letter, and in correcting one mistake another is sometimes made elsewhere in the line. The author should therefore examine carefully all lines affected.

Long and short pages. A new page should not begin with a short tag-end remnant of a paragraph (called a *widow*); if asked to do so, the author should try to bring forward one or more full lines from the preceding page. The last page of a chapter should have at least four lines on it; a heading should always have at least two lines under it on the same page. It is not always possible to secure pages of the right length without changing the wording slightly. If the makeup man in the compositor's office finds it impossible to adjust the length by changing the spacing around headings, he may mark the page "long" or "short"; in a short page he may insert a turned slug at the end of a paragraph. This will show in the proof as a wide black line. Again, if asked to do so, the author should add enough near the end of the paragraph to "gain a line"; the correction of the short page will then be simplified. Similarly, a page marked long can sometimes be corrected by crowding back a word or two. The author will make the compositor's work easier and save time and expense if he will make some change near the end of the paragraph that will "gain a line" or "lose a line" as needed, but only if he is asked on the proof to do so.

Illustrations, tables, and figures. The compositor will have inserted the illustrations and the larger tables in the pages as near to the point designated by the author or editor as the type will allow. The author should verify their position and their titles and any phrases in the text referring to them as "in the figure above," "on the next page," "opposite," "below." When a narrow illustration is inserted in a type page it is desirable to have the type form an even frame around it. Sometimes a paragraph ends beside the illustration so that the even outline is broken. The author can then add a word or two to fill the blank space and thus improve the appearance of the page. (For printing conventions relative to the placement of illustrations on the page, see p. 524.)

Front matter. When the page proofs are received, all folios can be filled in. Therefore the table of contents and the list of illustrations can be completed. All front-matter pages should be read with particular care, especially the title page and the copyright page.

Return of proofs. It is advisable to keep all the page proofs until the index has been prepared, work which should be started as soon as the first page proofs are received (see Indexes, which follows). Inconsistencies of spelling or statement often come to light during the index work, and if all the page proofs are at hand, corrections can be indicated. As promptly as possible the complete set of proof should be returned to the publisher.

Return of the page proofs indicates to him that the book is "O.K. for press." The author should not thereafter ask for any further changes.

INDEXES

The value of any book of nonfiction will be enhanced by a well-prepared index; without one, although the text may contain a wealth of information, the subject matter is largely inaccessible.

Most publishing contracts place the responsibility for the index upon the author, who must either prepare it himself or have one prepared at his expense by a professional indexer. Theoretically the author should prepare his own index, as he is obviously the best authority on his book. However, every author has his enthusiasms, modes of expression, and preferences in terminology, which, if he is inexperienced in indexing, may actually be a disservice to him. He may overlook some synonyms that might occur to the ordinary reader when searching for a topic. Or he may treat the index almost as an afterthought, hastily prepared to meet a deadline. A good professional indexer, who must have a wide vocabulary as well as skill in the mechanics of indexing, can usually interpret the material to the benefit of both the author and his readers.

If the author has the time and feels that he is best prepared to make the index for his own book, or if the index is being prepared by someone who is not a professional indexer, the following general principles can be used as a guide.

The index cannot be made until page proof is received, and then it should be done as promptly as possible so that production will not be delayed. Even publishers who have allowed a generous amount of time for index preparation in their original production schedule are often forced to make up time by reducing the original allotment. Indexing should not be begun in galleys, however, in hope of saving time later. The time-consuming task of cross-checking galleys with page proofs will offset any initial saving of time.

Procedure. First, read over the pages and mark the entries by underscoring the key word on the duplicate set of page proof. Write in the margin any wording that is not in the text (see p. 80). When all entries have been marked, write them, with their appropriate page numbers, on index cards. (Slips of paper are not efficient.) Each main entry, subentry, or cross-reference should be entered on a separate card. After all main entries, subentries, and cross-references have been entered, type the index on regular typing paper, double-spaced, one column to a page. When the publisher does not specify a particular line length, set the typewriter for 40 characters per line.

Professional indexers develop their own techniques—for example, typing entries on perforated adding-machine tape or using a dictating machine—but these are for experienced, skilled individuals; the inexperienced should use the index-card method as just outlined.

If the indexing is being done by a professional indexer, who is probably not expert in the subject matter of the text, the author may go over the page proof first and do his own underlining before the pages are sent to the indexer. The author should write in the margin of the proof synonymous terms, not mentioned in his text, for cross-references (a useful code from author to indexer is "X-R" to indicate cross-referred terms). The indexer must still read over each word of the text, but when his vocabulary and interpretations are added to the author's, the usual result is a better index.

Page references. There are several ways of presenting the page references (or folios) for an index entry. The simplest, but least desirable, is to give only the first page on which a discussion begins. A better method is to give inclusive page numbers for each entry. Perhaps the best method is to give inclusive page numbers for up to three pages but use the first page number of the discussion with *ff.* ("folios") for longer discussions. For example, a page reference such as "264–66" may be meaningful, but a reference such as "157–92" does not mean much to the reader except that a long discussion begins on page 157; the same information is conveyed more concisely with "157ff." Where there are numerous subentries in a complicated index, the page on which a major discussion begins can be indicated with *ff.* at the main entry, with subentries to lead the searcher to each particular topic. Also, provided there are subentries, *ff.* (as "42ff.") may be used with discretion to avoid unwieldy inclusive page references, such as "42–47, 48–49, 50, 51–58." Note that *ff.* should be set roman in the index.

Boldface and italic type are sometimes used for special references— to illustrations, tables, major discussions—with a note at the beginning of the index such as "Boldface page numbers indicate material in tables or illustrations." Proper use of *ff.* will often eliminate the need for boldface for major discussions; the use of a different typeface for illustrations or tables will depend upon the nature of the book.

A page reference to a footnote should be followed by a small *n.*—for example, "133n."

Inclusive page numbers may be given in full ("413–415") or elided ("413–15"); either way, when type is set an en dash, not a hyphen, should be used. There are several styles of elision; whichever is chosen, the indexer should be consistent. The following is a suggested way: Omit from the second number the digit(s) representing hundreds, except when the first number ends in two zeros, in which case the second number should be given in full. If the next-to-last digit in the first number is a zero, only one digit is necessary after the en dash.

8–10, 22–23, 100–102, 107–9, 119–21, 133–34, 1074–76

INDEXES BY NAMES

Most books require only one index, the so-called dictionary type, listing names, titles, and subjects in one alphabetical sequence. In special cases, however, more than one index may be desirable. For instance, for a book containing selections from many different authors or one that is heavily referenced, an index of authors may be useful; a book with many legal citations could have a table of cases. An anthology of poetry may require one index of authors and titles and another of first lines; on occasion there may be a need for a third, of subject matter, but usually the subject matter and author-title indexes are combined.

Index of authors. An index of authors requires only the transcribing of the names with the surname written first, and then an arrangement of the names in strict alphabetical order. The most efficient way to arrange the cards is first to sort them into alphabetical groups and then to arrange each group in order by the second or third letter as may be necessary, *a, ab, aba, ac,* and so on. For indexing names with prefixes, Spanish names, and pseudonyms, see pages 83–86.

In all types of indexes, initials in names should be closed up, as an economy of space (for example, *Adams, J.Q.*).

Table of cases. Index simple citations, such as *Trevett* v. *Weeden,* under both names. Even though the names may be italicized within the text, roman type is preferable for the table of cases.

Trevett v. Weeden, 49
Weeden, Trevett v., 49

Invert phrases like *In re, Ex rel., Ex parte, In the matter of, Estate of,* and *U.S.* (use the abbreviation in the index even if *United States* is spelled out in the text).

Adams, Abigail, Estate of, 96
Addington v. Baker, In re, 195
Adler, John, In the matter of the
    petition of, 140
Adler, Robert, U.S. v., 43
Admont, Harold, Ex parte, 78

For corporate names beginning with initials, index under both the first initial and the surname.

P.N. Jones Co., U.S. v., 133
Jones Co., P.N., U.S. v., 133

Index of titles. An index of titles requires the listing of the titles in alphabetical order, word by word, inverting an initial article—*A, An,* or *The.* The same rule for inverting articles applies to titles in a foreign language, but since many readers may not be familiar with a particular language, it is often wise to index under the article as well as the noun. For example, a reader might look under the *D*'s for *Das Kapital* but under the *M*'s for *La Misère de la philosophie;* therefore the first should be under both the *D*'s and the *K*'s, and the second could be placed only under the *M*'s. When in doubt, put the title in both places.

It is a good practice to place the author's surname in parentheses following a title, especially when the title is general, as *An Autobiography* or *Travels in Greece*. For a common surname it is sometimes well to use identifying initials also—for example, *American, The* (Adams, J.T.).

Titles should not be inverted (except for initial articles) and entered as if they were subject-matter entries. For example, *Fundamentals of Electrical Engineering* should not be entered either as *Engineering, Fundamentals of Electrical* or as *Electrical Engineering, Fundamentals of*. If both the title and the subject matter are to be indexed, the entries should be as follows:

> Engineering, electrical, 57
> *Fundamentals of Electrical Engineering*
> (Wilton), 57

Possible: Electrical engineering, 57

The following is a sample of how some titles might be alphabetized in an index:

Iacchus (Hystonis), 13
I Am a Camera (Isherwood), 49, 53
If All These Young Men (O'Brien), 122
I Pagliacci. See *Pagliacci.*
Iron Heel, The (London), 91, 101
It Is Never Too Late to Mend (Reade), 14

Ladies!, The (Adams, L.), 275
Lady of Belmont, The (Ervine), 93
L'Allegro (Milton), 158
Lenin (Ottmark), 184, 185
Les Misérables (Hugo), 142–44
Levanna Overcome, La (Farkle), 351

Note that *L'Allegro* and *Les Misérables* should also be indexed as *Allegro, L'* and *Misérables, Les.*

COMPILING A DICTIONARY-STYLE INDEX

The general index, in contrast with the foregoing kinds, is often exceedingly hard to compile, for compilation not only involves a transcription of titles and names, with an occasional inversion, and arrangement according to set rules, but it also requires discretion and judgment in choosing what is to be included and in placing the best word in the key position.

The indexer must first decide what to put on the index cards, then how to write the entry. Only after his cards are all written does the question of order arise. The sections immediately following, therefore, concern choice of entries and the form in which they should be written. The alphabetical arrangement of the cards is discussed on pages 90–93.

Choice of key words. The bulk of the entries in an index should directly relate to the subject matter of the text, so before any key words are selected, the indexer must have a clear idea of the central concerns of the text and from this must decide on the scope and kind of entries to be selected. For example, the general policy in indexing is to give entries for most proper names, titles, and other capitalized terms. But, if a book on motion pictures discusses the scene in Alfred Hitchcock's *Saboteur* (1942) in which Robert Cummings signals a message using a book titled *Escape,* Alfred Hitchcock, *Saboteur,* and Robert Cummings would be indexed, *Escape* probably should not be indexed.

Proper names, capitalized events and eras, titles of books or articles, and definitions are easily picked up in choosing key words. Choosing a key term from a general discussion is more difficult, and selecting synonyms for a key term is perhaps the most difficult of all. It is here that a wide vocabulary and an acquaintance with the subject matter are most important. The indexer must consider what the general reader might look for, not just the expert. Also, the author may use different key words for the same topic in various places in the book, but they all must be put together and cross-referred in the final editing of the index. For example, in one section "Congress" may be the key word, in another section "Senate," and in other places "House committees," "representatives," "bills," "legislation," "elections," and so on. These must all eventually be consolidated or cross-referred. If "Spanish architecture in the Southwest" is the topic, make entries for the three obvious words and also immediately consider "buildings," "churches," and "homes." In addition to the two key terms in "heart attacks in the elderly," also make entries for "cardiovascular disease," "aged," "coronary artery disease," and possibly "death rate" or "mortality," "blood vessels," "arteries," and "circulation." The number of additional entries made for any subject will depend upon the importance of the discussion in the context of the whole book and the readers for whom the book is intended.

Once the key word has been selected, write it on the index card, inverting qualifying words if necessary and punctuating carefully so that everything reads back to the key word or term. Each subentry is entered on a separate card and must be keyed back to the main entry so that it will accompany the main entry when the index is assembled.

Compound nouns. Distinguish compound nouns, properly indexed like simple nouns, from noun-and-adjective combinations, which should be inverted and indexed under the noun.

fund raising	figurines, earthenware
Triple Entente	granaries, public
water supply	tablets, ancestral

If there is a question whether certain words form a compound, make entries in both ways, and if there are only a few page numbers, give them with both entries. (This is called double entry.) The page numbers take less space than a cross-reference, and the searcher is saved time.

The following shows a failure to bring the significant words to key positions; ordinarily an adjective should not be used as a key word.

Poor: *Better:*

Divine, art, 000; art and the, 00; the,   Art: divine, 00; divine origin of, 00
   and earthly beauty, 00; the, as ma-   Divine, the: art and, 00; earthly
   terial, 00; origin of art, 00; purpose   beauty and, 00; as material, 00
   explains beauty, 00; purpose explains   Divine purpose: explains beauty, 00;
   pleasures of imagination, 00   explains pleasures of imagination, 00

Impure, art, 00; ideational style, 000;   Architecture, inner character of, 00
   music, 000; sensate style, 00   Art, impure, 00
Inner, aspect of culture, 00; character   Culture, inner aspect of, 00
   of architecture, 000; character of   Experience, inner, 00
   literature, 00; character of music, 00;   Literature, inner character of, 00
   experience, 00; culture of mentality,   Mentality, inner culture of, 00
   00   Music: impure, 00; inner character of,
                                                00
                                             Style: impure ideational, 00; impure
                                                sensate, 00

Phrases. Arrange phrase headings logically, with the significant word in the key position.

dress, forms of   Bill of Rights
humor, sense of   Elixir of life
Labor, Department of   Society of Jesus

Do not use descriptive phrases as headings, but invert to bring to the key position the noun denoting the topic of the paragraph or article indexed.

Wrong: *Better:*

Agriculture   Agriculture
   formation of prorate districts, 1   date palms, quarantine, 3
   regulating livestock grazing on U.S.   director of, salary of, 5
      public lands, 2   eggs, storage, 4
   relating to quarantine on date palms   grapevines, transportation regula-
      of Africa, 3   tion of, 6
   relating to storage of eggs, 4   livestock grazing on U.S. public
   salary of director of, 5   lands, 2
   transportation regulation of grape-   prorate districts, formation, 1
      vines, 6

Highways   Highways
   establishing an additional state high-   additional state, establishing, 11
      way, 11   appropriation for construction of, 12
   making an appropriation for the con-   width of, 13
      struction of, 12
   relating to the width of highways, 13

Inversions. When the natural order of closely connected words is changed to bring the important word to the key position, place such words as closely together as possible.

Brown, James, Jr. [*not* Brown, Jr., James)
Homer, poetry of, popular theology in
James I, King, anecdote of [*not* James I, anecdote of King]

Apply the rule with discretion and avoid unnecessary inversion.

Poor: *Better:*

Great Britain, isolation of, German policy of, 106 / Great Britain, German policy of isolation of, 106

Hoare, Sir Samuel, Laval and, peace plan of, 28 / Hoare, Sir Samuel, peace plan of Laval and, 28

Schuschnigg, Dr. Kurt von, Government of, overthrow of, 70 / Schuschnigg, Dr. Kurt von, overthrow of government of, 70

If subentries require *with, and, vs.,* and so on, invert when possible so that the key word appears first.

Poor: *Better:*

Dash, em, 00 / Dash, em, 00
  with comma, 00 /   comma with, 00
Heredity, 00 / Heredity, 00
  vs. environment, 00 /   environment vs., 00
Obesity, 00 / Obesity, 00
  and diet, 00 /   diet and, 00

Conciseness. Words and phrases like *concerning, regarding, relating to* have no place in an index. Omit prepositions, conjunctions, and articles that are not necessary to clearness. The articles *a* and *the* are seldom required in an inverted entry[1] unless they are the first word of a capitalized title; even then they may sometimes be omitted.

Not concise: *Better:*

Hosea, 38; conditions of, 39; marriage of, 40; message of, 46; person and life of, 43; teaching of, 49 / Hosea, 38; conditions, 39; marriage, 40; message, 46; person and life, 43; teaching, 49

Macmillan Company, The, 168 / Macmillan Company, 168
See also Ascot; Derby, The; Grand National / *See also* Ascot; Derby; Grand National
Society of Jesus, the, 96 / Society of Jesus, 96

---

[1]An example of an article necessary to meaning is shown on page 81; divine, the: art and, 00; earthly beauty and, 00.

Take care not to omit a preposition that is necessary to the meaning. For instance, *Acceptance, benefits* is ambiguous; a preposition is needed to make the meaning clear: *Acceptance, benefits of,* or *Acceptance, of benefits.* So, also, the following:

Exchange, 30
  restrictions on, 32
Inoculation, 19–21
  by nurses, 20

Geographical names. Index under the specific word unless the parts are inseparable or they compose the name of a city.

Fundy, Bay of   Cape Coast, Ghana
Horn, Cape   Lake of the Woods
Olympus, Mount   Mount Vernon, N.Y.

Names of persons. Observe established usage in the indexing of proper names. Persons referred to in the text by surname only should be further identified in the index. Ordinarily it is not the purpose of the index to supply information that cannot be found within the text, but an exception should be made to provide forenames or initials for names of persons. This is necessary to avoid possible problems with entries such as "Washington" (George, D.C., or state?), or "Booth" (Edwin, John Wilkes, or telephone?).

If no forename or initials appear in the text, there may be some question about the identity of the person. For example, in a discussion of the period just preceding the Revolutionary War, a mention of General Howe probably refers to William but could mean George Augustus. The author must be queried in such cases. If the author checks the page proofs before they go to the indexer, he should note in the margin the forename or initials next to entries about which confusion could arise.

English compound names. Index under both parts.

Campbell-Bannerman, Sir Henry, 76
Bannerman, Sir Henry Campbell-, 76

Spanish compound names. Index under the first part.

Avila Camacho, Manuel
Cervantes Saavedra, Miguel de
Moreto y Cabaña, Agustin
Vega Carpio, Lope Félix de

Noblemen. Index noblemen under their most familiar designation, usually the highest title or the family name. In most cases, double entries should be made: if there are only a few page numbers, give them in both places; otherwise make a cross-reference. When the title is the entry, the family name should follow in parentheses.

> Buckingham, Duke of (George Villiers)
> Chesterfield, Earl of (Philip Dormer
>   Stanhope)
> Disraeli, Benjamin
> Pitt, William

Cross-references to the preceding would read as follows:

> Beaconsfield, Earl of. *See* Disraeli,
>   Benjamin.
> Chatham, Earl of. *See* Pitt, William.
> Stanhope, Philip Dormer. *See* Chester-
>   field, Earl of.
> Villiers, George. *See* Buckingham,
>   Duke of.

Sovereigns, princes, writers. Index under the first name if the person is generally so known. (See also Pseudonyms, below.)

> Bonaparte. *See* Napoleon.   Twain. *See* Mark Twain.
> Ward, Artemus. *See* Artemus Ward.   Henry, O. *See* O. Henry.

Index under the full name if the person is habitually so spoken of.

> Kemal Ataturk
> Mark Antony
> Omar Khayyam

Pseudonyms. Index under the pseudonym if that is better known than the real name; otherwise under the real name with a cross-reference from the pseudonym. Use a double entry if desirable.

> Elia. *See* Lamb, Charles.
> Eliot, George, 42.
> Evans, Mary Ann. *See* Eliot, George.
> Lamb, Charles, 96.

Names with prefixes. Prefixes are difficult to handle, as they should properly be indexed according to the bearer's personal preference or in the form traditionally most familiar. Since this is often difficult to determine, the indexer must use his judgment about making double entries. The following very general rules can be used as a guide to solve most problems.

For an Englishman or an American retain the prefix:

> De Forest, Lee   La Follette, Robert
> De Quincey, Thomas   Van Buren, Martin

For a Frenchman or a Belgian retain the prefix when an article is involved:

La Fayette, Comtesse de   La Bruyère, Jean de
L'Enfant, Pierre Charles   Du Guesclin, Bertrand

but invert a pure preposition:

Balzac, Honoré de   Sévigné, Madame de
Grasse, Comte de   Tocqueville, Alexis de

For a Spaniard (except compound names) or a Dutchman retain the *de:*

de Sitter, Willem   De Soto, Hernando

For a German or a Dutchman invert *von, van*:

Beethoven, Ludwig van   Hoff, van't
Bernstorff, Johann, Graf von

In doubtful cases (Van Eyck, Della Robbia) list under the specific name and the preposition also.

Arabic names. In Arabic names *abu* (father), *umm* (mother), *ibn, bin* (son), and *ahu* (brother) are considered to begin the surname and therefore determine the alphabetical position.

abu-Bekr   ibn-Saud, Abdul-Aziz

Articles—*al-, ad-, ar-, as-, at-, az-*—are disregarded in determining the alphabetical position but are not inverted.

al-Masudi—under M

Oriental names. In indexing names of Orientals take care not to mistake titles for names. The following are titles: *Babu, Bey, Emir, Pasha, Pundit, Sri.*

Ibrahim Pasha—under I

In Chinese names the surname always comes first; therefore there is no inverting when such names are indexed, though a cross-reference may occasionally be necessary.

Chiang Kai-shek   Sun Yat-sen

Japanese, Indian, and Vietnamese names are written in the same order as English names.

Gandhi, Indira   Noguchi, Hideyo
Ky, Nguyen Kao   Thieu, Nguyen Van
Nehru, Jawaharlal   Yamagata, Prince Aritomo

Saints. Saints referred to as personages should be indexed under the personal name. In the names of places, churches, universities, and so on, *St., Saint,* or *Sainte* should be placed in the *S*'s; the same holds for names of persons other than saints (see also p. 116).

Agnes, Saint, 00   Sainte-Marie, Buffy, 00
Paul, Saint, 00   St. Paul's Bay, 00
St. Agnes Street, 00   St. Petersburg, Fla., 00
St. Denis, Ruth, 00   Saint Xavier College, 00
Sainte-Étienne, 00

Alternative foreign names. The indexer must keep in mind that he is dealing only with the information contained within the particular book, so that entries must be worded as they appear there. Therefore cross-references are not usually made for alternative forms such as the following unless they are specifically requested:

Antwerp *or* Anvers   Columbus, Christopher, *or*
Brussels *or* Bruxelles   Colón, Cristóbal
Cologne *or* Köln   Raphael *or* Raffaelo
Florence *or* Firenze   Wilhelm *or* William
Nuremberg *or* Nurnberg

The careful writer may inform his readers that *Suomi* is the Finnish name for Finland or that Oslo was formerly called Kristiania (or Christiania). In this case, both names should be indexed.

Words with more than one meaning. Make a separate entry for each subject. *Form,* for instance, can mean *a body, a social convention, physical and mental condition, a rank of students,* or *a printing frame.* Make a separate entry for each separate meaning. If headings spelled alike name persons and places as well as subjects, the correct order is person, place, subject, title.

cell, Communist, 177   London, Jack, 97
cell, electrical, 65   London, England, 48–50
cell, nucleus of, 312   London, Treaty of, (1913), 140
cell, prison, 36   *London* (Johnson, S.), 51
cell division, 496
cell structure, 495

Words with different uses. The following example incorrectly combines under the heading *water* several other distinct topics besides *water.* The noun *water* is one subject. Where *water* is combined with another word to form a compound noun a new main heading is needed.

Wrong:	*Right:*
Water, 00; beetles, 00; bodies of, 00; ducts, 00; evaporation of, 00; gradation agent, 00; horse (hippopotamus), 00; meter, 00; moccasin, 00; pimpernel, 00; properties of, 00; running, 00; spaniel, 00; supply, 00; table, 00; tubes, 00; vapor, 00; vascular system, 00; as wonder worker, 00; work of, 00.	Water, 00; bodies of, 00; evaporation of, 00; gradation agent, 00; properties of, 00; running, 00; supply, 00; as wonder worker, 00; work of, 00 Water beetles, 00 Water ducts, 00 Water horse (hippopotamus), 00 Water meter, 00 Water moccasin, 00 Water pimpernel, 00 Water spaniel, 00 Water table, 00 Water tubes, 00 Water vapor, 00 Water vascular system, 00

Singular and plural forms. In alphabetizing an index, the singular and plural forms of a word might be widely separated and one group of references overlooked as a consequence. The simplest way to deal with this problem is to add an *s* in parentheses following the singular: for example, *eye(s)*. This avoids separating the singular, *eye*, from its plural, *eyes*, by such terms as *eyeball, eyebrows, eyeglasses,* and *eyelashes*. When a plural is formed other than by adding an *s*, give the full plural in parentheses after the singular; for example, *foot (feet)* or *artery (arteries)*. In each of the preceding examples the matter in parentheses is disregarded in alphabetizing, so that *eye(s)* precedes *eyeball*, *foot (feet)* comes before *football*. Except in indexing exact titles, it is usually possible to group all references under one entry.

Subentries. Break up with subentries a comprehensive heading followed by many page references, so that the searcher will be able to quickly find the material he is looking for. There can be nothing more exasperating than an entry like the following:

Lighting, 298, 299, 314, 315, 317, 344, 372, 381, 398–99, 413, 416.

If an entry-a-line style of index (see pp. 93ff.) is used, an entry such as the following is much more useful:

Lighting
    direct, 314, 315, 317
    electric, 398–99
    gas, 413
    indirect, 344
    kerosene, 413
    oil, 416
    methods of, 298, 299
    semidirect, 372, 381

Or if paragraph style (see pp. 93ff.) is used:

    Jehovah: fatherhood, 00; glory, 00;
      holiness, 00; love, 00; majesty, 00;
      name, 00; nature and character, 00;
      righteousness, 00; sole deity, 00

The arrangement may be alphabetical, as in the preceding examples,
numerical (by page numbers), or chronological, when a time sequence is
important, as in histories or biographies. However, the numerical sequence
has several drawbacks. In the first place, most entries are clearer in alphabetical sequence, but if several sets of entries in an index are numerical,
then they all must be, for the sake of consistency. In the second place,
strict numerical order cannot usually be adhered to, as in the following
example, where the beginning of Rome is again discussed with its downfall.

    Rome
      beginning of, 137-38, 157
      growth of, 144–46
      in conflict with Carthage, 145–47
      laws of, 152, 155
      slaves in, 154
      downfall of, 157

And in the third place, even where time sequence might seem important,
the entries can usually be arranged so that there is little danger of confusing the reader, as in the following:

    Franklin, Benjamin, 196 ff.
      birth, 196
      boyhood, 197–98
      at Continental Congress, 199, 202
      death, 214
      diplomatic career, 201–3, 209
      as governor of Pennsylvania, 212
      inventions of, 207–8

    If the numerical order is used, place after the main heading all page
references to which no subentry is attached.

Wrong: *Right:*

Calhoun, John C.: Sectional Southern   Calhoun, John C., 189; Sectional
  party plan, 74, 77; 189; James Mur-   Southern party plan, 74, 77; James
  ray Mason and, 244   Murray Mason and, 244

Gropius, Walter   Gropius, Walter, 67
  director of Bauhaus, 58–64   director of Bauhaus, 58–64
  67   Architects' Collaborative, 70–72
  Architects' Collaborative, 70–72

Cross-references. Cross references are a vital part of any index. A topic
may be thoroughly covered by various entries and subentries, but if the
reader has difficulty finding them, they are nearly useless. Therefore, a
sufficient number of synonyms for a topic should be entered to guide the

average reader. For example, if *cancer* is a major entry with numerous sub-entries, there should be cross-references from *carcinoma, sarcoma, tumors, malignancies,* and *metastasis.* The following is the form to use for cross-references:

Coins and coinage. *See* Money.
Currency. *See* Money.
Money, 42–45, 60, 72.
    Bonds and, 53–55
    Coins and coinage, 73–77
    Currency, 78–83
    Flow of, 36–41

Jesuits. *See* Society of Jesus.
Society of Jesus (Jesuits), 9, 26

Use a *See also* cross-reference from an entry listing page references to another on a related subject.

Alliances, 98. *See also* Coalitions;
    Treaties.

Gambling, 100 ff. *See also* Horse rac-
    ing.
    morality of, 133
    taxation of, 125–26

In the entry-a-line form of index the cross-reference is placed immediately following a main entry or subentry (and any page references given with the main entry or subentry).

Shock, 323–25, 585
    anaphylactic, 432
    electric, 32. *See also* Electroshock
        therapy.
    insulin, 35–36

Turkey. *See also* Ottoman Empire.
    civil war in, 126–27

In the paragraph-style index the cross-reference is placed at the end of the paragraph.

Classwork: oral, 5, 13; written, 53,
    380. *See also* Recitations.

If an entry is in italics, and it has a cross-reference to another italicized entry, the *See* should be roman (applying the rule that what is italic in roman text is roman in italic text), as in the following example:

Bacillus botulinus. See *Clostridium botulinum.*

ALPHABETICAL ORDER

Two systems of alphabetizing are in use, the dictionary (or letter-by-letter) method and the word-by-word method. Present usage is toward adopting the dictionary practice, which is simpler, for all index work. Therefore, the examples that follow are ordered letter by letter unless otherwise noted.

Dictionary method. Dictionaries, encyclopedias, and atlases list entries alphabetically letter by letter, considering all letters in sequence up to a comma. If letter sequences are identical, common nouns precede proper nouns. Articles and prepositions within titles are considered in determining order.

Adams, John	East Africa
Adam's apple	Eastbourne
Adams-Stokes breathing	East Cape
Ad amussim	Easter Island
Adaptable	East Indies
Coffee	New Orleans
Coffee, Jacob	Newport
Coffee cake	New Rochelle
Coffeehouse	Newton
Coffee in History	
Coffee table	*School for Scandal, The*
	Schoolmaster, The
Duke	"School of Athens, The"
Duke, Doris	
Dukedom, sizes of	Vanadium
Duke of Albany	Van Briggle pottery

Letter-by-letter ordering is followed only up to the first comma, as exemplified by the following:

Citizens, influence of	Pasteur, Louis
Citizenship	Pasteur Institute
Index, price	Police, state
Index numbers	Police courts

Word-by-word method. In this method only the first element of a compound term is considered in the alphabetical order, as shown in the following examples:

Book lists	New Orleans
Book reviews	New Rochelle
Booklets	Newport
Bookplates	Newton

Hyphenated compounds. In either method of alphabetizing, hyphenated compounds are treated as one word, so that the following order is correct:

Self-control   high relief
Selfhood   highlight
Self-seeking   high-minded

Titles. Articles beginning a title are inverted but prepositions are not. A comma within a title is disregarded in determining alphabetical order.

Observer, The   Ratsbane
Offenbach, Jacques   *Rats, Lice and Men*
Off the East Dock   Rat snake
Of Human Bondage   Rattlesnake

First lines. In indexing first lines, an initial article is not inverted.

A bird came down the walk, 261
Advance is life's condition, 1130
Alter! When the hills do, 557
A wounded deer leaps highest, 120

Apostrophes. In any style of index an apostrophe does not affect the alphabetical position of an entry.

Oerlikon   *Boy's King Arthur, The*
O'er Our Sea   *Boys of Yesterday*
Oersted, Hans   *Boys' Workshop*

Abbreviations. Abbreviations may quite usefully appear in scientific and technical works and are occasionally necessary in general texts. For example, if *St. Louis* is the form used for the city in the text, that is how it should be in the index, not *Saint Louis*. For simplification, abbreviations or shortened forms (as Mlle and Mme), may be considered as falling into two categories. (1) Those for titles of persons (*Dr., M., Mme., Mlle., Mr., Mrs., Ms., St.*) and for geographical names (usually *Ft.* and *Mt.*) are alphabetized as they are to be read—for example, *Mr.* as *Mister, St.* as *Saint.* Titles of persons, except *St.,* will appear in an index only in the title of a work, such as a book or a play (e.g., the play and movie *Mr. Roberts*). *Mrs.* is to be read as *Mistress,* and presumably *Ms.* as *Miz.* (2) All other abbreviations fall into this second category and are alphabetized letter by letter, as they appear to the eye. For example, *RNA* should come after *rm* and before *ro* and should be cross-referenced to the spelled-out term as follows:

RNA. *See* Ribonucleic acid (RNA).

The following excerpts from a hypothetical index show the proper order for abbreviations of these two arbitrary categories; rarely would so many abbreviations be encountered in such a short space in a real index. Note that lowercase abbreviations retain that form even when the style is to capitalize full words.

Doberman pinscher
Dr. Kildare
dr. *See* Debtor (dr).
Drag racing

FM. *See* Frequency modulation (FM).
Foot-pound (ft-lb)
Forestry
Fortifications
Ft. Monmouth
Fortune Society
Ft. Wayne
Frequency modulation (FM)
Ft-lb. *See* Foot-pound (ft-lb).

Madame Butterfly
Mme. DuBois
Madan, Martin
Mlle. Jones
Magnetic north (MN)
Melting point (mp)

Mirrors
Mr. Deeds Goes to Town
Mr. Hulot's Holiday
Mistflower
Mistresses
Mrs. Wiggs
Ms. Eunice
MN. *See* Magnetic north (MN).
Molds
Mountain Brook, Ala.
Mountaineers, The
Mt. Carmel
Mt. McKinley
Mount Sinai Hospital
mp. *See* Melting point (mp).

St. Agnes Street
Saint Basil's College
Sainthood
St. Louis
Sb. *See* Antimony (Sb).

Mac, Mc, M'. Names beginning with *Mac* and its variations *Mc* and *M'* should logically be indexed in letter-by-letter order, as shown in the left-hand column below. However, there is a strong tradition to alphabetize them all as if they were *Mac*, as shown in the right-hand column. Present usage slightly favors the right-hand column.

MacDonald, Ramsay
Machado, Gerardo
Machinery
MacLeish, Archibald
Magazines
McClure, S.S.
McKinley, William
M'Culloch, John

McClure, S.S.
M'Culloch, John
MacDonald, Ramsay
Machado, Gerardo
Machinery
McKinley, William
MacLeish, Archibald
Magazines

When there are a number of names beginning with *Mac, Mc,* and *M'*, these are sometimes grouped in a separate section before the beginning of the regular *M* section.

Numerals. Numerals, as main entries, are alphabetized as if they were spelled out.

Forenames
42nd Street
Fossils
4-H Clubs
Fourteen Points

As subentries, numbers are sometimes more appropriate in ascending order, as in the following example. Each such case should be considered individually.

Crusades, 00ff.
  First, 00
  Second, 00
  Third, 00
  Fourth, 00
  Fifth, 00

Insignificant words. Articles, conjunctions, and prepositions should be disregarded in determining the order of subentries. Formerly such words were sometimes considered in letter-by-letter alphabetizing, but this practice is now seen less and less; it is troublesome for a reader to have to guess what preposition an indexer might have chosen to precede the key word he is looking for. It is best to consider only the key word when alphabetizing.

Angle, 00ff.
  of any magnitude, 00
  construction of, from function, 00
  cosecant of, 00
  cotangent of, 00
  of depression, 00
  of elevation, 00
  generation of, 00
  tangent of, 00

TYPOGRAPHY OF INDEXES

There are two forms of index: the run-in (paragraph) style and the entry-a-line (indented) style. The former is somewhat more economical in terms of space and is sometimes preferred for histories and biographies. The latter is more suitable for complex indexes and is preferable for most indexes for one important reason: vertical scanning of the entry-a-line form is easier than horizontal scanning of lines that have a variable number of complete and incomplete entries.

Indexes that occupy the full page width (called full-measure indexes) or that use leader dots from the entry to the page reference are not recommended.

The following samples demonstrate the difference between the same entries done in run-in style (on the left) and in entry-a-line style (on the right):

Constantinople: alien rule of, 66ff., Arab invasion of, 56, 68; Crusades and, 67ff., 142; founding of, 14, 16ff.; Golden Age of, 46ff.; Greek takeover of, 81–82; Ottoman conquest of, 87–89; Persian attack on, 55–56; renamed "Istanbul," 102-3; Venetian sack of, 94, 96, 97. *See also* Istanbul.

Constantinople. *See also* Istanbul.
  alien rule of, 66ff.
  Arab invasion of, 56, 68
  Crusades and, 67ff., 142
  founding of, 14, 16ff.
  Golden Age of, 46ff.
  Greek takeover of, 81–82
  Ottoman conquest of, 87–89
  Persian attack on, 55–56
  renamed "Istanbul," 102-3
  Venetian sack of, 94, 96, 97

Sub-subentries are awkward in paragraph indexes, as can be seen in the following comparison with an entry-a-line index. Sometimes sub-subentries in a paragraph index may be set off by em dashes, but the only real solution, short of eliminating them, is to use the entry-a-line style.

Genomes: bacterial, 106, 109–10; phage, 141–42, 831–32; spore, 84, 86; viral, 349, 350, 715, 716; base composition of, 741, complementation, 796, genetic relatedness of, 741, 746, information encoded, 833–35

Genomes
  bacterial, 106, 109–10
  phage, 141–42, 831–32
  spore, 84, 86
  viral, 349, 350, 715, 716
    base composition of, 741
    complementation, 796
    genetic relatedness of, 741
    information encoded, 833–35

Indention. The paragraph form of index is compact, using only one indention—for turnover lines—and is therefore easy to set.

In the entry-a-line form the main entries (the headings) are set flush and each subentry begins a new line, the relation to the main heading being shown by indention of one em. Sub-subentries are indented two ems, and turnlines are always indented, throughout the index, one em more than the largest indention otherwise used: that is, if there are no subentries, all turnlines are indented one em; if there are subentries but no sub-subentries, all turnlines are indented two ems; if there are sub-subentries, all turnlines are indented three ems. A third subentry (a sub-sub-subentry?) is rarely required; these and even sub-subentries can usually be eliminated by careful wording of subentries.

Biliary tract, 1123ff. *See also*
    Gallbladder; Liver.
  anesthesia, 1141–42
  colic, 998, 1001
    appendicitis vs., 433
  fistula, 1127
  pancreatitis and, 655, 666–69,
    680–82, 1126
  radiography of, 907, 1004. *See also*
    Cholangiography.
  vagus nerve and, 1124

The practice of running the first subentry in to the main entry when there are no page references given for the main entry is to be discouraged—too often confusion results, as in the following example.

Turkey, architecture of, 101
  reforms in, 125

[Does the subentry refer to "reforms in architecture" or "reforms in Turkey"?]

Sometimes dashes are seen in indexes preceding subentries, as in the following example, but their use is not recommended, as they are no improvement over mere indention:

```
Acidosis, 00ff.
—metabolic, 00
— —treatment of, 00
—respiratory, 00
— —diagnosis of, 00
— —treatment of, 00
```

Dashes may sometimes be used in a combined form of entry-a-line and paragraph styles. Subentries all begin a new line, preceded by an em dash, but sub-subentries are run in.

```
Education, 85ff.
—colleges: enrollment, 95, 97; first,
  91, 92; state, 94–95
—colonial, 79–80, 88ff.: free schools,
  88–89, 232; literacy rate, 91, 92;
  science, 93
—early Republic: academies, 111, 112;
  financing of, 110, 115; transportation
  and, 120, 123
—secondary school. See Secondary
  schools.
—Southern: New South, 123–25;
  Northern compared with, 119;
  separate-but-equal doctrine, 144,
  146
```

Punctuation. Periods are not necessary at the ends of lines in an index unless the line ends with a cross-reference.

```
Old Jerusalem, 175                Levites: early status of, 124; sons of
  guide to, 177                     Zadok and other Levites distinguished,
Omar, Mosque of. See Dome          128; status of, in Code of Deuteron-
  of the Rock.                      omy, 130
```

A comma should be used between two prepositions to stand for the inverted subject of the first preposition.

```
Amylase, level of, in blood, 60
Amyl nitrite, use of, in angina
  pectoris, 75
```

Capitalization. Initial words of main entries are often capitalized. Subentries begin with a lowercase letter, unless, of course, the initial word is a proper noun.

```
Costume, 139ff.                   Dances, 230ff.
  accessories for, 198              country, 252–54
  armor, 206                        Elizabethan, 234–35
```

Chapter and other topical headings in the book should not be capitalized if they appear in an index, but anything else capitalized in text should appear in the index in the same style.

Alphabetical divisions. Unless an index is very short and simple, the beginning of a new alphabetical group should be marked in some way, usually by inserting extra space, preferably two blank lines. (Use of A, B, C, . . . in the space is outmoded.) The first word of each group may be set cap-and-small or boldface. When special types are so used, the whole heading up to the first comma should be set in this type, not a part only.

Daight, L.   **Gadsden Purchase**
                                              [*not* **Gadsden** Purchase]
East, E. M., quoted

Checkup of copy. The work of the indexer can be checked by looking over the cards with the following questions in mind:

Has a wise choice of headings been made?

Is the information given under the headings the searcher would be most likely to look for first, and have cross-references been made from other possible headings?

Are the headings nouns or substantive phrases, not adjectives or descriptive phrases?

Are phrase headings indexed logically, inverted if necessary to bring the significant word to the key position?

Have headings been sufficiently divided by subheadings to enable the searcher to find quickly the material for which he is looking?

Are the subheadings correct subdivisions of the key word?

Do the subheadings read back properly to the key word?

Are the subheadings arranged logically—alphabetically or chronologically—and consistently?

Have all citations on a given subject been given under a single heading, not divided between singular and plural forms of the key word?

Are the entries concise, containing no such phrases as *concerning, relating to,* and no unnecessary articles and prepositions?

Have the rules for alphabetical order been accurately and consistently followed?

Bibliography. Several books on indexing cover basic principles for both beginners and experienced indexers:

Indexing Books, by Robert L. Collison. New York: De Graff, 1962. *Surveys stages in preparation of indexes, giving fundamentals and practical advice on materials—pens, paper, files—needed.*

Indexes and Indexing, 3rd rev. ed., by Robert L. Collison. New York: De Graff, 1969. *Discusses problems such as layout, organization, style, length, and technical difficulties as well as indexing materials other than books.*

Training in Indexing: A Course of the Society of Indexers, edited by G. Norman Knight. Cambridge, Mass.: MIT Press, 1969. *Series of lectures on fundamentals of indexing and on specialized indexing—medical, legal, scientific.*

PART III.   COPY-EDITING STYLE

STYLE

Style implies, first of all, the literary manner in which the author expresses his thoughts. A second meaning, upon which the following pages concentrate, might be qualified as *copy-editing style:* accepted usage in the treatment of abbreviations, numbers, italics, capitalization, punctuation, and hyphenation of words. In either meaning, the best style is that which the reader is least aware of, so that his attention may center on what the author has to say about his subject matter.

Formerly, when printers were responsible for many matters now handled by copy editors, the terms *printing style* and *press style* were generally used. *House style* refers to the usage preferred by individual publishing companies; some houses issue a style guide in pamphlet or booklet form for the use of their authors and editors.

Every manuscript must be copy-edited at some point before publication, and the publisher assumes this responsibility. The style of the writer who is aware of acceptable usage and is consistent in its application is willingly followed by the copy editor. However, even the finest writers have occasional lapses, and errors in spelling or inconsistencies in capitalization or usage of numbers creep in. The publishing house owes to the writer and to its own reputation a conscientious effort to eliminate inconsistencies and careless errors, which only distract the reader and lower the overall quality of a work.

If there is a point of primary importance in copy-editing style, it is consistency. There are many cases in which acceptable usage is merely a matter of personal preference, and so long as an author is consistent, the reader scarcely notices whether it is *percent* or *per cent, freelancer* or *free-lancer.* Ideally everyone would agree on the mechanical details of written communication, but this not being the case, the consistent application of good style will at least allow the reader to concentrate on the substance and literary quality of the writing.

The following books are useful to anyone concerned with copy-editing style:

The Elements of Style, 2nd ed., by William Strunk, Jr., revised by E.B. White. New York: The Macmillan Company, 1972.

A Manual of Style, 12th ed. Chicago: The University of Chicago Press, 1969.

The MLA Style Sheet, 2nd ed. New York: Modern Language Association of America, 1970.

The New York Times Style Book, edited and revised by Lewis Jordan. New York: McGraw-Hill Book Company, 1962.

A Practical Style Guide for Authors and Editors, by Margaret Nicholson. New York: Holt, Rinehart and Winston, 1967.

Proofreading and Copy-Presentation, by Joseph Lasky. New York: Agathon Press, 1971.

Style Manual, rev. ed. Washington, D.C.: United States Government Printing Office, 1973.

ABBREVIATIONS

The proper use of abbreviations depends upon the nature of the text. They are rarely appropriate in formal or literary writing, but many may be used with propriety in nonliterary tests and in technical and scientific works. Abbreviations avoid distracting the reader with a cumbersome repetition of words and phrases and are particularly desirable, for economy of space and conciseness, in tabular matter, footnotes, bibliographies, and lists of various kinds.

An abbreviation that can be used in certain branches of writing might be poor form in others. For example, *versus* should be abbreviated in citations of legal cases (*Smith* v. *Jones*) but the abbreviation would be inappropriate in such an expression as "the buyers versus the sellers." The following sections present general rules governing the use of abbreviations that should be consistent in all forms of writing. Dictionaries should be consulted for the form and meaning of abbreviations not included here. The section on composition of foreign languages in Part IV contains lists of abbreviations in French, German, Hebrew, Italian, Russian, and Spanish. (See pp. 288ff.)

Most terms to be abbreviated, except familiar forms for weights, measures, and time, should be given in full at their first appearance, followed by the abbreviation in parentheses. The author should be told that if he has any doubt he should spell out the term; it is easier later (and results in cleaner copy) for the copy editor to abbreviate a spelled-out term than vice versa, if a house style must be conformed to.

The following are nearly always abbreviated: *Mr.*, *Messrs.* (or *MM.*), *Mrs.*, *Ms.*, *Jr.*, *Sr.*, *Esq.*, *Dr.*, *M.D.*, *Litt.D.*, *M.P.*, A.M., P.M., B.C., and A.D.

Concerning the usage of periods in abbreviations, the trend is to eliminate them, especially in units of all kinds (*mg*, *mph*, *lb*, *hr*) and in expressions whose elements are all capitalized (*NATO*, *TVA*, *AWOL*), unless confusion might arise (for example, *in.* and *no.*, which spell words). Periods are retained, in the examples that follow, in abbreviations of most words other than units and where there is a strong tradition for their use. When internal periods are used there should be no space after the period (*M.D.*, *e.g.*, *Ph.D.*, *Litt.D.*, A.M.). When letters within a single word are used as an abbreviation they are capitalized but no periods are used (*TV*, *TB*, *IV*).

Acronyms and shortened forms. Acronyms[1] (words formed from the initial letter or letters of successive parts of a term) and shortened forms derive from informal speech and should not contain, or be followed by, periods.

AFTRA	COBOL
AWOL	CORE

[1]For thorough coverage of the subject, see *Acronyms and Initialisms Dictionary* (Detroit: Gale Research Co., 1970).

NATO   WAC, WAF, WAVES[2]
SEATO   ZIP
UNESCO

Some acronyms have become commonly used words: *laser, radar, scuba, snafu, sonar.*

Shortened forms should never be followed by a period: *ad, auto, co-op, exam, gym, lab, math, memo, phone, photo, prof, pub, steno, stereo, typo.*

Agencies and organizations. Government agencies (see pp. 159ff.), service organizations, fraternal societies, unions, and other groups are frequently designated by their initials. The full name should usually be spelled out at its first usage, followed by the abbreviation in parentheses.

AFL-CIO	CAB	NAACP	TVA
AMA	FTC	NLRB	UN
ANA	GATT	OAS	USMC
BPOE	HUD	OAU	YMHA
BSA	ILO	SBA	YWCA

Beginning a sentence. No sentence, except in a footnote, should begin with a symbol or an abbreviation (except Mr., Dr., and so on).

The Greek letter μ is the symbol for micron. [*Not:* μ is . . .]
Number 3 mound was the next excavated. [*Not:* No. 3 mound . . .]

The Bible. Wherever abbreviations are appropriate, the following forms may be used for the various versions and sections of the Bible:

Apoc.	Apocrypha
ARV	American Revised Version
ARVm	American Revised Version, margin
AT	American Translation
AV	Authorized (King James) Version
DV	Douay Version
ERV	English Revised Version
ERVm	English Revised Version, margin
EV	English Version(s)
JB	Jerusalem Bible
LXX	Septuagint
MT	Masoretic text
NAB	New American Bible
NEB	New English Bible
NT	New Testament

---

[2]WAC stands for *Women's Army Corps*, WAF for *Women's Air Force*, and WAVES for *Women Accepted for Volunteer Emergency Service*. Nonstandard forms in such expressions as "The Wacs were stationed nearby" or "She's a Waf (or a Wave)" are acceptable in informal writing.

OAB   Oxford Annotated Bible (RSV)
OT   Old Testament
RSV   Revised Standard Version
RV   Revised Version
RVm   Revised Version, margin
Syr.   Syriac
Vulg.   Vulgate

In references to passages of Scripture the names of the books of the Bible may be abbreviated when they are used with numerals designating chapter and verse.

I Cor. 1:16–20

When reference is made to an entire chapter, it is better in text to spell out the name of the book.

In the thirteenth chapter of First Corinthians...
In the First Epistle to the Corinthians, chapter 13, ...
In the Second Book of Kings, the second chapter,...
The first chapter of the Gospel according to John...

The following lists present accepted forms of abbreviations of the names of books of the Bible for five standard versions of the Bible. Many writers prefer, however, to spell out all names containing six or less letters. Either arabic or roman numerals may be used for the number of the book; the forms given are those most often used for the various versions.

Note: Alternative names of books of the Bible are given in parentheses accompanied by an asterisk; abbreviations in the versions that use these alternative names are indicated by asterisks.

OLD TESTAMENT

	AV	RSV	OAB	JB	NAB
Genesis	Gen.	Gen	Gen.	Gn	Gn
Exodus	Ex./Exod.	Ex	Ex.	Ex	Ex
Leviticus	Lev.	Lev	Lev.	Lv	Lv
Numbers	Num.	Num	Num.	Nb	Nm
Deuteronomy	Deut.	Deut	Dt.	Dt	Dt
Joshua	Josh.	Josh	Jos.	Jos	Jos
Judges	Judg.	Judg	Jg.	Jg	Jgs
Ruth	Ruth	Ruth	Ru.	Rt	Ru
I Samuel	I Sam	1 Sam	1 Sam	1 S	1 Sm
II Samuel	II Sam.	2 Sam	2 Sam.	2 S	2 Sm
I Kings	I Kings	1 Kings	1 Kg.	1 K	1 Kgs
II Kings	II Kings	2 Kings	2 Kg.	2 K	2 Kgs
I Chronicles	I Chron.	1 Chron	1 Chr.	1 Ch	1 Chr
II Chronicles	II Chron.	2 Chron	2 Chr.	2 Ch	2 Chr
Ezra	Ezra	Ezra	Ezra	Ezr	Ezr
Nehemiah	Neh.	Neh	Neh.	Ne	Neh
Tobit				Tb	Tb

OLD TESTAMENT—*cont.*

	AV	RSV	OAB	JB	NAB
Judith				Jdt	Jdt
Esther	Esther	Esther	Est.	Est	Est
1 Maccabees				1 M	1 Mc
2 Maccabees				2 M	2 Mc
Job	Job	Job	Job	Jb	Jb
Psalms	Ps.	Ps	Ps.	Ps	Ps(s)
Proverbs	Prov.	Prov	Pr.	Pr	Prv
Ecclesiastes	Eccles.	Eccles	Ec.	Qo	Eccl
Song of Solomon (Song of Songs*)	Song of Sol.	Song	S. of S.	Sg*	Sg*
Wisdom				Ws	Wis
Ecclesiasticus (Sirach*)				Si	Sir*
Isaiah	Isa.	Is	Is.	Is	Is
Jeremiah	Jer.	Jer	Jer.	Jr	Jer
Lamentations	Lam.	Lam	Lam.	Lm	Lam
Baruch				Ba	Bar
Ezekiel	Ezek.	Ezek	Ezek.	Ezk	Ez
Daniel	Dan.	Dan	Dan.	Dn	Dn
Hosea	Hos.	Hos	Hos.	Ho	Hos
Joel	Joel	Joel	Jl.	Jl	Jl
Amos	Amos	Amos	Am.	Am	Am
Obadiah	Obad.	Obad	Ob.	Ob	Ob
Jonah	Jon.	Jon	Jon.	Jon	Jon
Micah	Mic.	Mic	Mic.	Mi	Mi
Nahum	Nah.	Nahum	Nah.	Na	Na
Habakkuk	Hab.	Hab	Hab.	Hab	Hb
Zephaniah	Zeph.	Zeph	Zeph.	Zp	Zp
Haggai	Hag.	Hag	Hag.	Hg	Hg
Zechariah	Zech.	Zech	Zech.	Zc	Zc
Malachi	Mal.	Mal	Mal.	Ml	Mal

NEW TESTAMENT

	AV	RSV	OAB	JB	NAB
Matthew	Matt.	Mt	Mt.	Mt	Mt
Mark	Mark	Mk	Mk.	Mk	Mk
Luke	Luke	Lk	Lk.	Lk	Lk
John	John	Jn	Jn.	Jn	Jn
Acts (of the Apostles*)	Acts*	Acts*	Acts	Ac	Acts*
Romans	Rom.	Rom	Rom.	Rm	Rom
Corinthians	I & II Cor.	1 & 2 Cor	1 & 2 Cor.	1 & 2 Co	1 & 2 Cor
Galatians	Gal.	Gal	Gal.	Ga	Gal
Ephesians	Eph.	Eph	Eph.	Ep	Eph
Philippians	Phil.	Phil	Phil.	Ph	Phil
Colossians	Col.	Col	Col.	Col	Col
Thessalonians	I & II Thess.	1 & 2 Thess	1 & 2 Th.	1 & 2 Th	1 & 2 Thes
Timothy	I & II Tim.	1 & 2 Tim	1 & 2 Tim.	1 & 2 Tm	1 & 2 Tm
Titus	Titus	Tit	Tit.	Tt	Ti
Philemon	Philem.	Philem	Philem.	Phm	Phlm
Hebrews	Heb.	Heb	Heb.	Heb	Heb
James	James	Jas	Jas.	Jm	Jas
Peter	I & II Pet.	1 & 2 Pet	1 & 2 Pet.	1 & 2 P	1 & 2 Pt
John	I,II,III John	1,2,3 John	1,2,3 Jn.	1,2,3 Jn	1,2,3 Jn
Jude	Jude	Jude	Jude	Jude	Jude
Revelation	Rev.	Rev	Rev.	Rv	Rv

The following list covers all the books of the Old Testament as given in the Douay Version of the Bible, in which a number of the names differ from those used in other versions.

DOUAY VERSION

Books	Abbreviations	Books	Abbreviations
Genesis	Gen.	Canticle of Canticles	Cant.
Exodus	Ex./Exod.	Wisdom	Wis./Wisd.
Leviticus	Lev.	Ecclesiasticus	Ecclus.
Numbers	Num.	Isaias	Isa.
Deuteronomy	Deut.	Jeremias	Jer.
Josue	Josue	Lamentations	Lam.
Judges	Judges	Baruch	Bar.
Ruth	Ruth	Ezechiel	Ezech.
I Kings	I Kings	Daniel	Dan.
II Kings	II Kings	Osee	Osee
III Kings	III Kings	Joel	Joel
IV Kings	IV Kings	Amos	Amos
I Paralipomenon	I Par.	Abdias	Abdias
II Paralipomenon	II Par.	Jonas	Jon.
I Esdras	I Esdras	Micheas	Mich.
II Esdras	II Esdras	Nahum	Nah.
Tobias	Tob.	Habacuc	Hab.
Judith	Judith/Jth.	Sophonias	Soph.
Esther	Esther	Aggeus	Aggeus
Job	Job	Zacharias	Zach.
Psalms	Ps.	Malachias	Mal.
Proverbs	Prov.	I Machabees	I Mach.
Ecclesiastes	Eccles.	II Machabees	II Mach.

In the Douay Version, the names and abbreviations of the books of the New Testament are the same as those in the Authorized (King James) Version, except that the Revelation of St. John is called the Apocalypse of St. John in the Douay Version, and is abbreviated Apoc.

Books of the Old Testament Apocrypha are abbreviated as follows:

APOCRYPHA

Esdras	I and II Esd.
Tobit	Tob.
Judith	Jth.
The parts of Esther not in the Hebrew or Aramaic	Rest of Esther
The Wisdom of Solomon	Wisd. of Sol.
The Wisdom of Jesus, son of Sirach, or Ecclesiasticus	Ecclus.
Baruch	Bar.
The Song of the Three Holy Children	Song of Three Childr.
The History of Susanna	Sus.
The History of the Destruction of Bel and the Dragon	Bel and Dragon
The Prayer of Manasses, king of Judah	Pr. of Man.
Maccabees	I, II, III, and IV Macc.

APOCALYPTIC BOOKS

Books	Abbr.	Books	Abbr.
Book of Enoch	En.	Testaments of the Twelve Patriarchs	XII P.
Sibylline Oracles	Sib. Or.		
Psalms of Solomon	Ps. Sol.	Assumption of Moses	Asmp. M.
Book of Jubilees	Bk. Jub.	Apocalypse of Baruch	Apoc. Bar.

Compass directions. In certain kinds of printed matter compass directions are regularly abbreviated, using capital letters, without space between, and no periods.

N   E   S   W   SE   NNW   SSW

Dates and time. When time of day is expressed in numerals, abbreviations are used, usually set in small capitals (A.M. and P.M.) without space. Some publishers prefer lowercase (a.m. and p.m.) or capitals (A.M. and P.M.). The lowercase or uppercase style is seen almost invariably in newspapers and periodicals.

A.M. (*ante meridiem*), before noon
P.M. (*post meridiem*), after noon
M. (*meridies*), noon (12:00 P.M., midnight)

For the use of numerals with the preceding abbreviations, see page 135.

The names of days of the week and of months should not be abbreviated in straight text, but the following abbreviations may be appropriate for extensive tabular matter and bibliographies:

Sun.   Mon.   Tues.   Wed.   Thurs.   Fri.   Sat.

Jan.	Apr.	July	Oct.
Feb.	May	Aug.	Nov.
Mar.	June	Sept.	Dec.

May, June, and *July* are not abbreviated except in the following much shortened forms used when space is very limited:

Ja	Ap	Jl	O
F	My	Au	N
Mr	Je	S	D

For the standard units of time, the following represent both the singular and the plural. As the usage of these abbreviations is restricted to statistical text and tabular matter, periods are omitted.

sec, second(s)   wk, week(s)
min, minute(s)   mo, month(s)
hr, h, hour(s)   yr, year(s)
da, d, day(s)

When year numbers are given, by far the most common chronological designations are A.D. (*anno Domini,* in the year of the Lord) and B.C. (before Christ). Again only very strong tradition requires the use of small capitals rather than regular capitals (A.D. and B.C.). Less common abbreviations follow. In all cases, abbreviations beginning with A precede the year number; the others follow it. For further discussion of A.D. and B.C., see pages 128–129.

A.A.C. (*anno ante Christum*), in the year before Christ
A.C. (*ante Christum*), before Christ
A.H. (*anno hegirae*), in the year of the Hegira (A.D. 622, the Mohammedan era)
A.H.S. (*anno humanae salutis*), in the year of human salvation
A.M. (*anno mundi*), in the year of the world
A.P.C.N. (*anno post Christum natum*), in the year after the birth of Christ
A.P.R.C. (*anno post Roman conditam*), in the year after the building of Rome
A.R. (*anno regni*), in the year of the reign
A.R.R. (*anno regni regis, or reginae*), in the year of the king's, or queen's, reign
A.S. (*anno salutis*), in the year of salvation
A.U.C. (*ab urbe condita*), from the founding of the city (of Rome)
B.C.E., before the common era (Jewish)
B.P., before the present
C.E., common era (Jewish)

Dialogue. In dialogue and colloquial narrative, abbreviations may be used that are to be read as they stand, not amplified into the words they represent. Technical terms in their proper setting may also be so presented.

AWOL	OK'd, OKs[3]	Btu
DTs	on the q.t.	amps
GHQ	TWA	caps

"He is working for his Ph.D."
"It was a TKO in the third round."
"Send it to RFD number three."

Mr., *Mrs.*, and *Ms.* should be abbreviated in dialogue except as follows:

"Call me *Mister* Johnson" (for emphasis)
"Mister Roberts..." (in U.S. Navy contexts)

Doctor and *Saint* are sometimes spelled out for special reasons, particularly to avoid the plural (*SS.*) of the latter; but *Dr.* and *St.*, especially for place names (*St. Louis*), are acceptable in dialogue.

The forms A.M., P.M., B.C., and A.D. are correct in dialogue, and it is unnecessary to resort to coinages such as "in the ayem."

The ampersand (&) should never be substituted for *and* in dialogue.

"He bought a hundred shares of A and P stock."

Firm names. The abbreviation *Bro.*, *Bros.*, *Co.*, *Inc.*, and *Ltd.* and the symbol & are frequently used in the names of firms and corporations; each firm name should be printed as the name stands on letterheads and other official documents. In text matter of books these abbreviations are out of place, and the words should be spelled in full, except that *Inc.* and *Ltd.* are usually dropped.

Ginn and Company   Little, Brown and Company

[3]For plurals of abbreviations, see pages 115–116.

In notes, bibliographies, lists, and general business printing, abbreviations and shortened forms are better.

Ginn & Co.   Little, Brown & Co.

The word *Company* should not be abbreviated in the names of theatrical organizations: Chicago Civic Opera Company, Times Square Stock Company.

Some firms may be designated by their initials and an ampersand: A&P, AT&T.

In straight text the words *Associates, Corporation, Manufacturing*, and *Railroad* should be spelled out in firm names.

Geographical terms. In tabular matter and lists of addresses the abbreviations Ave., Bldg., Blk., Blvd., Pl., Sq., St. are used freely. Abbreviations of the names of the states may be used, but names having five letters or less are better spelled in full. None of these abbreviations should be used without a place name preceding. (The Government Printing Office rules that the names of the states may be abbreviated after geographical terms such as names of counties, mountains, national forests, national parks, navy yards, reservations, and the like, but this is not good style for book work nor for periodicals that make any claim to literary merit; it is better to spell the state name in full or, in some cases, to omit it.)

In text, bibliographies, and footnotes, reference to any of the following cities is sufficiently clear, in most instances, without accompanying designation of the state:

Akron	Grand Rapids	Richmond
Atlanta	Hartford	St. Augustine
Atlantic City	Honolulu	St. Louis
Baltimore	Indianapolis	St. Paul
Boston	Jersey City	Salt Lake City
Brooklyn	Los Angeles	San Antonio
Buffalo	Memphis	San Diego
Chattanooga	Milwaukee	San Francisco
Chicago	Minneapolis	Savannah
Cincinnati	Nashville	Scranton
Cleveland	New Orleans	Seattle
Colorado Springs	New York	Spokane
Dallas	Oakland	Tacoma
Dayton	Oklahoma City	Tampa
Denver	Omaha	Toledo
Des Moines	Philadelphia	Toronto
Detroit	Phoenix	Trenton
Duluth	Pittsburgh	Tulsa
Fort Wayne	Providence	Wheeling

In geographical names *Saint* is ordinarily abbreviated (*St.*). *Fort* and *Mount* are preferable to *Ft.* and *Mt. Port* and *Point* are always spelled out.

The following are the official abbreviations of the Post Office Department for states, territories, and possessions of the United States. Note that where there are internal periods, the abbreviation is closed up. The two-letter form following each abbreviation is the ZIP (Zone Improvement Program) code. (The abbreviations in parentheses are not sanctioned by the postal authorities but are used by many printers and in legal citations.)

Ala., AL	Ky., KY	Okla., OK
Alaska, AK	La., LA	Oreg. (Ore.), OR
Ariz., AZ	Maine (Me.), ME	Pa. (Penn., Penna.), PA
Ark., AR	Md., MD	P.R., PR
Calif. (Cal.), CA	Mass., MA	R.I., RI
C.Z.	Mich., MI	Samoa
Colo., CO	Minn., MN	S.C., SC
Conn., CT	Miss., MS	S.Dak. (S.D.), SD
Del., DE	Mo., MO	Tenn., TN
D.C., DC	Mont., MT	Tex., TX
Fla., FL	Nebr. (Neb.), NB	Utah, UT
Ga., GA	Nev., NV	Vt., VT
Guam, GU	N.H., NH	Va., VA
Hawaii, HI	N.J., NJ	V.I., VI
Idaho (Ida.), ID	N.Mex. (N.M.), NM	Wash., WA
Ill., IL	N.Y., NY	W.Va., WV
Ind., IN	N.C., NC	Wis. (Wisc.), WI
Iowa (Ia.), IA	N.Dak. (N.D.), ND	Wyo., WY
Kans. (Kan.), KS	Ohio, OH	

The names of the Canadian provinces are commonly abbreviated as follows:

Alberta	Alta.	Northwest Territories	N.W.T.
British Columbia	B.C.	Ontario	Ont.
Cape Breton	C.B.	Prince Edward Island	P.E.I.
Manitoba	Man.	Quebec	Que. *or* P.Q.
New Brunswick	N.B.	Saskatchewan	Sask.
Nova Scotia	N.S.		

The initial abbreviations used in addresses for London designate the old postal districts. Britain now has a Postal Code, similar to the ZIP Code, which follows the initials (for example, London W1X 8AB).

Eastern	E	South-Eastern	SE
Eastern-Central	EC	South-Western	SW
Northern	N	Western	W
North-Western	NW	Western-Central	WC

Countries. Names of countries are nearly always spelled out in text, with the exception of the Soviet Union, which is often designated *USSR.* Where abbreviations are appropriate (in statistical text, tabular matter, and so on), *UAR* (United Arab Republic), *UK* (United Kingdom), and *UN* (United Nations) may be used without periods. However, there is a strong tradition for retaining periods in abbreviating the United States, *U.S.,* in such contexts. As modern usage inclines toward eliminating abbreviating periods whenever possible, it seems logical that someday the common preference will be *US,* without periods. (*The American Heritage Dictionary* lists *US* and *U.S.* as equivalent alternates.) When several abbreviations appear together, consistency is the ruling factor, and either all should appear with periods or all without: *U.A.R., U.K., U.N., U.S., U.S.S.R.* or *UAR, UK, UN, US, USSR.*

In the most formal writing, *United States* should always be spelled out; in other works, *U.S.* is gaining currency as an adjective when preceding a government agency, department, or organization, or the name of a government vessel.

U.S. Civil Service Commission	U.S. Geological Survey
U.S. Commission of Fine Arts	U.S.S. *Saratoga*
U.S. Department of Agriculture	U.S. monitor *Nantucket*

When used as an adjective with general terms, *United States* should be spelled out.

United States economy	United States government
United States foreign policy	United States space program

If *U.S.* in copy is circled to be spelled out, thought should be given to the form, which should in most cases be *the United States*. One exception is *Bank of United States*.

Latin words and phrases. The following abbreviations may be used in all except literary and formal texts, although each of them has a reasonably short and simple English equivalent that would in many instances be preferable. Note that the modern trend is to set them roman type in text.

i.e. (*id est*), that is
e.g. (*exempli gratia*), for example
etc. (*et cetera*), and others, and so forth
c., ca., or circ. (*circa, circiter,* or *circum*), about (used with dates denoting approximate time)
q.v. (*quod vide*), which see, *pl.,* qq.v. (*quae vide*)
viz. (*videlicet*), namely

It is not correct to use *e.g.* as it is used in the following sentence:

Wrong: Socioeconomic level, e.g., is a factor to be considered.
Right: Socioeconomic level, for example, is . . .

When other abbreviations are used freely in a work, however, it is acceptable to *introduce* a word or a phrase with *e.g.*

In evaluating an IQ score several factors, e.g., socioeconomic level, must be considered.

The abbreviation *etc.* should be used as little as possible, and it should never be preceded by *and* (*and etc.*). Neither *etc.* nor its equivalent *and so forth* should be used after examples preceded by the words *such as.*

Wrong: Many animals, such as rabbits, horses, etc., shed their heavy winter coat at the approach of warm weather.
Right: Many animals, such as rabbits and horses, shed their heavy winter coat . . .

The following Latin abbreviations, including some seen only in older works, are not often appropriate to text except parenthetically but are useful in footnote material:

ad fin. (*ad finem*), at the end
ad inf. (*ad infinitum*), to infinity
ad init. (*ad initium*), at the beginning
ad int. (*ad interim*), meanwhile
ad lib. (*ad libitum*), at will
ad loc. (*ad hunc locum*), on this passage
aet. (*aetatis*), aged
cf. (*confer*), compare
con. (*contra*), against
et al. (*et alii*), and others[4]
et seq. (*et sequens*), and the following (rarely used *pl.*, et sqq.)
f. (*folio*), on the following page; *pl.* ff., on the following pages (used in indexes)
fl. (*floruit*), flourished
f. v. (*folio verso*), on the back of the page
ibid. (*ibidem*), the same[5]
id. (*idem*), the same
inf. (*infra*), below
inst. (*instans*), the current month
i.q. (*idem quod*), the same as
loc. cit. (*loco citato*), in the place cited
m.m. (*mutatis mutandis*), necessary changes being made
MS *or* ms (*manuscriptum*), manuscript; *pl.* MSS, mss
n. (*natus*), born
N.B. (*nota bene*), mark well
non seq. (*non sequitur*), it does not follow
op. cit. (*opere citato*), in the work cited
pass. (*passim*), throughout, here and there
prox. (*proximo*), next month
Q.E.D. (*quod erat demonstrandum*), which was to be demonstrated
sc. *or* ss. (*scilicet*), namely (in law)
sup. (*supra*), above
s.v. (*sub verbo*), under the word; (*sub voce*), under the title
ult. (*ultimo*), last month
u.s. (*ubi supra*), in the place above mentioned
v., vs. (*versus*), against[6]
v. (*vide*), see
v.l. (*varia lectio*), a variant reading; *pl.* vv. ll.
v.s. (*vide supra*), see above

For abbreviations of Latin phrases used in expressing dates and time, see pages 105–106.

[4]For heavily referenced works, where many authors' names appear in the text, the citation "Jones et al." may be used for three or more authors. (Some publishers prefer "and others.")
[5]For the use of ibid., op. cit., and loc. cit., see pages 29–30.
[6]For use of v. and vs., see page 120.

Latitude and longitude. The words *latitude* and *longitude* are always spelled out in general text. In statistical text and tabulations of coordinates, one of the following styles may be used (the preferred is on the left):

lat. 40°20′ N	lat. 40-20 N
long. 24°15′30″ W	long. 24-15-30 W

Laws, constitutions, bylaws. In setting laws, bylaws, and similar subject matter divided into articles or sections it is customary to spell out ARTICLE I or SECTION I and abbreviate ART. II or SEC. 2 and following articles or sections.

Measures, weights, and other units. Units of measure and weight may be abbreviated in technical copy when they are accompanied by a numeral; never abbreviate units when they follow a spelled-out number. Note that all the following abbreviations are identical in both the singular and the plural and that periods are usually not used (except for *in.*):[7]

Å, angstrom unit
ac, alternating current
AF, audiofrequency
amp, ampere
atm, atmosphere
at wt, atomic weight
avdp, avoirdupois
bbl, barrel
bd ft, board foot
bev, billion electron volts
bp, boiling point
Btu, British thermal unit
bu, bushel
c, curie
cal, calorie
Cal, large calorie (kilocalorie)
cc, cubic centimeter (for gas
    volume only)
cgs, centimeter, gram, second
    (system of units)
cp, candlepower
cps, cycles per second
cu ft, cubic foot
cu in., cubic inch
cwt, hundredweight
da *or* d, day
db, decibel
dc, direct current
doz, dozen
dr, drachma
dwt, pennyweight

emf, electromagnetic force
F, Fahrenheit
FM, frequency modulation
fps, foot, pound, second (system
    of units)
ft, foot
ft-lb, foot-pound
g, gram *or* gravity
gal, gallon
gr, grain *or* gross
hhd, hogshead
hp, horsepower
hr *or* h, hour
Hz, hertz
in., inch
IQ, intelligence quotient
IU, international unit
K, carat
kc, kilocycle
kph, kilometers per hour
kv, kilovolt
kw, kilowatt
kwh, kilowatt hour
lb, pound
ma, milliampere
mc, millicurie
Mc, megacycle
mev, million electron volts
min, minim *or* minute
mo, month
mol wt, molecular weight

[7]The American National Standards Institute (formerly the American Standards Association) and *The American Heritage Dictionary* endorse the elimination of periods from abbreviations of all types of units. However, some publishing house styles retain the periods, in which case they should be used consistently. For an exhaustive listing of scientific abbreviations, see the *American National Standard Abbreviations* (ANSI Publication Y1.1), New York: American Society of Mechanical Engineers, 1972.

mp, melting point
mph, miles per hour
msec, millisecond
oz, ounce
pk, peck
ppm, parts per million
psi, pounds per square inch
qt, quart
r, roentgen
RF, radiofrequency
rpm, revolutions per minute
SD, standard deviation
SE, standard error of the mean

sec, second
sp gr, specific gravity
sq, square
t, ton
tbsp or T, tablespoonful
tsp, teaspoonful
UHF, ultrahigh frequency
v, volt
VHF, very high frequency
w, watt
wk, week
yd, yard
yr, year

Except in highly technical work *cycle, dyne, erg, farad, henry, joule, lambert, volt,* and *watt* are not abbreviated, but in such work the first letter of each may be used to represent the word. Units of measurement for which there is no abbreviation are *gauss, gilbert, lumen, maxwell, mho, mil, ohm,* and *phot.*

Metric measurement. The three principal units of measurement in the international metric system are the meter, the gram, and the liter.

Length	*Weight*	*Capacity*
mm, millimeter	mg, milligram	ml, milliliter[8]
cm, centimeter	cg, centigram	cl, centiliter
dm, decimeter	dg, decigram	dl, deciliter
m, meter	g, gram	l or L, liter[9]
dam, decameter	dag, decagram	dal, decaliter
hm, hectometer	hg, hectogram	hl, hectoliter
km, kilometer	kg, kilogram	kl, kiloliter
mym, myriameter	myg, myriagram	myl, myrialiter
μ, micron	q, quintal	μl, microliter
mμ, millimicron	ng, nanogram	
	μg, microgram	

For the following surface and volume abbreviations, terms of square and cubic measurement are best stated as *sq mm, cu mm,* rather than *mm²* and *mm³*, to avoid confusion in works where superscript numbers indicate references and footnotes. (The same rule applies, of course, to *sq in., cu ft,* and so on.)

Surface	*Volume*
sq mm or mm², square millimeter	cu mm or mm³, cubic millimeter
sq cm or cm², square centimeter	cu cm or mm³, cubic centimeter
sq dm or dm², square decimeter	cu dm or dm³, cubic decimeter
sq m or m², square meter	cu m or m³, cubic meter
sq dam or dam², square decameter	cu dam or dam³, cubic decameter
a, are	s, stere
ha, hectare	ks, kilostere

[8]For liquid capacity, always use *ml*; for gases, use the equivalent unit of volume, *cc*.
[9]It is usually best to use the capital *L,* to avoid confusion with the numeral one (1), unless there are many other metric measures nearby, as in statistical matter.

Abbreviations such as these should be used only with a numeral.

Poor: In this test an ml of salt solution is injected.
Better: In this test a milliliter (*or* 1 ml) of salt solution is injected.

Medical terms. Again, when there is any doubt about the familiarity of a term, spell it out first, and follow it with the abbreviation in parentheses.

In binomial nomenclature, the genus may be abbreviated after it is first given in full: *Escherichia coli*, then *E. coli*.

For the abbreviation of medical journal titles, the best source is the National Library of Medicine's compilation, *List of Journals Indexed in Index Medicus*, obtainable from the Superintendent of Documents, Government Printing Office, Washington, D.C. 20402.

Many abbreviations used in medical writing are discussed elsewhere (e.g., see the section on metric measurements on p. 112 and the general abbreviations on pp. 111–112). The following are specific to medicine:[10]

ACTH, adrenocorticotropic hormone
ADH, antidiuretic hormone
ADP, adenosine diphosphate
Ag-Ab, antigen-antibody
AHF, antihemophilic factor
AP, anteroposterior
ATP, adenosine triphosphate
AV, atrioventricular
BCG, bacille Calmette-Guérin
BMR, basal metabolic rate
BSP, bromsulphalein
BUN, blood urea nitrogen
CBC, complete blood count
CF, complement fixation
CFA, complement-fixing antigen
CNS, central nervous system
CoA, coenzyme A
CSF, cerebrospinal fluid
CVA, costovertebral angle
d-, l-, dextro-, levo- (optical rotation)
D-, L-, *dextro-, levo-* (configuration)
D & C, dilatation and curettage
DNA, deoxyribonucleic acid
DNase, deoxyribonuclease
ECG, electrocardiogram
ECHO virus, enteric cytopathogenic human orphan virus
EEG, electroencephalogram
ESR, erythrocyte sedimentation rate
FSH, follicle-stimulating hormone
GFR, glomerular filtration rate
GH, growth hormone

GSH, glutathione
G6PD, glucose-6-phosphate dehydrogenase
h, Planck's constant
Hb, hemoglobin
HE, hematoxylin and eosin
HGF, hyperglycemic factor
ICSH, interstitial-cell–stimulating hormone
IM, intramuscularly
IP, intraperitoneally
IU, international unit
IV, intravenously
LD_{50}, lethal dose (for 50% of inoculated group)
LOA, left occipitoanterior
LSD, lysergic acid diethylamide
LTH, luteotropic hormone
M, mole
MCHC, mean corpuscular hemoglobin concentration
MCV, mean corpuscular volume
MED, minimum effective dose
mEq/L, millequivalents per liter
m*M*, millimole
MSH, melanin-stimulating hormone
N, normal
NF, *National Formulary*
NPN, nonprotein nitrogen
OD, optical density
PKU, phenylketonuria
PMN, polymorphonuclear neutrophilic leukocytes

[10]For an encyclopedic listing of chemicals and drugs, see *The Merck Index*, 8th ed. (Rahway, N.J.: Merck & Co., 1968).

PPD, purified protein derivative
PPLO, pleuropneumonialike organisms
RBC, red blood cells; red blood (cell) count
RNA, ribonucleic acid
RNase, ribonuclease
RQ, respiratory quotient
s, sedimentation constant
S, Svedberg unit
SA, sinoatrial

SGOT, serum glutamic oxalacetic transaminase
SK-SD, streptokinase-streptodornase
STH, somatotropic hormone
TMV, tobacco mosaic virus
TSH, thyrotropic hormone
USP, *United States Pharmacopeia*
WBC, white blood cells; white blood (cell) count

Electrocardiographic terms are preferably written as follows:

aVF, aVL
leads I, III, V_6
QRS complex

P-R interval
S-T segment
T wave

Following are a few other useful symbols in medical writing:

F_1, F_2 generations
genotypes *AaBB, AAbb*
MNSs blood type

pH, pO_2, pCO_2, pK
Rh negative
R_0 antigen

T_2 phage
X chromosome

Prescriptions. Many abbreviations are used in writing prescriptions. The following are common:

aa (*ana*), of each alike
a.c. (*ante cibum*), before meals
aq. (*aqua*), water
b.i.d. (*bis in die*), twice a day
ea., each
gt. (*gutta*), drop (*pl.*, gtt.)
ol. (*oleum*), oil
p.ae. (*partes aequales*), equal parts
p.c. (*post cibum*), after meals
p.r.n. (*pro re nata*), as necessary
q.h. (*quaque hora*), every hour; q.h. 2, every two hours
q.i.d. (*quater in die*), four times a day
q.l. (*quantum libet*), as much as you please
q.pl. (*quantum placet*), as much as seems good
q.s. (*quantum sufficit*), a sufficient quantity
q.v. (*quantum vis*), as much as you will
S. (*signa*), mark (indicates directions to be written on the package)
sc., scruple
sol., solution
t.i.d. (*ter in die*), three times daily
ut dict. (*ut dictum*), as directed

℥	ounce	Ði	one scruple
℥i	one ounce	Ðss	half a scruple
℥ss	half an ounce	f℥	fluid
℥iss	an ounce and a half	fʒ	fluid dram
℥ij	two ounces	℞ or ℞	minim, or drop
ʒ	dram	℔	pound
ʒi	one dram	M̶	mix
Ð	scruple	℞	(*recipe*) take

"Number." In tabular matter, footnotes, and references, and occasionally parenthetically in text, *number* may be abbreviated *No.*, capitalized and followed by a period. This is one of the few abbreviations that adds an *s* in its plural, *Nos.*, as in "Publ. Nos. 2-653 and 2-655." Awkward usage such as "in exercise No. 6" can be avoided by using "in exercise 6."

To indicate the number of individuals in a test group (in tables or parenthetically in text) an italicized *n* is preferable to a small cap N (for example, $n = 781$).

Personal names. It is not good form to abbreviate personal forenames in straight reading matter, even if the person himself does, or did, sometimes use that form. George Blank may sign his name *Geo. Blank,* and Charles Smith may habitually write *Chas. Smith,* but the full forms *George* and *Charles* should always be used in writing about them. Care must always be taken, however, not to spell out a name that looks like an abbreviation but is in reality the full name. For instance, *Edw.* is plainly an abbreviation, but *Fred* might or might not be.

The use of initials for the first and middle names is allowable whenever brevity is in keeping with the nature of the text. When there are only two initials, regular spacing should be maintained in straight text matter (G. A. Howe, S. P. Farkle); three or more initials should be closed up (P.G.T. Beauregard). However, in lists of names, bibliographies, and catalogue entries even two initials should be closed up (Smith, J.A., and Jones, P.R.). Some publishers, particularly of scientific works, advocate elimination of any unnecessary punctuation in bibliographies (Smith JA and Jones PR), but this practice has not yet been widely adopted.

No periods are used when a person has become well known by his initials only: FDR, JFK, LBJ. In dialogue a person may be familiarly called by his initials, which should be closed up: "Come here, J.A.!" or "J.A. wants to see you in his office."

Philological terms. The following philological terms are used without periods by most publishers:

AF, Anglo-French	MHG, Middle High German
AN, Anglo-Norman	ML, Middle, or Medieval, Latin
AS, Anglo-Saxon	NL, New Latin
EE, Early English	OE, Old English
IE, Indo-European	OF, Old French
L, Latin	OHG, Old High German
LG, Low German	ON, Old Norse
ME, Middle English	

The following require more than initials, and periods are used:

Eng., English	Gr., Greek
Fr., French	Mod.E., Modern English
Gael., Gaelic	N.Gr., New Greek
Ger., German	O.Gael., Old Gaelic

Plurals. To form plurals of abbreviations that contain no periods an *s* is added without an apostrophe: IQs, MPs, OKs, YMCAs. For plurals of abbreviations that use internal periods, add an apostrophe and *s*: M.S.'s, Ph.D.'s. A few abbreviations (usually preceding a number) that use an end period form plurals as follows (see also next section): Exs., Figs., Nos.

Abbreviations for units (weight, measure, time) — that is, for any amounts or quantities — are identical in form for both singular and plural: ml, ft, doz, lb, wk.

Some Latin abbreviations have their own special forms: MSS, ff. Also commonly used are pp. (pages) and *Ss* (subjects).

References. In footnotes, bibliographies, and similar material — but not in text except parenthetically for cross-references — words designating a part of a work are abbreviated when accompanied by a numeral. Capitalization is usually as here shown. A common error is the use of singular abbreviations with numerals denoting more than one.

App., Appendix	No. (*numero*), number (*pl.*, Nos.)
Art., Article (*pl.*, Arts.)	NS, New Series
Bk., Book	OS, Old Series
Bull., Bulletin	p., page (*pl.*, pp.)
Ch. or Chap., Chapter; *in law*, c. (*pl.*, Chaps.)	par., paragraph
col., column, (*pl.*, cols.)	Pl., Plate
Div., Division	Pt., Part
Ex., Example	sec., section; *in law*, s. (*pl.*, ss.)
Fig., Figure (*pl.*, Figs.)	ser., series
fol., folio	st., stanza (*pl.*, sts.)
f., the following (*pl.*, ff)[11]	Supp., Supplement
l., line (*pl.*, ll.)	v. or vs., verse (*pl.*, vv. or vss.)
n., note (*pl.*, nn.)	Vol., Volume (*pl.*, Vols.)

"Saint." *Saint* should be spelled out when preceding the name of a saint, except in tabular and other tight matter. Omit *Saint* before the names of apostles.

In Saint Jerome's revision . . .	Matthew was the first . . .
According to Saint Thomas Aquinas . . .	

For female saints the following forms are correct: Saint Agnes, Saint Anne, Saint Joan. The French feminine *Sainte* (*Ste.*) is seen only in place names and personal names.

When abbreviated, *St.* is the singular and *SS.* the plural. Some house styles (following the French and British practice) omit the period after *Ste, St,* and *SS;* however, in American practice the consistent use of the end period predominates: *St., SS., Ste.*

In geographical place names and in the names of churches, streets, and squares, *Saint* should be abbreviated even in regular text matter.

St. Andrews	Sault Ste. Marie
St. Augustine	St. Mark's Square
Lake St. Clair	St. Patrick's Cathedral
Church of SS. Constantine	St. Petersburg
and Helena	St. Philip Street

[11]Before numbers with f. or ff. use a plural noun: pp. 50 f., 50 ff.

In French place names and church names , the saint's name is hyphenated: Rue St.-Honoré, St.-Germain-des-Prés.

In personal names *Saint* is treated according to the bearer's preference: Ruth St. Denis, Arnold Saint-Subber, Susan Saint James, Buffy Sainte-Marie.

Scholastic, military, and civil honors. Designations of rank or position, membership in monastic or secular orders, academic degrees, military or civil honors, are commonly abbreviated when they are used after a proper name. Most of such abbreviations consist of capital letters, and the prevailing practice is to set them without spaces between the letters.

B.A. *or* A.B., Bachelor of Arts	LL.B., Bachelor of Laws
B.S. *or* S.B., Bachelor of Science	M.A. *or* A.M., Master of Arts
D.D., Doctor of Divinity	M.D., Doctor of Medicine
D.D.S., Doctor of Dental Surgery	M.P., Member of Parliament
D.S.O., Distinguished Service Order	M.S. *or* S.M., Master of Science
Esq., Esquire	Ph.D., Doctor of Philosophy
F.R.S., Fellow of the Royal Society	R.N., Registered Nurse
Kt., Knight	S.J., Society of Jesus
Litt.D., Doctor of Letters	U.S.N., United States Navy

Note also two italicized abbreviations used following names:

Cantab. (*Cantabrigiensis*), of Cambridge
Oxon. (*Oxoniensis*), of Oxford

In an informal context the abbreviations of academic degrees are sometimes allowable without an accompanying name.

Originality is highly desirable in the M.A. thesis.
Six years later the university gave him a B.A. degree.

Temperature and gravity. When temperature and gravity are expressed in figures with the degree sign, the name of the thermometer or hydrometer is usually abbreviated: 68 degrees Fahrenheit, 68°F (note that the degree sign and the "F" are closed up to a thin space).

abs *or* T, absolute temperature	Bé, Baumé
C, centigrade *or* Celsius	Tw, Twaddell
F, Fahrenheit	API, American Petroleum Institute
K, Kelvin	
R, Reaumur	

Time. See page 105.

Titles. The abbreviated forms *Mr., Mrs., Ms., Messrs.* (or *MM.*), and *Dr.* are always used when the titles occur before proper names;[12] also *Esq., Sr.,* and *Jr.*[13] following names. The adjectives *Reverend* and *Honor-*

---

[12]For usage of *Mr., Mrs., Ms.,* and *Dr.* in dialogue, see page 106.
[13]*Senior* and *junior* used informally in narrative should be spelled out. "It was all right for Thomas senior to take part in an affair of this sort, but quite another matter when Thomas junior was involved."

able, either in full or abbreviated, should never be used with a surname alone.

> *Wrong:* Rev. Black; Hon. White; Reverend Black

If the article *the* is used, *Reverend* and *Honorable* should be spelled in full.

> *Right:* the Reverend A. B. Black; the Honorable H. A. White; the Reverend Mr. Black.

If the article is not used, the abbreviated forms *Rev.* and *Hon.* are correct form.

> *Right:* Rev. Alfred Black, Rev. A. B. Black, Hon. Harold White, Hon. H. A. White

Two titles of the same significance should not be used, one preceding the name and the other following.

> Dr. Walter Franklin *or* Walter Franklin, M.D.
> Mr. Frank Milton *or* Frank Milton, Esq.
> the Reverend Dr. Black *or* the Reverend A. B. Black, D.D.

But titles of different significance can be so used.

> Professor Theodore Howard, D.D.S.

The titles *Mr., Ms., Dr., Professor,* and the like may properly be omitted from the names of authors cited, except in the case of living authors, where the title is often used as a matter of courtesy when the surname alone is given. If mention of a person is personal, as in acknowledgment of courtesies or services, such titles should be used.

> In this connection Arnold has stated . . .
> . . . for Joseph Heller and Doris Lessing. As Mr. Heller and Ms. Lessing both . . .
> The method suggested by J. W. Butler . . .
> Calculations made in collaboration with L. A. Kilgore show . . .
> The committee elected Dr. A. E. Kennelly its honorary president. Dr. Kennelly has served . . .

The French titles *Monsieur* (pl. *Messieurs*), *Madame* (pl. *Mesdames*), and *Mademoiselle* are abbreviated before names when they are mentioned in the third person.

> M. Monie is . . .
> MM.——and——entered the room.
> Mlle. Rose took the children away.

Some publishers, following the French practice, omit the period after

Mme and *Mlle* but retain it for *M.*; for most American publishers it is usual to retain it for all, as for *Mr., Mrs.,* and *Ms.*

If these French titles are used with the name in the second person, they should be spelled out.

> "How are you today, Monsieur Monie?"
> "Madame Curie, will you accept this . . ."
> "Mademoiselle Rose, will you please take the children away."

Monsieur, Madame, and *Mademoiselle* should be spelled out and capitalized when they are used alone in the third person as a substitute for a name.

> When this untoward event took place, Madame went with all haste to tell her husband; but Monsieur only shrugged his shoulders.

When they are used without a name as substantives in direct address, they should be spelled out and not capitalized.

> "Oui, madame."
> "It shall be done at once, monsieur."
> "Can we help you, mademoiselle?"

The Spanish *Señor, Señores, Señora,* and *Señorita* and the Italian *Signor, Signora,* and *Signorina* are seldom abbreviated, but Sr. (*Señor*), Sra. (*Señora*), and Srta. (*Señorita*) may be used like Mme., Mlle., and M.

Titles denoting position or rank should usually be spelled out in general text. Occasionally abbreviations may be appropriate (for example, in tabular matter and in military texts). Abbreviations should not be used when only the surname is given (*General Washington*). Whenever there is any doubt about the propriety of using abbreviated forms of titles, they should be spelled in full. Titles like *Secretary of State* and *Secretary* when it is the title of a Cabinet member should never be abbreviated; *Amb., Sen.,* and *Rep.* are allowable only in tabular matter.

Amb., Ambassador	Pres., President
Atty., Attorney	Prof., Professor
Atty. Gen., Attorney General	Rep., Representative
Dist. Atty., District Attorney	Sec., Secretary
Gov., Governor	Sen., Senator
Gov. Gen., Governor General	Supt., Superintendent
Lt. Gov., Lieutenant Governor	Treas., Treasurer
Msgr., Monsignor, Monseigneur	

Adj.	Asst. Surg.	Col.
Adj. Gen.	(Assistant Surgeon)	Comdr.
Adm. (Admiral)	Brig. Gen.	Cpl.
A1c. (Airman,	Bvt. (Brevet)	CWO (Chief Warrant
First Class)	Capt.	Officer)

Ens. (Ensign)	Maj.	Sgt.
1st Lt.	Maj. Gen.	Sgt. Maj.
Gen.	M. Sgt.	S1c. (Seaman,
Hosp. Sgt. (Hospital	Pfc. (Private,	First Class)
Sergeant)	First Class)	S. Sgt.
Insp. Gen.	PO (Petty Officer)	Surg. Gen.
Judge Adv. Gen.	Pvt.	T2g. (Technician,
Lt.	Q.M. Gen.	Second Grade)
Lt. Col.	Q.M. Sgt.	T. Sgt.
Lt. Comdr.	Rear Adm.	Vice Adm.
Lt. Gen.	2d. Lt.	WO (Warrant Officer)
Lt. (j.g.)		

"Versus." Versus should be spelled out in general text but may be abbreviated, *vs.*, in tight matter. For legal cases only, the preference is *v.* (rather than *vs.*), set roman between italicized names of the parties involved: *Smith* v. *Jones*.

SYMBOLS AND SIGNS

A symbol represents elements, relations, quantities, operations, or certain other qualities; symbols are usually letters alone or combined with signs or numbers. Signs are any sort of notation other than regularly printed letters. Symbols and signs never use periods in the sense of conventional punctuation.

Book sizes. A symbol indicating book size is not followed by a period.

f°, folio	16mo, sextodecimo
4to or 4°, quarto	18mo, octodecimo
8vo or 8°, octavo	32mo, thirty-twomo
12mo, duodecimo	

Chemical elements and compounds. In general text the names of chemical elements and compounds should be spelled out (lowercase), but in scientific and technical writing symbols for elements are used freely, especially in equations and formulas. Periods are never used in this form of abbreviation.

Actinium	Ac	Bromine	Br	Dysprosium	Dy
Aluminum	Al	Cadmium	Cd	Einsteinium	Es
Americium	Am	Calcium	Ca	Erbium	Er
Antimony	Sb	Californium	Cf	Europium	Eu
Argon	A	Carbon	C	Fermium	Fm
Arsenic	As	Cerium	Ce	Fluorine	F
Astatine	At	Cesium	Cs	Francium	Fr
Barium	Ba	Chlorine	Cl	Gadolinium	Gd
Berkelium	Bk	Chromium	Cr	Gallium	Ga
Beryllium	Be	Cobalt	Co	Germanium	Ge
Bismuth	Bi	Copper	Cu	Gold	Au
Boron	B	Curium	Cm	Hafnium	Hf

Helium	He	Nitrogen	N	Silicon	Si
Holmium	Ho	Nobelium	No	Silver	Ag
Hydrogen	H	Osmium	Os	Sodium	Na
Indium	In	Oxygen	O	Strontium	Sr
Iodine	I	Palladium	Pd	Sulfur	S
Iridium	Ir	Phosphorus	P	Tantalum	Ta
Iron	Fe	Platinum	Pt	Technetium	Tc
Krypton	Kr	Plutonium	Pu	Tellurium	Te
Lanthanum	La	Polonium	Po	Terbium	Tb
Lawrencium	Lw	Potassium	K	Thallium	Tl
Lead	Pb	Praseodymium	Pr	Thorium	Th
Lithium	Li	Promethium	Pm	Thulium	Tm
Lutetium	Lu	Protactinium	Pa	Tin	Sn
Magnesium	Mg	Radium	Ra	Titanium	Ti
Manganese	Mn	Radon	Rn	Tungsten	W
Mendelevium	Md	Rhenium	Re	Uranium	U
Mercury	Hg	Rhodium	Rh	Vanadium	V
Molybdenum	Mo	Rubidium	Rb	Xenon	Xe
Neodymium	Nd	Ruthenium	Ru	Ytterbium	Yb
Neon	Ne	Samarium	Sm	Yttrium	Y
Neptunium	Np	Scandium	Sc	Zinc	Zn
Nickel	Ni	Selenium	Se	Zirconium	Zr
Niobium	Nb				

Isotopes should be written with the mass number as superscript preceding the element symbol.

$$^{198}\text{Au} \qquad ^{131}\text{I}$$
$$^{60}\text{Co} \qquad ^{32}\text{P}$$
$$^{51}\text{Cr} \qquad ^{99m}\text{Tc}$$

When isotopes are spelled out in general text, they should be written ^{198}gold, ^{131}iodine, and so on.

The following are some symbols used as prefixes for certain chemical compounds (for example, *m*-sulfanilic acid, *n*-propyl alcohol):

α-, alpha-	N-, indicates nitrogen attachment
β-, beta-	*nor-*, indicates parent compound
cis-, stereochemical opposite of *trans-*	*o-, ortho-*
d-, dextro(rotatory)	*O-*, indicates oxygen attachment
D-, *dextro* (configuration)	*p-, para-*
l-, *levo*(rotatory)	*S-*, indicates sulfur attachment
L-, *levo* (configuration)	*sec-*, secondary
m-, meta-	*tert-*, tertiary
n-, normal	*trans-*, stereochemical opposite of *cis-*

In writings on biochemistry, organic chemistry, and medicine, various expressions for the carbon atom are encountered. The following is a suggested treatment:

C_{14} indicates a compound with 14 carbons
C-14 refers to the position of the fourteenth carbon atom
^{14}C indicates an isotope of carbon

Chess. Terms used in chess are represented by the following symbols.

B, bishop	P, pawn
K, king	Q, queen
KB, king's bishop	QB, queen's bishop
KKt, king's knight	QKt, queen's knight
KP, king's pawn	QR, queen's rook
KR, king's rook	R, rook
Kt, knight	

Degree mark. See pages 117, 130.

Mathematical signs. See pages 123–124.

Prime marks. The single prime ('), not an apostrophe, is used for feet or minutes; the double prime ("), not quotation marks, is used for inches or seconds.

$$25'8'' \qquad\qquad \text{latitude } 40°30'30''N \qquad\qquad 20°30' \text{ (of arc)}$$

Times sign. The times sign (or multiplication sign) is often used for magnification, for crosses in genetics and breeding, and sometimes for dimensions.

magnification: $\times$ 100
$DD \times dd$ produces an F_1 all of the dominant phenotype.
14×28

Trigonometry. Abbreviations of trigonometric terms are symbols.

cos, cosine	log, logarithm
cosec *or* csc, cosecant	sin, sine
cot, cotangent	tan, tangent

Standard signs and symbols. A special kind of abbreviation is the sign, also called a symbol. The following list contains signs and symbols frequently used in specialized matter.

ACCENTS
⁄ acute
˘ breve
₎ cedilla
∧ circumflex
‥ dieresis
＼ grave
⁻ macron
∼ tilde

BULLETS
● solid circle; bullet
• bold center dot
• movable accent

DECORATIVE
✚ bold cross
✛ cross patte
✖ cross patte
❖ cross patte
✿ (184 N)
⚷ key
⚱ (206 N)

ELECTRICAL and CHEMICAL
ℜ reluctance
↔ reaction goes both right and left
↕ reaction goes both up and down
↕ reversible

⇵ exchange
↑ gas
→ direction of flow; yields; direct current
⇋ reversible reaction; alternating current
⇌ reversible reaction; alternating current
⇌ reversible reaction beginning at left
⇋ reversible reaction beginning at right
Ω ohm; omega
MΩ megohm; omega
μΩ microhm; mu omega
ω angular frequency, solid angle; omega
Φ magnetic flux; farad; phi
Ψ dielectric flux; electrostatic flux; psi
γ conductivity; gamma
ρ resistivity; rho
Λ equivalent conductivity; lambda
HP horsepower

MATHEMATICAL

÷ geometrical proportion
—: difference, excess
‖ parallel
‖s parallels
∦ not parallels
| | absolute value
· multiplied by
: is to; ratio
÷ divided by
∴ therefore; hence
∵ because
:: proportion; as
> greater than
⊏ greater than
≥ greater than or equal to
≧ greater than or equal to
≷ greater than or less than
≯ is not greater than
< less than
⊐ less than
≶ less than or greater than
≮ is not less than
≤ less than or equal to
≦ less than or equal to
≲ equal to or less than
≂ equal to or less than
≳ equal to or greater than
⊥ equilateral
⊥ perpendicular to
⊢ assertion sign
≐ approaches
⪯ equal angles

≠ not equal to
≡ identical with
≢ not identical with
𝗛𝗛 score
≈ or ≑ nearly equal to
= equal to
~ difference
≈ perspective to
≅ congruent to, approximately equal
≏ difference between
⇌ equivalent to
(included in
) excluded from
⊂ is contained in
∪ logical sum or union
∩ logical product or intersection
√ radical
√ root
$\sqrt[2]{}$ square root
$\sqrt[3]{}$ cube root
$\sqrt[4]{}$ fourth root
$\sqrt[5]{}$ fifth root
$\sqrt[6]{}$ sixth root
π pi
e base (2.718) of natural system of logarithms
ϵ is a member of; dielectric constant; mean error; epsilon
∈ element of
+ plus
+ bold plus
− minus
− bold minus
/ virgule; solidus; separatrix; shilling
± plus or minus
∓ minus or plus
× multiplied by
= bold equal
number
⅋ per
% percent
∫ integral
| single bond
\ single bond
/ single bond
‖ double bond
≏ double bond
⫽ double bond
∂ or δ differential; variation
∂ Italian differential
→ approaches limit of
~ cycle sine
⌒ horizontal integral
∮ contour integral
∝ variation; varies as
Π product

Σ summation of; sum; sigma
! or ∟ factorial product

MISCELLANEOUS
℀ account of
℅ care of
¶ paragraph
Þ Anglo-Saxon
₵ center line
℞ recipe
Ο or ☉ or ① annual
☉☉ or ② biennial
♃ perennial
ƒ function
φ diameter
☍ opposition
⊠ station mark

MUSIC
♮ natural
♭ flat
♯ sharp

PLANETS
☿ Mercury
♀ Venus
⊕ Earth
♂ Mars
♃ Jupiter
♄ Saturn
♅ Uranus
♆ Neptune
♇ Pluto
☊ dragon's head, ascending node
☋ dragon's tail, descending node
☌ conjunction
☍ opposition
☉ Sun
♨ Sun's lower limb
☌ Sun's upper limb
① solar corona
⊕ solar halo
● Moon
● new Moon
☽ first quarter
◑ first quarter
◐ third quarter
◐ last quarter
◐ last quarter
☾ last quarter
○ full moon
⊛ full moon
⊖ eclipse of Moon
♉ lunar halo
∪ lunar corona
⚳ Ceres
⚵ Juno

SEX
♂ or ♂ male
□ male, in charts
♀ female
○ female, in charts
⚥ hermaphrodite

SHAPES
◆ solid diamond
◇ open diamond
▲ solid triangle
△ triangle
□ square
■ solid square
▭ rectangle
▤ double rectangle
★ solid star
☆ open star
∟ right angle
∠ angle
√ check
✔ check
☛ solid index
☞ index

WEATHER
⊤ thunder
⦦ thunderstorm; sheet lightning
↓ precipitate
⦶ rain
← floating ice crystals
↔ ice needles
▲ hail
⊗ sleet
∞ glazed frost
⊔ hoarfrost
∨ frostwork
✳ snow or sextile
⊠ snow on ground
⊹ drifting snow (low)
≡ fog
∞ haze
⌓ aurora

ZODIAC
♈ Aries; Ram
♉ Taurus; Bull
♊ Gemini; Twins
♋ Cancer; Crab
♌ Leo; Lion
♍ Virgo; Virgin
♎ Libra; Balance
♏ Scorpio; Scorpion
♐ Sagittarius; Archer
♑ Capricornus; Goat
♒ Aquarius; Water bearer
♓ Pisces; Fishes

NUMBERS

In deciding whether to use numerals or spell out numbers, the nature of the writing should be considered and literary style distinguished from technical and scientific style. At one extreme is the formal style of the Bible, proclamations, and similar writings, in which all numbers, even of years, are spelled out; at the other extreme is the statistical style, in which all numbers are expressed in numerals. The following sections indicate usage for most types of writing (except for dialogue; see page 275), with the caution that authors and editors should keep consistency uppermost in mind when determining style.

As a very broad rule, numbers under 101, round numbers (for example, *about two hundred years ago . . .*), and isolated numbers are expressed in words in general text matter; in scientific and technical writing the rule is to use numerals for all physical measures and for most quantities and qualities of 11 and over. Another rule of thumb is that when several numbers appear in the same context the style for the larger numbers governs that for the smaller.

> *Poor:* There were 132 men but only ten women interviewed.
> *Better:* There were 132 men but only 10 women interviewed.

Large numbers. *Specific* numbers over one hundred, except isolated numbers, are set in numerals. This does not apply to *round* numbers, which are expressed in words.

> There were 1,240 people in the town.
> There were about twelve hundred people in the town.

If specific numbers and round numbers appear in the same sentence or paragraph, however, the specific number governs the style.

> There were only 1,240 people in the town in 1950, but the population was expected to increase to about 9,000 by 1970.
> Last year we had 125 registered in the course, but next year there should be about 200.

In writing round or approximate numbers of thousands, the following forms are recommended:

> a thousand (or one thousand)   seven thousand
> fifteen hundred   seventy-five hundred

Whenever confusion might arise, use numerals even for round numbers.

> *Vague:* He said he could defend against two or three hundred.
> *Better:* He said he could defend against 200 or 300.

When *isolated* numbers (that is, numbers that appear only rarely in a manuscript) are spelled out, it is unnecessary to use *and* following the word *hundred* or *thousand* (although in formal literary and legal contexts *and* may be used).

two hundred fifty-six   six thousand nineteen
seven hundred forty   forty-six thousand two hundred seventy-two

Uneven isolated numbers over one hundred thousand should usually be set in numerals: *two hundred thousand* or *five hundred thousand* may be written out, but it is better to use numerals for uneven thousands: "between 270,500 and 572,200."

In scientific, technical, and statistical writing, a hybrid form of expressing very large numbers has come into acceptable use, but only for numbers of six or more ciphers. Numerals are used in combination with spelled-out words for any number of millions or more[14] that are rounded to hundred thousands.

. . . 6 million cells per cubic millimeter.
The gross national product was $860.7 billion.
. . . in which $5 million par of preferred stock controlled $1.5 billion capitalization of subsidiary companies.
The 1970 population of Japan was 101,090,000.

For informal, general writing use the rule of spelling out numbers under 101 and even hundreds *(eight million, sixty-two billion, three hundred million,* but *122 million)* except in mixtures such as the following:

From $121 million in sales came $9 million in profit.
He got only 650,000 of the 4 million votes cast.
India has 525 million people as opposed to about 100 million in Pakistan.

Small numbers. Numbers under 101 are usually written out in informal, general text; but where there are frequent numbers, only those under 11 are expressed in words. Small numbers are set in numerals when they occur in a group with larger numerals and refer to similar things.

One booklet contained 36 pages, one 72 pages, and the third 108 pages.
Of the total of 358 landholder members there were 4 Irish peers, 98 sons of peers, and 155 near relatives of peers.

In dealing with more than one series of quantities it is usually best to set them all in numerals.

The number so employed in the Lynn district was 68, as compared with 110 in the 28 other districts.
Besides 300 men and 275 women there were 48 children under 12.

[14]In the United Kingdom *billion* is used as equivalent to *trillion* in the United States.

A small number referred to as a number and not as a quantity is set in figures.

> Put water into a graduated beaker until it just reaches the 40 mark.
> Change sentences 3 to 8 as numbers 1 and 2 are changed.
> Count by 2s to 20.

When two separate numerals appear together, use an alternate style for each or recast the sentence.

> six 8-cent stamps [*not* 6 8-cent stamps]
> 112 eight-cent stamps
> 12 four-foot boards
> sixty-eight 33-rpm records
> In 1973 there were 313 cases of . . . [*not* In 1973, 313 cases of . . .]

Numbers with abbreviations. Regardless of size, numbers used with abbreviations and symbols are always set in numerals.

No. 3	9 cu ft	35°C
5 mg	75%	84th Cong., 1st sess.

Ages. Ages of persons are spelled out in literary and informal writing, but in technical works ages are usually set in numerals. Casual references to decades in all types of text matter are written out, as "in his thirties"; but where frequent ages are given in numerals, "in his 30s" is better.

Nonscientific:	*Scientific:*
aged forty-five	aged 45
six-year-old	6-year-old
eleven months old	11 months old

> Between ages 40 and 65 tolerance to pain decreases significantly; men over 65 tolerate pain less well than men in their 20s.

Beginning a sentence. A numeral should not stand at the beginning of a sentence. A number so placed should be spelled out or the sentence reworded to bring the number elsewhere in the sentence.

> *Wrong:* $36,000 was the amount of the debt.
> *Right:* The amount of the debt was $36,000.
>
> *Wrong:* 1941 saw our entry into the war.
> *Right:* The year 1941 saw our entry into the war.

If a large number at the beginning of a sentence is spelled out, editors sometimes repeat the number in figures, enclosed in parentheses. A common error is the misplacing of the gloss. For instance, "six thousand ($6,000) dollars" should be either "six thousand (6,000) dollars" or "six thousand dollars ($6,000)." The latter form is preferable.

If two related numbers occur at the beginning of a sentence, only the first need be spelled out:

Fifty or 60 steers, 20 cows, and 150 sheep.

Use of commas. With the exceptions to be noted, commas should be used to point off numbers of five digits or more. It is recommended that four-figure numbers be set with commas (8,540), although some house styles differ (8540). Where the comma has been eliminated in text matter, it should nevertheless be used for four-digit numbers in tabular matter for alignment.

$$9,850$$
$$10,560$$
$$12,322$$

Commas are not used in years,[15] page numbers, binary digits, radio-frequency designations, serial numbers, telephone numbers, degrees of temperature, or to the right of a decimal point.

2500 B.C.	serial number 650321
page 1372	555–1212
00110101	3071°F
1550 kilocycles	0.31417

In Great Britain and in foreign languages, the practice is sometimes to use periods or spaces instead of commas (46.862 or 46 862). This is not American usage.

Dates. The number of the year is never spelled out except in very formal writing, such as proclamations, and occasionally in dialogue.

nineteen hundred seventy-three, or nineteen seventy-three
Very formal: one thousand nine hundred and seventy-three

In referring to decades, *the sixties* or *the 1960s* is generally preferred (not *'60's, '60s, 60's,* or *60s;* the last form is used occasionally for ages of persons). A few cases in which it is unnecessary to use the full number of the year in informal contexts are *the class of '65, the spirit of '76,* and *the gold rush of '49.*

In referring to centuries, words are preferred to numerals: *nineteenth century, twelfth century, twenty-first century.*

The abbreviations A.D. and B.C. have practically become symbols that are used with year numbers without conscious thought of the words

[15]Commas are used for year numbers of five or more digits: 36,000 B.C.

for which they stand,[16] and are therefore commonly placed after the number. Nevertheless, careful writers still place the A.D. before the number, B.C. after the number, except for centuries expressed in words.

31 B.C.   fifth century A.D.
A.D. 195

Occurring in lowercase roman text, A.D. and B.C. are commonly set in small caps without space. In a line of capitals and lowercase, such as a heading or a displayed line, A.D. would be preferable. With italic lowercase, italic caps should be used.

The rule of Augustus continued to A.D. 14.
The First Council of Nicaea, A.D. 325
Kalidasa, Fifth Century (?) A.D.

Day of the month. The day may be expressed by a spelled-out ordinal preceding the month or by a cardinal numeral after the month. In military texts and in many foreign countries a cardinal numeral precedes the month: 15 January 1972. However, for most writing the preferable form is as follows:

second of January   October 12
third of June   January 1, 1973

Do not use *-nd*, *-rd*, *-st*, or *-th* with the day when given in numerals.

Once a specific date has been indicated, spelled-out ordinals may be properly used.

Wilson arrived on June 9 but did not sign the agreement until the tenth.

A comma should not be omitted following the year in such expressions as the following:

On July 15, 1962, there was . . . (*not* On July 15, 1962 there was . . .)

When only the month and the year are given, commas are unnecessary.

He began writing in May 1971 and finished in April 1972.

Numerical dates. If a date is written numerically in copy and circled to be spelled out, care should be taken to do so correctly. In the United States the month is usually written first; therefore 2/12/70 means February 12, 1970. In the United Kingdom the first figure indicates the day; therefore 2/12/70 means December 2, 1970, according to that practice.

---

[16]For the meanings of these and less common chronological abbreviations, see pages 105–106.

Decimals. For quantities of less than one, a zero should be set before the decimal point except for quantities that never exceed one. Also, when units are not abbreviated the singular is used for quantities of one or less. Align decimal numbers on the decimal point in tabular matter.[17]

0.32 second	.62 coefficient of correlation
0.75 grain	probability of .85
2.75 grains	.44 caliber revolver[18]
0.26 part	

Degrees. Temperature readings and degrees of latitude and longitude are usually set in numerals.

temperature of 52 degrees	latitude 49 degrees N
52°F	longtitude 24 degrees west
60°–75°	lat. 40°20′ N

Dialogue. With few exceptions, as year dates, numbers in dialogue should be spelled out.

"I paid two dollars and forty-nine cents for it."
"Harry got a six-percent loan in 1973."
"She said that of a hundred twenty-five invited, all but three showed up."
"No, it was April third, not May third."

Conversation or direct discourse, however, as in a novel or biography, should be distinguished from quotations from a formal speech or from matter that may have been previously written. ("Dr. Smith said" could mean he said it in a letter or an article rather than orally.)

In quotations that might have been written in a letter or an article or delivered in a speech, numbers should be treated as previously discussed for general or scientific text matter.

Dr. Smith said, "Of 325 animals inoculated with the virus in 1969, there were 292 alive in 1973."

Dimensions. Dimensions may be expressed in words or numerals, depending upon the nature of the text. In technical matter dimensions under 11 are spelled out unless numerals are used nearby for other measures. The word *by* is usually preferable, but the symbol × or x may sometimes be more appropriate, especially when the sign for inches or feet is used; when the sign is used it should appear with both numerals.

two by four inches	five feet four inches
three- by five-inch cards	5 ft 4 in.
64 feet 10 inches	24″ × 3″

He was six feet four inches tall (or six feet four, or six four) and weighed 175 pounds.

[17]In British usage a decimal is a raised point (2·75), and in Continental practice a comma represents a decimal (2,75).
[18]In informal writing this may be written out as "a forty-four revolver."

Divisions. Numbers of divisions (parts, chapters, paragraphs, sections, rules) are usually set in numerals, either arabic or roman; also numbers denoting lines, notes, verses, figures, and plates.

Psalm 23	Item 6	question 2	verses 6 to 8
Hymn 49	Figure 30	column 3	line 10

In cross-references take care to make the style agree with the specifications for the particular manuscript. For example, the sixth chapter may be designated Chapter 6, Chapter Six, or Chapter VI.

Elision. If two year numbers are connected, the hundreds may be omitted from the second unless the first number ends in two ciphers, when the full number must be repeated. In some styles, only one numeral is needed after the dash.

1775–79	1800–1801	1901–2	1453–7

The same principle may apply to inclusive page numbers, especially in indexes.

8–10, 22–23, 100–103, 107–9, 119–22, 1074–76

In such elisions an en dash should be used; never use an en dash when the numbers are preceded by *from* or *between*.

from 1892 to 1898   between 1955 and 1960

For year numbers with A.D. or B.C. do not elide.

440–421 B.C.   A.D. 133–156

End-of-line division. If necessary a number of five numerals or more may be divided at the end of a line, using a hyphen as in a word. The division should always be made at a comma, and the comma should be retained before the hyphen.

Fractions. Do not use *th* and *ths* after fractions.

$3/100$, *not* $3/100$ths $1/25$, *not* $1/25$th

A fraction expressed in figures should not be followed by *of a*, *of an*.

$3/8$ inch, *not* $3/8$ of an inch

If the sentence seems to require *of a*, the fraction should be spelled out.

Wrong: Fractions are first introduced as expressions of part of a whole—half a pie, ¾ of an hour, and so on.
Right: . . . half a pie, three quarters [*or* three fourths] of an hour.

If the writer of the example above wished for special reasons to use the form ¾, he should, for the sake of consistency, have used ½ also.

Fractions should be expressed in words or as decimals where possible; typesetting is easier and the appearance is better. The fraction ¹/₁₀₀ written in words is better made *one hundredth* rather than *one one-hundredth*, ⁷/₁₀₀ as *seven hundredths* rather than *seven one-hundredths*; or *0.01* and *0.07* may be equivalently used.

> 0.5 percent, *not* ½ of 1 percent
> 5.5 million, *not* 5½ million

Fractions expressed in words are governed by the same rules as other compounds: as nouns they are not hyphenated, as adjectives they should be.

Noun	*Adjective*
three fourths of the total	three-fourths share
two thirds of the members	two-thirds majority
one fifth of a mile wide	one-fifth mile wide

House and room numbers. Except in dialogue, house and room numbers are set in numerals.

> He lived at 249 Western Avenue.   He was sent to room 46.

Money. An isolated mention of a sum less than a dollar should be expressed as cents and spelled out.

> eighty cents, *not* $0.80
> fifty cents per thousand cubic feet, *not* 50¢ per thousand cubic feet

A sum less than a dollar that requires more than two numerals to express it should be set with dollar sign and cipher; and in statistical matter a uniform style should be followed.

> $0.7525 per ounce
> The cost is not high (about $0.02 per ton) and is distributed as follows: depreciation (10 percent), $0.003; power, $0.006; maintenance, $0.012.

A sum of dollars with no cents expressed is usually set without the decimal point and ciphers; but if there is in close proximity a sum of dollars and cents, the ciphers may be used with the sum of even dollars.

> $3, $486 $5–$10 [note two signs]
> One owed $5.00, the other $5.45.

An isolated mention of a sum of dollars that can be expressed in one or two words should be spelled out:

> forty-five dollars   three thousand dollars
> He subscribed $250 to the fund.

If several sums are mentioned in a short space, all should be in numerals, even if some are only dollars or cents.

For treatment of millions and billions of dollars, see page 126.

Foreign money. Sums of money in foreign currency are usually expressed in much the same way as money of the United States. As world monetary policies change, the trend is toward using the decimal system universally. For example, in the United Kingdom under their previous system the style was £6 8s. 7d. (6 pounds, 8 shillings, 7 pence); an equivalent amount is now expressed as £6.43. (They no longer use shillings and there are 100 pence to the pound.) Symbols preceding amounts of foreign currency are closed up, as in the following examples:

Canada:	Can$100	100 dollars
France:	fr100	100 francs
Japan:	¥100	100 yen
Mexico:	Mex$100	100 pesos
West Germany:	DM100	100 deutsche marks

Ordinals. Ordinals are usually spelled out in text matter. In tabular matter, footnotes, and references the hybrid form of a numeral and suffix (*-st, -nd, -rd, -th*) are used; no period follows the suffix.

tenth day	twentieth century
twelfth grade	twenty-fifth of June
3rd ed.	5th printing

A few expressions are conventionally written in the hybrid form even in general text matter: 99th percentile, 38th parallel.

The following are ordinals that are capped:

Fifth Dynasty	Twenty-fourth Amendment
Twenty-third Psalm	Fourth Congressional District,
Eighty-first Congress, second session	fifth ward
Fourth of July	Twenty-first Assembly District

Street and avenue numbers are written in full below 100th Street unless there are frequent other numerals in the text, in which case the hybrid numeral and suffix are used for all streets and avenues.

| Fifth Avenue | Forty-second Street | 112th Street |

Decades of streets follow the same general rule: *in the East Fifties, in the Seventies, in the 140s.*

Some writers feel that the hybrid forms are impure or bungling and that "51st Street" is no better than "51 Street" (to be read as an ordinal). This latter form is being increasingly used for street addresses; and in medical writing the preferred form for the tenth cranial nerve, for example, is "X cranial nerve."

In designating military units the following style may be used: roman numeral for the corps number; spelled-out ordinal for army, fleet, or air force; and hybrid ordinals for divisions, regiments, wings, and other special units. (However, Government Printing Office style is that military units, except *Corps*, are expressed in hybrid ordinals when not the beginning of a sentence.)

V Corps
Seventh Army, Fleet, or Air Force (GPO style: 7th Army)
3rd or 111th Division, Regiment, or Wing
98th Field Artillery, 13th Engineers, 10th Cavalry, 1st Coast Artillery

For hyphenation of ordinals in compound adjectives, see page 232.
Page numbers. Numerals are always used for page numbers:

On page 5 he tells us . . .

Percentage. In literary works percentage numbers are spelled out: "fifty percent." In technical and scientific writing, numerals always precede the word *percent*, with the single exception that isolated references to *one percent* may be spelled out. In statistical material and where other numerals appear frequently with abbreviations in scientific copy, the percent sign (%) may be used in text matter. Always use the sign (%) in tables, and never use a fraction before the word *percent* or the percent sign.

10 percent   2.25 percent, *not* 2¼ percent
12.3 percent   0.5 percent, *not* ½ of 1 percent

Roman numerals. The letters used in Roman notation are I, V, X, L, C, D, and M. Repeating a letter repeats its value. A letter placed after one of greater value adds to it. A letter placed before one of greater value subtracts from it.

Any number of thousands is expressed by a line drawn over any numeral less than one thousand. Thus $\overline{V}$ denotes 5,000, $\overline{LX}$, 60,000. So likewise $\overline{M}$ is one million and $\overline{MM}$ two millions.

As a medieval roman numeral Y stands for 150, $\overline{Y}$ for 150,000.

1	I	15	XV	300	CCC		
2	II	20	XX	400	CD		
3	III	21	XXI	500	D		
4	IV	25	XXV	600	DC		
5	V	30	XXX	900	CM		
6	VI	40	XL	1,000	M		
7	VII	50	L	2,000	MM		
8	VIII	60	LX	3,000	MMM		
9	IX	70	LXX	4,000	M$\overline{V}$		
10	X	80	LXXX	5,000	$\overline{V}$		
11	XI	90	XC	10,000	$\overline{X}$		
12	XII	100	C	100,000	$\overline{C}$		
13	XIII	200	CC	1,000,000	$\overline{M}$		
14	XIV						

Monarchs, popes, and family members with identical names are identified by roman numerals.

Henry VIII	Paul VI
Louis XVI	Oscar Hammerstein II
Philip IV	Loudon Wainwright III

Musical chords are denoted in roman numerals: I, IV, V, V^7 (dominant seventh).

Lowercase roman numerals (i, ii, iii) are generally used for page numbers in the front matter of a book to distinguish them from the arabic-numbered pages in the rest of the book. One reason for this is that parts of the front matter, the table of contents, for instance, cannot be set until the remainder of the book is paged; also it is not unusual for the front matter to require last-minute revisions.

Subscripts and superscripts. Subscripts and superscripts are numbers in a smaller typeface set below or above the line. Commonly used subscripts are those indicating the numbers of atoms of an element in a molecule (H_2CO_3) and the base for a number system (25_8). Superscripts are used for exponents (2^3), for the mass number of isotopes (^{235}U), and for footnote or bibliographic references (Smith[9, 10]); note in this last case that where a comma is needed, the following number is closed up to a thin space.

Superscripts for footnote or bibliographic references are set outside commas, colons, and periods; all other superscripts and subscripts are set inside the punctuation.

The variable is x^2.   See Brown's article.[2]

Time of day. In straight text matter it is usually better to express time of day in words whenever the expression is simple.

four-thirty	quarter to three
five o'clock	half past one
twelve noon	quarter past ten

For exact time use numerals, including ciphers for even hours, and A.M. and P.M. as required.

12:00 M. (noon)
12:00 P.M. (midnight)
The meeting will take place at 5:00 P.M.
On Monday afternoon at 2:35 we were at Joan's.
Take the 11:28 from Rockville Centre.

Never use A.M. or P.M. with *in the morning, in the afternoon,* or *o'clock.* In the military time system no punctuation or abbreviations are used.

Reveille was at 0630, breakfast at 0645.

In works of fiction, it is acceptable to use either "two P.M." or "2 P.M."

Votes and scores. For tabulations of votes and for scores of various kinds, use numerals.

yeas 67, nays 42   Chicago 3, Philadelphia 2
vote of 74 to 28   98–97 margin of victory

Ashe defeated Okker in the finals, 10–8, 6–3, 6–4.
Nicklaus's birdie 3 on the par-4 sixth put him 4 strokes ahead.
He scored 18 on a 20-point scale.
The Mets went into the bottom of the ninth behind, 4–3.

Note that an en dash is used between scores (it represents "to").

ITALICS

Italics are used to distinguish letters, words, or phrases from the rest of the sentence so that the writer's thought or the meaning and use of the italicized words will be quickly understood. In italic text, roman is used for words that would be italic in a roman text. If a word ordinarily in italic is used in text set in small caps or cap-and-small, it is usually quoted instead of italicized.

THOSE WHO PASSED THE "STUDENTEXAMEN" IN 1971

Italics are little used in newspapers. When the meaning might be ambiguous without some differentiation from surrounding words, quotation marks are used. The following rules, therefore, do not apply to newspaper work.

For the typeface of punctuation following italicized words, see pages 172–173.

Book titles. Titles of books (including essays, cycles of poems, long single poems usually printed separately, and works issued in microfilm) are generally set in italics.

Cat's Cradle   Beowulf
The Wealth of Nations   Nibelungenlied
Axel's Castle   Canterbury Tales

Following a possessive an article beginning a title may be dropped. If a work is merely referred to as already mentioned or as familiar to the reader, or if the title is abridged, an initial article may be omitted or left outside the italics.

He read Dickens's Tale of Two Cities.
The statement is made by Adam Smith in his Wealth of Nations.
Gibbon published the Decline and Fall between 1776 and 1788.

Be careful about the use of italics for a title after a preposition in such constructions as the following:

Incorrect: In Jackson's book on *Immunity to Parasitic Animals* . . .
Correct: In Jackson's book on immunity to parasitic animals . . .
   or In Jackson's book, *Immunity to Parasitic Animals,* . . .

The following are not italicized:

Bible	Corporation Manual
books of the Bible: Genesis, . . .	Revised Statutes
Koran, Talmud, Mishnah, . . .	Social Register
Prayer Book	Blue Book
Book of Common Prayer	Constitution (of U.S.A.)
ancient manuscripts: Codex Sinaiticus,	Declaration of Independence
Book of Kells, Ratisbon MS.	Gettysburg Address

The italic style for book titles is generally followed in footnotes and bibliographies, but in book lists and book reviews these titles may be differentiated in other ways, often by using caps and small caps.

Characters in books and plays. When characters in books or plays are referred to, it is unnecessary to italicize or quote their names.

He was excellent as Hamlet, Othello, and King Lear.
He performed well in *Hamlet, Othello,* and *King Lear.*
Tristram Shandy thought his father was a bigot.

Differentiation. When a word is used not to represent the thing or idea it usually represents, but merely as the word itself, it should be italicized. Also, italics or quotation marks (p. 219) should be used following *term* or *word* when a definition is being given.

Inexperienced writers are likely to use too many *and*'s and *but*'s.[19]
Bromine takes its name from its odor. *Iodine* comes from the Greek word for *violet.*
Lord is the general term used to describe a landowner, usually a nobleman, who had a group of persons dependent upon him. *Vassal* is the name given to all persons dependent upon a lord or person of higher rank.
The word *sensitivity* connotes responsiveness.
When they came to the word *schlaf,* "sleep," their eyes would pop wide open.

This rule should be kept clearly in mind when writing sentences like the following. Acts and words can be said to have meaning, but objects and things cannot.

Wrong: What is the meaning of a plebiscite? From what word in Roman history is it derived?
Right: What is the meaning of *plebiscite*? [Do not use an article before a word used in this way.]

[19]Note that the *s* in plurals of italic words as words is roman: *and*'s, *but*'s.

Emphasis. Italics may be used to emphasize a word or phrase. (When used too freely for this purpose, however, they lose their force.)

> We are in danger of being satisfied with the *forms* of knowledge without its *substance.*
> We not only *know* the various objects about us through sensation and perception, but we also *feel* while we know.
> *Force lost: If* the last letter of the *stem* of a weak verb is *t* or *d*, the letter *e* is *inserted* between the *final letter* of the stem and the *t* which precedes the personal endings of the *imperfect tense.*

Occasionally italics may be desirable, particularly in scientific and technical writing, as a sort of heading in long paragraphs on related topics. For example, in a discussion of drugs the word *tranquilizers* may be italicized in one paragraph, followed by a paragraph in which the topic *sedatives* is italicized, then *narcotics*, and so on. This practice is not recommended for short paragraphs or where several items are covered in a single paragraph.

Foreign words. Foreign words or phrases that have been adopted into the English language are set in roman type. (See pp. 482–487 for a list of such words and phrases.) The prospective readers of a work should be considered, and any words or phrases in a foreign language that might be unfamiliar to them should be italicized. Most dictionaries no longer give guidance in the choice of roman or italic for foreign words.

Italics should not be used for foreign proper names (churches, streets, institutions, titles, organizations, places). The following are examples that should be set in roman type:

Abbé——	Légion d'honneur
Arc de triomphe	Mar Toma
Banco de la República	Musée de Louvre
Bibliothèque l'Arsenal	Place Vendôme
Comédie Française	Prado
Croix de guerre	Prix de Rome
l'Étoile	Santa Maria degli Angeli
Hôtel de Ville	Unter den Linden

Foreign money is not italicized: lira, escudo, peso, franc.

It is not necessary to italicize foreign language names of art objects, titles of poems, articles, and so forth, that are enclosed in quotation marks.

> Cicero's "De Senectute"
> Kuenen, "Über die Männer der grossen Synagoge," in his *Gesammelte Abhandlungen.*

An unmistakably foreign word that occurs repeatedly in an article or a book may be italicized the first time it occurs, usually accompanied by a definition of its meaning and use, and thereafter set roman.

Care should be taken to distinguish between an English word and a foreign word that have the same spelling but different meaning, as, for instance, *exit*, "he or she goes out," which set in roman means "a way of egress"; or *item*, "also," "likewise," which means "a separate thing" when it is set in roman.

Many publishers prefer to set Latin abbreviations in roman rather than italic (see pp. 109–110).

Some Latin phrases that have come into common use do not need italicization: de facto, mea culpa, ad hoc, non sequitur, sine qua non, quid pro quo, status quo.

Legal citations. The names of the parties in the titles of court cases are usually italic but may be roman. Such phrases as *et al.* and *ex parte* are italic if the rest of the title is roman; roman if the rest of the title is italic. (See pp. 281–283 for a fuller presentation of legal styles and p. 120 for the use of the abbreviation of *versus*.)

In re *Wight* v. *Baker* the court ruled . . .
Pagano *et al.* v. Charles Beseler Co., 234 Fed. 963 (1916)
Burleson v. United States *ex rel.* Workingmen's Cooperative Publishing Association, 274 Fed. 749, 260 U.S. 757 (1921)
Wright, appellant, v. Johnson, respondent

Letters. Single letters are italicized when they are particularly referred to.

the *i*'s and *o*'s   the *n*th power

In an italic context, italic rather than roman is still the choice of most editors.

A curve whose equation is of the nth degree is called a curve of the nth degree.

When a letter is used to designate shape, tradition calls for a gothic letter, but roman is now more often used. Italic is not satisfactory for this purpose.

a U-shaped tube   an S-shaped cross section
a great V of foam   a T-square

Letters in algebraic equations are italicized (see also p. 283).

$$ab + bc - cd = 62$$

Also, letters are italicized in legends or text referring to corresponding letters on geometrical figures or illustrations, whether italic in the illustration or not.

We reach any point P of the plane by going from O a certain distance $x = OM$ along OX.

If then two propellers A and B, absorbing a given power but A turning at a speed of rotation twice that of B, are compared, a little computation shows that the top speed of the geared propeller B is about 30 percent less than that of A.

Chemical formulas, however, should be set roman: Fe_2O_3, also sin, tan, log, and so on, in trigonometric equations.

Letters in parentheses to indicate subdivisions are usually roman, as are letters after the number of verse or page to indicate a fractional part: Luke 4:31a.

Musical compositions. Titles of songs and short musical selections are usually set in roman quoted; operas and oratorios are italicized; selections named by number or key are sufficiently distinguished by roman caps and lowercase without quotes.

"The Star-Spangled Banner"	*The Messiah*	Quartet in A minor
the "Marseillaise"	*Lohengrin*	Fifth Symphony
"Sonata Appassionata"	*Peer Gynt Suite*	Opus 147, in B

Newspapers and periodicals. The names of newspapers and periodicals are usually italicized.

The New York Times	the New York *Daily News*
the London *Times* or *The Times* (London)	the St. Louis *Post-Dispatch*

The names of magazines sometimes seem to need to be followed by the word *magazine*, to distinguish them from newspapers and journals. In such cases, unless it is a part of the official title, *magazine* should be set lowercase roman.

Cue magazine	*New York* magazine
Newsweek magazine	*The New York Times Magazine*

Strictly speaking, an initial article should be italicized if it is part of the official title; but the fact is often difficult to verify, and if many titles appear in a work, some of which have *The* and some have not, an apparent consistency of treatment is gained by setting all with a lowercase roman *the*. Official titles of newspapers are especially difficult to verify.[20] Some have the article and the name of the place of publication, some have the place and not the article, some have neither. In general, when in doubt italicize only the distinctive part of the name.

If the name of a publication is used within itself, it is usually set cap-and-small.

[20]The official title is the name on the masthead or the statement of ownership found usually in the upper corner of the editorial page. The title at the top of the first page may vary. *The Boston Globe,* for example, has *The Boston Globe, The Boston Evening Globe,* and *The Boston Sunday Globe* at the top of its various editions.

Plays, movies, television series. The titles of plays (including one-act plays), motion pictures, and television series are italicized; the title of an episode in a television series is set roman in quotes.

The Dock Brief	*All About Eve*	*The Waltons*
Medea	*The Seventh Seal*	*Ironside,* "The Ghost of
Two Gentlemen of Verona	*Casablanca*	the Dancing Doll"

Resolutions and legislative acts. *Resolved,* in resolutions, legislative acts, and the like, is always italicized. The word *provided* in the body of such matter is also usually set in italics.

Resolved, That the Secretary of the Senate be authorized to . . .
Resolved by the Senate of the State of California, That the Judicial . . .
. . . either on its title page or on the first page of music: *Provided,* That . . .

Scientific names. Taxonomic names of genera, species, and varieties are italicized, but roman type is used for any larger taxon: phylum, class, order, family, and subphylum, infraclass, superfamily. The name of the founder of a species and the abbreviations for *species* (singular *sp.,* plural *spp.*) and *variety* (*var.*) are set in roman. The kingdoms—animal kingdom, mineral kingdom, and plant kingdom—should be set in roman lowercase except in tabular classifications, where they may be capped.

The branches of the banyan tree (*Ficus bengalensis*) send out numerous trunks.
    Some *Ficus* spp. may be dwarfed.
. . . the house mouse (*Mus muscularis,* family Muridae, order Rodentia, class
    Mammalia, phylum Chordata)
sweet corn (*Zea mays* var. *rugosa*)
Rickettsia prowazekii da Roche-Lima
Toxoplasma strain BK

The genus name appearing alone should be italicized. In many cases the genus name is the same as that for an individual organism, so that care must be exercised in cases such as the following:

Epidemic strains of *Streptococcus* are common, and when the streptococcus
    invades the heart valve, rheumatic fever results. Penicillin is effective against
    most streptococci, but . . .
Entamoeba histolytica is the most important of the amebae.
Most strains of *Salmonella* are motile; the flagella of a salmonella . . .

Sounds. A combination of letters used to represent a sound is usually set in italics. If an onomatopoeic word is used, italics are not necessary, though occasionally they are used.

"*Urr-aw! Urr-aw!*" suddenly barked an automobile horn.
The *dit-dar* of the Morse was growing stronger.
a long *whooooo*

the steady drip, drip of the rain
"Boom, bo-o-om," roared the surf.
Suddenly the bell rang—ding, ding, ding-ding-ding.

Thoughts. Unspoken thoughts, which might appear in context with dialogue, are often italicized rather than set roman within quotation marks.

Will Hank be on time for a change? he thought. *Sure he will.*

Vessels, airplanes, spacecraft. The names of vessels, submarines, airships, airplanes (but not the type of plane), spacecraft, and trains are italicized.

the carrier *Hornet*	the *ZR-5*
H.M.S. (or HMS) *Ajax*	the *City of New Orleans*
the *Spirit of St. Louis*	the *Sunset Limited*
Boeing 747	lunar excursion module *Eagle*
Super Caravelle	moon rocket *Apollo 12*

CAPITALIZATION

Although there are many uses of capitals about which no one requires guidance, the question of what to capitalize ("cap") and the necessity of following a consistent style of capitalization often prove troublesome to the writer, the copy editor, and the proofreader. The following rules are generally accepted.

CAPITALS TO MARK BEGINNINGS

Sentences. The first word of a sentence is ordinarily capped.

A sentence set in small caps for emphasis should begin with a cap and all proper nouns should be capped.

American names such as *du Pont*, *van Dyke*, and others that usually begin with a small letter are capped if they begin a sentence but not within a sentence.

A sentence that is one item of a series should begin with a cap.

Examples of easy colloquial speech are such words as *folks; he don't* or *it don't; Stop the car, I want out; He comes of a Sunday; This is all the farther the road goes; . . .*
Case has plenty of legitimate uses, as in *Circumstances alter cases* and *In case of fire, give the alarm.*

Exceptions. The two following sentences, taken from a foreign language textbook, illustrate a kind of sentence that may legitimately begin with a small letter.

rr is forcibly trilled in Spanish.
a, o, and **e** are strong vowels.

A sentence enclosed in parentheses inside another sentence need not begin with a cap.

> That overstocking does not occur is due to ruthless fishing (man is a severe enemy of fish), to pollution, and to the many other enemies that surround fish.
> These can be drawings, pictures, pressed flowers, sunflower seeds (have you ever eaten sunflower seeds?), Indian relics . . .

Partial sentences. The word following an exclamation point or a question mark is not always construed as beginning a new sentence requiring a cap. If the matter following is closely connected with what precedes, completing the thought or making the meaning clearer, no cap is necessary.

> Progress where? or, even more fundamentally, progress whence?
> What a piece of work is man! how noble in reason! how infinite in faculty! —
> SHAKESPEARE.
> How is the gold become dim! how is the most fine gold changed!
> Lord, who shall abide in thy tabernacle? who shall dwell in thy holy hill?

In a series of questions not in complete sentence form, such as are often used in school textbooks at the end of chapters, the usual practice is to cap after the question mark, for here the part following the question mark does not form an integral part of what precedes but is a new question requiring a separate answer. Some writers, however, consider the cap unnecessary here, also.

> What part did Philip I play in the history of Holland? Of England?
> What effect might that have upon the market price of trees? Upon the desirability of going into the business?

Although lowercase may be as good usage as caps in some of the foregoing instances, there is no justification for lowercase in sentences such as the following:

> *Wrong:* Do you wear out a few words? which ones?
> *Right:* Do you wear out a few words? Which ones?

The use of (*a*), (*b*) before questions does not make it necessary to cap as for a new sentence.

> *Cap unnecessary:* Who stands the loss if the car burns and freight is destroyed (*a*) As the result of being struck by lightning? (*b*) Because a match is carelessly thrown inside the car?
> *Better:* Who stands the loss if the car burns and freight is destroyed (a) as the result of being struck by lightning? (b) because a match is carelessly thrown inside the car?

If each part introduced by a letter in parentheses begins a new line, caps are still unnecessary, but some writers and editors object to the appearance of a lowercase letter in this position.

> 1. What methods of travel
>   (a) were used by your grandparents when they were your age?
>   (b) have been added since the year 1900?

> *Or:*

> 1. What methods of travel
>   (a) Were used by your grandparents when they were your age?
>   (b) Have been added since the year 1900?

Whether to use caps or lowercase is a matter of choice also in such instances as the following, but a consistent style should be followed throughout:

> Factors Influencing Creative Work
>   I. *The teacher should*
>   1. Be creative herself
>   2. Possess a rich cultural background
>   3. Plan for many worthwhile experiences

Poetry. In poetry, the convention is to start each line with a cap; however, since some modern verse does not follow this convention, the original must always be followed.

> "Like fearful shadows, / Slowly passes / A funeral train."
> "Say goodby to black / black scrub on fire."

Direct discourse. A cited speech in direct discourse should begin with a cap, as should a direct thought; when intermixed with spoken dialogue, thoughts are often better expressed in italics, without quotes.

> At this moment John said, "It's seven o'clock."
> I glanced at the clock and thought, "Can it be seven already?" (or *Can it be seven already?*)

Direct questions. The first word of a direct question may or may not be capped; the choice is the author's.

> There is one question the reader should always ask himself, and that is, is the meaning clear?
> We may ask, how can it best be understood?
> We understand this question to mean, What number divided by 5 gives 6 as a quotient?
> To introduce this topic ask pupils, Who are the owners of the vast grazing lands described on pages 110–13?

In sentences like the following, capping a single interrogative word and using a question mark after it is too cumbersome:

> *Poor:* The subject is the word that answers the question Who? or What?
> *Better:* The subject is the word that answers the question *who* or *what*.

Direct quotations. A quotation complete in itself, formally introduced, and not grammatically joined to what precedes, should begin with a cap. However, a quotation may be so closely woven into a sentence that no cap is required. (See also Excerpts, pp. 17–19.)

> Golden Rule Jones probably did as well as anyone could with the baffling problem of defining law when he said that "law in the United States is anything that the people will back up." Emerson also observed to the same effect, "The law is only a memorandum."
> "It was a major victory for the Administration," states the Boston *Transcript,* with the conclusion that "it will be a blow to the utility interests of the entire country."

Rules, slogans, mottoes. The first word, and only the first, of rules, slogans, and mottoes should be capped.

> Another rule is, Divide the volume in cubic inches by 2150.
> All for one, one for all.
> Write up cases and incidents from your own experience that illustrate the truth of these sayings: Haste makes waste; A stitch in time saves nine; Better late than never.

Following a colon. The first word after a colon usually should be capped when it begins a complete sentence.

> In conclusion I make this prophecy: The coming year will show greater advance on this line of research than have the past ten years.
> They can be summarized as follows: (1) All trade is essentially an exchange of goods and services. (2) Goods are given . . .
> The following were elected: president, William Jones; vice-president, Frank Smith.
> You need tools to carry on in a science course: not just workbench tools and laboratory test tubes and beakers, but books, magazines, visits to shops and museums.
> . . . two kinds of information: (1) the results of research that has been completed; (2) the results of applied methods in the same field as the problem chosen.

Letters. In the salutation of a letter only the first word of the salutation and the name of the person addressed or the noun used in place of the name are capped. (See also p. 281.)

> Dear Sir: My dear Cousin,
> My dearest Emily,   Most honorable Sir:

Subjects for debate, legislative acts. The first word following *Resolved* (or any words in italic that follow *Resolved*) should be capped; as should also words following introductory italicized phrases in documents and the like.

> *Resolved,* That woman suffrage has been of benefit to the country.
> *Resolved by the Senate, the Assembly concurring,* That . . .
> *Know all men by these presents,* That . . .
> *To whom it may concern,* Greeting: Know that . . .

The word WHEREAS, if set cap-and-small introducing a paragraph, should be followed by a comma and a capped word. If *whereas* is not so distinguished from the rest of the sentence, no comma should follow and the next word should be lowercase.

> WHEREAS, Substantial benefits . . .; now, therefore, be it
> Whereas the Constitution provides . . .; and . . .

The vocative "O." The vocative *O* in English is always capped; but *oh* is capped only when it begins a sentence or stands alone.

> Thy ways, O Spirit, are unconfined.
> For if you should, oh! what would become of it?

In Latin sentences cap *O* only at the beginning of a sentence.

> Hoccine seclum! o scelera! o genera sacrilega! o hominem impium!
> —TERENCE, *Adelphi,* III, ii, 6.

TITLES, HEADINGS, LEGENDS

English titles of books, pamphlets, newspapers, magazines; of parts, chapters, sections, of a work; of poems, articles, lectures, should be distinguished by capitalization. In such titles, in headings, and in similar matter set cap-and-lower or cap-and-small the first and the last word should be capped, regardless of what part of speech they may be. Also cap all important words, no matter how short; that is, cap nouns, pronouns, verbs, adjectives, and adverbs. Articles, coordinate conjunctions, and prepositions are generally considered unimportant and are not capped.[21]

What Is a Law?	Books to Have on Hand
Inside, As It Is to the Worker	The Captain Who Stood By
Determining Per Capita Costs	Should a Pupil Add Up or Down a Column
Percent of Urban Population	Procedure If Cash Does Not Prove

[21]The library style of capping only the first word and proper names of book titles is occasionally followed in bibliographies, but it is not generally favored by American trade book publishers. However, in many foreign languages it is the preferred style, in which case the original must be followed.

Prepositions. Many publishers cap all prepositions of four or more letters. Whether to follow this rule should be determined at the outset, and the style chosen observed throughout the book.

Days Before History   All Through the Night
From the War Into Peace   Shout All Over God's Heaven

Prepositions that are an inseparable part of a verb should be capped.

Getting Down to Business
How to Lay Off a Line Through a Given Point That Shall Be Parallel to a Given
    Line
Things to Do With and Find Out About a Book

The usual form for infinitives is *to Be, to Do, to Go.*

If office practice is to lowercase all prepositions, a problem sometimes arises when prepositions are accompanied by an intimately qualifying word, forming, in effect, a compound preposition. An exception may well be made of such instances and caps used for the sake of appearance.

According to   Except for   Instead of
Apart From   from Among   Just Before

Such problems are usually avoided by capping all prepositions of four or more letters.

Articles. In longer titles cap an article, preposition, or conjunction that immediately follows a marked break indicated by a punctuation mark.

Woodrow Wilson: The Man, His Time, His Task
Business, Politics, Diplomacy—An Autobiographical Biography
Looking Them Over—A Glimpse of the New Machine at the Show
The Psychopathic Employee: A Problem of Industry

Hyphenated compounds. Usage varies greatly in the capitalization of hyphenated compounds in titles and headings. The following general rules should cover most situations.

The first element of a hyphenated compound should always be capped (*Un-Americans in Congress*). Cap both elements of a temporary compound (that is, a unit modifier that would not be hyphenated otherwise) and both elements of a coordinate term (*Good-Evil Continuum*).

English-Speaking Union   Complement-Fixing Antibody
Short-Story Writers   Time-Consuming Chores
Twentieth-Century Man   Blue-Green Leaves
Well-Known Authors   Acid-Base Balance

If the compound would ordinarily be hyphenated, even if it were not a unit modifier, the second element (and third, as in *Son-in-law's Dilemma*) should be lowercase.

Twenty-fifth Anniversary	Merry-go-round Tactics
Absent-mindedness in Washington	Medium-sized Warships
Cul-de-sac Infections	The Co-author's Role
Self-conscious Students	So-called Successes

The words in a longer hyphenated phrase should be treated as in any other heading (*All-or-Nothing-at-All Attitude*).

Cross-references. A cross-reference to a heading within a work should be styled as the heading itself. Such cross-referred headings may appear in roman type (as in the following examples), be italicized, or be set within quotation marks; whichever style is selected should, of course, be consistent throughout the work. Also be careful to distinguish between references to the heading itself and to the subject matter discussed under it.

> See the section on life among the primitives.
> *Or:* See the section Life Among the Primitives.

> See the discussion of present-day styles.
> *Or:* See under Present-Day Styles.
> *Or:* See Present-Day Styles.

PROPER NOUNS AND ADJECTIVES

Cap proper nouns and proper adjectives and words used as proper nouns.

Boston	Tennessee	Shakespeare
Bostonian	Tennesseean	Shakespearean

Proper names and proper adjectives used to designate a particular kind or variety of the common classification tend to become lowercase (as *panama hat, graham cracker, turkish towel*), as do verbs derived from proper nouns (*pasteurize*). However, defining the distinction between proper adjectives with a *proper* meaning and derivatives (including verbs and nouns) with a *common* meaning is sometimes difficult. The following are a few specific examples:

Proper	*Common*
Afghan hound	afghan (lap robe)
Alaskan pipeline	baked alaska
Brussels griffon	brussels (*or* Brussels) sprouts
Italian industry	italic typeface
Puritan migration	puritanical ethics
Roman civilization	roman type
Russian troops	russian (*or* Russian) dressing

Anglicize, Anglophile, and *Francophile* are usually capped. Copyrighted trade names should always be capped.

The following are derivatives of proper nouns with common meanings and may be set lowercase:

alexandrine	delftware	macadamize
ampere	derringer	madras cloth
angora wool	draconian	manila paper
angstrom	fletcherize	maxwell
apache dance	french dressing	melba toast
astrakhan fabric	gauss	merino sheep
benday process	gothic novel	morocco leather
bessemer steel	herculean job	oxford shoe
bohemian set	hessian fly	parkerhouse roll
bologna sausage	india ink	petri dish
bowie knife	jamaica ginger	plaster of paris
braille	japan varnish	portland cement
brewer's yeast	joule	quisling
bunsen burner	klieg light	roentgen
burley tobacco	kossuth hat	roman numeral
cashmere sweater	kraft paper	russian bath
china silk	lambert	rutherford
congo red	leghorn hat	shanghai
cordovan leather	levantine silk	simon pure
coulomb	lilliputian	spanish omelet
cuban heel	london purple	timothy grass
curie	lynch law	venturi tube
daguerreotype	lyonnaise potatoes	zeppelin

Medical terms retain the capital in possessives, but adjectives derived from proper nouns are usually lowercase.

Gasser's ganglion	gasserian ganglion
Gram's stain	gram-negative bacteria
Hunter's sore	hunterian chancre
Müller's duct	müllerian duct
Parkinson's disease	parkinsonian tremor
Achilles' tendon	cesarean section
Hansen's disease	fallopian tube
Hurler's disease	graafian follicle

Words used as proper nouns. No one has any difficulty in recognizing a simple proper name but even experienced copy editors often become confused in determining when a common noun is used as a proper noun or is part of a compound proper noun and should be capitalized. For instance, is the correct form *King of England* or *king of England, Book of Genesis* or *book of Genesis, the Government* or *the government, my Uncle Charles* or *my uncle Charles*? Few would hesitate on *Massachusetts General Court,* but many are confused when the term *legislature* is used: *the California Legislature* (see p. 160).

Just as some writers consider many marks of punctuation necessary for clarity where others would dispense with them, so some writers seem to believe that caps are necessary when others find the meaning is clear with lowercase. British writers use caps much more freely than Americans do, and French writers much less. Most publishing houses have established an office style for capitalization, which they follow if the author has no objection. Some book publishers like a free use of caps (up-style), while others prefer few caps (down-style); most magazines follow an up-style, while newspapers generally follow a down-style.

The rules that follow are generally acceptable to most book publishers.

GEOGRAPHICAL TERMS

Divisions of the earth's surface. Cap the names of the great divisions of the earth's surface; also the names of distinct regions or districts.

Arctic Circle	Great Divide
North Pole, the Pole	Tropic of Cancer
Torrid Zone	North Temperate Zone

Antipodes (when specifically referring to Australia and New Zealand)

Lowercase: antipodean, equatorial, polar regions, the arctics, the tropics, the temperate zone.

Divisions of the world or of a country. Cap names of the divisions of the world or of a country.

the Old World, the New World	Lower California (Mexico)
the Near East, the Far East	North Atlantic States[23]
the Far North	New York State
the Middle East, the Mideast	the Atlantic Coast
the Continent [continental Europe][22]	the Pacific Coast

Lowercase: state [used in a general sense and when it does not follow a proper name: state of New York, state of Oklahoma] and coast [when the meaning is the shoreline rather than the region: Pacific coast, Atlantic coast].

When referring to the hemispheres the following forms are usually preferred:

Eastern Hemisphere: Eastern(er), Orient(al)
Western Hemisphere: Western(er), Occident(al)
Northern Hemisphere: Northern(er)
Southern Hemisphere: Southern(er)

Names of points of the compass and adjectives derived from them are capped when they are part of a name established by usage as the designation of definite area, but not when they denote simply direction or compass points.

[22]When the adjective alone is used referring specifically to continental Europe it should be capped: *Continental tastes, Continental customs, Continental cuisine,* but *European continent.*

[23]The divisions of the United States used by the United States Census are New England States, Middle Atlantic States, East North Central States, West North Central States, South Atlantic States, East South Central States, West South Central States, Mountain States, and Pacific States.

the East, eastern(er)	the North, northern(er)
East Africa	North Africa, central Africa
the Far West,	the South, southern(er)
far western(er)	Southeast Asia, southeastern
the Midwest or Middle West,	Asia, central Asia
midwestern(er)	the West, western(er)

Lowercase: traveled west (or south, north, east); southern California; northern Michigan, eastern Gulf states; western canners; northern manufacturers; southern aristocrats; a western (movie).

The following are treated as political divisions when referring to specific periods in history:

Northern(er), Southern(er): American Civil War period
Western Europe, Eastern Europe: World War II to present
Central Europe (*or* Powers): World War I period

Regions and localities. Cap popular appellations for regions and localities.

the Levant	the Eastern Shore (Chesapeake Bay)
the Promised Land	the West Coast
the Eternal City	the Upper Peninsula (Michigan)
the Dark Continent	the Golden State
the Spanish Main, the Main	the City of Churches
the Gold Coast	the Battery
the Great Plains	East Side, West Side
the Badlands	the City, West End (London)
the Corn Belt, the Black Belt	the Loop, the Stock Yards (Chicago)
Down East	the National Capital
the Deep South	

Lowercase: fatherland, ghetto, no-man's-land.

Rivers, lakes, mountains. Cap generic terms like *river, lake, mountain,* and many others when they are used with a proper name and form an organic part of it. Newspaper usage is divided, some papers preferring lowercase for the common-noun portion of the name unless it comes first, as in *Mount Everest.*

Penobscot River	Rocky Mountains
Tombigbee River	Bay of Naples
Casco Bay	City of Mexico
the Coast Range	Lake Michigan

Varying styles are used in cases that are more confusing than those just mentioned. The following rules provide for most situations wherein a generic term passes from proper usage into common usage, and therefore should be lowercase: (1) When a plural generic term follows two or more proper names it should be lowercase. (2) When a generic term precedes proper names it is capped unless the generic term itself is preceded by the article *the*. (3) When a generic term (such as *valley* or *range*) follows a capped generic term—that is, is used descriptively—it should be lowercase.

the Ohio and Mississippi rivers	the Rocky Mountains
the Himalaya and Andes mountains	the Rocky Mountain range
Lakes Huron and Michigan	the Ohio Valley
Mounts Whitney and Rainier	the Ohio River valley
the river Nile	the Mississippi Valley
the rivers Danube and Elbe	the Mississippi River valley

The following geographic names contain foreign words that are the equivalents of generic terms:

Fujiyama	Rio Verde
Mauna Kea	Sahara
Mauna Loa	Sierra Madre
Rio Grande	Sierra Nevada

In technical, scholarly, or formal usage, these words should not be accompanied by a generic term. In informal usage, however, such forms as *Sahara Desert* or *Sierra Nevada Mountains* are often used.

Make a distinction between nouns that have become established by usage as parts of proper names, and nouns used casually before proper names.

Isthmus of Panama	city of Chicago
Isle of Wight	island of Cuba

When such generic terms are used alone for the whole name, they should be lowercase even if the meaning is specific, except in a few instances such as the following:

the Canal (the Panama Canal)
the Channel (the English Channel)
the Cape (Cape of Good Hope)
the City (the financial district of London)
the District (District of Columbia)
the Falls (Niagara Falls)
the Gulf (Gulf of Mexico)
the Horn (Cape Horn)
the Islands (Philippine or Hawaiian Islands)
the Isthmus (Isthmus of Panama)
the Sound (Long Island Sound)
the Street (Wall Street)

Some office style books list other similar terms of a local application, as, for instance, *the Hills,* meaning the Black Hills, and *the Falls,* meaning the Menomonee Falls.

Deities. Cap all names or appellations of a supreme being, of prophets, and of the Virgin Mary.

Adonai	Hara	the Prophet
Allah	Heaven [personified]	the Paraclete
the All-Wise	the Holy Family	the Savior
the Almighty	the Holy Spirit	Siva
Buddha	Jehovah	the Son of Man
the Child Jesus	King of Kings	the Supreme Being
the Christ Child	the Logos	the Way
the Comforter	Lord of Lords	the Word
Divine Providence	Mahadeva	Yahweh
El	the Messiah	the Blessed Virgin
Elohim	Mohammed (or Muhammad)	Madonna
the Godhead	the Omnipotent	Mater Dolorosa

Lowercase: fatherhood, godlike, godly, messianic, omnipotence, sonship.

Except in extracts from the Bible or the Book of Common Prayer, cap pronouns referring to the foregoing if necessary to avoid ambiguity, but lowercase when their antecedent closely precedes. The relative pronouns *who, whom,* and *whose* are ordinarily lowercase.

> For God was a loving Father to whom his people could always go for comfort in their troubles.
> God in his mercy . . .
> Christ chose his apostles.
> When Christ saw the multitude, He went up into the mountains.
> Live in the Spirit of One who says first of all, "Come unto me," and then, "Go ye into all the world."

The Bible. Cap all names for the Bible and its books and divisions; also the names of versions and editions of the Bible.

the Good Book	Acts of the Apostles, the Acts
Holy Writ	Sermon on the Mount
(Holy, Sacred) Scriptures	Golden Rule
Old Testament	Lord's Prayer
Pentateuch	Ten Commandments, the Decalogue
Exodus, or Book of Exodus	Apocalypse
First Book of Kings	parable of the Sower
Wisdom Literature	King James Version
Major and Minor Prophets	Revised Version
Priestly Code	American Translation
Apocrypha	Douay Version
Synoptic Gospels	Septuagint
Gospel of John	Vulgate
First Epistle of Paul to the Corinthians	Polychrome Bible

Lowercase: biblical, scriptural, apocryphal; gospel [except when reference is directly to one of the four books so named]; epistles.

Ancient manuscripts. Cap the titles of ancient manuscripts.

Codex Alexandrinus   Codex Canonicianus

Lowercase: the Alexandrian codex.

Creeds. Cap the names of creeds and confessions of faith.

the Apostles' Creed   the Augsburg Confession
the Creed of Chalcedon   the Westminster Confession

Denominations, monastic orders, movements. Cap the names of
religious denominations, monastic orders, movements, and their adherents. The word *church* is capped when it forms a part of such names or
of names of particular edifices, but not when it stands alone unless it is
used to denote a religious organization of the whole world or of a particular
country. This form is usually found in contradistinction to *State:* the Church
and the State.

Protestantism   the Society of Friends
Catholicism   the Plymouth Brethren
Baptist   the Doukhobors
Fundamentalist   St. James's Church
Modernist   Friends' Meeting House
Mormonism   the First Church of Christ, Scientist
Latter-day Saints   High Churchman
Seventh-day Adventists   Black Friars
Protestant Episcopal Church   Carmelites
the Mother Church   Church Fathers
the Established Church   Papist
the Church of Rome   Agnosticism
High Church, Low Church   Gnosticism
the Oxford Movement   Scholasticism (Schoolmen)
Nonconformist   Tractarianism

Lowercase: spiritualist, theosophist, agnostic [in common-noun sense].

Theological terms. Cap theological terms used with definite application.

Ark of the Covenant, the Ark   Ecce Homo
the Chosen People   Day of Judgment
Annunciation   the Disciples (meaning the Twelve)
Immaculate Conception   the Apostle Paul, Paul the Apostle
Nativity   Pharisees
Last Supper   Gentile (*noun*)
the Magi   Satan, Beelzebub, the Enemy
Crucifixion   Paradise (meaning the Garden of Eden)
Resurrection   Hades
Ascension   Gehenna

Lowercase: heaven [meaning a place], hell, purgatory, paradise, nirvana, happy
hunting ground; devil [unless used as a specific name for Satan]; gentile (*adj.*);
scribes; for heaven's sake.

Ecclesiastical terms. Such ecclesiastical terms as the following should be capped.

Eucharist	the Sovereign Pontiff
Lent, Lenten	the Consistorial
Epiphany	Sacred College of Cardinals
Septuagesima	Sacred Congregations
Advent	Sacred Heart
Ember Days	Pontifical High Mass
Holy Grail	Nuncio
Holy of Holies	Papacy[24]
Holy Orders	House of Bishops
Holy See, See	General Synod
Holy Father (the Pope)[24]	General Conference

Lowercase, in secular writing: father, friar, brother, sister [except when used with name]; benediction, holy communion, mass, high mass, confessional; papal [when not part of a title].

Capped: Christian, non-Christian, Christianity, Christianize, Christendom, Christology; but: christological, unchristian.

Additional religious terms. Cap the names of other religions, their heads, their followers, and their sacred writings.

Buddhism	Koran	Talmud
Confucianism	Lamaism	Taoist
Dalai Lama	Mishna	Upanishads
Dhammapada	Parsee	Veda
Hinduism	Peshitto	Zen Buddhism
Islam	Shinto	Zend-Avesta
Jewish	Sufism	Zoroastrian

Lowercase: Olympic gods, the sun god, the god Baal; koranic, talmudic, vedic.[25]

Quaker dates. Capitalize Quaker dates: *First-day, Eighth Month.*

HISTORICAL TERMS

Historical epochs. Cap the generally accepted names of historical periods and movements.

Bronze Age	Great Depression
Christian Era	Industrial Revolution
Colonial days (U.S.)	Iron Age
Counter-Reformation	Middle Ages
Crusades	Norman Conquest
Dark Ages	Paris Commune, the Commune
the Exile	Reconstruction (U.S.)
the Flood, the Deluge	Reformation
Golden Age of Greece	Stone Age

[24]*Pope* and *papacy* are usually lowercase in secular writing.
[25]*Vedic* is capped in its other connotations: *Vedic culture, Vedic language.*

Lowercase: twentieth century; medieval, medievalism; space age, nuclear age, ice age; cold war, gold rush, a depression.

Periods in the history of language, art, literature. Cap the names of periods in the history of language, of art, and of literature.

Age of Reason	Elizabethan Age	Old English
Age of Pericles	Enlightenment	Renaissance
Alexandrian Era	Middle High German	Revival of Learning

Lowercase: postexilic writings, baroque period

Philosophical and cultural movements. The names of philosophical schools are capped when they derive from a proper name.

Aristotelian	Kantianism
Cartesian	Neoplatonism
Epicurean	Platonism
Gongorism	Thomism
Gregorian	

Lowercase: epicurean tastes, platonic relationship.

The following philosophical, cultural, and stylistic terms are properly lowercase, as are their derivatives (for religious terms, see pp. 153–155):

classicism	hedonism	neoclassicism
constructivism	humanism	nihilism
determinism	idealism	nonobjectivism
eclecticism	imagism	philosophe
existentialism	miracle plays	physiocracy
fatalism	morality plays	rationalism
formalism	mystery plays	realism
functionalism	mysticism	transcendentalism
futurism	naturalism	utilitarianism
gymnosophy		

Some such terms are more troublesome, as they are used in several senses. The following are capped when applying to a doctrine or school of thought, but are lowercase in a common, general sense:

Cynic, Cynicism	a cynical attitude, a cynic in playwriting
Peripatetic	peripatetic habits
Romantic, Romanticism	romantic fiction, romantic love
Sophists (*but* sophistry)	sophistic argument, a sophist in a crisis
Stoic, Stoicism	a stoical mother, a stoic facing troubles
Utopian, Utopianism	utopian island, utopian plans

Artistic terms with particular application to painting are sometimes the most difficult to distinguish when deciding whether to cap or set lowercase. In the following list the right column gives terms that should be capped when referring to a historical movement, school, or theory; terms in the left column either should always be lowercase or (if there is a right-column counterpart) should be lowercase when referring to a style or technique or when used in any sense other than for painting:

Style or Technique	*Movement, School, or Theory*
abstract impressionism, -ist, -istic	Cubism, -ist, -istic
baroque	Dadaism, -ist, -istic
cubism, -ist, -istic	Fauvism, -ist, -istic
expressionism, -ist, -istic	Impressionism, -ist, -istic
impressionism, -ist, -istic	Luminism, -ist, -istic
luminarism, -ist, -istic	Mannerism, -ist, -istic
mannerism, -ist, -istic	Neoimpressionism, -ist, -istic
naturalism, -ist, -istic	Pre-Raphaelite
pointillism, -ist, -istic	Renaissance
postimpressionism, -ist, -istic	Symbolism, -ist, -istic
rococo	
surrealism, -ist, -istic	
symbolism, -ic, -ist, -istic	

The following terms are always capped when referring to architecture:

Doric	Ionic
Gothic	Romanesque

Lowercase: gothic proportions, gothic novel.

Treat revivals and adaptations as follows:

Neo-Darwinism	Neo-Hegelianism
Neo-Freudian	Neo-Hellenism
Neo-Gothic	Neo-Thomism

Lowercase: neoclassicism, neoromanticism, neonaturalism, neofascism, neo-humanism, neopaganism.

Exceptions: Neoimpressionism, Neoplatonism.

Congresses, councils, expositions, fairs. Cap the names of congresses, councils, expositions, and similar proper names.

the First Congress of Races	Passion Play (of Oberammergau)
the All-Russian Congress of Soviets	Illinois State Fair, the state fair
Olympic Games	the Council of Trent
Tournament of Roses	Louisiana Purchase Exposition
the Mardi Gras	

Important events. Cap the names of important events.

American Revolution, the Revolution   Vietnamese War
Civil War (American)   Battle of the Bulge
War of 1812   Fall of Rome
World War I, World War II[26]   Reign of Terror
Thirty Years' War   Whisky Insurrection
Korean War   Louisiana Purchase

Lowercase: Spanish civil war; the two world wars; prewar, postwar, before the war.

Historic documents. Cap the names of historic documents.

Declaration of Independence   Bill of Rights
Constitution (U.S.)   Petition of Right
Magna Charta   Turkish Capitulations
the Edict of 1807; the great reform edict of 1807

Treaties. Cap the word *treaty* only when it is given as part of the exact title of a treaty of historical importance.

Treaty of Ghent   the treaty at Versailles
Versailles Treaty   the Franco-German treaty
Treaty of Paris   the Cuban
Jay's Treaty   reciprocity treaty

MILITARY TERMS

Officers. Cap the following titles of high-ranking officers before or after a name or standing alone. Titles of other officers are lowercase unless they precede a name (*Captain Smith*) or are used in direct discourse (*"Yes, sir, Lieutenant"*).

Commander in Chief   The Adjutant General[28]
Supreme Allied Commander   the Inspector General
Chairman, Joint Chiefs of Staff   the Judge Advocate General
Chief of Staff   the Paymaster General
General of the Army   the Surgeon General
Admiral of the Navy, Fleet Admiral[27]

Lowercase: the general, the admiral, the colonel, the captain, and so on.

Organizations. Cap the names of organizations of the military. *Army, navy,* and *air force,* used as noun and adjective, are not usually capped (*army post, navy patrol, air force blue*), unless they refer to the official organization of a country (*Army policy, Navy officer*) or if the name of the

[26]Usage is divided on capitalization of the preceding ordinal, but convention suggests *First World War* and *Second World War.*
[27]Cap the British titles *First Lord of the Navy* (parliamentary chief of the navy) and *First Sea Lord* (professional chief).
[28]Capitalization of *The* prescribed by law.

country precedes (*United States Army, British Navy*). (For use of ordinals preceding a military unit, for example, *Third Army*, see p. 134.)

(For use of ordinals preceding a military unit, for example, *Third Army*, see p. 134.)

General Staff	Signal Corps
General Board (of the Navy)	Corps of Engineers
Headquarters of the Army	Coast Guard
Battle Force	B Company, Company B
Naval Reserve	National Guard, the Guard
Air Service	Regular Army, a Regular

Lowercase: national guardsman, artilleryman, an infantry sergeant, an army captain.

Combatants. Cap the names of the parties participating in wars.

Federals (the North)	the Triple Alliance
Confederates (the South)	the Central Powers
Allies	the United Nations
Insurrectos	the Axis

Lowercase: Italian front, western front.

Decorations. Cap the names of war decorations.

Congressional Medal of Honor	Victoria Cross
Distinguished Service Medal	Iron Cross
Navy Cross	Légion d'honneur
Purple Heart	Croix de guerre

GOVERNMENTAL AND POLITICAL TERMS

Political divisions. Cap the names of political and administrative units and subdivisions.

Holy Roman Empire, the Empire	the Colonies, Colonial (U.S.)
United Kingdom	West Riding
French Republic	Essex County
Irish Free State	Evanston Township, the township
the Republic (the United States)	Ward 2
the Union (the United States)	Precinct 4
the Dominion or Provinces (Canada)	Eleventh Congressional District

Lowercase: the first ward, the second precinct.

Judicial bodies. Cap the name of an international court or the word *Court* used alone referring to it; a national, district or state court of law whenever it is named in full. Names of city and county courts are lowercase.

the World Court, the Court
the International Court of Arbitration, the Court
the Court of Impeachment (the U.S. Senate), the Court

the United States Supreme Court, the Supreme Court
the Circuit Court of the United States for the Second Circuit, the circuit court, the court
United States Commerce Court, the court
Court of Appeals of the State of Wisconsin, the court of appeals
Court of Claims, the court
Court of Customs and Patent Appeals, the court
Court of Private Land Claims, the court
District Court of the United States for the Eastern District of Missouri, the district court, the court

Lowercase: the Lynn municipal court, the Boston police court, in justice court, federal court, the juvenile court, the land court.

In England the names of all courts are capped.

Her Majesty's High Court of Justice, the High Court
the Court of Chancery
the Court of Session

Legislative bodies. Cap the names of legislative bodies.

Congress	House of Commons, the Commons
the Senate	Chamber of Deputies, the Chamber
House of Representatives, the House	Texas Legislature[29]
Parliament (United Kingdom)	House of Delegates
House of Lords, the Lords	Chicago City Council

Lowercase: the lower house, the two houses, the state senate, the state legislature, the city council, parliamentary law, congressional, senatorial.

The following are examples of some national legislative bodies:

Althing (Iceland)	Great People's Khural (Outer Mongolia)
Bundestag; Bundesrat (West Germany)	Knesset (Israel)
Chamber of Deputies (Luxembourg)	Majlis (Maldives)
Congress (Brazil, Colombia, Mexico)	National People's Congress (China)
Constituent Assembly (Thailand)	Oireachtas: Seanad Éireann, Dáil
Cortés (Spain)	Éireann (Ireland)
Council of Ministers (Burundi, Cuba)	Panchayat (Nepal)
Diet (Japan, Liechtenstein)	Parliament (Belgium, Canada, France,
Duma (czarist Russia)	Italy, Kenya, Nauru, New Zealand)
Eduskunta (Finland)	People's Chamber (East Germany)
Federal Assembly (Yugoslavia)	Rajya Sabha; Lok Sabha (India)
Folketinget (Denmark)	Riksdag (Sweden)
Grand National Assembly (Romania)	Sejm (Poland)

[29]Unless the exact designation is used there is no reason for capping: Massachusetts legislature, New York legislature. Many states officially designate their legislative body the Legislature; but almost as many call it the General Assembly or the Legislative Assembly, and in Massachusetts it is the General Court. The *World Almanac* and the *Information Please Almanac* give the exact titles of state legislative bodies.

Ständerat; Nationalrat (Switzerland) Supreme Council (Yemen-Aden)
States-General (The Netherlands) Verkhovny Sovet (USSR)
Storting (Norway) Yuan (Taiwan)

Civil and noble titles. Cap *President* and *Vice-President* when referring to the incumbent official in the United States or when preceding the name of an individual. Lowercase *presidency, presidential, vice-presidency, vice-presidential.* All other civil and noble titles should be lowercase unless preceding the name of the individual, in which case they should be capped.

secretary of state   emperor
attorney general   king
senator   queen
congressman (preferred to   dictator
  "representative")   kaiser
majority leader   sultan
speaker of the House   chancellor
chief justice   premier
associate justice   prime minister
ambassador   governor general
director   foreign minister
governor   minister
mayor   member of Parliament
assemblyman
councilman

Exception: Chief Justice, of the U.S. Supreme Court.

When the article *the* precedes a title before a name, the title becomes a common noun in apposition to the personal name and is lowercase: *the emperor Napoleon, the young king Louis.*

The following titles of nobility, in order of rank, are treated as civil titles — that is, lowercase except when preceding a name:

prince, princess   baron, baroness
duke, duchess   baronet
marquess (or marquis), marchioness   lord, lady
earl, count, countess   knight
viscount, viscountess

The Turkish titles *bey, pasha,* and *hakim* should be capped when immediately following a name: *Kemal Pasha.*

Departments, bureaus, offices, services. Cap the names of United States departments, bureaus, offices, services, and so on.

Department of the Interior   General Education Board
Bureau of Pensions   Marine Hospital Service
Census Bureau   Interstate Commerce Commission
General Land Office   Tennessee Valley Authority
the Cabinet   Civil Service Commission
Central Intelligence Agency   National Labor Relations Board
National Security Council   Peace Corps

Lowercase: agency, board, commission, bureau, department, authority, council, used alone in place of the full name.

Do not cap *government,*[30] *federal, administration,* or *executive branch* except when used in the name of a governmental agency or organization.

The federal government exercises tremendous power through the Internal Revenue Service.
The Food and Drug Administration has become quite active.
The Nixon administration at times seems unaware of the large landholder in political office.
The government of Mexico replied quickly to the statement.
The executive branch grows ever more powerful.
The expenses of government have mounted rapidly.
The Federal Communications Commission has revised its license renewal policies.

Laws. Cap *act* and *law* only in the formal titles of bills that have become laws.

Volstead Act	Aldrich-Payne Tariff Act
Sherman Antitrust Act	the Eighteenth Amendment

Lowercase: the prohibition amendment, the tariff act, the revenue act of 1921; established by an act of Congress; public acts, private acts, pending labor bill.

Political parties. Cap the names of political parties (but usually not the word party) and their adherents.

Democratic party	Socialist Labor party
British Labour party	Republicans
Mugwumps	Communist bloc
the Solid South	Opposition (the party out of power
Black Shirts	in a foreign country)

Lowercase: anarchist, bolshevism, communism, facism, nihilism, socialism, socialistic, left wing, farm bloc, opposition (U.S.).

MISCELLANEOUS

Academic degrees. Cap names of academic degrees and scholastic and military honors and decorations when they are used following a proper name, whether they are written in full or abbreviated. In the expression "the degree of Doctor of Laws" either caps or lowercase may be used, the nature of the publication determining the preferable form.

[30]In Great Britain *government,* applied collectively to the cabinet and ministers holding office at any given time, is construed as plural and is usually capped: "His Majesty's Government are resolved to assist." In the United States, *the government* is construed as singular.

Walter Blank, Bachelor of Science
Charles Black, LL.D.
Frank White, Fellow of the Royal Geographic Society
Frank White, F.R.G.S.

Buildings and organizations. Cap the names of buildings, organizations, institutions, and so on. (Newspaper usage is divided, some papers preferring lowercase for the common-noun portion of the name unless it comes first, as in *University of Chicago, Hotel Roosevelt.*)

the White House	Independent Order of Odd Fellows,
the Blue Room	an Odd Fellow
the Mansion House (official residence	Boy Scouts, a Boy Scout, a Scout
of the Lord Mayor of London)	Camp Fire Girls
the Vatican	Red Sox, Cardinals, Mets
the Plaza Hotel	Penn Central Transportation Company
Hull House	Meredith Corporation
the Sphinx	Dartmouth College
the Pyramids	Grolier Society
Tomb of the Unknown Solider	Chicago Civic Opera Company
Rose Bowl	Little Theater
Masonic Order, a Mason, a Freemason	Boston Symphony Orchestra
American Legion, the Legion	Royal Northwest Mounted Police

Lowercase: ladies' aid society, parent-teacher association, chamber of commerce [unless they are accompanied by a proper name]; legionnaires.

Common-noun elements of these proper names, such as *club, college, company, hotel, railroad,* and *society,* should not be capped when they are used in the plural with two or more proper names.

the Union Pacific and Southern Pacific railroads
Simmons and Radcliffe colleges
the Warner and Paramount theaters

If the common noun is used alone, it should not be capped, even if the reference is specific, unless a cap is necessary to avoid ambiguity. A few such nouns that *are* capped when used alone are the following (they should be lowercase unless they have the meaning here specified):

the Bank (Bank of England)	the Legion (American Legion or
the Garden (Madison Square Garden)	French Foreign Legion)
the Yard (Scotland Yard)	the Street (Wall Street)

Epithets. Cap epithets used as parts of proper names or as substitutes therefor.

Charles the Bold	the Emancipator
Coeur de Lion	the Tiger of France
Billy the Kid	the First Lady

When epithets and names of groups of people, parties, or localities have lost their original identity and have taken on the nature of institutions with a well-defined significance, they should be capped.

Four Hundred   Little Italy
Old Guard   Great White Way

Estates. Capitalize the names of houses and estates.

The Croft   The Maples

Family appellations. *Father, Mother, Uncle,* and so on, when used as names, should be regarded as proper nouns and written with a capital letter; otherwise they are lowercase.

I have had a letter from Mother.
I have had a letter from my mother.
Yesterday Uncle John came.
Did you know my uncle had come?
She became a mother at twenty.

Note that when a possessive pronoun is used before an appellation and a personal name, the appellation becomes a common noun in apposition and is lowercase.

You should have tasted my grandmother Maude's pizza.
My uncle John had a great influence on me.

Do not cap *brothers* in a sentence such as the following:

The James brothers were notorious outlaws.

Flags. Cap names of national flags, emblems, and college colors.

the Star-Spangled Banner   Great Union
Old Glory   Union Jack
the Stars and Stripes   Tricolor
the Lilies of France   the Crimson

Games. Names of games are not capped unless they are trademarks.

golf   Monopoly
dominoes   Ouija
hide and seek   Scrabble
blindman's buff   Ping-Pong
musical chairs   Mille Bornes

In the game of chess all terms, unless proper names, should be lowercase. For example: Ruy Lopez opening, Philidor's defense, king's bishop's gambit, queen's gambit declined, giuoco piano opening, zugzwang.

Never cap the pieces: king, queen, bishop, knight, rook (or castle), pawn; but when moves are given by initials, set: K–Kt3, Kt–K6, K–R sq. P(Kt2)xP, KxB, Q – Kt8 ch.[31]

Geological names. Cap the accepted names for eras, periods or systems, and epochs or series.

Cenozoic era	Tertiary period
Recent epoch	Quaternary period
Silurian system	Acadian epoch

No difficulty will be experienced with those terms whose proper-noun character is obvious. The situation is different with *upper, middle,* and *lower,* which are properly capped only when they are part of an accepted name for a period. In the following list cap any of the adjectives in the left column when preceding the terms in the right column (these are the official names suggested by the U.S. Geological Survey):

Upper	Cambrian
Late Middle	Cretaceous
Middle Lower	Devonian
Late Lower	Jurassic
Early	Mississippian
	Ordovician
	Pennsylvanian
	Permian
	Silurian
	Triassic

The adjectives *upper, late middle, middle lower, late lower,* and *early* should be lowercase (as merely descriptive) when appearing before any of the following terms: *Miocene, Oligocene, Paleocene, Pliocene,* and *Precambrian.*

Many dictionaries contain a table of geological times and formations.

Coal measures, calciferous, lignitic, magnesian, and *red beds* may also be used as common nouns or adjectives. The forms *glacial, preglacial,* and *postglacial, Postpliocene, Precambrian,* are correct.

Such nouns as *anticline, dome, formation, group, member, syncline, terrace,* and *uplift* should not be capped when preceded by a proper name.

The following ages are capped as official geological periods:

Age of Copper	Bronze Age
Age of Fishes	Ice Age[33]
Age of Mammals	Iron Age
Age of Man[32]	Stone Age
Age of Reptiles	

[31]Rule from *The New York Times Style Book.*
[32]The present geological period.
[33]Cap only when referring to the Pleistocene glacial epoch; otherwise it is lowercase.

The names of the twenty-four great soil groups are capped.

Alpine Meadow	Laterite	Sierozem (Gray)
Bog	Pedalfer	Solonchak
Brown	Pedocal	Solonetz
Chernozem (Black)	Podzol	Soloth
Chestnut	Prairie	Terra Rossa
Desert	Ramann's Brown	Tundra
Gray-Brown Podzolic	Red	Wiesenboden
Half-Bog	Rendzina	Yellow

Heavenly bodies. Cap the names of the stars, planets, and constellations.

North Star, Polaris	Cassiopeia
Milky Way	Halley's Comet
the Dipper	Leonids
the Southern Cross	Great Nebula
the Great Bear	Charles's Wain
Venus	Jupiter

The words *sun, moon,* and *earth* are capped only when used in context with other astronomical terms; and *earth* is never capped when preceded by the article *the.*

four corners of the earth
salt of the earth
I would go anywhere on earth to find it.
The moon seemed to shed more light than the sun last weekend.
The atmosphere of Mars or Venus could not support the creatures of Earth.

Holidays. Cap the names of holidays, festivals, and other special days.

New Year's Day	Passover
the Fourth of July, the Fourth	Thanksgiving Day
Veterans Day	Shrove Tuesday
Mother's Day	Holy Week, Passion Week
Commencement Day, Commencement	the Feast of St. John
Christmas Eve	Yom Kippur, Rosh Hashana

Lowercase: noel, D day.

Nouns with numerals. Nouns or abbreviations used with a letter or numeral to denote a place in a numbered series are usually capped. Such words are often lowercase in parentheses, footnotes, references, and similar matter; and house style sheets often note certain words as exceptions. *Line* (l.) and *page* (p.) are always set lowercase. (Do not abbreviate these words in straight text.)

Act I	Volume II	Chapter 2
Article IV	Part I, Part A	Division 4
Book III	Plate II, Plate B	Figure 48

Lowercase: room 8, grade 5, grade V tumor, question 14, item 3, adenovirus type 8, blood factors II, III, VIII.

Used in the plural such words are usually capped.

Refer to Tables 20 and 21.
Illustrated in Figures 1, 3, 4.
See Chapters XV, XVII, XX.
Especially Opera 127, 131, 132, 135.

For the usage of *Number* or *No.* see page 115.

Parts of books. Cap *Preface, Contents, Index,* and other names of parts of a book when they are specifically referred to within the same book.

The meanings of technical terms can be found in the Glossary.
See the Table of Food Values on page 87.
The interested student will find in Chapter XI, "The Press and Its Working," a complete exemplification of the subject.

Peoples, races, tribes. Cap names of peoples, races, tribes, and other groupings of mankind.

Aryan	Kaffir	Negro, Negroid
Asian	Magyar	Nordic
Caucasian, Caucasoid	Mongol, Mongolian,	Oriental
Cliff Dwellers	Mongoloid	Pygmy
Indian	Negrillo, Negrito	Sioux

Lowercase: redskins, whites, blacks, colored, quadroons, octaroons, mongolian and mongoloid [in medical sense referring to mongolism], pygmy [simple reference to a short person].

Personal names with "de," "le," "von." Articles and prepositions (particles) in the names of Englishmen and North Americans are usually capped, unless a preference for the lowercase form has been shown by the families themselves.

Thomas De Quincey	Martin Van Buren
Theodore L. De Vinne	Robert La Follette
Henry du Pont	

In French, Italian, Spanish, Portuguese, German, and Dutch names these secondary elements are not usually capped if a forename or title precedes.

Catherine de' Medici	Ludwig van Beethoven
M. de Tocqueville	Herr von Ribbentrop
Lucca della Robbia	Hendrik Willem van Loon
Leonardo da Vinci	René le Bossu
Eamon de Valera	Miguel de la Torre

Here again family preference may determine the form.

Maxime Du Camp   Edmondo De Amicis

In French names *Le, L', La,* and *Les* are usually capped; *de* and *d'* are never capped.

Comtesse de La Fayette   Philibert de L'Orme
Jean de La Fontaine   Charles Le Brun
Alexis de Tocqueville   Charles de Gaulle

Other than the foregoing guides, personal usage and seemingly competent authority vary so widely in the treatment of names with foreign particles that only a very general course can be suggested. If any sort of title (or a forename) is used, lowercase the particle (except *Le, L', La* and *Les*): *Comte de Grasse, General von Kleist, Duc de La Rochefoucauld, Mme. du Barry, Count von Moltke.* If possible, drop the particle: *Beethoven, Cervantes, Ribbentrop, Leeuwenhoek, Torquemada, Maupassant.* Where the particle seems required, and no title or forename precedes, cap the particle: *Da Vinci, Van Gogh, La Fayette, De Valera, Della Robbia;* also *Ibn-Saud, Abu-Bekr.*

The particle is capped at the beginning of a sentence regardless of its usage within the sentence.

Personification. Personifications of abstract ideas or objects are occasionally capped. Always cap *Mother Nature* but cap *Nature* only when the personification is as clear as in the following:

The great artist, Nature, had finished her masterpiece.
We ask you, Reason, to rule in our judgments.
But: We ask you to appeal to reason in making your judgment.

Prefixes. Prefixes added to proper nouns or adjectives are usually lowercase.

ante-Norman   inter-American   pre-Islamic
anti-Catholic   mid-Atlantic   pro-German
anti-English   non-Euclidean   semi-American
anti-Semite   post-Victorian   un-American
ex-President   pre-Columbian

Some prefixes are capped as part of a specific name.

Anti-Lebanon Mountains   Neo-Hebraic   Pan-Slavic
Neo-Christianity   Pan-American   Semi-Pelagian

Other prefixes have been incorporated as one word *(Antiremonstrant, Precambrian, Neoplatonism),* and are capped; still others incorporated as one word *(transatlantic, transpacific, postexilic)* are lowercase. Since there is no general rule for cases of this kind, consult a dictionary.

Be careful of certain pairings that might have different meanings, as *unplatonic* and *un-Platonic*.

Note that an en dash, rather than a hyphen, should follow a prefix to an open compound: *pre–Revolutionary War*.

Scientific names. In botany, paleontology, zoology, cap the names of all divisions higher than species, that is, genera, families, orders, classes, and phyla; but not adjectives and English nouns derived from them. (For the use of italics for the name of the genus and for further discussion of the lowercasing of the genus, see p. 141.)

Chiroptera, chiropter	the genus *Hydra*; a hydra
Coelenterata, coelenterate	the tall buttercup (*Ranunculus acris*)
Protozoa, protozoon	the harp seal (*Phoca graenlandica*)

Parasitism is emphasized in the study of Sporozoa, Platyhelminthes, and Nemathelminthes.

Note that such terms as *Metazoa* and *Protozoa* are properly capped as divisions of the animal kingdom; however, the plurals of the lowercase common-noun designations *metazoon* and *protozoon* are *metazoa* and *protozoa*, so that when used with the article *the* (that is, when speaking of members of the division and not the division itself), the proper style is the *metazoa* and the *protozoa*. Also sometimes confusing is *Primates*, capped as the name of an order, and *primate* or *primates*, lowercase as members of the order.

Seasons. The names of the seasons are not capped unless they are personified or to avoid ambiguity.

Hail, Autumn, with thy joyous harvests!
The water is cold in the Spring.

Signs. Cap the wording of signs as titles are capped.

Go to the door marked Exit. (*But* Go to the exit door.)
The sign said Do Not Enter.
On his sign was Save the World for Me.

"The." Cap *The* when it is part of a name.

The Hague	The Weirs (N. H.)
The Dalles (Oregon)	The Buttes (Cal.)

The Adjutant General [prescribed by law]

Lowercase: the Netherlands, the Bronx.

It is not necessary to cap *the* because the name following is capped or capped and quoted. (Note the position of the quotation marks below and see also p. 217.)

the Virgin Mary	called "the North Star State"
the Tiger of France	the Joneses (*not* The Joneses)

For usage regarding *the* in titles of books and newspapers see pages 136, 140.

The word *the* used before the name of an Irish or Scots clan designates the chieftain of the clan. It is sometimes capped but more often not.

the McManus   the MacGregor

Thoroughfares, parks, squares, airports. Cap the names of streets, avenues, parks, squares, airports, stadiums, and the like; such words are considered an organic part of the name.

Boston Common	Unter den Linden	Friendship Airport
Prospect Park	Trafalgar Square	Love Field
Champs Elysées	Mitchel Field	Shea Stadium

Lowercase in the plural: Union and Market streets.

Always cap *Wall Street, Fleet Street, Downing Street,* and *Main Street* when they are used with special significance beyond that of a street name. When the word *street* or *avenue* forms part of the name of a building or organization, it should, of course, be capped.

Fifth Avenue Presbyterian Church
Fourth Street Boys' Club

Titles. Cap civil and religious titles when they precede a name. Capitalization of such titles used after the name is definitely governed by the nature of the publication. Some kinds of printed matter might require capitalization of all titles denoting position or rank when used following a name.

Superintendent Wagner	____ ____, governor of Massachusetts
Deacon White	____ ____, parole officer
General Manager Howard	E. P. Wagner, superintendent of schools

Distinguish a title from a substantive and an appositive.

the general manager, John Howard   the papal legate, Riorio

When titles are used alone referring to specific persons, they are usually lowercase: *the bishop, the duke, the king, the general, the pope.* Cap titles of honor, nobility, and respect.

His Holiness Paul VI	Her Grace
His Grace the Duke of Devonshire	Your Excellency
His Excellency	Your Honor

Cap titles used in direct address as synonyms of proper names.

Mr. Mayor, Ladies and Gentlemen: "Come this way, Doctor."
"I am, Captain, your . . ."   "How are you, Pop?"
"Well, General, what is the next step?"   "I'll be there, Mumsie."

But: "Yes, miss." "Oh, no, sir." "Yes, my lord." "No, madam."

Any common noun used as part of or in place of a proper name may be capped in a publication if the noun is highly specific for the audience addressed.

the Dean, the Prex (in a school publication)

Trademarks. A *trade name* is the name under which a concern does business. A *trademark* is a proprietary name protected by law. Popular usage of a trademark may reduce it to a common noun unless its owner makes an effort to seek out improper lowercase uses. Thus many firms insist on capitalization and also on certain phrasing of the trademark plus a generic term. To avoid having to use such bulky expressions as *BAND-AID Brand Adhesive Bandages,* the author or editor is better advised to say *adhesive bandage* and forget the trademark. Some brand names are difficult to express concisely in generic terms: *Frisbee, Muzak, Land-Rover, Ouija, Dacron, Orlon, Monopoly, Pyrex.* The following are some trademarks (as of 1973) followed by an alternate generic term in parentheses; use only the generic term when possible:

Alka-Seltzer (antacid tablet)
Baggies (plastic bags)
Ben-Gay (analgesic ointment)
Benzedrine (amphetamine sulfate)
Chap Stick (lip balm)
Coca-Cola, Coke (cola or soft drink)
Demerol (meperidine)
Dexedrine (dextroamphetamine)
Dictaphone (dictating machine)
Dramamine (dimenhydrinate)
Ex-Lax (laxative)
Fiberglas (glass fibers or fiber glass)
Fig Newtons (fig cakes)
Formica (laminated plastic)
Frigidaire (refrigerator)
Jell-O (gelatin dessert)
Keds (canvas shoes, sneakers)
Kleenex (tissues)
Kodak (camera or film)
Kool-Aid (soft-drink mix)
Levi's (jeans)
Masonite (hardboard)
Merthiolate (thimerosal)
Mixmaster (food mixer)
Naugahyde (vinyl-coated fabric)

Nembutal (pentobarbital)
Novocaine (procaine)
Pentothal (thiopental sodium)
Ping-Pong (table tennis)
Plexiglas (acrylic plastic)
Polaroid (camera or self-developing film)
Prestone (antifreeze)
Pyrex (heat-resistant glass)
Q-Tips (cotton swabs)
Saran Wrap (plastic wrap)
Sanka (decaffeinated coffee)
Scotch tape (cellophane tape, plastic tape)
Seconal (secobarbital)
Styrofoam (plastic foam)
Sucaryl (noncaloric sweetener)
Tabasco (pepper sauce)
Technicolor (movies in color)
Teflon (nonstick coating)
Thermo-Fax (copying machine)
Touch-Tone (pushbutton dialing)
Vaseline (petrolatum or petroleum jelly)
Xerox (copier, duplicator, photocopier)

When a trademark is used, do not cap the common-noun portion.

Campbell's soup   Eagle pencil
Doublemint gum   Ford station wagon

The following former trademarks are now unprotected and have passed into the public domain, so that they are properly lowercase:

aspirin	kerosene	nylon
calico	lanolin	raisin bran
cellophane	linoleum	trampoline
escalator	milk of magnesia	yo-yo
hansom	mimeograph	zipper

PUNCTUATION

Some writers use punctuation skillfully, giving their words the greatest possible clearness, emphasis, and value. Some disregard accepted usages completely, using whatever punctuation occurs to them at the moment. The great mass of printed matter, however, is written by persons of neither of these sorts. They use punctuation sometimes inaccurately, sometimes awkwardly, usually reasonably well, though seldom so well that a capable copy editor cannot find faults to correct.

Punctuation should primarily prevent misunderstanding of thought or expression and should secondarily facilitate reading: the best punctuation is that which the reader is unaware of.

Since violations of good usage are exceedingly annoying to many readers, copy editors must give careful attention to punctuation if they are to prevent unfavorable criticisms of their company's publications.

Most presentations of the principles and rules of punctuation are not sufficiently comprehensive and do not go deep enough into usage and into the difficulties of applying the rules to another person's written work to meet the needs of the copy editor. He needs to know not only the rules and the principles underlying them but also the modifications allowed in different contexts, and the details subject to the dictation of the publisher. He needs to know also which rules can be safely followed and which must always be left to the judgment of the author. The following pages on punctuation have been compiled especially for copy editors, with their problems and difficulties uppermost in mind. Many of the examples are taken from actual writing, and most of them were incorrectly punctuated when they came to the copy editor's attention.

Typeface. Commas, colons, and semicolons are set in the typeface (italic or boldface) of the preceding word. Quotation marks, exclamation points, question marks, and parentheses are set according to the overall context of a sentence.

One species, *B. anthracis,* is highly virulent.
What was the correct length of *AC?*
"But now," he said, "I *know*!"
Harris thumbed down several, including *Away We Go* (which became the huge
    success *Oklahoma!*).
Wolf said about *Sounder:* "Beautifully thought-out and directed."

A single italic letter preceding any punctuation mark does not justify
setting the mark in italic type.

in Figure 6*a;*

PERIOD

Sentences. Use a period after a complete declarative sentence or words
standing for such a sentence — that is, a virtual sentence.

It is a pleasant day. Yes, indeed.

Sentences containing an imperative may be marked by a period or
by an exclamation point. The author must decide which to use, since the
choice depends upon the degree of emphasis.

Come at once.   Be careful!

Do not use a period after a complete sentence enclosed in parentheses
or in quotation marks and interpolated within another sentence.

Make composition a class exercise by having students suggest a topic (the
    teacher writes it on the board) and then develop the theme in outline form.
John's confession, "I did it because if felt good," stunned the court.

Abbreviations. The trend is away from using a period in abbreviations,
particularly for units of any kind (see pp. 111–112), but abbreviations for
many words properly retain it.

etc.   Mrs.   Co.   St.   pp.   LL.D.

Some abbreviations are termed symbols and use no period (see
pp. 120–122).
In some proper names initials are used without periods.

AE   Maurice L Rothschild
W J McGee   Charles C Thomas
Harry S Truman   Daniel D Tompkins

The French and some British publishers use no period after an abbreviation ending with the last letter of the word for which it stands (see pp. 118–119). The practice is seldom followed in the United States.

Contractions. Do not use a period after a contraction, an abridged form of a word using an apostrophe for omitted letters.

m'f'g   sec'y   Ass'n

Shortened forms. Use a period after abbreviations of names, such as Thos., Chas., Benj., but not after shortened forms of names.

Alex   Ed   Nell   Pete   Will

Some shortened forms in common use or appearing in a localized setting are not abbreviations (see p. 101).

Shortened forms of the names of English counties are not followed by an abbreviating period.

Berks (Berkshire)   Lincs (Lincolnshire)
Hants (Hampshire)   Salop (Shropshire)

Letters. Use no period after letters used in place of names unless they are the initials of the name.

A said to B   Mr. A. (for Mr. Abbot)
Mr. A told Mr. B   Mr. B. (for Mr. Baker)

Numerals. Do not use a period after the ordinals 1st, 2nd, 3rd, and so on; they are not considered abbreviations.

Do not use a period after roman numerals except in enumerating items.

Charles I   Louis XIV   Vol. II   IV. Protozoa

Headings, captions, entries, lists. Never use periods after centered heads and running heads. They are not necessary after topical side heads set on a line by themselves, after captions or legends not in complete sentence form, after items in a column, at the end of entries in an index. Items in a list should ordinarily be parallel in construction—that is, all phrases or all sentences—but if one item in the list is a complete sentence and therefore requires a period, then each item in the list, including phrases, should end with a period. Otherwise no periods are necessary in lists of phrases.

Use with other marks. The period may be used in conjunction with quotation marks, parentheses, and brackets, but not with other marks unless it points an abbreviation.

If an exclamation point or a question mark comes at the end of a declarative sentence, omit the period.

Have you ever eaten in a big city hotel and thought to yourself, "When I am earning enough money of my own, this is where I shall live"?

He was holding one end of the heavy shirt of mail which an Elliott had worn many a time when answering the call of his king and country, "To arms!" [*Not:* . . . country, "To arms!".]

From Hamilton, "At last!" [*Not:* From Hamilton, "At last!"."]

However, when a period points an abbreviation preceding a question mark or exclamation point, both marks are used.

When is it proper to use the abbreviation *pp.?*

With quotation marks. When the period is used with closing quotation marks, place it inside (see also p. 222).

The other was "The Old Folks at Home."

"The other was 'The Old Folks at Home.'" [*Not:* . . . Home'."]

Do not mistake an apostrophe for a single quotation mark in sentences like the following:

"He'll never hurt ye at a'."

"Thank you kindly, ma'am, for that second helpin'."

With parentheses. When the last words of a sentence are enclosed in parentheses, place the period marking the end of the sentence after the closing curve (called the "close paren"). (The preceding sentence illustrates the correct form.) The period is properly placed inside a parenthesis only when the matter enclosed is not part of the preceding sentence but is an independent sentence beginning with a capital. Note that in such constructions there is regularly a period before the parenthesis.

Sometimes the mouth-to-nose method is used. (See Mouth-to-Nose Resuscitation, p. 51.)

In enumerations when parentheses are used with letters or numbers, periods are unnecessary after the parentheses.

Winter may be defined as (a) a season, (b) a cold period, (c) a barren time, (d) old age.

In lists numerals with periods are preferable to numerals with parentheses.

1. Call for help.
2. Stop the bleeding.
3. Keep the patient warm.

EXCLAMATION POINT AND QUESTION MARK

Exclamations. The exclamation point is used after a command, an expression of strong feeling, or an otherwise forceful utterance. At the close of a sentence beginning with an interjection an exclamation point is used if the emotion is strong enough to justify it, a matter to be determined by the author.

> Hurrah!
> How beautiful on the mountains are the feet of him that bringeth good tidings!
> Oh, Anna, please hurry!
> Alas, they were too late!

In the example below, the exclamation point was incorrectly placed. This sentence scarcely needs an exclamation point, but if one is used, it should be placed at the end of the sentence.

> *Wrong:* Oh! yes, it will be hard for us to think of school in terms of growing self-discipline.
> *Right:* Oh, yes, it will be hard for us to think of school in terms of growing self-discipline!

Emphasis or irony. An exclamation point is also used to point an emphatic or ironical comment and sometimes after what in form is a direct or indirect question, to indicate strong feeling.

> I remember a statute passed in one of our middle states, I believe, to the effect that two trains approaching an intersection must both come to a full stop, and neither may start again until the other has passed!
> You don't mean it!
> Oh, dear, what now!

Direct questions. Use a question mark after a direct question.

> Where is he now?
> It is well to ask, can this be done without harm to anyone?
> In the sentence "How do beavers build dams?" which words name things?
> Nor was it disclosed—why need it have been?—that John had taken the case.
> These two boats (can you find them?) gave the world not only steel fighting ships but the turret.
> How is the resistance of a wire affected by doubling its length? by doubling its diameter? by halving its length? by halving its diameter?

Some writers prefer to use only one question mark with a series of questions such as those in the last example, and this form is preferable whenever the construction is like the following:

> What is meant by the terms (a) atom of an element, (b) molecule of an element, (c) molecule of a compound?

The following sentences illustrate a use of interrogative words, essentially questions yet so short that the capital and question mark would be cumbersome. Italics may well be used in such sentences.

> He must take the verb and ask the question *who* or *what* to find the subject.
> They are easily recognized by asking the questions *how, when, where,* or *why.*

Place the question mark at the end of the question, not elsewhere.

> *Poor:* Will she fight Victor's case, I wondered?
> *Better:* Will she fight Victor's case? I wondered.

> *Wrong:* "It's a mighty hard pull, though, isn't it? when people don't have confidence in you."
> *Right:* "It's a mighty hard pull, though, isn't it, when people don't have confidence in you?"

Sometimes a sentence declarative in form is to be read as an exclamation or a question and is punctuated accordingly.

> Surely it cannot be that you are so utterly barren of manhood that you will deny me what you promised!
> Suppose I introduce you to the Burleighs?

It is for the writer to decide whether he will cap the first word of a direct question. If he intends it formally (as in addressing an audience), it is better to cap; informally, lowercase is acceptable.

> The question arises, What is the role of government in medical care?
> We must ask ourselves, How can inflation be stopped?
> Ask yourself the question, How can the study of trigonometry be of use in my future?
> Before proceeding ask yourself, is it worthwhile?
> I wondered, should I wait for Jane?

Indirect questions. Do not use a question mark after indirect questions.

> Artemas had wondered why.
> He asked how old John was.
> How to live—that is the essential question for us.

The difference between a direct and an indirect question may be slight.

> *Direct:* It is well to ask, what is the precedent in such and such a case?
> *Indirect:* It is well to ask what the precedent is in such and such a case.

There is a form of interrogation in which the exact words of the speaker are not given, nor is the sentence worded like a statement, as in the indirect questions given above. In such sentences the question mark is necessary, but quotation marks are not used.

> Could it be possible? he asked himself.
> When will it be? everyone wondered.

Requests. A period, not a question mark, may be used after a sentence which is interrogative in form but is in reality a command.

> Will you please close the door.
> As you read the following selection will you write down each word whose mean-
> ing you do not already know.

When a sentence of this type is a request rather than a command, however, the question mark should be retained.

> Will you kindly return this as soon as possible?

When such sentences are intended as definitely interrogative they should be punctuated as questions.

> Will you please close the door?

Similarly, what may appear to be a question may really be a statement of fact. It should be followed by a period, not by a question mark.

> "Do you know," she said, "I am writing a story about Donald Kane."

Doubt. A question mark, usually in parentheses, may be used to express a doubt about what precedes.

> Edmund Spenser (1552?–1599)
> Nest(?) of Nighthawk. The eggs are laid on a bare hillside with no effort to build
> a nest.
> On March 2(?), 1542.

Use with other marks. The exclamation point and the question mark may be used in conjunction with quotation marks, dashes, parentheses, and brackets. Usage varies about the need or correctness of using these marks with commas, periods, and each other.

With quotation marks and parentheses. When either the exclamation point or the question mark is used in conjunction with closing quotation marks or a closing parenthesis, it should precede the quotation marks or the parenthesis if it is a part of the quoted or parenthetical matter; otherwise it should follow. This rule is invariable.

> He cried out, "Wake up!"
> How absurd to call this stripling a "man"!

"Did you memorize 'The Chambered Nautilus'?"
Ask the pupils, What did Whittier mean by "It sank from sight before it set"?
Another familiar idiom is "How do you do?" For this the French idiom is *Comment vous portez-vous*? (How do you carry yourself?) and the German idiom, *Wie befinden Sie sich*? (How do you find yourself?).

With a period. If an exclamatory or an interrogative sentence and a declarative sentence end at the same time, retain the exclamation point or the question mark and omit the period.

But does not one sometimes say, "I don't feel quite sure about the first important problem in percentage"?

With each other. If a question or an exclamation occurs within a question, both ending at the same time, retain the stronger mark. It is often hard to say which is the stronger mark, but the following sentences illustrate acceptable forms:

"Has it ever occurred to you that she might retort, 'Dangerous for whom'?"
"How about 'Where Are You Now, Old Pal of Mine'?"
Did you hear somebody yelling "Fire! Fire!"
How many of you have heard the question, "Which is the more important for a person, his heredity or his environment?"
Haven't you noticed the appeal, "Ten dollars down and ten dollars a month"?
What would he think if we should turn to him and say, "Old man, just think, you're responsible for all this!"

With a comma. If a question or an exclamation ends within a sentence and the construction calls for a comma at the same point, the comma should in most instances be omitted. A comma and an exclamation mark or a comma and a question mark used together are so unusual that they attract the attention of the reader without in any way clarifying the sentence.

Poor: The curriculum problem is clarified when we consider that it raises the question, not so much, What has? or What is?, or even What will be?, but rather, What should be?

No confusion of meaning results when the commas are omitted.

The curriculum problem is clarified when we consider that it raises the question, not so much What has? or What is? or even What will be? but rather, What should be?
Voices rose in a tumult of shouting: "Free her! Free her!" "Hang the bloody captains!" "Hang the villains!" "They're the murderers!"
"Are Languages Practical?" *F. R.*, Vol. 5 (Nov. 1931), pp. 141–45.

Some persons prefer to drop the question mark instead of the comma.

One who asks, "What is money, anyway," would be quick to kick if short-changed.

With a semicolon. When the sentence construction calls for a semicolon where a question or an exclamation ends, the semicolon is usually retained, in spite of its obviousness.

> ...cultivate an interest in the following topics: "How Old Is the Earth?"; "The Beginnings of Life"; and "Human Beings."
> ...the mottoes of which were the same: "Enjoy life, for it is short!"; "Wine, women, and song"; or, as in advertising: "Unhappy?—Buy a Chevrolet!"; "Buy Swift's ham and be happy!"; ...

Titles of works. When an exclamation point or a question mark is part of the title of a work, such as a play or a book, it must be retained, and in these rare instances double punctuation may have to be used (with colons, semicolons, commas, and each other, but not with periods), but try to avoid such situations by rewording. The following exemplify correct usage:

> Could it possibly be as successful as *Hello, Dolly!*?
> Many would consider Rodgers and Hammerstein's top three musicals as *Oklahoma!*, *The King and I*, and *South Pacific*. [Or reword if possible: ... *The King and I*, *South Pacific*, and *Oklahoma!*]
> "Wasn't his first big role in *Fiorello!*?" she asked.
> The following students were in the sketch from *M*A*S*H**: Tim Wilson, ...

In the following example a successive question mark may acceptably be omitted:

> Wasn't his last book *What Do You Say After You Say Hello?*

COLON

The colon is used after a word, phrase, or sentence to introduce something that follows, such as a formal question or quotation, an amplification, or an example. It may be said to replace the words *that is* or *for example*, but should not be used after these words or after *such as*, *namely*, or *for instance* unless a complete sentence follows. (See p. 145 for capitalization following a colon.)

> Bacteria are everywhere: in the air, in water, in milk, in dust, in soil, in the mouth, and on the hands.
> Of both centuries, meanwhile, two things are true: neither in itself presents much literary variety, and most of what was published in each has already been forgotten.
> Cooking has various purposes: (1) to make food more completely digestible; (2) to make food better flavored; (3) to make ...
> In the *Dictionary of Graphic Arts Terms* we may read: "Imposition is the operation of ..."

Do not use a colon before a list unless the items are in apposition to an introductory word. Do not use a colon between a verb or preposition and its direct object.

> *Wrong:* ...produced in large quantities, such as: cacao beans, coffee, fruits, and Panama hats. [The colon should be omitted.]
> *Wrong:* The principal criteria determining the selection of the site were: location, initial cost, suitability, and convenience to transportation. [The colon should be omitted.]

Before a direct quotation or question a comma is often sufficient.

> The question may be raised, Is the reaction really a reaction between the molecule of barium nitrate and the molecule of sodium sulfate?

Use neither colon nor comma before an indirect question or quotation.

> *Wrong:* The problem is: How to conserve it for plant use.
> *Right:* The problem is how to conserve it for plant use.
> *Or:* The problem is, how can it be conserved for plant use?

Occasionally in formal literary use, a colon may be used to join two almost equal sentences that are saying essentially the same thing.

> The Lord bless thee, and keep thee: The Lord make his face shine upon thee: The Lord lift up his countenance upon thee and give thee peace.
> —Numbers 6:24–26

Lists. A colon is used after an introductory statement that contains the words *as follows* or *the following*; either a colon or a period may be used after other statements introducing lists.

> We will discuss the following types of psychotherapy:
>   1. Client-centered therapy
>   2. Rational-emotive therapy
>   3. Behavioral therapy
>   4. Psychoanalysis

When the introduction is not a complete sentence and one or more of the items of the list are needed to complete it, no colon or dash should be used.

> Two types of psychotherapy are
>   1. Client-centered therapy
>   2. Rational-emotive therapy

An exception to the foregoing is that a colon may be used to replace a comma when the statement preceding a displayed list ends with such words as *for example* or *that is*.

We have thus far covered two types of therapy, that is:
  1. Client-centered therapy
  2. Rational-emotive therapy

Salutations. Use a colon after the salutation of a letter — except an informal note — and of an address.

My dear Mr. Barr: GENTLEMEN:
Mr. President, Ladies and Gentlemen:

Scripture references. References to passages in the Bible usually use the colon after the chapter number. (See also p. 279.)

I Corinthians 13:4–7   2 Chronicles 1:3–5

Side heads. A colon is sometimes used after a side head, although a period is usually sufficient.

Conciseness. Before beginning to write, know...
Instruction: Stand squarely before the mirror,...

Time. In expressions of time the colon is used.

at 10:20 A.M.   the 7:40 train

Use with other marks. Whenever a colon is used in conjunction with closing quotation marks, place it after the quotes. It should also follow a parenthesis.

He began with "The Village Blacksmith": "Under the spreading chestnut tree..."
There are several advantages to growing the red mulberry (*Morus rubra*): Its fruit is edible; ...

The combination of a colon with a dash was once popular but is rarely used today; a colon is sufficient.

SEMICOLON

The semicolon indicates a more definite break in thought or construction than a comma would mark, and calls for a longer pause in reading. It is used wherever a comma would not be sufficiently distinctive.

There is no undergrowth, no clinging vines, no bloom, no color; only the dark, innumerable tree trunks and the purplish, scented, and slippery earth.
Preliminary practice was given as follows: in the formboard, by two trials with eyes open; in the star tracing, by fifteen minutes of practice.

Coordinate clauses without a conjunction. When the parts of a compound sentence are not connected by a coordinating conjunction, they are best separated by a semicolon, especially if the parts express a contrast or an antithesis.

> The intention is excellent; the method is self-destructive.
> An engineer may use this construction to lay out a large circle; a navigator may use it to avoid a dangerous reef or a submerged rock; a surveyor may utilize it in locating a new point on a tract that has been partly mapped.

Coordinate clauses with a conjunction. When the clauses of a compound sentence are long, involved, or internally punctuated, use a semicolon between them.

> The governments of Europe adopted the parliamentary type which was developed in England through many centuries; and even the countries with dictators have more or less held on to the parliamentary forms of government.
> We have seen how the people of this state work, how they earn their living; now let us see how they play, how they use their leisure.

If the clauses of a compound sentence are joined by one of the conjunctive adverbs—*so, therefore, hence, however, nevertheless, moreover, accordingly, besides, also, thus, then, still,* or *otherwise*—a semicolon between the clauses is usually better than a comma; but before *so, then,* and *yet* a comma is often sufficient, especially when the subject is omitted (as understood) in the second clause.

> I saw no reason for moving; therefore I stayed still.
> The importer pays the tax and then sells the articles at an advanced price; thus the consumer is really the man who pays the tax.
> He took his seat, then waited to be summoned.

Series. If one or more items of a series contain commas, semicolons are often necessary between the parts to make the meaning clear.

> These were located in Newbury, Cambridge, Saugus, Watertown, and Rowley, Massachusetts; Portsmouth, New Hampshire; and Lincoln, Rhode Island.
> The children were drawn from three schools: the Hebrew Orphan Asylum, New York City; School 38; and the Normal Practice School, Buffalo, New York.

> *Clear with commas:* Mr. W. A. Stauffer, principal, Mr. Leonard McKaig, instructor, Mr. Stanley Hawthorne, instructor, Jay Bonnet, winner of the state oratorical contest, and the members of the civics class of the Ripon High School were extended the privilege of the floor of the Assembly for this day.
> *Clear with commas:* An ellipsis, a two-letter syllable, a two-letter word (unless quoted or carrying closing quotes), or a dash must never be used as the concluding line of a paragraph.

Resolutions. In a series of *whereas* clauses in a resolution each clause is usually followed by a semicolon.

Whereas in virtue of Article 3, paragraph 3, of the Covenant, the Assembly . . . ;
and whereas, according to Part IV, section III, of the report . . . ; the assembly
decides . . .

Use with other marks. Whenever the semicolon occurs in conjunction
with closing quotation marks, place it outside. If it is used with parentheses,
always place it after the closing parenthesis.

Pronouns referring to the Deity should not be capitalized except to avoid
ambiguity, as "We owe all to Him who made us"; but in a case like "The Lord
pitieth them that fear him," the capital is not necessary.
Dates given without the month are designated as "the eighteenth"; dates with
the month in this style: January 18.
Don't use "at the corner of" (say Michigan Avenue and Second Street); "put in
an appearance"; apt, for likely.
Historic connotes something famous or memorable ("this historic document
marked the end of the age"); *historical* indicates something having to do
with history ("historical documents of that age show . . .").

Note. Some writers contend that semicolons or colons should be placed
inside quotation marks if the semicolon or colon was in the original printing of
the matter quoted. Such an arrangement leaves the sentence in which the quotation appears without the punctuation called for by its construction. Even if the
semicolon or colon was used in the original it has no significance in the new
sentence unless it is necessary outside of the quotation marks.

COMMA

A comma should be used only if it makes the meaning clearer or enables the reader to grasp the relation of parts more quickly. Intruded
commas are worse than omitted ones, but keep in mind at all times that
the primary purpose of the comma is to prevent misreading.

In informal writing and in fiction, authors should be allowed wide
latitude in their preferences for comma placement (or nonplacement).
However, in formal, scientific, and technical writing and in works of high
literary quality the standard rules for the comma should be observed.

Open and close punctuation. The terms *open punctuation* and *close*[34]
punctuation are often used with reference to commas and mean, respectively, using only as much punctuation as is absolutely necessary for clearness, and using as many marks as the grammatical construction will justify.
Close punctuation often results in jerkiness without making the meaning
clearer.

Too closely punctuated: The soil, which, in places, overlies the hard rock of
this plateau, is, for the most part, thin and poor.
Better: The soil, which in places overlies the hard rock of this plateau, is for the
most part thin and poor.

[34]Klos, tight; not kloz.

Compound sentences. The coordinate clauses of a compound sentence are usually separated by a comma, though a semicolon may be called for by complexity of thought.

> The whole judicial selection process requires fundamental overhaul, but the state legislature remains derelict in its failure to act on the present amendment.
> During the prohibition era, wine-making became an art practiced in many homes, and America's consumption of wine actually increased in those years, despite the ban on alcoholic beverages.

Note that in the following compound sentence the first several words of the second clause might be misread until the verb is encountered:

> Carlos was proclaimed king by the clerical party, and the pretender Dom Miguel of Portugal recognized him immediately.

The comma may be omitted if the clauses are short and closely connected in thought, especially if the connective is *and*; but again take care not to omit a comma that is needed to prevent the subject of the second clause being read, at least momentarily, as a part of the object of the first clause.

> *Faulty:* Zinc oxide is substituted for calcium oxide and selenium and charcoal are added.
> *Clear:* Zinc oxide is substituted for calcium oxide, and selenium and charcoal are added.
> *Faulty:* The bland flavor of the egg white does not call for the addition of lemon juice and more delicate flavors are used.
> *Clear:* The bland flavor of the egg white does not call for the addition of lemon juice, and more delicate flavors are used.

Clauses joined by *but* regularly take a comma; a semicolon may be used when the contrasted clauses are complex or when one clause is balanced against two others.

> Copy editors may try to make a sentence clear by punctuation, but they cannot so clarify one that is badly worded.
> We know what he did and what he said, and we know what interpretations his friend put upon his words and actions; but we can only guess at his ulterior motives.

A clause introduced by the conjunction *for* should be set off by punctuation to prevent misreading of *for* as a preposition. Either a comma or a semicolon may be correct.

> He must have been in a great hurry, for his steps had been from twenty to thirty inches long.

> *Confusing:* The last two have much promise for the future for those fields have only recently come into prominence.
> *Clear:* The last two have much promise for the future, for those fields have only recently come into prominence.

A clause beginning with *so, then,* or *yet* should usually be separated from a preceding clause, and a comma is often sufficient for this purpose (see p. 183).

Compound predicate. Compound predicates are usually not separated by punctuation.

> It begins with a discussion of different types of soil and goes on to explain the care of various kinds of flowers.
> These people have been dominated by foreigners for centuries but have still kept their language, culture, and racial integrity.

A comma is sometimes needed to prevent misreading.

> *Faulty:* He stifled the cry that rose to his lips and lay motionless.
> *Better:* He stifled the cry that rose to his lips, and lay motionless.

Punctuation of sentences like the following seems to trouble many writers and readers. It is best in such sentences to observe the rule of not breaking a compound predicate, even though if the sentence were read aloud there might be a natural pause before the conjunction. In other words, in following rules for punctuation the written word always takes precedence over the oral word.

> She clapped her hands and, after bowing once more to her father, ran to find her mother.
> He escaped from the guard having him in charge and, crossing the frontier of the Papal territory, embarked at Naples.

No rule is disregarded in punctuating the following; the comma before the conjunction separates the parts of a compound sentence, the other comma separates a dependent clause or phrase from the main clause following it.

> Riga is icebound for a shorter time, and except for a few days each winter, icebreakers can keep its harbor open.
> His wealth is entirely in the service of employment and production, and if it were taken from his management, his only loss would be the opportunity to serve.

Series. Two parallel elements connected by a conjunction require no comma between them.

> *Commas incorrect:* Through the courtesy of Mr. West, Miss June Hucke, and members of the eighth grade of the Pleasant Grove School, were extended the privilege of the floor of the Assembly.
> *Better:* Through the courtesy of Mr. West, Miss June Hucke and members of the eighth grade of the Pleasant Grove School were extended the privilege of the floor of the Assembly.

Letters or figures in parentheses make no difference in the rule.

Wrong: South America may be divided into two sharply contrasted regions;
namely, (1) the Pacific, and (2) the Atlantic drainage basins.
Right: South America may be divided into two sharply contrasted regions,
namely (1) the Pacific and (2) the Atlantic drainage basins.

In a series of the form *a, b, and c* or *red, white, and blue* most publishers prefer a comma (sometimes called the serial comma) before the
conjunction, whether the items of the series are words, phrases, or clauses.
The consistent use of this comma is recommended, for in many sentences
it is essential for clarity.

Motion pictures, the press, and the radio have tremendous influence.
Errors in spelling, typesetting, punctuation, or sentence construction lead to
humorous statements.

In the examples to follow, the use of the serial comma is necessary to
prevent misreading.

A series of personal names should have the comma before the conjunction to prevent the first name of the series being construed as a noun in
direct address.

"Anna, May and Walter are here." [Might be read as telling Anna that May and
Walter are here.]
"Anna, May, and Walter are here." [Can mean only that three persons are here.]

In a series of phrases the comma is necessary before the final phrase
whenever the preceding phrase of the series contains a conjunction.

... for punctuation, choice and spelling of words, and good diction.
"Strike-anywhere" matches are tipped with a mixture of phosphorus sulfide
and some oxidizing material such as potassium chlorate, powdered glass or
some other material to increase the friction, and glue to bind the ingredients
to the match.

When the phrase preceding the conjunction contains a preposition, a
comma is needed to prevent misreading the last phrase as part of the object
of the preposition.

The electrochemical industry, battery action and cell action, the smelting of ores,
the corrosion of metals, and many other important types of chemical changes
are examples of the transfer of electrons.

Obscure: The green trees, the sight of the hills and fresh raw meat had quickly
revived the beast.
Clear: The green trees, the sight of the hills, and fresh raw meat had quickly
revived the beast.

In a series of three or more clauses clarity almost invariably requires the comma before the conjunction.

> A pencil placed in a glass of water appears to be bent, the moon appears to be larger at the horizon than when it is overhead, and lake water appears to be green or blue depending on the depth, the position of the observer, and other factors.

If one or more members of the series contain commas, semicolons may be necessary between the parts to make the meaning clear (see p. 183).

Inserting a comma after the last member of the series, thus bringing a comma between subject and verb, is contrary to modern American style.

> *Not American usage:* Camels, llamas, elephants, and carabaos, are as valuable in tropical countries as horses in America.
> *Not American usage:* A strong unreasonable prejudice, class or personal hatred, or insane fear, bears its frightful fruits of death, destruction, and persecutions.

In a series in the form *a, b, c* (no conjunction), preceding a verb or clause that stands in the same relation to each member of the series, use a comma after each member.

> Poise, dignity, the peace of a man at one with his soul, shone from his face.
> We noticed the misery, the suffering, the hardships, that lay hidden in the neighborhood.

Writers should avoid careless use of this construction, for in many instances a conjunction is preferable to the awkward comma.

> *Poor:* Flies, insects, buzzed about.
> *Better:* Flies and insects buzzed about.
>
> *Poor:* Potatoes, rice, spaghetti, barley, may be added for bulk and thickening.
> *Better:* Potatoes, rice, spaghetti, or barley may be added for bulk and thickening.

No comma is needed in a series in the form *a and b and c* (conjunction between each two).

> The figures and postures and expressions of the Madonna, of the saints, of the prophets and sages, were less stern.
> She meant she wasn't thinking of changing her house or job or husband *right now.*

Coordinate adjectives. Two or more adjectives modifying the same noun should be separated by commas if they are coordinate in thought — that is, if *and* could be used between them without changing the meaning; but if a compound is formed by one or more adjectives included in thought with the noun modified, an adjective modifying this compound should not be separated from it by a comma.

a rotund, cheery little man	a huge boxlike affair
fine tall, straight timber	a titanic flaming-red torch
cool, humid climate	cold corned beef
broad, shallow rivers	big white plantation house

Snowflakes are beautiful complex six-sided figures.

Wrong: On a little, native, banana plantation
Right: On a little native banana plantation

Wrong: From the many, quaint, old fishing villages have come...
Right: From the many quaint old fishing villages have come...

Nonrestrictive and restrictive modifiers.[35] A nonrestrictive phrase or clause is one that could be omitted without changing the meaning of the principal clause; it should be set off by commas. A restrictive phrase or clause, on the contrary, so qualifies or limits the word it modifies that it could not be omitted without affecting the meaning of the sentence, and it should *not* be set off.

Nonrestrictive: Many animals, such as rabbits and horses, shed their heavy winter fur at the approach of warm weather.
Restrictive: Animals such as the beaver, seal, rabbit, fox, and cat are protected against cold by a soft silky covering called fur.
Nonrestrictive: Bobcats make their home in a hollow log or cave in the rocks, where they raise from two to four young.
Restrictive: Cheetahs live in various regions in Africa and Asia where they are able to find deer and antelope.
Nonrestrictive: It was improved very little until the middle of the nineteenth century, when matches were invented.
Restrictive: We should not use formal language when simpler expressions would serve.

A sentence containing a restrictive or nonrestrictive relative clause may be so worded that only by the punctuation can the reader tell the meaning. The writer should always check the punctuation of such sentences with special care. In scientific work, the terminology of which may be unfamiliar to a reader, this is particularly necessary. (The preceding sentence is itself one in which the relative clause might be construed as either restrictive or nonrestrictive. If it were intended restrictively it would be better to reword: "In scientific work whose terminology may be...")

Nonrestrictive: They passed on to the next victim, who seemed to be suffering greatly. [The clause here gives us the information that the next victim seemed to be suffering greatly.]
Restrictive: They passed on to the next victim who seemed to be suffering greatly. [Here we understand that the victims suffering greatly were attended before those suffering less severely.]

[35]Some writers on punctuation prefer the terms *nonessential* and *essential*, *nondefining* and *defining*, or *nonlimiting* and *limiting*.

Whether or not to use a comma preceding such clauses often depends on the usage of *that* or *which*. Ideally, as a relative pronoun *that* should always be used restrictively, whereas *which* is usually nonrestrictive but may be restrictive (as in clauses introduced by *of which* and *that which*). Therefore if *that* and *which* are used correctly, the problem of insertion or omission of a comma concerns only clauses beginning with *which*.

Nonrestrictive: He was greatly disturbed by the letter, *which* he received this morning. [This clause gives incidental information about one of several letters he may have received this morning.]
Restrictive: He was greatly disturbed by the letter *that* he received this morning. [Here he received one particular letter and it disturbed him. Note that in this usage the *that* may be omitted: "the letter he received."]

Restrictive: Those are the streets on *which* the traffic practically moves backward but *which* the mayor claims are a fine achievement. [The *on which* clause fairly defines the streets, so that the clause following *but* has lost some of its definitiveness; although the second clause would require *that* if it stood alone, *which* is preferable here.]

A restrictive relative clause often incorrectly set off is one that does not immediately follow the word it defines or limits. Usually the problem of this kind of sentence arises from an unnecessary attempt to avoid ending the sentence with a preposition (see p. 381).

Wrong: They spoke of old battles, in which they had fought together.
Right: They spoke of old battles in which they had fought together.
Or: They spoke of old battles that they had fought together in.

Wrong: It is this William, for whom the rock in Salem harbor called Bowditch's Ledge is named.
Right: It is this William for whom the rock in Salem harbor called Bowditch's Ledge is named.
Or: It is this William whom the rock in Salem harbor called Bowditch's Ledge is named for.

Essential phrases, as well as clauses, should be recognized as restrictive and not be set off by commas.

Wrong: The learning and staging of operettas, dramatically insipid and musically void, may vitiate a year's work in music. [This states that all operettas are dramatically insipid and musically void. The commas should be omitted, making "dramatically insipid and musically void" a phrase limiting "operettas."]
Right: The learning and staging of operettas dramatically insipid and musically void may vitiate a year's work in music.

Wrong: The Ngauruhoe volcano was smoking lazily, like some old man, deep in reflection. [As punctuated, this states that the Ngauruhoe volcano was deep in reflection.]
Right: The Ngauruhoe volcano was smoking lazily, like some old man deep in reflection.

Wrong: It was a peaceful migration, rather than a military conquest, such as that of the German invaders of the Roman Empire.
Right: It was a peaceful migration, rather than a military conquest such as that of the German invaders of the Roman Empire.

Wrong: An exercise, like C6, should be done orally in class.
Right: An exercise like C6 should be done orally in class.

Appositives. Set off words in apposition by commas.

He was replaced by a German leader, Odoacer, and thus a ruler from the barbaric tribes was recognized in Rome.
Kid leather, the skin of young goats or sheep, is one of the finest of leathers.

Simple appositives like the preceding examples are usually recognized and properly punctuated, but when the appositional phrase is longer, the second comma is often overlooked.

Wrong: Emporia, the prophet's own county with its 10,000 inhabitants bought more than 2,500 copies.
Right: Emporia, the prophet's own county, with its 10,000 inhabitants, bought more than 2,500 copies.

Use dashes to set off an appositive whenever they would make the meaning and construction more quickly understood.

They determined the stimulus—sound or an electric current—that would just suffice to waken an individual.
All of the active members—students and teachers—attended the concert.

Appositional "or." Words or phrases in apposition are often introduced by *or.* Such words should always be set off when they explain or define a noun, but the commas may sometimes be omitted about a phrase that defines an adjective. Alternative *or* and appositional *or* should be carefully distinguished and no commas omitted that are necessary to clearness of meaning.

Flying Mammals, or Bats
Connotation, or the Suggestive Power of Words
Destruction of the Original, or Virgin, Forests
They are of the mixed or teratomatous variety.

Restrictive appositives. A restrictive appositive is one used to distinguish its principal from other persons or things of the same name, group, or class. Such as appositive, which could not be omitted without robbing the sentence of meaning, should not be set off.

the poet Longfellow   his brother Will
Mary Queen of Scots   the yacht *Sally Jane*
the grand duke Edward   my friend Pat

Explain how the poem "Louisiana Night" brings to mind a railroad train.
The expression "Where am I at?" is a provincialism.
The threefold division *antiquitas, media aetas, recens aetas* becomes common.
Based on *U.S. Census Reports on Population* covering the period 1790–1970.

The most frequent failure to recognize the restrictive nature of an appositive occurs in sentences like the following.

Wrong: The Greek philosophers, Leucippus, Democritus, and Epicurus, advanced a doctrine . . . [There were other Greek philosophers.]
Right: The Greek philosophers Leucippus, Democritus, and Epicurus advanced a doctrine . . .

Wrong: The early-nineteenth-century American publisher, Robert Bonner, laid down a set of rules . . . [This conveys the impression that Robert Bonner was the one and only publisher of the early nineteenth century.]
Right: The early-nineteenth-century American publisher Robert Bonner laid down a set of rules . . .

Parenthetical expressions. Set off a parenthetical phrase or clause that interrupts the even flow of a sentence and could be omitted without altering the meaning of the sentence. A parenthetical sentence requires dashes or parentheses; commas are not strong enough.

Even heavy-soled shoes, to say nothing of thin-soled ones, are not proof against continued wetting.
The great event in typography, so far as Western civilization is concerned, occurred about 1450 in Germany.
From the Amarillo field, in a pipeline eleven miles long, are carried the natural gases from which helium is made.
At times—there were more than a few—their situation seemed hopeless.

One of the most frequent mistakes in punctuation is omission of one of the two marks required to enclose an interrupter. "One to separate, two to enclose" should be kept constantly in mind.

Absurd: Senators must have been citizens of the United States for at least seven years prior to their election, and each must be living at the time of the election, in the state from which he or she is elected.
Right: . . . each must be living, at the time of the election, in the state from which he or she is elected.

In the following sentence there should be two commas or none.

Possible: Watch someone, unobtrusively but closely, for five minutes.
Better: Watch someone unobtrusively but closely for five minutes.

The parts of a compound subject consisting of only two members are often separated by a comma. The second part should then be treated as a parenthetical element requiring a comma after it also.

The prominence given these proposed changes in tariff policy, and the sharp
division of opinion fostered by the activities of the Anti-Corn-Law League,
inevitably dominated the campaign.

The subsequent arrest of the pope by Guillaume de Nogaret, and the compul-
sory removal of the popes to Avignon, marked clearly this turn.

Note. These examples illustrate the use of commas to indicate the manner
in which the writer himself would read the sentence. Since the grammatical
construction does not require any punctuation, the writer should assume full
responsibility for its use. On the other hand, the copy editor should recognize
this use of commas to indicate emphasis and inflection and should not delete
commas in such cases. It should, of course, be recognized that so punctuating
a compound subject does not make the subject singular—subject and verb are
both plural. The construction should not be confused with the sentence in this
pattern: "The captain, as well as the sailors, is . . ."

Interpolations and transition words. Set off an expression such as
I think, to tell the truth, to say the least, in short, besides, unless it is so
close as practically to limit the accompanying words. A word and short expression that does not require a pause in reading need not be so set off.

He will come, I think, on the noon train.
To tell the truth, I'm rather tired.
This was not military work, anyway.
There were two, in fact.
Of course I will. Indeed she may. I too must follow.
Whereupon the President adjourned the meeting.

Possible: Pitiful, indeed, is the person who perpetually apologizes for himself.
Better: Pitiful indeed is the person who perpetually apologizes for himself.

The word *therefore* is not necessarily a signal for punctuation.

Commas necessary: An easy answer would be, therefore, to say . . .
They will understand, therefore, why . . .
Commas unnecessary: It is not so noticeable and therefore not so objectionable.
It is the logical method and it therefore appeals to the rational person.
Commas may or may not be used; the question is one of emphasis:
They may therefore reasonably assume . . .
It must therefore be less . . .

Use of the comma after *accordingly, consequently, yet,* and *hence* is a
matter of choice.

Distinguish carefully between words used for transition and the same
words used as adverbs.

It is very small, however.
However small it may be, it can be used.
Thus has the earth been peopled with a wealth of life forms.

"Do you think." One of the commonest of errors comes about through
failure to recognize *do you think (do you suppose)* as an interpolation. We

find such obviously incorrect questions as "Why do you think they are carrying a cross?" under a picture of men carrying a cross; or "Why do you think he is called Thrifty?" immediately following a statement that he *is* called Thrifty. The primary error in such sentences is faulty word order. The questions should be: "Why, do you think, are they carrying a cross?" and "Why, do you think, is he called Thrifty?" When the words are correctly arranged, the nature of *do you think* is immediately apparent.

> *Wrong:* For what do you think the woman in the picture has been using her rake?
> *Right:* What, do you think, has the woman in the picture been using her rake for?

The phrase *do you think* is not always an interpolation, however, but may be the main verb of the sentence. The difference in use is clear in the following:

> Why do you think he has gone? [Meaning what reasons have you for thinking he has gone?]
> Why, do you think, has he gone? [Meaning he has gone; what reasons do you think he had for going?]

In similar questions introduced by *what* no difficulty arises; the phrase *do you think* can often be construed either as a main clause or as an interpolation, and the sentence punctuated accordingly.

> What do you think is a fair way to have television used in political campaigns?
> What, do you think, is a fair way to have television used in political campaigns?

Introductory words. Use a comma after a word like *yes, no, well, why,* introducing a sentence. Take care to distinguish between *well, why,* and similar words used as mere introductory words and the same words used as adverbs.

> Yes, I'll go.   No, he is not here.   Oh, yes, to be sure, sir.

> Well, I may.   Well I may.
> Why, can't you?   Why can't you?
> Still, the flag is waving.   Still the flag is waving.
> Now, there are three who...   Now there are three who...

Observe the same care when words of this sort are used at the end of a sentence.

> What's to be done, then?   What's to be done then?

For example, that is, and similar phrases introducing an illustration or an explanation should usually be set off from the rest of the sentence. The comma is sometimes omitted after *thus, hence, namely,* and the abbreviations *e.g.* and *i.e. For instance* or *for example* used after the illustrative phrase may need no comma before it. Clearness may require setting off the illustration or explanation, together with the introducing phrase, by dashes or parentheses.

> These sepals serve as a protective covering for the rest of the flower in the early stages of its development, that is, when it is a bud.
> To obey the general will, says Rousseau, is to obey the enlightened self. Thus, when a policeman rightfully arrests a burglar, the latter is self-arrested and afterward self-condemned.
> Soils may be put into three groups, namely clay, sand, and loam.
> Other birds, the great horned owl for instance, are considered beneficial in some sections and harmful in others.
> They are getting the best kind of exercise obtainable—namely, exercise that is enjoyed.
> The President may veto an act of Congress—that is, forbid it to go into effect—or the courts may set aside a law.
> ...who would argue that if what was "real" (that is, what existed) was reasonable, then the opposition or contradiction called forth by that which existed was likewise reasonable.

Adverbial phrases. Do not use a comma after an adverbial phrase beginning a sentence when it immediately precedes the verb it modifies. The comma should be omitted in all these sentences:

> *Wrong:* Against all these evils, is the overwhelming benefit which comes from the American system of government.
> *Wrong:* To the west of No. 1, is another enclosure.
> *Wrong:* Among these lines, may be seen the refracted rays.
> *Wrong:* Between Mars and Jupiter, are a large number of small bodies called asteroids.

After an adverbial phrase that is not independent of the part of the sentence that follows, the comma is not necessary unless confusion of meaning would result from its omission (as it would in this sentence).

> On our way home we met the returning fisherman.
> In many parts of the world the wind has an important part in soil-making.
> In the year 1900 our exports were ...
> To the altitude flyer mere height ...

Whenever such a phrase ends with a verb or a preposition, use a comma before a following noun to prevent misreading.

> At the parties at which the dances were performed, the parents were invited guests.
> On the sandy shores beneath, fishermen spread their nets to dry.
> Soon after, their first settlement was started.

Writers often punctuate incorrectly when such phrases as the foregoing stand at the beginning of the second part of a compound sentence. The coordinate parts of the sentence should be separated, and then if necessary for clarity or ease of reading a comma may be inserted after the phrase. Very seldom should there be a comma before the phrase, setting it off like a parenthetical element.

> *Wrong:* Stars are quite punctual and, by use of a transit telescope, their passage can be accurately noted.
> *Right:* Stars are quite punctual, and by use of a transit telescope, their passage can be accurately noted.
>
> *Wrong:* He had come alone to the party, and, as he made his way through the crowd, he saw no one he knew.
> *Right:* He had come alone to the party, and as he made his way through the crowd he saw no one he knew.

Adverbial clauses. As a rule, use a comma after an adverbial clause preceding its principal clause. There is a strong tendency to omit the comma if the clause is short and no misreading would result from the omission.

> When a sufficient supply of oxygen is provided, any further increase has little effect.
> Before the individual tries to write a question on the board, let the group offer questions.
> After some years had passed, the situation changed.
> Were there space in this book we could cite hundreds of instances.

In the following sentences a comma is necessary after the adverbial clause to ensure correct reading.

> As soon as he stepped in, the elevator fell.
> As the pond fills, the plants are killed off.
> When the glaze is worn off, the inner surface becomes absorbent.
> When we shoot, our bowstrings give a twang that's heard but a little way off.

Always use a comma after an adverbial clause introduced by *as*, *since*, or *while* if it is intended to express any idea of cause or condition; without the comma these conjunctions express time only.

> As we flew over the lake we could see the cottages bordering it.
> As the plane was flying low, we could see the cottages distinctly.
> As we read the stories of the westward movement we shall see . . .
> As we learned before, from the time of the explorer Coronado the Spaniards had tried to establish settlements north of the Rio Grande.

Do not use a comma before an internal adverbial clause, preceding the main clause on which it depends, unless the clause is clearly nonrestrictive and can be read as a parenthetical element.

> *Wrong:* Dark walls of rock rise steeply from the shores, and, when it is calm, snowcapped peaks and blue glaciers are mirrored in the water of the strait.
> *Right:* Dark walls of rock rise steeply from the shores, and when it is calm, snowcapped peaks . . .

Wrong: This means that, if the cylinder is full of gas when the piston is at the bottom, the gas will occupy only one fifth of the volume . . .
Right: This means that if the cylinder is full of gas when the piston is at the bottom, the gas will occupy only . . .

Punctuation of an adverbial clause following the main clause depends upon its nature: a nonrestrictive clause, which merely gives additional information, should be preceded by a comma; a restrictive clause, which limits the action of the main verb to a particular time, manner, or circumstance, should not be preceded by a comma.

Clauses introduced by *though* or *although* are always nonrestrictive.

Such forms are still living in the Australian region, although they have become extinct elsewhere.

Clauses introduced by *if* are restrictive.

Publication can be hastened if the paper is made as concise as possible.
If a certain state charges $16 for each auto license, how many registered cars were there in the state if the amount of the license fees was $4,757,216?

Clauses beginning with *because* are usually restrictive, but they may be nonrestrictive.

Millions of stars cannot be seen with the eye because they are too far away to supply the necessary light for vision.
The Austrian government proposed a new statute prohibiting the breeding of carrier pigeons, because carrier pigeons might be used by unauthorized persons for the purpose of smuggling military or industrial information out of Austria.

Clauses introduced by *unless* and *except* are usually restrictive.

The tense and the voice of the verbs in a composition should not be changed unless the meaning demands it.
Do not give space to unimportant negative findings except when convinced that they add to the force of the argument.
Intention is of no avail unless it is stated at the time of the contract.
She must have reached Stamford by now, unless she took the wrong train.

Do not use a comma before a clause beginning with *before, when, while, as,* or *since* restricting the time of the action of the principal verb.

You should make it a point to master each of these facts before you attempt to write the assignment.
The Indians became more and more alarmed as the white settlements appeared farther and farther westward.

When a clause introduced by *as, while,* or *since* does not restrict the verb but expresses cause or condition, use a comma before it.

Reed did not play, since he had injured his ankle.
The United States was concerned with wages and prices, while Japan concentrated on productive quality and efficiency.

Use a comma before a clause of result introduced by *so that* but not before a clause of purpose so introduced.

This alloy possesses the peculiar property of not shrinking when it cools, so that the case is a sharp and accurate reproduction in relief of the letter cut in the matrix.
Can you construct a line *MN* through a given point *P* so that it is parallel to a given line *CD*?

Absolute phrases. Set off an absolute phrase.

That being so, there was no binding contract.
The sun having risen, they packed the tents and sleeping bags into the car and took to the road again.
The vote taken, the committee proceeded with other business.

Take care that no comma precedes the participle in absolute expressions like the following:

The commission having been reduced to an innocuous statistical and research agency, the problem of effective regulatory procedure was returned to a perturbed and disconcerted Congress.
The offer being to ship not later than May 15, the buyer had the right to fix the time of delivery at any time before that.

Infinitive phrases. Do not set off an infinitive phrase used as the subject of a sentence.

Wrong: To think of our solar system as a part of the Milky Way, gives us a diminished conception of our importance.
Right: To think of our solar system as a part of the Milky Way gives us a diminished conception of our importance.

Wrong: To neglect the integrity of the family or the prosperity of any considerable social class, will sooner or later injure society as a whole.
Right: To neglect the integrity of the family or the prosperity of any considerable social class will sooner or later injure society as a whole.

Participial phrases. Set off a participial phrase unless it is restrictive or is used in place of a noun.

The temperature of plants changes rapidly, depending on the amount of external heat they receive.
In 1888 Roux, working in Pasteur's laboratory, found that the diphtheria germ produces a toxin which causes symptoms of the disease.
Making use of the maps and tables answers the questions and supplies the missing words.
Valued at only $15,000 in 1822, it was estimated at $450,000 by 1845.

Nonrestrictive: Some atoms broke up spontaneously, forming atoms of other elements.

Restrictive: Associated with the two dominant trees are ash, elm, walnut, linden, and a wealth of smaller trees and shrubs forming a lower layer under the higher trees.

Restrictive: Refer to the graph showing the price of wheat.

Restrictive: Make a collection of pictures showing the great extremes of land surface in our country.

Wrong: Describe a scene trying to convey an impression of heat, cold, restfulness, confusion, fear, horror . . .

Right: Describe a scene, trying to convey an impression of heat . . .

Direct questions or quotations. Usually set off by a comma a short quotation, maxim, or similar expression; before a long, formal quotation or question use a colon (See p. 180.)

Is has been well said, "The tongue is a little member and boasteth great things."

The question is, How shall we know what are good books?

It was Congreve who wrote, "Love's but a frailty of the mind."

Quotations as parts of speech. A quoted word or phrase that constitutes the subject or object of a sentence should not be set off by commas.

He had a habit of saying "Understand?" at frequent intervals.

"If he says 'I'll do it,' what then?"

Miss Johnson's topic was "Adjustment Problems and the Visiting Teacher."

"A soldier is no better than his feet" is an old saying that is true for all of us.

I'm not sure that the mother bird said "Thank you."

One of the important rules of singing is "Sing as you speak."

If a slogan must be carried in mind, perhaps "No calories without vitamins" is as good a precept as any.

Note: The writer and the copy editor may sometimes disagree about the application of this and the preceding rule. One often sees sentences in which one person would use a comma before a quotation and another would not. The copy editor should be more circumspect about making changes.

A quotation immediately preceded by the conjunction *that* should not be separated from *that* by any punctuation.

Wrong: The law stipulated that, "Employees shall have a right to bargain collectively."

Right: The law stipulated, [or:] "Employees shall have a right to bargain collectively."

Or: The law stipulated that "employees shall have a right to bargain collectively." (See also p. 145.)

Phrases used with quoted conversation. Mistakes are often made in punctuating descriptive phrases following a quotation. A participle and an adverb should be distinguished; a participle should always be set off, an adverb generally not.

"That will never do," he said, laughing.
"That will never do," he said laughingly.

Sometimes a quoted speech is followed by such an expression as *he smiled*. In formal writing this is not joined to the quotation as if it were a verb of speaking, but is set as a separate sentence. In informal writing and popular fiction, however, it may be joined to the quotation, for stylistic reasons.

"You're a very attractive young lady." He smiled.
"Haven't we met somewhere before?" he leered.

When the words interrupting a speech are not words of speaking or saying, it is better to use dashes to separate them from the quotation (see also p. 208).

"I can't help thinking"—he smiled at her suddenly—"that part of it is me."

Antithetical elements. Set off an antithetical clause following the main clause on which it depends.

The function of the heading is to tell the facts, not to give the writer's comment on the facts.
Deeds, not words.

An antithetical phrase introduced by *not* and followed by *but* may be set off by commas if it could be omitted without destroying the grammatical completeness of the sentence or changing the meaning. Many such phrases are so short, however, that commas are not necessary.

The way that is at once easiest and most honorable is not to be silencing the reproaches of others but to be making yourselves as perfect as you can.
Hence the diagnosis may be made, not by laboriously pondering the data, but by a computer.
His interest dwindled not in observation but in participation.
It was his mental not his physical qualifications that worried me.
What he offered me was not hospitality but detention.
Not more machinery but more intelligence is needed.

Correlative phrases. Do not confuse a correlative phrase and an antithetical phrase. A correlative phrase is not usually set off by commas.

They listen for the voice of God not alone in the records of the past but in the stillness and silence of their own souls.
It has become the chief trading center not only for the northwest section of the state but also for parts of Montana and Canada.

The phrases in the following sentence are nonrestrictive and should

be set off from the main clause, the meaning of which would remain un-
changed were they omitted.

> The landscape varies greatly, not only from place to place but from season to
> season.

Omission of a common verb. When two or more clauses of a sentence
require the same verb, the omission of the verb in clauses following the
first should usually be marked by a comma unless they are short.

> One aspect of the United Nations might be entitled Cooperation in the Works of
> Peace; the other, Cooperation for the Prevention of War.
> Within this area Mohammedans have occupied the desert of the northwest;
> Hindus, the Deccan plateau and the upper and middle Ganges valley; and
> Buddhists, the Burman and Ceylon areas.
> Tories become conservatives, and Whigs liberals.
> Thus evil is made to seem good, and good evil.

Elements common to more than one phrase. Use a comma before an
element that belongs equally to two or more phrases but is expressed only
after the last.

> Just as none of us are physically, so few are mentally, fit all the time.
> They may play a secondary role in reinforcing or weakening, accelerating or re-
> tarding, disturbances.
> He was aware of, but did not concern himself about, the problem.

> *Note.* Many sentences requiring the application of this rule could be re-
> phrased more clearly and smoothly.

Commas are not needed when a conjunction is used between brief
modifiers or coordinate elements.

> Many infectious diseases have been partly or almost entirely conquered.
> They can, ought to, and most assuredly will win.

Repeated words. Separate repeated words in the same construction by
commas. Repeated words in different constructions often need separation
by a comma for the sake of clearness.

> The mere knowledge of what this substance is, is of great value.
> What money there was, was steadily drained away.
> A dosage was arrived at, at which each bird was found to maintain constant
> weight.

Repeated *that*'s in different constructions do not require a comma.

> . . . and it was his desire to achieve that that impelled him to enter the campaign.

Residence, position, title. Set off by commas phrases indicating residence, position, or title, following a person's name.

> Mr. and Mrs. Peters, of Portland, Maine, were . . .
> William Russell, dean of the College of Liberal Arts, was . . .
> Donald Brown, Jr., is . . .
> Barbara Marshall, chairwoman of the Finance Committee, spoke . . .

> *Note.* There is a tendency, especially in newspaper work, to omit the comma before *of* phrases, as in *John Smith of Chicago, Harold Jones of Northwestern University,* and before *Jr.* and *Sr.*

If such a phrase has practically become a part of the person's name, do not separate it from the rest of the name.

> Lord Curzon of Kedleston   Timour the Tartar
> King George of England

Specifying phrases. Set off a phrase, a name, or a number that makes a preceding reference more specific. It is in effect a parenthetical element, which cannot be read without a slight pause before and after it.

> The president declared the provisions of Article IV, Section 15, of the Constitution suspended.
> line 20, page 43, to line 17, page 44.
> In Figure II, column 1, are given the square Hebrew characters.
> *Exploring Books, II,* presents a series of problems.

Geographical names.

> He set up his shop in Cambridge, Massachusetts, in 1638.

> *Wrong:* Mail and stagecoach lines were established, traveling from St. Joseph, Missouri and Atchison, Kansas over the Oregon Trail to California.
> *Right:* Mail and stagecoach lines were established, traveling from St. Joseph, Missouri, and Atchison, Kansas, over the Oregon Trail to California.

Dates.

> His death on May 20, 1506, attracted little attention.

A comma is unnecessary when only the month and year are given; however, many writers and editors use commas to set off the year number. Whichever style is used, be consistent. When the name of a holiday or other special day is given, set off the year number.

> It was in July 1943.
> On Thanksgiving Day, 1969, we . . .

Addresses. Do not use a comma in addresses between the number of the house and the name of the street, except in British addresses, and even in these the comma is not always used.

412 Fifth Avenue, New York, New York
We arrived at 525 East Fiftieth Street, New York City, at eight o'clock.

Distinct proper nouns. Separate consecutive proper nouns referring to different individuals or places.

To Elsie, Robert seems perfect.
To Europe, America is the land of opportunity.

Numbers. Do not use a comma in page or date numbers (except for dates of five figures or more) or after a decimal point. Other numbers of four figures usually take a comma after the digit representing thousands. (See also p. 128.)

1500 B.C.   p. 3675   0.65842   35,000 B.C.

For the sake of clarity, separate two unrelated numbers coming together by a comma, or reword the sentence if possible.

In 1971, 250,000 more were sold than in 1970.
In 1971 the company sold 250,000 more than in 1970.

Dimensions, weights, and measures. Commas are unnecessary in the punctuation of phrases denoting dimensions, weights, and measures.

five feet seven inches   4 lb 3 oz   5 hr 10 min
His age is 6 years 4 months 12 days.

Direct address. Set off proper names and substantives used in direct address.

How are you, Mother?   Come again, Uncle.   Yes, sir.   No, miss.
I move, Mr. Chairman, that the meeting adjourn.

Wrong: Awake, my little ones and fill the cup...
Right: Awake, my little ones, and fill the cup . . .

"O" and "oh." Do not use a comma immediately after the vocative *O*, as from its very nature it requires another word or words to complete it. After *oh*, on the contrary, use a comma if other words follow.

O lovely goddess!
Oh, Robert, how could you!
Oh, for Pete's sake!

Interrogative phrases. An interrogative phrase that changes a statement into a question should be preceded by a comma.

He hasn't it, has he?   You will go, won't you?

Inflection or emphasis. The copy editor should not forget that a comma represents a slight pause in reading, and that a careful writer may often use it to indicate the way in which he would like his sentence read.

It would have been better for them if they had become good farmers rather than poor doctors, or skillful mechanics rather than blundering engineers.
Though the sailing vessel is still used to some extent, it is too much subject to the vagaries of the weather, and too slow, to meet man's general needs.
That is important, and interesting, too.
Most of the people in the past have been like that, too.

Construction likely to be misread. Use a comma to separate any sentence elements that if not separated might, in reading, be improperly joined or be misunderstood. (See also pp. 195, 196.)

To the courageous, men turn with respect.
Wherever practicable, translations of unusual German phrases have been included in the vocabulary.
Two new university extension courses, Singing for Recreation, and Business Law, will be started in this city Tuesday night.

Unnecessary comma punctuation. Do not separate a subject from its verb even when the subject is fairly long. Exceptions are when there are repeated words (see p. 201) and when ambiguity might occur.

Establishment of a citizens' monitoring committee to check progress toward full implementation of the legislative reforms was urged as a first step.
What we seem to see is one continuously changing picture.
Whether any particular expression is suitable depends upon the purpose . . .
Prunella up against it would be dangerous to whoever threatened her security.
Entering the controversial debate on abortion when he did not have to, cost him thousands of votes. [Many writers would not use even this comma.]

Do not separate a verb and its object in a simple sentence like the following:

Wrong: It is only a means to an end and has for its purpose, (1) wise expenditure and (2), an accumulation of savings.
Right: It is only a means to an end and has for its purpose (1) wise expenditure and (2) an accumulation of savings.

Do not use a comma before or after the conjunction *that* in sentences constructed like the following:

Wrong: But the author believes that, with average high school classes it will take three recitations.
Right: But the author believes that with average high school classes it will take three recitations.

Wrong: It can be easily seen that, when such amounts are fixed too high, owners of property can be crippled and ruined by such high taxation.
Right: It can be easily seen that when such amounts are fixed too high, owners of property can be ruined by such high taxation.

Do not set off a phrase separating the parts of a compound conjunction.

Wrong: There are grounds for believing that no earlier period exhibits *such* a long-continued decline in the actual price, *as* occurred in the nineteenth century.
Right: There are grounds for believing that no earlier period exhibits such a long-continued decline in the actual price as occurred in the nineteenth century.

Wrong: Such is the connection between words and things, *that* a thorough study . . .
Right: Such is the connection between words and things that a thorough study . . .

Use with other marks. The comma may be used in conjunction with quotation marks and parentheses, but not ordinarily with other marks (see p. 179).

When a comma and closing quotation marks fall alongside, set the comma inside. This practice represents the preference of practically all American publishers and printers. (The British practice is to set a comma inside closing quotation marks only if it is part of the quotation.)

If a sentence containing parentheses would require a comma were the parenthetical matter omitted, place a comma *after* the closing parenthesis.

Excerpts from the works of other authors (when they are more than a phrase or a sentence), problems, examples, and test questions are generally set in smaller type than the body of the text itself.

In the following sentences no comma is needed after the parenthesis.

Two admonitions are usually needed in preparing tables: (1) always put the title of a table above it . . . ; and (2) never put more than one kind of data in a table.
In the years after the Civil War both Seward (who had remained Secretary of State under President Johnson) and President Grant tried to buy the islands.

A comma may precede an open parenthesis only if the word or words in parenthesis clearly limit a following word.

Cheese, (full) cream

> Adventitious, (of roots) those that arise from any structure other than a root; (of buds) those that arise other than as terminal or axillary structures

The combination of a comma and a dash is unnecessary; either one or the other alone will make the meaning clear.

> Ultraviolet and other short rays—too short for the eye to perceive—are especially active.
> Our everyday speech is full of them, often too full, but we must keep the worst of them out of our written work.

EM DASH

The em dash is properly used to mark a suspension of the sense, a faltering in speech, a sudden change in the construction, or an unexpected turn of the thought.

The following sentences are properly punctuated with dashes.

> I—I think so.
> No doubt he could see—as who could not—that . . .
> But the jewels—if they are sold they cannot be replaced.
> Instead of which—but let me quote his own words.
> The school will be constantly dealing with the live questions and problems of today—social, economic, political, what you will.
> How did Larry get hurt? Fall over a cliff—stampede, maybe?
> A tree, a potato, a cabbage growing in the garden, a daisy—all are alive.

In the following sentences other punctuation may be substituted for the dash in most cases; however, some contexts may require the dash.

> "What is it?—please tell me."
> *Or:* "What is it? Please tell me."

> He glanced at his watch again—it was five minutes of one.
> *Or:* He glanced at his watch again; it was five minutes of one.

> The question is—will they make trouble for you?
> *Or:* The question is, will they make trouble for you?

> *Wrong:* . . . all sorts of devices: filter papers, double perforations—one large and one small—, pots of china, enamel, glass.
> *Right:* . . . all sorts of devices: filter papers, double perforations (one large and one small), pots of china, enamel, glass.

Appositives. Use dashes to set off an appositive whenever a comma might be misread as a series comma.

> The frontier—upland and mountain regions to be settled later in the eighteenth century—offered much more uniform conditions.

If commas are used to mark minor divisions within an appositive, dashes are generally needed to set off the whole appositional phrase.

A crate full of Murano glass—iridescent goblets, lovely shells of spun glass, beautifully shaped chalices—a tremendous English-Latin lexicon, . . .

Bismarck immediately demanded of the rulers of the larger North German states—Hanover, Saxony, and Hesse-Cassel—that they stop their warlike preparations.

The solid structure of trees—that is, the dry matter in the roots, stalks, and branches—is about 95 percent carbon and oxygen.

Do not use em dashes to chop up sentences like the following, which have short appositional phrases. Parentheses are better.

Choppy: When age was the parameter, the younger group—those under 40—and the older group—those over 40—performed equally well.

Better: When age was the parameter, the younger group (those under 40) and the older group (those over 40) performed equally well.

Confusing: Waltz in A flat—Brahms, "Skaters' Waltz"—Waldteufel, Minuet—Mozart, Lullaby—Brahms, "Rock-a-bye, Baby"—traditional, . . .

Clearer: Waltz in A flat (Brahms), "Skaters' Waltz" (Waldteufel), Minuet (Mozart), Lullaby (Brahms), "Rock-a-bye, Baby" (traditional), . . .

A comma is not strong enough to mark the end of an appositive or a parenthetical expression that is preceded by a dash. Unless the expression ends the sentence, another dash or a semicolon is necessary for clearness.

Wrong: Hampton Roads, which actually embraces five cities—Norfolk, Portsmouth, Newport News, Hampton, and Suffolk, provides anchorage for giant vessels.

Right: Hampton Roads, which actually embraces five cities—Norfolk, Portsmouth, Newport News, Hampton, and Suffolk—provides anchorage for giant vessels.

Wrong: Paracelsus held that the three elements—earth, air, and water, were represented by salt, sulfur, and water, respectively.

Right: Paracelsus held that the three elements—earth, air, and water—were represented by salt, sulfur, and water, respectively.

The combination of a dash followed by a semicolon is incorrect.

Wrong: Observe (a) kinds of food purchased—e.g., fresh fruit, baked goods, meat—; (b) buying habits—e.g., use of list, businesslike attitude—; (c) shopping courtesies; (d) sales talk; (e) brands sold.

Right: Observe (a) kinds of food purchased—e.g., fresh fruit, baked goods, meat; (b) buying habits—e.g., use of list, businesslike attitude; (c) shopping courtesies; (d) sales talk; (e) brands sold.

Right: First, he presented his qualifications—legal, moral, and practical; then his objectives were put forward.

Parenthetical expressions. Use dashes to set off a parenthetical expression whenever commas are needed for minor divisions within the expression.

> The face—thin, harsh, cold, and forceful—was deeply lined.
> Not only was the fur trade of great importance in providing the settlers with much of their heavy clothing—a portion of the trade, it may be noted, that would not swell the export figures—but it was also one of the many minor ways that the settlers had of eking out a living.

Commas are not strong enough to set off a complete sentence interpolated within another; dashes or parentheses are required.

> *Wrong:* Ace, people who don't know him well call him "Goody," is aided by a natural sense of humor.
> *Right:* Ace—people who don't know him well call him "Goody"—is aided by a natural sense of humor.

Put the dash in the right place.

> *Wrong:* A sight—a sound may be supernatural—that is from the romanticist's standpoint—but not a cough.
> *Right:* A sight, a sound, may be supernatural—that is, from the romanticist's standpoint—but not a cough.

> *Wrong:* I cannot remember them all now, but two do stand out clearly from the rest—Thenaud's, *Notes on the Detection of Hidden Spaces*—and Wilson's, *Studies of Ancient Architecture.*
> *Right:* I cannot remember them all now, but two do stand out clearly from the rest—Thenaud's *Notes on the Detection of Hidden Spaces* and Wilson's *Studies of Ancient Architecture.*

> *Wrong:* Robert Haven Schauffler, in a charming little book—*The Joyful Heart* maintains that a joyful heart comes only to those whose life—whether in business or in study—is so organized that a surplus of energy remains at the end of the day's work, energy to be expended in recreation and intercourse with others.
> *Right:* Robert Haven Schauffler, in a charming little book, *The Joyful Heart,* maintains that a joyful heart comes only to those whose life, whether in business or in study, is so organized that a surplus of energy remains at the end of the day's work—energy to be expended in recreation and intercourse with others.

Divided quotations. When a quoted sentence in direct discourse is interrupted by phrases that are not words of speaking or saying, use dashes before and after the interrupting phrase.

> "My dear little girl"—his tone was all concern—"I'm so sorry."
> "At nine o'clock I gave her ten grains of Trional in hot milk"—the doctor flickered an eyelid at me. "Umpf," said he—"and she took it without any trouble."

Distinguish sentences like the preceding from those in which the dash is part of the quotation and would be used if the quotation were not interrupted at that point. The dash should then be placed within the quotation marks.

> "Because—" reasoned Jo, "because nobody . . ."
> "Perhaps I can guess what's on your mind," he ventured, "—that in some way you are . . ."

Unfinished sentences. The sudden breaking off of a sentence is marked by an em dash. No period should be used after the dash. (If the sentence is interrupted by another speaker, a 2-em dash is preferred by some publishers, but a 1-em dash seems sufficient.)

> "But if we—" he began.
> "Then how can—?" he asked.
> "It'll never—" He stopped suddenly.

> "I will go to the King of Poland and I will tell him—"
> "The King of Poland has no need of such as you."

> "Now, if I can get two mo—"
> We heard no more. He had been shot.

> "If only I had—"
> "The money?"
> "—the support of the farmers."

Dash or colon? A dash or a colon may be used with equal correctness in several situations. In plays the names of the speakers are often followed by a colon, but in court reports, interviews, proceedings of public bodies, and the like a dash is usually preferred to a colon after the names of the speakers.

Similarly:

> Q.—
> A.—

Neither a dash nor a colon should be used in lists such as the following (for the use of colons with displayed lists, see pp. 181–182):

> In it the bookkeeper records (a) the account to be debited, with the amount; (b) the account to be credited, with the amount; (c) a complete explanation of the transaction.
> In it the bookkeeper records
> (a) the account to be debited, with the amount
> (b) the account to be credited, with the amount
> (c) a complete explanation of the transaction

In lists, between a word and its definition a comma, a colon, or a dash may be used: a comma when the definition is a simple appositive, a colon for a more involved construction. A dash is often used in a glossary.

Drop folio—a page number placed at the foot of a page instead of in a running head at the top of the page.

Mechanical uses. Besides the foregoing uses as a punctuation mark with more or less definite significance, the dash is a convenient mark of separation, as before a credit. Here the dash may be omitted if the credit is in a type style that differentiates it clearly from the quotation.

Science is, in its source, eternal; in its scope, unmeasurable; in its problem, endless; in its goal, unattainable.—VON BAER.

The en dash and em dash are used in presenting golf scores and the like. (Tennis scores are usually set with en dashes: 5–4, 6–3.)

72–69—141

In a vague date or in an open-ended date, as in biographical data, an em dash is used in place of numerals.

in 18—   1905—

EN DASH

To represent *to* between figures or words an en dash is used. (Some publishers accept a hyphen but those who use the en dash feel that a hyphen is a poor substitute and should be reserved for its own distinct uses, as a connector in compound words and as a separator in showing syllabication.) An em dash is sometimes used with this meaning when an en dash might be confusing, as in Scripture references extending over more than one chapter.

the years 1970–73   I John 1:3–5
pages 5–15   I John 1:1—2:5
the New York–Chicago bus
Berlin–Bagdad Railway

The word *to*, not an en dash, must be used if the numbers are preceded by the word *from*.

Wrong: Completion of the transcontinental railroad from 1869–1885.
Right: Completion of the transcontinental railroad from 1869 to 1885.
Or: Completion of the transcontinental railroad, 1869–1885.

Wrong: Chief among these were the two governors, George Clinton (from 1777–95) and DeWitt Clinton (from 1817–22 and 1824–28).
Right: Chief among these were the two governors, George Clinton (1777–95) and DeWitt Clinton (1817–22 and 1824–28).

An en dash cannot be substituted for *and*.

Wrong: Between 1923–29 Mussolini had transformed this.
Right: Between 1923 and 1929 Mussolini had transformed this.

Use an en dash instead of a hyphen in a compound when one of the components contains a hyphen.

English–Scotch-Irish parentage
Cambrai–St.-Quentin direction

In a hyphenated word set in caps use an en dash instead of a hyphen.

ANGLO–SAXON GRAMMAR

TWO-EM DASH

The 2-em dash is used to indicate the omission of part of a name or other word. In this use the dash is set close to the letter preceding it, but takes after it the regular spacing of the rest of the line.

I saw Mr. D—— and Miss M——.

PARENTHESES

Parentheses[36] are used to enclose expressions that are of such a nature that they would not be sufficiently set off by commas or dashes. Many times dashes or parentheses are equally good style, but parentheses should preferably be reserved to enclose matter having no essential connection with the rest of the sentence in which it occurs. (See also p. 208.)

Dry out each substance in an oven under slow heat (do not char).
In 1732 Joist Hite (he spelled his name in a variety of ways) entered the valley with sixteen families.
Most took a "newspaper" (thirty days), a few a "magazine" (sixty days), and one gangster had pleaded guilty to carrying concealed weapons and had taken a rap for a "book" (a year and a day), rather than be held for investigation on a major charge.

If matter enclosed in parentheses is a complete declarative or imperative sentence, the period is omitted; but any other mark of punctuation called for by the construction may be used within a parenthesis.

[36]The terms *curves* and *round brackets* are never used in composing rooms for parentheses. The term there used is *parens,* separately designated *open paren* and *close paren.*

Being inferior in strength (for who could be equal to the strength of Atlas?), he cried: . . .

Investigations in McCollum's laboratory (Orent and McCollum, 1931, 1932) and at the University of Wisconsin (Kemmerer, Elvehjem, and Hart, 1931; Skinner, Van Donk, and Steenbock, 1932; Van Donk, Steenbock, and Hart, 1933), while differing somewhat in detail . . .

When the letters *a*, *b*, *c*, and so on, marking divisions of an enumeration, are to be enclosed in parentheses, use both open and close parentheses — (*a*), (*b*), (*c*). (Note that when letters or figures are enclosed in parentheses, no period or comma is needed.)

Declarer may (a) accept the lead, (b) forbid the lead of that suit, or (c) treat the card as a penalty card.

Type. Italic or boldface parentheses are rarely used in a roman text; when an italic letter or word in roman text is to be enclosed in parentheses, roman parentheses should be used, not italic, and the punctuation following should be roman.

In this instance paper slips (*flyers*) are used.
Work (foot-pounds) = force (pounds) × distance (feet)

BRACKETS

Brackets[37] are used to enclose comments, explanations, queries, corrections of error, or directions inserted in a quotation by some person other than the original writer. The matter enclosed may be wholly independent of the text, or it may be words supplied to secure complete and understandable sentences, as in the last of the group of examples that follow.

"Let them [all the sons who have abandoned the paternal house] return."

". . . to write with the Pen then [*sic*] to work with my Needle."

"Nay, the most honest among them would hardly take so much pains in a week as now [after Dale's changes] for themselves they will do in a day."

"On October 11 [12] Columbus discovered the New World."

"Of the lighter kind, we have no poem anterior to the time of Homer, though many such in all probability there were [possibly on the lower planes of literature]. . . . [The first of these forms, developed later into tragedy,] originated from the dithyrambic hymn, the other [comedy] from those phallic songs. . . ."

The stage directions in plays may be enclosed in either parentheses or brackets (see pp. 277–278).

[37]The word *bracket* signifies only one thing to a compositor. Asking him to use a "square" bracket is unnecessary.

Sal: [*to the Commons at the door*] Sirs, stand apart; the king shall know your mind. [*He comes forward.*]

In reprints of early manuscripts, brackets are used to enclose passages whose authenticity is doubtful.

Parenthetical matter. Use brackets to enclose parenthetical matter within matter already included in parentheses. Brackets should be within the parentheses, not outside, except in algebra. Sometimes it may be just as easy to use em dashes along with parentheses, so that brackets are unnecessary.

> This Po River country was called Cisalpine Gaul (Gaul on this side [i.e., Italian side] of the Alps) because there was another Gaul on the other side of the Alps (Transalpine [trans·al'pin] Gaul).
> Bleeding is a real danger (half of the infants had to be evacuated [Smith, 1970]), and 30 percent developed wound infection.
> *Or:* Bleeding is a real danger—half the infants had to be evacuated (Smith, 1970)—and 30 percent . . .

Use with other marks. No punctuation is used with brackets except what is required by the matter bracketed. If bracketed matter is inserted in a quotation, it is unnecessary to use quotation marks before and after.

> "I could perceive [writes Warwick] he was very apprehensive."

QUOTATION MARKS

The main function of quotation marks, as their name indicates, is to mark words as spoken in direct discourse or as written at some time previous to the text of which they now form a part. In addition, quotation marks are used, more or less interchangeably with italics, to differentiate words or phrases from surrounding text in order to make their meaning or use clearer.

Direct quotations. Direct quotations, which may be of any length, from a word to an extract running through several paragraphs, should be enclosed in quotation marks.

> According to one recent writer, the "working principle" of punctuation is emphasis.
> When Tansey said "ironclad," I knew he meant ironclad, and not rusty iron, either.
> "That's it," he murmured. "'Pro and con' were the champ's last words."
> Some of the phosphorus now in our bodies may once have "formed part of huge reptiles, living millions upon millions of years ago."
> "United States casualties, one officer killed, ten enlisted men wounded." Contained in the "one officer killed" statement was the grimmest part of it all to us.
> The question has often been asked, "What does aviation offer to women?" or "Is there an opportunity for women in this field?"

A short quotation in a foreign language, set in italics, requires quotation marks.

> Cato ended all his speeches to the Senate by saying, "*Carthago delenda est.*"

A common error is misplacing of the quotation marks when the quotation is introduced by *that*.

> *Wrong:* This is done as a courtesy to the applicant, who may be discouraged and disheartened by a direct statement "that the office cannot even consider him as an applicant."
> *Right:* . . . a direct statement that "the office cannot even consider him as an applicant."

> *Wrong:* He received such a shock "that" in his own words, "He would not take another for the kingdom of France."
> *Right:* He received such a shock that, in his own words, he "would not take another for the kingdom of France."

Equally frequent is unnecessary use of *that*.

> *Wrong:* There is doubtless much truth in the dictum of Plato that, "Ruin comes when the trader, whose head is lifted up by wealth, becomes ruler."
> *Better:* There is doubtless much truth in the dictum of Plato, "Ruin comes when the trader . . . becomes ruler."
> *Or:* There is doubtless much truth in the dictum of Plato that "ruin comes when the trader . . . becomes ruler."

Excerpts. All direct quotations from the work of another should be enclosed in quotation marks unless they are set as extract, in type smaller than the text type, or set solid within a leaded text. If the quotation includes more than one consecutive paragraph, opening quotes should be used at the beginning of each paragraph and closing quotes at the end of the last paragraph. (See also pp. 18–19.)

A quotation within a quotation should be enclosed in single quotation marks. If this, in turn, contains a quotation, this last should be double quoted. One within this should be single quoted.

> The student who answered the question, "Who said, 'See that thou fallest not by the way'?" with "Elisha to Elijah, when the latter started up to heaven in the chariot," was evidently airminded.
> "This he said many times. Then he walked away and stood and talked to himself, and I heard him say: '*He* said, "Unless you repent, you shall die on a dark night, in a lonely spot, with no one nigh."' And he kept repeating, 'On a dark night, in a lonely spot, with no one nigh.' And then he would look around him at the trees and the mountains and the solitary shores. . . ."

Direct discourse. In discourse the exact words of the speaker should be enclosed in quotation marks.

> "Yes, wasn't it strange?" Jo said softly.

Do not punctuate sentences of separate speeches as if they were a single speech.

> As the two gentlemen approached they could hear shouts of "Open the bank!" "Let us in!" "When does the bank open?" and a medley of a similar tenor. "Nothing doing!" and "We wouldn't take your own word against yourself!" the officers reassured her.

The use of dashes between the words of different speakers does not always make it quite clear that several persons are speaking. (See also p. 209.)

> *Obscure:* "Oh, la, la, la! — Something of damage? — An accident, eh? — Has any-one been hurt? — Oh, la, la, la!"
> *Clearer:* "Oh, la, la, la!" "Something of damage?" "An accident, eh?" "Has anyone been hurt?" "Oh, la, la, la!"

No quotation marks are used with direct discourse in the older editions of the Bible.

> Thine ear shall hear a word behind thee saying, This is the way, walk ye in it — when ye turn to the right, and when ye turn to the left.

No quotation marks are necessary in interviews and dialogues when the name of the speaker is given first, or in reports of testimony when the words *Question* and *Answer* (or Q. and A.) are used.

> Mayor James M. Gordon — I believe the ordinance should be revised.
> William Jones — I am not in favor of revision at this time.
> Q. — Did you see the defendant in the room?
> A. — I did.

A quoted speech is often interrupted or followed by a phrase like *he said* or *replied Janet*. This phrase should not be included in the quotation and should be separated from it by some kind of punctuation.

> *Wrong:* "I choose this man" he says "to be my friend."
> *Right:* "I choose this man," he says, "to be my friend."
> *Or, occasionally:* "I choose this man [he says] to be my friend."

Direct thoughts. Direct thoughts may be italicized (see p. 142) or enclosed in quotation marks. Some publishers suggest quotation marks if the thought is accompanied by words like *he said to himself* or *muttered to himself,* rather than *he thought.* Others feel that neither italics nor quotes are necessary for direct thoughts. Whichever style is selected, it should be consistent throughout a single manuscript.

> "I had a fortune almost in my fingers," he said to himself.
> "So are we," he muttered to himself, "but we're still in the race."

What I won't do when I get in there! he must have thought to himself.

Direct and indirect discourse are sometimes confused as follows:

Wrong: "Carlyon would not harm a woman," Andrews thought. "It is only a trap to catch me." But then, was it likely that they would plan such a trap for me, a coward? They could not expect to do anything but repel him by danger.

In the sentence after the quotation marks, *was* indicates indirect discourse and *me* could be only direct discourse. The sentence should read thus:

But then, was it likely that they would plan such a trap for him, a coward?
Or: "But then, is it likely that they would plan such a trap for me, a coward?"

If an unexpressed thought and a spoken thought occur together, both enclosed in quotation marks, take care to use the marks about each in such a way as to avoid any confusion.

"Some quarryman out of work" was his unspoken thought. "What does he want with me?" "Well," he said sharply.

Indirect discourse. Do not enclose indirect discourse in quotation marks.

Wrong: She signed herself, "gratefully his, Edith Kilgallen."
Right: She signed herself gratefully his, Edith Kilgallen.
Right: She signed herself "Gratefully yours, Edith Kilgallen."

Wrong: He told them to "Charge it to the county."
Right: He told them to charge it to the county.
Right: He told them, "Charge it to the county."

Wrong: "How had that happened?" I finally wanted to know.
Right: How had that happened? I finally wanted to know.
Right: "How did that happen?" I finally wanted to know.

Wrong: "Had he bungled? Was he hit?" were his first thoughts.
Right: Had he bungled? Was he hit? were his first thoughts.
Right: "Have I bungled? Am I hit?" were his first thoughts.

Some uncertainty seems to exist about how to construe *yes* and *no* in such sentences as the following, whether as direct or indirect quotations. *Yes* and *no* should be quoted only as part of direct discourse. These words are properly lowercase in all the following examples:

A single answer is given and that is no.
He could not say no to him.
Frances had said yes to Stanley's suggestion.
Tell him yes. [Just as one would say: Tell him I will.]
Answer yes or no.
He said yes, he would.

In the following, *yes* is specified as direct discourse:

He answered sharply, "Yes!"
On his way out he said, "Answer 'Yes' for me."

Direct questions. Do not enclose direct questions in quotation marks unless the words have been borrowed. (See also pp. 213–215.)

The question arises, What is this book about?
The question will arise, What does *reconstructed* mean?

After "entitled," "signed." Use quotation marks to enclose words or phrases following *entitled, marked, endorsed, signed,* except, of course, where the words would regularly be set in italics, as in the fourth example below.

They were told to mark the case "Handle with care."
The note was signed "Chas. Arnold, Assessor."
The bill was entitled "An act to provide county library systems."
A novel entitled *For Whom the Bell Tolls.*

After "called," "known as." The meaning and use of words or phrases following *termed, called, so-called, known as* are usually clear without quotation marks.

The black soil we call humus.
Intensive sales effort, the efficiency manager calls it.
A secretary, or clerk, as he is called, acts as chairman.
This kind of policy, termed major medical, covers all such instances.
It is now known as an acceptance.
The so-called Alabama Claims
There grew up in France what is called the Napoleonic legend.
It has also been given other names, as the North Star State, the Land of Ten
   Thousand Lakes, the Bread and Butter State, and the Playground of the Nation.

Quotes are frequently used, however, as a substitute for *so-called.*

These data provide illustrations of the "principle of limits."
Or: These data provide illustrations of the so-called principle of limits.

What is the difference between overhead cost and "fixed charges"?
It is believed that insects perceive moving objects quickly; they also appear to
   "remember" landmarks. Bees see colors familiar to us and probably others
   beyond the violet end of the "human" spectrum.

Differentiation. Quotation marks (or italics) are frequently required to make clear the meaning or use of words or phrases. For instance:

Wrong: The Duke of Windsor had not been used as a title since the Black Prince
   used it in the fourteenth century.

Right: "Duke of Windsor" had not been used as a title since the Black Prince used it in the fourteenth century.

Wrong: Northern Ireland is the name of a distinct political division.
Right: "Northern Ireland" is the name of a distinct political division.
Better: Northern Ireland is a distinct political division.

Technical words. An unusual or technical word presumably unfamiliar to the reader may be enclosed in quotation marks. In a book of a serious nature such an unusual word or expression is often quoted — or italicized — only the first time it is used.

A high leak resistance may cause the tube to "block."
There comes a time, if the wind is not too variable, when the kite "stands," neither rising nor falling.

A word or phrase, however, which has long been familiar in the meaning required by the context should not be quoted.

The coffee and raised biscuits melted away.
I was having the creeps like an old woman.
The city's Four Hundred.
He was in the Little Lord Fauntleroy tradition.
One of the high spots of the meeting was the debate.
He has never felt that Latin is a dead language.

Technical terms should not be quoted in a text addressed to persons to whom the terms are familiar.

Quotes unnecessary: When "setting up" a "take" of *Republican* copy marked for an initial letter, allow the following indention on the first line, and an extra "en" space on the next three lines. This applies to "7-point" type only.
Better: When setting up a take of *Republican* copy marked for an initial letter, allow the following indention on the first line, and an extra en space on the next three lines. This applies to 7-point type only.

Explanations. Enclose in quotation marks a word or phrase to which attention is called for the purpose of definition or explanation. If the word or phrase has not been previously used, italics are usually preferred. (See also p. 137.)

By "federal" is meant a government with a strong central power.
What do we mean by "air conditioning"?

If the article *the* forms a part of the phrase referred to, take care to include it within the quotes.

What is meant by "the Settlement of 1815"?

The four phrases "the Cultural Heritage," "the Basic Dividend," "the Unearned Increment," and "the Just Price" were the abracadabra of Mr. Aberhart's necromancy.

Or: The four phrases "Cultural Heritage," "Basic Dividend," "Unearned Increment," and "Just Price" . . .

A profusion of quotation marks mars the appearance of a printed page. If following the rule above results in such a profusion of marks, use italics.

Most languages have only six words for comparison: *small, smaller, smallest;* and *great, greater, greatest.* Therefore verbal quantitative specifications like *great and small, more or less, increase and decrease, rise and decline, growth and decay,* are limited in meaning . . .

Wrong: Use such descriptive words as "considerate, cooperative, well-poised, cheerful, assertive, diffident."

Spotty: Use such descriptive words as "considerate," "cooperative," "well-poised," "cheerful," "assertive," "diffident."

Better: Use such descriptive words as *considerate, cooperative, well-poised, cheerful, assertive, diffident.*

Right: Such terms as *antecedent, auxiliary, correlative, coordinate, active, passive,* were so much gibberish to the average boy.

Commas are not an acceptable substitute for italics or quotation marks following the word *term.*

Wrong: The term, materials for research, is here understood to refer to materials of three types.

Right: The term "materials for research" is here understood to refer to materials of three types.

Right: The term *materials for research* . . .

Definitions. Words defining another word or words are quoted or italicized; sometimes neither quotation marks nor italics may be considered necessary.

I use the word *static* in the sense of "unchanging."

It was an Indian name signifying *he who splits the sky.*

Our word "priest" comes from the Latin word *presbyter,* meaning elder.

Chrestus, the Greek word for Messiah

Corporation (from Latin *corporare,* to form into a body)

If quotes are used and synonyms are given defining a word, quote each separately.

Lace at first retained the meaning of its Latin original, "noose," "snare," or "net."

The Latin *camera,* "vault," "arch," later "chamber," became *chambre* in French.

Translations. The English translation of a foreign word or phrase is usually quoted.

Although the etymology of the word *interlude* is clear (*ludus*, "a play," and *inter*, "between" or "among"), the application of the term is uncertain.
Zeitgeist, "spirit of the times."
He sent home the shortest of all famous dispatches: *Veni, vidi, vici*, "I came, I saw, I conquered."
Julius Caesar began his *Commentaries* with the well-known sentence, *Gallia est omnis divisa in partes tres* ("Gaul as a whole is divided into three parts").

Articles, essays. Titles of books are usually set in italics, but titles of parts of books, articles in magazines, essays, poems, sermons, unpublished manuscripts, and similar titles are roman quoted. Titles of series may be quoted, but usually cap-and-lower is sufficiently distinctive. In referring specifically to the Preface, the Glossary, and so on, it is not necessary to quote; capitalization is sufficient.

Part I, "The Manufacture of a Book"   English Men of Letters Series

An essay may sometimes be referred to by its subject matter rather than by its title; quotes and caps are not then used.

essay on how to read
essay on the poet

Art objects. Names of paintings, sculptures, and other works of art are often quoted; however, some house styles prefer the use of italics.

Claude Monet's painting "Antibes"
two Gothic sculptures, "Bull Fight" by Goya and . . .
What is the background of Da Vinci's "The Last Supper" and of Michelangelo's "The Last Judgment"?
In what year was Pollock's *Mural on Indian Red Ground* completed?

Musical compositions. Names of songs and of short musical selections are usually quoted. (See also p. 140.)

"Begin the Beguine"   "I Am Woman"
"Stars and Stripes Forever"   "Maxwell's Silver Hammer"
"Some Enchanted Evening"   "The 59th Street Bridge Song"

Buildings, thoroughfares. Names of public buildings, hotels, parks, streets, and the like do not need quotation marks.

Berlin's famous avenue Unter den Linden . . . the Brandenburg Gate at the entrance of the Tiergarten, a large park.

Nicknames and sobriquets. Whether a nickname should be enclosed in quotation marks depends largely upon how and in what context it is used. It would, for instance, be quite unnecessary to use quotation marks in naming Babe Ruth, Teddy Roosevelt, or FDR. In almost any context in which the use of nicknames would be appropriate, a nickname would be recognized as such without enclosing it in quotes, as Kit Carson, Dizzy Dean, Bing Crosby. If the nickname is given together with the real name, as a further identification of the person under discussion, quoting the nickname would be appropriate.

Casey Stengel	Charles Dillon "Casey" Stengel
Yogi Berra	Lawrence "Yogi" Berra
Buffalo Bill Cody	William Frederick "Buffalo Bill" Cody

Clearness is of course the criterion. *Nicknamed* is like *called* or *christened,* and its object is sufficiently set off by caps; but if the word *nicknamed* does not appear, quotes may be required as a substitute, just as noted about *so-called,* page 217.

He had been nicknamed Thomas the Sudden.
Thomas "the Sudden" was there too.

Slang. Slang should not usually be quoted. Using quotation marks around a slang word or phrase as a sort of apology for using the expression only distracts the reader.

He had to lay out more bread for the earphones.
That cat totaled his car.
John used pull to get his traffic ticket fixed.

Horses, dogs; rifle matches, races, stakes, cups; estates, cottages. It is unnecessary to quote names of these, but they should be capitalized.

Sea Biscuit	the Derby	Davis Cup
Man o' War	Preakness	The Oaks

Unnecessary quoting. Do not use quotation marks without a reason. In the following sentences, for instance, none are needed:

A font contains . . . most c's and a's; fewest z's and q's.
The ballots were marked in a special way: those with one hole meant nay, and those with two holes yea.
The letter N must precede the license symbol.
Shop now for the holiday over the Fourth.
Old Faithful geyser
As question 4 should have suggested . . .

Single quotation marks. Single quotes are properly used to enclose quoted matter within a quotation (see p. 214). Other uses are more or less arbitrary and unusual. They may be used to save space when quotations are numerous, or to differentiate certain terms from others similar.

29. **dissilire:** 'is split,' as we speak of "a splitting headache."

A once-common practice, now usually confined to etymological texts, was the use of double quotes only for extracts and dialogue and single quotes in all other cases where quotes were called for.

Use with other marks. Set quotation marks outside of periods and commas.

Quotation marks in conjunction with colons and semicolons should be set inside, because the colons and semicolons are sentence punctuation, not part of the quotation.

Set quotation marks outside of exclamation points and interrogation points that are part of the quotation, inside of points that are not (see pp. 178–179).

When quotation marks are used in conjunction with points of ellipsis or with *etc.*, care should be taken to see that the quotation marks are placed so that they indicate clearly whether the omitted matter is part of the quotation or not.

The audience quieted as he began to speak: "Ladies and gentlemen, . . ."
The topics were "How to Fix Up Old Cars," "Where to Buy Pre-1940 Models,"
    etc.

When a possessive must be formed for a quoted word or phrase, the apostrophe and *s* are placed outside the closing quotes; but try to avoid this construction by rewording.

"Stardust"'s first line is . . .
Better: The first line of "Stardust" is . . .

APOSTROPHE

The apostrophe is used to indicate the possessive case of nouns and the plural of letters, figures, signs (see pp. 478–481). It is also used to denote the omission of a letter or letters. In this last usage mistakes often occur, sometimes in misplacing the apostrophe or omitting it, sometimes in misspelling the word in which it belongs: as, *its* for *it's, your* for *you're, where're* for *where'er.*

An apostrophe may also mark the omission of figures, as in *the class of '92.*

The apostrophe is omitted in such shortened forms as *bus, cello, possum, phone, cab, squire,* and similar commonly used words. Never use an apostrophe before *round* and *till,* which are full words, not shortened forms.

Do not use the apostrophe in place of a single turned comma in Scotch names like M'Gregor.

HYPHEN

Division. The hyphen is used at the end of a line to show that part of the word or number has been carried over to the following line. (See the rules relating to division of words, pp. 237–239.)

In manuscript a hyphen occurring at the end of a line should be marked by the author or copy editor if it is to be retained. For instance, if the word *never-ending* were divided in the manuscript, *never-* on one line and *ending* on the next, the hyphen should be marked as a double hyphen to prevent the printer from setting *neverending*.

Syllabication. The hyphen is used to indicate syllabication. It is not needed with an accent mark. The space dot—a period cast in the center of the type body—is often used instead of the hyphen for this purpose.

lab'o-ra-to-ry   spec-tac'u-lar

Etymology. Used at the end of an etymological part of a word, the hyphen indicates a prefix; before an element it indicates a suffix.

The hyphen is omitted when *in-* and *un-* are prefixed to a word.
Names of metals and metallic radicals usually end in *-ium.*

Connection. The hyphen is used to join words to form compound words. For rules regarding this use, see the section on compounding of words, pages 226ff.

Prefixes. The modern tendency is to eliminate the hyphen between a prefix and a root unless the root is a proper noun or adjective, such as *un-American* (see pp. 168–169). Whenever there is any question as to the proper formation, a dictionary should be consulted, as usage differs. With common roots the following prefixes form solid words except on the occasions to be noted for homographs or when this would result in doubling a vowel. (*Co-, de-, pre-, pro-,* and *re-* may be set solid even when this forms a double vowel.)

anti-	infra-	non-	semi-
co-[38]	intra-	pre-	sub-
de-	macro-	pseudo-	supra-
hyper-	micro-	re-	un-
hypo-			

antiwar	intra-abdominal	pseudosophisticated
coauthor	macromolecule	reenter
defoliate	microorganism	semielliptical
hyperresonant	nonnative	subheading
hypoactive	nonviolent	supra-auditory
infrared	preeminent	unnamed

[38]An exception is the shortened form *co-op* (for *cooperative*), which is usually written with a hyphen.

In a middle ground between derivatives (formed with prefixes) and compound words are the following combining forms that are commonly used in other senses as separate words:

| extra- | over- | pro- | ultra- |
| out- | post- | super- | under- |

extra-articular	overripe	supermarket
extracurricular	postgraduate	superrefractory
outfight	postterm	ultra-atomic
outpoll	proamnion	ultrahigh (frequency)
overdecorated	proethnic	ultranationalist

Appropriate here, although more accurately defined as compound words, are the following, which are written with hyphens as noted:

all-: all-around, all-embracing (*but* allspice)
half-: half-asleep, half-blooded, half-dollar (*but* halfhearted, halfway)
quasi-: quasi-public, quasi-judicial [the preceding are adjectives and are always written with the hyphen; nouns are open compounds: quasi corporation, quasi scholar]
self-: self-conscious,[39] self-control, self-seeking (*but* selfhood, selfless, selfsame)

Homographs. Usually there is little chance for confusion when homographs are used in context; all of the following may ordinarily be written solid, without a hyphen. An obvious example is *overall*: it is difficult to imagine a context in which the adjective *overall* (meaning "comprehensive, general") could be confused with the noun *overall* (chiefly British usage, for a "loose-fitting protective garment"; American usage is the plural *overalls*). Possibly less distinct are *overage* ("surplus, excess") and *overage* ("too old"); *unionized* ("formed into a labor union") and *unionized* ("not separated into ions"); *redress* ("to requite for a loss") and *redress* ("to dress again"). Where confusion might arise, the forms that incorporate a prefix (the second meanings in the examples above) may be hyphenated.

The following words formed with the prefix *re-* may frequently require a hyphen for distinction from a solid homograph:

re-act	re-create	re-pose	re-sign
re-ally	re-form	re-present	re-solve
re-collect	re-lease	re-search	re-sound
re-cover	re-mark	re-serve	re-store

Only occasionally would hyphenation be required for *repress, reprobate, restrain, retreat.*

[39]*Webster's Third New International* gives *unselfconscious* as the form with the added prefix *un-*; but *unself-conscious* is also acceptable.

POINTS OF ELLIPSIS

Function. Three points, spoken of as points of ellipsis,[40] are properly used to indicate an omission, a lapse of time, or a pause too long to be indicated by a 1-em dash; if the omission occurs at the end of a sentence a fourth point is used—the period for that sentence. (For the use of points of ellipsis with excerpts, see pp. 20–21.)

> "All persons born in the United States . . . are citizens of the United States."
> "Hm-m! Yes, malaria . . . mosquito infection without a doubt . . . tricky thing, malaria."
> "I do not know that play. . . . And it is that play that matters."

In fiction, when points of ellipsis are used at the end of a sentence that trails off, the writer must decide whether the sentence is to be construed as incomplete (three points) or complete (period plus three points).

> "I didn't know you cared . . ."
> *Or:* "I didn't know you cared. . . ."

In telephone conversations where only one side is given, it is customary to indicate the pauses for the other side of the conversation by points of ellipsis, using quotation marks only at the beginning and the end of the conversation.

An omission of whole paragraphs or stanzas of poetry, a change of subject, or the lapse of time may be indicated by a line of points or a line of asterisks.[41] The points or asterisks are usually set about two ems apart, and three, five, or seven are used, depending on the width of the page. They should never extend beyond the longest type line.

Use with other marks. Although points of ellipsis are usually unnecessary at the beginning or the end of a quotation, if they are used they should be set inside the quotation marks. In the following example the author did not cap the first word of the quotation, as it did not originally begin a sentence, so that points of ellipsis should be used:

> In the early part of the eighteenth century Hawkins in his *Pleas of the Crown* stated: ". . . there can be no doubt, but that all confederacies whatsoever, wrongfully to prejudice a third person, are highly criminal at common law."

Punctuation should be used with points of ellipsis when it is felt the meaning will be clearer.

> "These associations must be accepted cautiously; . . . remissions have often occurred."

---

[40]Ellipsis points are usually spaced with en quads but in a closely spaced line the spacing of ellipsis points should be reduced.
[41]The use of three asterisks for an ellipsis within a sentence or paragraph is now rare, though formerly common practice.

COMPOUND WORDS

The term *compound word*[42] suggests to many persons a word made up of two simple words connected by a hyphen *(hyphenated compounds)*. The term is not, however, so restricted and includes also two words written separately *(open compounds)* or as one word *(solid compounds)* that together express a single idea.

Principles of compounding words. Many compound nouns have the same meaning whether written as one word or two, and good usage sometimes sanctions more than one form. Unfortunately the dictionary makers have not all followed the same general rules in deciding the question; and although most noun forms may be located in some dictionary or other, many adjectival and inflected forms of compounds are unlisted anywhere.

The preferred form depends primarily upon the stage of evolution the compound is in: it begins as an open compound, increasing usage sees it hyphenated, and eventually, when we no longer analyze it into its elements, it becomes solid, as *blackberry, postman, newspaper.* Many copy editors become unnecessarily concerned about the inconsistencies in the forms of compound words. Yet it is doubtful if anyone outside composing rooms and publishing houses is conscious of the form of compound words unless he comes upon some unusual form that is momentarily surprising because it is unfamiliar. A tendency among some writers and some dictionaries is to combine into a single word compounds that are clear and not ambiguous as two words, with the result that the reader must accustom himself to new forms. Compounding should be natural and should not anticipate usage too much. *Copyholder* and *proofreader* have been used so long in these forms that they are immediately clear to everyone; in logic, *copyeditor* is as valid, and some writers present it as one word, but many still prefer *copy editor* for the noun form, and *copy-edit* for the verb.

The present trend is to eliminate hyphens and write a compound solid as soon as popular usage indicates it will become a permanent compound. Far more important than the form of compound words is the use of the hyphen to make the meaning or syntactical connection of words clear.

Temporary compounds. Two words connected by a hyphen to show their syntactical relation to another word may be termed a temporary compound, in contradistinction to the permanent compounds whose form is determined by usage and is usually given in dictionaries. Compounds that change their form according to their position in the sentence present a problem. A rule of thumb is that compounds used adjectivally preceding a noun *(unit modifiers)* are often hyphenated to prevent misreading. For

---

[42]Not properly classified as compound words are derivatives from word-forming prefixes *(non-, un-, re-, semi-)*. These are discussed on pages 223–224.

example, is "an old book collector" an old person who collects books, or someone who collects old books? Depending upon the intention, this would be written "an old-book collector" or "an old book-collector."

On the following pages a few rules are given relating to permanent compounds, many more relating to temporary compounds.

Form dependent upon meaning. Some otherwise permanent compounds change form according to position and meaning. Sometimes *baldhead* may be a solid compound, but a distinction must be made between "He had a bald head" and "It was a baldhead." Similarly *hairbrush* may be solid, but "She used a camel's-hair brush" is certainly different from "She used a camel's hairbrush."

He had firsthand information.	He got the information at first hand.
Stay for a while.	Stay awhile.
He ran onto the field.	He went on to better things.
He was a fine ballplayer.	He was a fine baseball player.

Several other compounds that may require a change in form are the following:

anyone	Did you see anyone?
any one	Any one of these will be satisfactory.
everyone	Everyone will be there.
every one	Every one of the children will be there.
anyway	I'll come anyway.
any way	I'll go any way you like.
sometime	Sometime in May
some time	Some time ago
apiece	5 cents apiece
a piece	5 cents a piece
hothouse	They have a hothouse.
hot house	They have a hot house.
handwriting	The handwriting on the wall
hand writing	The hand writing on the wall
matter-of-fact	He is very matter-of-fact.
matter of fact	It is a matter of fact.
great-grandfather	He is a great-grandfather.
great grandfather	The boy thought him a great grandfather.

"Full" and "-ful." Distinguish phrases like *car full of people* and *carful of people.*

Wrong: The car full of people stood up. [This plainly says that the car stood up.]
Right: The carful of people stood up.

Wrong: Knead a cup full of rye flour into the dough.
Right: Knead a cupful of rye flour into the dough.

Wrong: Then came oars, sails, and steam; and finally the great ships that could take trainloads of freight and a city full of people from country to country.
Right: . . . and a cityful of people from country to country.

Names of kindred. The following are correct forms for the names of kindred.

godmother, goddaughter, godson
grandfather, grandson
stepdaughter, stepsister, stepparent
great-aunt, great-uncle, great-grandfather
half sister, half brother, half aunt
foster child, foster brother
second cousin

Two nouns of equal value. Use a hyphen between two nouns used together to indicate that the person or thing referred to partakes of the character of both nouns.

man-child	priest-ruler	boy-king
wolf-lion	soldier-statesman	bridge-tea
city-state	secretary-treasurer	dinner-dance

Compounds referring to nationality may be more perplexing. One would hardly hyphenate *American Baptist* or *Scotch whisky*, as the proper adjectives are obvious. However, *Spanish American* refers to an American of Spanish ancestry; *Spanish-American* refers to the two countries, Spain and the United States. Similarly *French Canadian* refers to a Canadian of French ancestry; *French-Canadian* to the two countries. The difficulty lies in determining the equality of the elements of the compound: *French-Canadian* is properly hyphenated if it denotes a child whose father is French and mother Canadian. (The language spoken by many French Canadians, incidentally, is properly *Canadian French.*)

Do not confuse the preceding with prefixes, which are always followed by a hyphen: *Anglo-American, Austro-Hungarian, Sino-Soviet.*

Noun and possessive. In most dictionaries the hyphen is used in those very numerous botanical names that are made up of a noun preceded by a possessive.

adder's-tongue	elephant's-ear	St. John's-wort
bird's-eye	Job's-tears	shepherd's-purse

Nouns of similar construction other than botanical names tend to become solid words without an apostrophe.

beeswax	ratsbane	townspeople	sheepshead (a *fish*)

Exceptions: bull's-eye, crow's-foot.

"Like." Adjectives ending in -*like* are usually one word except when the root word ends in two *l*'s or is a proper noun or adjective.

| childlike | businesslike | snaillike | ball-like |
| womanlike | fossillike | American-like | shell-like |

Do not combine adjectives with *-like* as in *globularlike;* use either *globular* or *globulelike.*

Add *-like* to an open compound (for example, *bone spicule*) as follows: *bone-spiculelike.*

"Self," "Half," "Quasi." See page 224.

Adjective and noun. When an adjective and a noun combined to form a compound noun have lost their separate meanings and equal stress and have become a combination with one accent and a single specialized meaning, they are usually written as a single word. Not infrequently the compound word written as a single word has a meaning different from that of the compound written with a space between the parts.

| aircraft | headway | seaweed | taxpayer |
| oilcloth | cornflower | lawgiver | bookkeeper |

Compounds of this nature are not inseparable in the sense that whenever they are used they must be written as a single word. For instance, *cornstarch* is often written as one word; *wheat starch,* on the contrary, is two words. In a discussion comparing corn and wheat starches the following would be correct form. (Note that *corn* and *starch* are equally stressed in reading this sentence.)

For corn starch the maximum is about 91°C, and for wheat starch it is about 95°C.

Dictionaries agree that *freshwater, saltwater,* and *seawater* are the preferred forms for the adjectives, but many recommend *fresh water* and *salt water* as open compounds in the noun usage and list *seawater* as a noun. Writers and copy editors can only be guided by their good sense in trying to be consistent.

Modified compounds. If a compound noun ordinarily written solid is preceded by an adjective which modifies the first part of the compound, separate the compound.

schoolboy	high-school boy	aircraft	lighter-than-air craft
ironworker	structural-iron worker	glassware	cut-glass ware
housekeeper	lodging-house keeper	grapevines	wild-grape vines
stockholder	common-stock holder	gristmill	grist and saw mills
taxpayer	income-tax payer	sawmill	

Wrong: The strong silent public schoolboys from Winchester and Eton.
Right: The strong, silent public-school boys from Winchester and Eton.

If a modifier is used before a compound noun written as two words, the noun must be joined to prevent misreading.

color filter   red color-filter   place names   foreign place-names
letter writers   public letter-writers   line design   fine line-design
[Note for the last: *fine-line design,* if a different meaning is intended.]

Similarly, a compound must be joined if a prefix or a suffix is added.

dessert spoon   dessert-spoonfuls   measuring cup   measuring-cupfuls

Adjectives. A compound adjective made up of an adjective and a noun in combination should usually be hyphenated.

cold-storage vaults   short-term loan
hot-air heating   small-size edition
toy-repair shop   different-size[43] prisms
ten-cent-store toys   present-day matches

The purpose of the hyphen in these modifiers and in those noted in other rules later is to prevent a misplacing of the stress or a momentary impression that the first word of the phrase modifies the third word instead of the second. For instance, *toy repair shop* might easily be read as if it were *toy repair-shop,* whereas *toy-repair shop* shows the correct reading immediately.

Following this rule strictly would result in the use of a great many hyphens. Therefore copy editors must recognize instances in which a hyphen might be more distracting than helpful, as in the following:

advertising agency employee   mechanical engineering school
appeals court ruling   sportscar racing strip
city hall news   stock exchange building
grand jury room   yellow fever outbreak

Compound chemical or scientific terms usually need no hyphen when they are used as modifiers.

a carbon dioxide extinguisher   hydrogen ion activity

A compound of adjective and noun need not be hyphenated when used as an adjective if both parts are capped.

a Class A member   Bronze Age tools
Safety First rules   the Open Door policy
Old World countries   Great Society programs

Verb and preposition or adverb. Compound verbs ending in a preposition or adverb are written as two words. The same two words used as nouns or adjectives should be written as a single word or hyphenated.

---

[43]Some prefer *different-sized.*

black out	blackout	lay out	layout
break down	breakdown	line up	lineup
break in	break-in	make out	makeout
break out	breakout	play off	playoff
break up	breakup	run off	runoff
build up	buildup	set up	setup
cave in	cave-in	stand by	standby
check up	checkup	stand in	stand-in
come down	comedown	take off	takeoff
come on	come-on	take out	takeout
fade out	fade-out	warm up	warm-up
hand out	handout	worn out	worn-out
lay off	layoff		

Adjectives similarly formed should be hyphenated when they precede the noun they modify.

Everything must be speeded up.	a speeding-up process
a log hollowed out by hand	a hollowed-out log

An adverb preceding such a compound adjective usually modifies the whole compound, not just the first part. Two hyphens should be used.

a solution much sought for	a much-sought-for solution
The dialogue was long drawn out.	a long-drawn-out dialogue

No hyphen sould be used between the adjective and the noun.

Wrong: He used the trade-in-value of his old car for the first payment on a new one.
Right: He used the trade-in value of his old car for the first payment.

Number and noun. Always hyphenate a compound in which one component is a cardinal number and the other a noun or adjective.

eleven-inch stick	a three-hundred-dollar clock
ten-pound bag	seven-pointed star

Exception: onesided affair

Strict observance of the rule is particularly important when the noun modified is in the plural, for without the hyphen the phrase might be misleading or at the least ambiguous. It is unsafe to depend upon the context to make the meaning unmistakable. Note the difference in meaning.

ten acre farms	ten-acre farms
two dollar tickets	two-dollar tickets

Usage is not strictly consistent in the number of the noun in the compound. Note, for instance: sixty-day note, six-months note, four-week period.

Use hyphens logically. In the following sentence neither hyphen is needed, as will readily appear if the numbers are spelled.

Wrong: A customer hands you a $5-bill for a 72¢-purchase.
Right: A customer hands you a $5 bill for a 72¢ purchase (a five-dollar bill for a seventy-two-cent purchase).

Similar compounds using ordinal numbers are hyphenated when they precede the word they modify.

second-story room	third-class coach
third-race entries	second-rate hotel

The suffix *-fold* is not hyphenated at the end of a spelled-out number; however, with numerals it is hyphenated.

tenfold	75-fold
twenty-fivefold	110-fold
hundredfold	

Observe, however, the following:

His office visits increased five to ten fold.

Compounds with *-score* are also one word if they are used as adjectives.

within twoscore years	Two score will speak.

Compounds of a number with *-odd* are hyphenated.

forty-odd	180-odd

Fractions. Always hyphenate fractions used as modifiers unless the numerator or the denominator contains a hyphen. Do not further hyphenate a fraction whose numerator or denominator contains a hyphen.

a one-third interest	four twenty-fifths part
one-half life size	twenty-nine fiftieths ownership
three-fourths inch	sixty-two hundredths point
two and one-eighth inches	
one-third lower	

The average is about one-half that of winter wheat.
It weighs one-half as much as cork.
Add three-fourths the amount of olives.

It is not correct to hyphenate fractions used as nouns; so used, the ordinal (or *half* or *quarter*) is a simple noun preceded by a number adjective.

His share was three fifths.
He owned a three-fifths share.
One half of the planet is perpetually frozen, while the other half has a temperature around 660° F.
Whatever it deposits during the first quarter of a cycle it removes in the next three quarters.
Its composition was three-fifths copper, two-fifths zinc.

Colors. A compound adjective denoting color whose first element ends in *-ish* should be hyphenated when it precedes a noun, but it need not be hyphenated when it follows the noun it modifies. When a noun is compounded with a color, or two colors are combined, they should always be hyphenated.

bluish-purple flowers	The leaves are reddish brown.
yellowish-green foliage	emerald-green
blue-green	snow-white
orange-yellow	coal-black
pink-lavender	iron-gray
sky-blue	

Compounds with present participle. Hyphenate compound adjectives made up of a noun, adjective, or adverb and a present participle when they precede the noun they modify. Many such compounds, including the following, have become permanent, taking the hyphen or being written solid regardless of their position:

breech-loading	backbreaking
earth-shaking	childbearing
far-reaching	copyholding
good-looking	easygoing
hair-raising	heartbreaking
nerve-racking	lifesaving
peace-loving	toolmaking

He was a horse-racing enthusiast.
He enjoyed horse racing.
The coal-mining industry must be ecologically responsible.
We must turn to increased coal mining.

If the compound is preceded by an adjective modifying the first word in the compound, omit the hyphen, or if necessary for clearness, use two hyphens.

beet-raising area	sugar beet raising area
rot-producing fungus	moldy-rot-producing fungus

Compounds with noun plus "-d" or "-ed." Ordinarily hyphenate compound adjectives of which one component is an adjective and the other a noun to which *-d* or *-ed* has been added.

able-bodied   old-fashioned
blue-eyed   dull-witted
middle-aged   dimple-cheeked
acute-angled   freckle-faced
ripple-edged   large-fruited

If the adjective so combined ends in *d,* omit this *d:* a child with *dimpled cheeks* is *dimple-cheeked,* not *dimpled-cheeked;* a boy with a freckled face is *freckle-faced,* not *freckled-faced;* a paper with a rippled edge is *ripple-edged,* not *rippled-edged.*

If the first part of the compound is qualified by a preceding adverb, omit the hyphen.

fine-grained sugar   extra fine grained sugar

The hyphenated form of modifier noted in the preceding rule is the correct form to use in describing, for instance, a man with a pleasant face. He is a *pleasant-faced man,* not a *pleasantly faced man.*

Wrong: She had permanently waved blue-white hair.
Right: She had permanent-waved blue-white hair.
Right: Her hair was permanent-waved.

Compounds with the past participle. Always hyphenate compound adjectives made up of a past participle combined with a noun or adverb when they precede the noun they modify. Many such have become established as hyphenated compounds regardless of their position in the sentence.

deep-seated   gold-filled   worm-eaten   poverty-stricken
tongue-tied   high-flown   hard-boiled   high-powered

Adverb-and-adjective compounds. Hyphenate a compound adjective of which one component is an adverb and the other an adjective or participle if it acts as a unit-modifier and the adverb could be misread as a modifier of the noun. When it follows the noun it may be set as two words unless it has been used so commonly that it has become established as a hyphenated compound.

best-informed man   man who is best informed
shows much-improved growth   growth is much improved
the most-prized furs   furs that are most prized
a long-desired outlet   an outlet long desired
the above-mentioned facts   the facts above mentioned
his so-called friends   his friends, so called

If the adverb ends in *-ly* or could not be misread as a simple adjective modifying the noun, the hyphen is not necessary.

equally effective cures	too ardent fisherman
newly found treasures	less rigid climates
poorly equipped factory	carefully thought out scheme
well organized and effectively managed life	

Since some adjectives end in *-ly*, adjectives and adverbs in this construction must be carefully distinguished.

a finely built, scholarly-looking man

Compounds with *well* and *ill* need not be hyphenated when they follow the noun modified, although the hyphenated form is often seen.

Well-bred cows well fed may turn out from four hundred to six hundred pounds.

If the compound adjective is preceded by an adverb modifying only the first word of the compound, omit the hyphen.

a well-organized program   a reasonably well organized program

Suspended compounds. When successive compound adjectives have one component the same in all, this component is sometimes omitted in all except the last. The best method is to retain the hyphen in each one.

...that cut across all such school- and book-imposed subject lines.

When during the Crimean War ships protected by iron or steel armor were first built in France, and almost immediately afterwards in England, a large number of adjectives, as the *Oxford Dictionary* tells us, were used to describe them: *iron-* or *steel-* or *armor-plated, -cased, -clothed, -sided,* and many others, and *iron-plated* was the official adjective until 1866.

He estimated that out of thirty tons hauled, twenty tons would be third- and fourth-class commodities and ten tons first- and second-class.

This is a much overused construction that writers should avoid when possible. Of the two forms of the following phrase, for example, the second is preferable and not cumbersome:

3-, 4-, and 6-inch mortars
3-inch, 4-inch, and 6-inch mortars

Undesirable: You can play an important part in developing such a program by studying in- and out-of-school facilities in Colorado.
Better: ... by studying in-school and out-of-school facilities.

Similarly, avoid expressions like the following.

given and surname   cattle and sheepmen

They should be written:

given name and surname   cattlemen and sheepmen

Before using hyphens to form a suspended compound, make sure the hyphens are needed. The following is a possessive construction (see p. 231):

> *Wrong:* Examine a series of cross sections of woody stems, of one-, two-, and three-years' growth.
> *Right:* Examine a series of cross sections of woody stems, of one, two, and three years' growth.

Phrases. There are many compounds made up of phrases and of words or syllables similar in form or sound that are quite familiar to everyone as hyphenated words, such as:

Nouns	*Adjectives*
aide-de-camp	happy-go-lucky
daughter-in-law	helter-skelter
hocus-pocus	higgledy-piggledy
hurdy-gurdy	namby-pamby
jack-o'-lantern	topsy-turvy
touch-me-not	wishy-washy

Do not hyphenate phrases used as nouns in regular grammatical construction.

> *Wrong:* Courses in how-to-read-and-study . . . Knowing how-to-study also connotes having right attitudes toward study.
> *Right:* Courses in how to read and study . . . Knowing how to study also connotes having right attitudes toward study.

Phrases used as attributive adjectives usually require hyphenation to make clear their relation to the noun they modify. (Many such have already been noted incidental to previous rules.)

the how-to-study area   a life-and-death struggle

If such a phrase modifier is hyphenated at all, it should be hyphenated throughout; but no hyphen should be used between the modifier and the noun.

> *Wrong:* low milk-and-cream yielding dam
> *Right:* low milk-and-cream-yielding dam
>
> *Wrong:* pay-as-you-go-plan
> *Right:* pay-as-you-go plan

If there is scarcely any possibility of misreading, hyphens need not be used.

half a dozen different shops	yellow pine timber belt
two and a half seconds	white crêpe de Chine dress
a story and a half house	a mahogany and leather chair
cream of corn soup	red leather pocketbook
chicken and tomato soup	two hundred and twenty horsepower
a thirty percent increase	radial
seven and one-half million dollars	

Writers should avoid unnecessary use of this construction. It is hardly fair to expect a copy editor to solve such a problem as is presented by the following:

three quarters of a billion dollar business

The thoughtful author would write:

a business worth three quarters of a billion dollars

Hyphenating a phrase used as an adjective makes it a single adjective. The writer of the following failed to see that he had two adjectives connected by *and,* each modifying the noun. This construction should be distinguished from the phrase requiring hyphens.

Wrong: Add more of the chilled-and-partly "set" broth.
Right: Add more of the chilled and partly set broth.

Long phrases like the following should be hyphenated.

never-to-be-forgotten friend
twenty-dollar-a-week clerk
sun-and-wind-browned cheeks
one of those God-help-me-why-am-I-here expressions
that chap's funny-at-all-costs type

A foreign phrase used as an adjective need not be hyphenated.

viva voce vote   a pro rata assessment

DIVISION OF WORDS

In order to secure that evenness of spacing which is essential to the good appearance of a printed page, it is often necessary to divide a word at the end of a line, carrying over one or more syllables to the beginning of the next line.

When a word is broken at the end of a line in manuscript, the copy editor should routinely mark whether the hyphen is to be kept or is to be deleted and the word closed up by the typesetter.

There should be little trouble with divisions in galley proof if the copy editor and the proofreader observe the following simple rules.

GENERAL RULES

A word may be divided only at the end of a syllable. If one is not sure of the syllabication, he should refer to the dictionary.

Words pronounced as one syllable should never be divided.

cracked   drowned   through   often

Words of two syllables of which one is a single vowel — or its equivalent — should not be divided.

around   aegis   even   over

A terminating syllable of only two letters should not be carried over unless the exigencies of very narrow measure make it necessary. Some printers assert that an initial syllable of two letters should not stand at the end of a line, but such a division is preferable to a line more widely spaced than adjoining lines.

A division before a single vowel that alone forms a syllable should be avoided except in the case of the suffixes -*able* and -*ible* and words in which the vowel is the first syllable of a root.

munici-pal	consider-able	inter-oceanic
privi-lege	remedi-able	un-equal
deroga-tory	dis-united	

There are many words ending in -*able* and -*ible*, however — in which the *a* or *i* does not by itself form a syllable — that may be divided after the *a* or *i*, as, for instance:

ame-na-ble	char-i-ta-ble	pos-si-ble
ca-pa-ble	gul-li-ble	ter-ri-ble

The terminations -*cial*, -*sial*, -*tial*, -*cion*, -*sion*, -*tion*, -*gion*, -*cious*, -*ceous*, -*tious*, and -*geous* should not be divided.

se-ba-ceous   con-sci-en-tious   coura-geous   ad-van-ta-geous

The letter *t* belongs with the following letters in words like

ad-ven-ture   for-tune   pre-sump-tu-ous   lit-era-ture

When the final consonant of a verb is doubled in forming the past tense or the participle, the second consonant belongs with the letters following it. Single or double consonants in the root word should not be carried over.

occur-ring	di-vid-ing	fore-stall-ing
for-get-ting	forc-ing	buzz-ing
re-gret-ted	trav-el-ing	dis-till-ing

When there is a distinction made in the pronunciation of a word to denote its part of speech, the word should be divided according to pronunciation.

proj'ect (*noun*)   pro-ject' (*verb*)   prod'uce (*noun*)   pro-duce' (*verb*)

Abbreviations such as M.A., Ph.D., FDA, YMCA, A.M., P.M., A.D., and B.C. should not be divided.

Do not put the initials of a name on different lines, and if possible avoid separating initials, titles (Mr., Rev.), and degrees (M.D.) from a name.

Avoid, if possible:

1. Dividing at the ends of more than two successive lines
2. Dividing the last word of a paragraph; if it is divided, carry over at least four letters
3. Dividing the last word on a right-hand page
4. Dividing a hyphenated compound word except at the hyphen (that is, avoid two hyphens)
5. Dividing numbers expressed in numerals; for numbers of five or more numerals, if they must be divided, divide only at the comma and retain the comma (4,656,-)
6. Dividing combinations like 98.6°F, 14 B.C., 6:30 P.M.
7. Separating (a) or (1) from the matter to which it pertains

Rules for the division of words of foreign languages will be found in the sections dealing with the respective languages (see pp. 288ff.). They should be studied for the guidance they give in the division of words and names of foreign origin, as, for instance:

sei-gneur	zem-stvo	Pa-gli-acci	To-ma-szow
lor-gnettes	se-ra-glio	Bu-czacz	Lasch-tschenko

PART IV.
TYPOGRAPHICAL STYLE

ELEMENTS OF TYPOGRAPHICAL STYLE

It has been noted in previous sections that the first task in starting a book manuscript on its way to manufacture is to decide questions of format, such matters as size, bulk, page dimensions, typefaces, illustrations, and binding. Next to be considered are elements of typographical style—leading, spacing, and indention—that the editor must relate to the problems of typographical style presented by each book. All of these aspects of typographical style are important in determining the appearance of the finished work. Even though some of the material given here may fall in the province of the production department or the designer, the more knowledge the editor has of the factors involved, the more guidance he can give toward producing a physical book in harmony with its contents.

LEADING

Leading, or line spacing in photocomposition, is the spacing set between lines of type. It is measured in points, or with some photocomposition in half-points, with 1-point and 2-point leading the sizes most generally used. When matter is to be leaded, it is specifically noted on the type specifications, as 9 on 11 (i.e., 9-point type with 2-point leading; often written 9/11). In linotype composition, 1-point or 2-point leading is usually cast on the body of the type. Matter set without any leading is called "solid," although it is not, of course, actually solid, since the body of the type is always larger than the typeface, thus automatically providing some amount of space between lines. In photocomposition it is possible with some type faces having short ascenders and descenders to use *minus-leading* of ½ point, as 10 on 9½ (i.e., the film will advance only 9½ points).

Besides giving a more pleasing appearance to the page, leading also increases legibility by keeping the eyes from being distracted by lines above or below and enabling eye movements to locate the correct line easily. The amount of leading desirable for any given work is dependent on a number of factors, including the size of the typeface, the length of the line, the size of word spacings, and the kind of material being set. Sir Cyril Burt, in *A Psychological Study of Typography*, considered several of these factors in relation to leading (see pp. 506ff.). In terms of type size, he found that the amount of leading needed related to the actual rather than the point size of the type, and determined that leading about one and one-half times the x-height provided good legibility. Thus, while for most 8-, 9-, or 10-point type, 2-point leading may be desirable, for some "small" typefaces in this range 1-point leading may be sufficient. In relation to line length, or *measure*, this study confirmed that wide measures require wide leading, with good legibility yielded by leading about ⅓₀ of the line measure. In most cases, leading of more than 3 points with a conventional type size cut down on legibility.

The kind of material being set also influences the amount of leading; for example, reference material that is read only in patches requires less leading than biography or fiction, which are read consecutively, while poetry, which must be read word-for-word, will require even wider leading. The Burt study revealed that children's books require generally wider leading to avoid confusion in locating the correct line. Another point to be considered is that readers will find most legible the width of leading to which they have become accustomed. With so many factors involved, the final decision on leading must be made in terms of what will be most easy to read, not in terms of general rules or formulas.

Certain stylistic points may require variations in the leading within a book. Two different sizes of type should always be separated by more space than is used between lines; this is especially important with footnotes. Generally, the smaller type size, as for extracts or footnotes, will require less leading than the book text; in fact, footnotes are often set solid. A heading within a text, whether centered or flush left, should be closer to the type below, to which it belongs, than to the type above it. Finally, spacebreaks, created by leading the equivalent of one or two extra, blank lines, are sometimes desirable for indicating elapsed time or a change in characters or scene or subject under discussion within a chapter (see p. 272).

SPACING

The spaces used to justify lines—that is, to spread out words to fill the line—are the following:

em quad	1 em
en quad (a "nut")	½ em
3-to-an-em (a "thick")	⅓ em
4-to-an-em ("thin space")	¼ em
5-to-an-em ("thin space")	⅕ em
hair space	less than ⅕ em

"Em" here, of course, means the square of whatever size type is being used, not a *pica* em. Thus a thick space of 12-point is 4 points thick, of 9-point it is only 3 points, and so on. A "pica thin space" is 3 points thick.

Text. The standards of good printing require that lines be spaced as evenly as possible and that spacing of a work be as nearly uniform as possible throughout. The space most used is the three-space, also called a thick or third space. Wider spacing is allowable in leaded matter than would be acceptable in solid matter. An occasional 5-spaced line is permissible in solid matter but not in leaded text. Spacing between words in small caps should be greater than between words in lowercase. No less than an en space should be used between words set in even caps.

A widely spaced line—an en quad between words
A normally spaced line—a thick space between words
A closely spaced line—a 4-space between words
A line spaced too tightly—a 5-space between words

The last line of a paragraph should be spaced about the same as lines above and below it, not widely spaced to fill the line or tightly spaced to secure open space at the end. However, it should not be allowed to stand with less than an em space at the end but should be either full width or have an em, preferably more, at the end. The last word of a paragraph should never be divided if such a division can be avoided; and in no case except in very narrow measure should less than four letters be carried over to stand alone in the line, even if wide or tight spacing is required to avoid it.

Punctuation. After a question mark or an exclamation point within a sentence the spacing should be less than that used after periods, usually an en if sentences are em spaced.

To arms! they come! the Greek! the Greek!
"Ha! There's some mischief afoot!" Raoul said to himself.

Following the period after a numeral that precedes a topic or a paragraph in a series an en space or two thick spaces may be used, in some cases an em quad. Consistency should be maintained.

1. How would you stimulate the imagination?
2. Explain the cause of such errors.
3. Classify the school subjects. [en space]

IX. DECISIONS
9. DECISIONS [2 thicks]

UNIT IV. THE REVOLUTION [em quad]

Between double and single quotation marks, a thin space should be inserted. Quotes and dashes are often separated from adjoining words by a thin space. References to footnotes look better thin-spaced away from the word preceding if no punctuation intervenes. In the footnote a thin space after the superior makes a better appearance than no space.

Some publishers use a thin space in contractions—except *don't*, *won't, can't,* and *shan't*—but this is not a general practice.

Letterspacing. Inserting thin spaces between the letters of a word is called letterspacing. Headings in caps or small caps are letterspaced sometimes to improve their appearance, or for the sake of a change from ordinary style. Space between words in a letterspaced heading should always be wide enough to separate them clearly: EDITOR'S PREFACE

Letterspacing in text is an expedient that should not be used if it can be avoided. In narrow measure, however, as in a column of a table or beside a text illustration, words may be letterspaced to fill the line. Whenever this is done, a short word should be spaced, not part of a long one.

INDENTION

The blank space left in setting a line in from the margin is called indention (never indentation), and everyone is familiar with indention as marking the beginning of a paragraph.

Paragraphs. The indention of paragraphs should be in proportion to the width of the page. An em space is enough for a page less than 27 picas—4½ inches—wide. An em and a half looks better for widths from 28 to 35 picas; 2 ems for 36 picas or more. In books set in two or more sizes of type it is desirable to have the paragraph indentions as nearly alike as possible. However, a page containing paragraphs of 10-point, 9-point, and 8-point type will look very well if the paragraph indentions are respectively a 10-point em, a 9-point em, and an 8-point em. More exact measurements are not practical in Linotype composition.

Outlines and similar matter. Indention is of greatest use in showing the relation of items to one another, in making clear the subordination or coordination of topics or statements.

Dairy Products:	Meat Products:
Milk	Beef
Butter	Lamb
Cheese	Pork

III. The legislative function
    1. Distribution of legislative powers between the House of Lords and the House of Commons
        a. Legislative powers of the House of Lords (see Chapter VIII, item VII, 2 and 3)
        b. Legislative powers of the House of Commons: the House of Commons possesses all powers of Parliament not belonging to the House of Lords, . . .
    2. Classification of bills

A. The problem and the assumptions
    1. Problem one of discovering how day-to-day prices result from the forces of supply and demand
    2. Approach by analysis of the actions of individual producers
        a. Assumptions: (1) Single homogeneous good, (2) aim to maximize revenue or minimize losses immediately or in the future, (3) full information, (4) freedom of access to market

The last two examples above illustrate different practices in regard to turned-over lines. The editor should clearly indicate his preference on the copy.

Headings and display. Indention is used in headings and display matter in several ways. Since paragraphs usually have the first line indented, *paragraph indention* means set with the first line indented and following lines flush. The amount of indention of the first line must be specified. *Reverse indention* is the opposite: first line flush and following lines indented; the more common term is *hanging indention,* and paragraphs set with hanging indention are called flush-and-hang paragraphs. Here, too, the amount of indention must be specified. The *block style* uses no indention for any line except the last, which is centered.

Diagonal, or drop-line, indention and inverted pyramid indention are used for display. The *drop-line indention* (also called echelon when more than two lines are involved) is often used for newspaper headlines, the first line flush at the left, the second flush at the right.

Rare Antiques Lost
      As Curio Shop Burns

Inverted pyramid indention, sometimes called half-diamond indention, is an arrangement for headings taking less than three full lines. Each line is centered and shorter than the line preceding. In book work in earlier days the last several lines of chapters were frequently arranged in this form.

Rare Antiques Lost/As Curio Shop/Burns

Rare Antiques Lost
As Curio Shop
Burns

PROBLEMS OF TYPOGRAPHICAL STYLE

Along with the basics of typographical style, the editor must consider the specific problems of typographical style presented by a particular book. These problems include decisions to be made on the forms for headings, extract, synopses, quotations, initials, footnotes, bibliographies, tables, and any other matter.

HEADINGS

Type. The book editor may use for headings in a book faces or letters quite different from those used for the text. If the same face is used for headings as for the text, there are several combinations of types that can be used to distinguish main headings from subheads. All the variations of which a regular font is capable—all caps, cap-and-small, cap-and-lower, italic, and boldface, with variations as needed or desired—may be used to distinguish headings from subheads.

Coordinate divisions should have headings of the same size and style. Part heads, for instance, should be in one size and form throughout a work, and no other headings in the work should be of this size and form. The size and form used for chapter titles are generally used also for the headlines of the preface, contents, list of illustrations, introduction, appendix, notes, glossary, bibliography, vocabulary, and index.

Arrangement. In the body of a book a centered head that can be set in less than three full lines should be arranged in inverted pyramid style, each line centered and shorter than the line above it.

STANDARDS FOR THE PRESENTATION OF OBJECTIVE DATA IN
REPORTS OF RESEARCH STUDIES

A longer heading, unless special directions are given by the editor to "pyramid," may be set in block style—that is, each line but the last flush left and right and the last line centered.

The Potential of Audio-Visual and Electronic
Media Aids in Unstructured Remedial Programs
for the Economically or Culturally Deprived
Preschool Child

Punctuation. In a heading set in caps, en dashes should be used for hyphens, and spacing between words should be not less than an en. The larger the type the greater should be the space between words.

ANGLO-SAXON GRAMMAR

Division. Dividing words and separating closely related terms or groups of words should be avoided whenever possible without marring the typographical appearance.

Numerals. The typographical appearance of the regular Old Style hanging numerals in a line of caps is unpleasing. This should be borne in mind in planning the type format of a book containing many headings of this sort. Many Old Style fonts now have both the hanging and the lining (modernized) numerals. (See p. 504.)

TURKEY SINCE 1900
FROM A.D. 449 TO THE NORMAN CONQUEST, 1066

TURKEY SINCE 1900
FROM A.D. 449 TO THE NORMAN CONQUEST, 1066

Small caps. Headings using small caps may be set cap-and-small or all small caps. (See pp. 146–148 for rules of capitalization in headings.)

These two forms are useful especially when several levels of subheads must be differentiated.

THE FAMILY ACCORDING TO FAULKNER

THE SARTORIS FAMILY

"Continued." If a heading is followed by the word *continued,* the form of this word should be indicated. It is usually italic lowercase, enclosed in parentheses or preceded by an em dash.

Side heads. Side heads may be set flush left with space above and below; or they may have regular paragraph indention with the text running in after a period and a space or a dash. They are usually boldface or italic, occasionally cap-and-small, rarely small caps. Side heads may be capped like centered heads or only the first word and proper nouns may be capped. For styles of side heads see pages 16–17.

Cut-in side heads are set in type other than the text type and inserted in a space left for them when the text type is set. Where they are used, they should be placed under the first two lines of the paragraph, with an even frame of type on three sides, and should be the same width throughout the book. (See p. 17.)

Marginal heads, or marginal notes, are also set in other than the text type and must be placed beside the text by hand. They should be set in the outside margin, with at least an en space between the text and the note, and should be placed as nearly as possible beside the line to which they apply. They are ordinarily set lowercase except the first letter and proper nouns; they may be set in justified lines, or align on only one side. (See p. 17.)

Running heads. As part of the page design, running heads serve both a decorative and an informative purpose. They are especially useful in textbooks and reference books, but may be omitted in books where they will have no real use, as in novels with untitled chapters or in books of poetry.

Position. There are a number of possible positions for running heads. The simplest style is with the head centered on the line, standing alone, with the folio at the foot of the page. However, in many books the running head and folio are given on the same line:

144   NEWSPAPER TYPOGRAPHY AND MECHANICS

TYPE FACES AND HOW TO USE THEM   145

Very often the head is centered, with the folio at the outside margin:

86   THE POTTERY PRIMER

SLAB BUILDING   87

The head may be flush with the margin instead of being centered:

164   ECOLOGY AND LIFE ON THE EARTH

THE LIVING COVER OF THE EARTH   165

At times, the head may be set on a line above or below the folio:

HEMINGWAY — CRITICAL INTERPRETATIONS — THE EARLY WRITINGS
122

                                                                BONFIRES IN EDEN
                                                                123

This last position has the advantage of allowing for lengthy running heads, when they are necessary, and also requires less setting and handwork than when folio and running head are on the same line.

    Composition. Running heads can be made up in a number of ways. Occasionally, the same running head, generally a section or chapter title, will be used on facing pages, but most often the facing pages will have different running heads, for example:

Left-hand page	*Right-hand page*
Author of book	Book title
Book title	Chapter title
Part or unit title	Chapter title
Chapter title	Subject matter on page
Author of article	Title of article

For books such as dictionaries or encyclopedias, the running head for each page will generally name the first and last entry on that page.

    While the more detailed running heads giving subject matter on the page may be most helpful in a technical book, it should be remembered that the more different heads set, the greater the expense. A number of different heads also requires very clear, detailed instructions to the compositor and careful checking at every stage through page proofs and repros.

    The running head should not run over one line, allowing at least an em space between the head and the folio. When a running head must be cut to fit, care should be taken that all essential information is retained, and if extensive cutting is involved, the author should be consulted before final decisions are made on the form.

INITIALS

THE FIRST WORD, phrase, or line of a chapter is often made prominent by setting it in caps or cap-and-small or by using an "initial"—a letter much larger than the text type, occupying the space of two or more lines. With any of these devices, the first line of the paragraph can be set flush left or indented a paragraph indention or more. If the chapter begins with a synopsis, a quotation, or a stanza of verse indicating the theme of the chapter, the initial would appear at the beginning of the text proper.

A DESCENDING, or dropped, initial should align at the top with the first line of the paragraph and at the base with the base of the last line abutting it. An ascending, or stickup, initial projects above the first line, aligning with it at the base.

Form and position of the initial word. The rest of the word, phrase, or line following the initial can be set in small caps, caps, or lowercase. If the first word is part of a proper name and is set in small caps or caps, the rest of the name should be set in the small caps or caps. This rule for capping after an initial is similarly applied to chapters beginning with cap-and-small.

> THE AMERICAN RED CROSS was on the scene within hours.
> PIPPA PASSES, in Kentucky, is the site of Alice Lloyd College.
> DEMENTIA PRAECOX is one of the most curious of all mental diseases.

In newspaper work, where the measure is narrow, the caps following an initial should not extend over one line.

> MR. AND MRS. JAMES EDWARD
> Montgomery were host and . . .

U NLESS THE INITIAL is the article *A*, the pronoun *I*, or the vocative *O*, the rest of the word should be set close to the initial. In metal type, if initials of a large size are used, the letters *A* and *L* are usually mortised (part of the body of the type is cut away) to allow the rest of the word to approach the face of the initial and thus avoid the white hole that there would be otherwise. Even *B, C, D, O*, and *Q* may be slightly mortised at the upper corner if they are of a comparatively large size. In photocomposition, letters can be set as close as if mortised, with no need for any hand-cutting of typefaces.

Abutting lines. The second and following lines that abut on a descending initial should be spaced away from the initial and should align with themselves. With the letters *F, P, T, V, W*, and *Y*, the abutting lines may be set flush, with the white space at the bottom of the initial providing sufficient spacing. The initials used in this section provide examples of various possible spacing methods.

W HEN the paragraph beginning with an initial is so short that it ends before the initial is completely surrounded by type, the evenness of the white frame is broken. This is a matter the author can remedy by adding a few words to avoid ending the paragraph at that point. The same difficulty of an even frame also arises when the initial is used to begin a poem that has alternate lines indented. An even frame for the initial is more important here than the indention of the lines.

Quotation marks. If quotation marks are used with initials, they should be of the text type, not of the type of the initial, and they should be set in the margin. Most compositors omit them before initials.

SYNOPSES

The chapter heading may be followed by a synopsis set in some manner distinct from the text. A synopsis may be a flush-and-hang paragraph in italic; or a regular paragraph in type smaller than text; or a small-type paragraph with the first line flush and the last line centered, the whole indented on both sides. It may be a list of the subheads, each on a separate line and the list centered as a whole.

CHAPTER IV

CONFIDENCE AND COLLAPSE—THE WORLD'S
FINANCIAL CRISIS

The crash in June 1931. The political causes. The Credit Anstalt. The German short-term debts. The Hoover Moratorium. The fall of the pound. The effect of the financial crisis upon the economic depression. The essence of the crisis; a "gap in the balance of payments unbridged by new credit." Its underlying causes: dead-weight debts, reckless lending, and high tariffs.

CHAPTER XIX   ..   THE ROLE OF THE MODERN HOME

  I The Function of the Home
 II The Economic Aspects of the Home
    1. Formerly the Center of Production
    2. Today an Agency of Consumption

III The Social Aspects of the Home
    1. The Home and the Community
    2. Vital Importance of Early Training
       (a) The Need for Durable Standards
    3. The Character of the Home Itself
       (a) Abnormal Home Life
       (b) The Rule of the Impersonal
       (c) Social Mobility

 IV Conclusions

CHAPTER I

OF THE DEFINITION OF CRIME, AND OF CERTAIN GENERAL PRINCIPLES
APPLICABLE THERETO

1. Crime Defined	35. Criminal Capacity
6. The Criminal Act	53. Intent in Statutory Crimes
26. The Criminal Intent	58. Justification for Crime

EPIGRAPHICAL QUOTATIONS AND CREDITS

Occasionally, a quotation appropriate to the theme of an article or chapter will be inserted between the heading and the text. This is always set in type smaller than the text itself. An initial, if used, should be at the beginning of the text type, not in the quotation.

Chapter IX

THE VISIT

Our souls sit close and silently within,
And their own webs from their own entrails spin.
Dryden: *Mariage à la Mode*

SUNSET sifted through the lace-hung windows, making orange moss of the space between the chairs. The light flowed gradually down the floor and over the edge of the carpet, then suddenly left, as if cut off by the ringing of the bell. . . .

For epigraphs, poetry is often centered, but may have a set indention throughout the book. A prose quotation may be set as a regular paragraph or a flush-and-hang paragraph, full measure or indented on both sides. Quotation marks are not necessary.

The credit line is usually aligned at the right of the quotation, generally on the line beneath the quote (see p. 21). However, if the quote ends in a short line, the credit may be run on the same line if it is short enough to fit completely on the line.

If your will want not, time and place will
be fruitfully added. *King Lear*
———————————

He that hath never done foolish things never
will be wise. — CONFUCIUS
———————————

He hath no leisure who useth it not.
— George Herbert

EXCERPTS

The author is responsible for careful indication of all matter quoted from the work of another. For that reason, detailed instructions about how excerpts should be distinguished from an author's own text were given in Part I (see pp. 17–22). The editor should be thoroughly familiar with the rules and practices there set down, for he will need to check carefully the author's accuracy and good judgment in following those instructions. The editor must decide upon the typographical form and carefully mark each excerpt that is to be reduced; the printer will take no responsibility for consistency of treatment. In addition, the editor should observe the following typographical practices:

Prose. Letters, excerpts from plays, and prose excerpts five lines or more in length are usually reduced. Extra leading should always be used above and below an extract. If an editor wishes to have less space above than below, he should mark on the copy the spacing desired, unless, of course, the printer has standing instructions on this detail. Quotation marks are not necessary, but if type the size of the extracts is used to any extent for matter other than quotations, quotation marks may well be used to avoid any possibility of misunderstanding.

Poetry. Excerpts from poetry are generally reduced. Each citation, or each group without intervening text, should be centered as a unit, regardless of other extracts falling on the same page. Poetry should be set line for line, with the alignment of the quoted lines following that in the original as closely as possible.

> My tender age in sorrow did begin:
> And still with sicknesses and shame
> Thou didst so punish sin,
> That I became
> Most thin.
> —George Herbert, *Easter Wings*

If a poetical extract begins with part of a metrical line, this line should be set to indicate the fact clearly.

> "long and level lawn,
> On which a dark hill, steep and high,
> Holds and charms the wandering eye."

FOOTNOTES

If the author has read and followed accurately the instructions for the arrangement of footnotes given in Part I, the editor's work will be light. The editor must, of course, decide upon the typographical style, whether 6-point, 7-point, or 8-point, solid, or leaded. Footnotes amplifying the text or acknowledging indebtedness for excerpts give little trouble. The editor must make sure, however, that all necessary credits are given and that the form of the credit remains the same even if the rest of the note is in shortened form (see p. 28). Footnotes of the bibliographical sort are so full of details that only an experienced person can maintain consistency on all points. They will therefore require careful editing. Models for such footnotes are shown on pages 26–28. Publishers may, of course, decide upon a different style. In publications of the United States Geological Survey, for instance, italics and quotation marks are not used, only the first word and proper nouns in titles are capped, and the punctuation is different from that shown in this book.

Copy check-up. A check-up of the copy of footnotes can be made by answering the following questions:

Have references to footnotes been placed properly and in correct order?

Have all essential data been given?

Can the name be supplied for any initial that stands alone before a surname?

Have notes been given in shortened form whenever that is desirable?

Have the names of periodicals that are abbreviated been given once in full and the same abbreviation used for all others?

Have *ibid., op. cit.,* and *loc. cit.* been used as they should be?

Have the names of publishers been given in consistent form in all notes?

In the page references are numbers set in the same way, all in full or all double numbers elided?

Have *vol.* and *p.* been omitted when possible?

Are the punctuation and capitalization consistent?

TABLES

The following pages are concerned primarily with typographical aspects of tabular composition; matters of construction and arrangement of data were considered in Part I.

Type and spacing. Tabular matter should be set off from reading text by using a smaller type, with extra space above and below the table. For ruled tables set in hot type, a solid set or a body larger in point size than the face is usually chosen—for instance, an 8-point face on a 9-point body, as is common. The purpose of this is to avoid having to cut leads; if leads were used between the lines, each one would have to be cut whenever it intersected a vertical rule. In photocomposition, ruled tables do not present this problem.

Lists in columns. The simplest kind of tabular matter is represented by lists in columns. If an uneven number of items is to be set in two columns, the first column may be one line longer than the second, or the last item may be centered separately under the two columns. If the number of items to be set in three columns lacks one of being a multiple of three, the first two columns should be one line longer than the third. If the number is one more than a multiple of three, the first or the second column should include one item more than the other two columns. This style cannot always be followed if there are turnover lines; in such cases as even an outline as possible should be secured.

The example at the left below shows alignment of periods and of right-hand digits in a column of dates. The example at the right shows alignment of the closing parentheses and the list of names, which results in equal spacing between the parentheses and the words.

i.	311 A.D.	(*l*)	William of Orange
j.	476 A.D.	(*m*)	Peter the Great
k.	44 B.C.	(*n*)	Magellan
l.	4 B.C.	(*o*)	Cortes
m.	325 A.D.	(*p*)	Xerxes
n.	1066 A.D.	(*q*)	Alexander the Great
o.	1519 A.D.	(*r*)	Marco Polo

Two-column tabulation. A simple two-column tabulation should be indented on both sides so that the columns are not too widely separated to be read easily.

1929	$5,471
1930–1932	5,500
1934	4,175

Whenever possible a long narrow tabulation should be set in half measure and arranged in two columns.

SEEDS PRODUCED BY A SINGLE AVERAGE-SIZE WEED

Dandelion	1,700	Burdock	24,500
Cocklebur	9,700	Russian thistle	25,000
Oxeye daisy	9,750	Purslane	69,000
Prickly lettuce	10,000	Crab grass	89,600
Beggar ticks	10,500	Willow foxtail	113,600
Tumbleweed	14,000	Tumble mustard	1,500,000
Ragweed	23,000	Worm seed	26,000,000

The stub. When the first vertical column of a table lists the items that are being dealt with in the table, it is called a stub. If complex matter is being dealt with, the stub may have boldface or italic subheads—centered or flush left—and several indentions. A colon or a dash should be used after lines that are used as side heads.

If an entry in the stub must be run over, or turned over, the first line should be run to the full measure of the stub, and the turned-over line should be indented about one em under the first line. The word *Total* can be omitted when the footing is obviously a total; when used it is generally indented at least one em more than any other line in the stub.

In some tables a row of dots, *leaders*, is used to emphasize the relationship of the stub entries to the following columns. Leaders are especially useful when several long lines in the stub force the shorter lines to be widely separated from the rest of the data. Careful use of leaders can make a table clearer, but their overuse results in a table unnecessarily "busy" and cumbersome, and often unduly separates the data to be compared.

Alignment. In tables, both horizontal and vertical alignments are important. Each entry in the tabular columns must align horizontally with the entry in the stub to which it refers. When the stub entry is several lines long, alignment is with the bottom line of the stub entry; if column entries are also several lines long, align the first lines.

In vertical alignment, columns of whole numbers are aligned on the right; decimals are aligned by the decimal points. Plus, minus, and equals signs should be aligned. Dollar signs and percentage signs should also be aligned, and in a full column in which all numbers denote either dollars or percentages, the signs are used with the first number in the column and after every break such as a rule or a heading.

Gold	Au	197.3		Shelter	$576–$1,200
Hydrogen	H	1.008		Food	480– 1,000
Indium	In	113.7		Clothing	408– 850

A column of words or dissimilar items may be centered or aligned at the left. Short items are better centered, longer items may be aligned at the left.

Tax Rate
11½ mills
?
$24.00
2.8%

Boston to Philadelphia
Chicago to Denver
Louisville to Minneapolis
Miami to New Orleans
Washington to San Francisco

Number columns. In unruled tables and those using braces to show the relation of headings to each other, the columns of numbers must always be centered under the headings. In book and magazine work, where tabular matter is used to amplify the text, columns of figures in ruled tables are centered in the width of the column by the longest number. In books of statistics like the *Statistical Abstract* or the Census reports, columns of numbers are set an en space (less if the table is crowded) from the rule at the right.

In a number column a nonentry indicating that no data is available can be represented by close leaders running either the full width of the column or as wide as the widest item, but at least three leaders wide. An em dash, centered in the column, is sometimes used for a nonentry in unruled tables. When a data column heading is not applicable to a particular stub item, the data column space relating to that item should be left blank, not filled with leaders or a dash.

It is desirable to secure as even an outline as possible in a column of figures, but addition of decimal ciphers to chemical or mathematical data should be made warily, and only with sure knowledge that the significance of the number is not affected by the addition.

Tables without figures. One-line entries in tables using words can be centered if they are very short or aligned on the left if they are longer. Entries of more than one line are often difficult to set with a pleasing appearance. A choice must be made between even spacing of words with a ragged outline on the right and uneven spacing with occasional letterspacing to secure an even outline. Vertical rules should not appear crowded, but should have at least two points of space on each side.

	COTTON	LINEN	WOOL	SILK
Practicability	Rather strong. May collect chalk and dust. Easily laundered.	Strong. Needs frequent pressing. Easily laundered. Sheds dust, chalk, lint.	Rather strong. Holds lint and ravelings, chalk, and dust. Usually needs dry cleaning. Soils easily but cleans easily.	Strongest. Wrinkles little. Sheds dust, chalk, lint, and ravelings.

Tables using words are most effective when they are kept in an uncomplicated form. They generally present fewer problems when set in an unruled table, in which ragged outlines and unjustified lines do not detract from appearance or comprehension. For example, the above table could be reset in the following unruled form:

Material	*Practicability*
Cotton	Rather strong. May collect chalk and dust. Easily laundered.
Linen	Strong. Needs frequent pressing. Easily laundered. Sheds dust, chalk, lint.
Wool	Rather strong.

Ditto marks. The use of ditto marks in tables should be avoided. Ditto marks can be eliminated either by repeating the words or by changing the form of the heading to include the words.

Captions. The descriptive caption, or title of the table, is most often set cap-and-small. The table number is usually set in caps if it is on a separate line above the caption, the same as the caption if it is run in. If there is a subhead on the caption, it is usually given in parentheses on a line below the caption and is often set in a different type style (e.g., cap-and-lower).

The table caption is placed above the table and is usually centered; over a narrow table, the caption should be set the width of the table, if possible. When a table is set broadside—reading across from the bottom to the top of the page—the table caption should also be set broadside, so that it will be at the top of the table when the book is turned sideways to read the table. If the tables in a book are set in two or three different type sizes in order to accommodate them to the measure of the book, the captions should all be in one size.

Column heads or boxheads. There are no fixed rules about the size and face of column heads in an unruled table. They may be cap-and-small, cap-and-lower, or italic cap-and-lower, in the same size as the table or smaller. In an unruled table, headings are preferably set so that they align across the bottom, and are centered over the columns to which they refer.

State	Wheat	Oats	Barley and Rye	Total All Grains
Colorado	0,000,000	0,000,000	000,000	00,000,000
Kansas	0,000,000	0,000,000	000,000	00,000,000
Nebraska	0,000,000	0,000,000	000,000	00,000,000

In a ruled table all column headings are centered vertically and horizontally in the boxes alloted to them. The stub may or may not have a heading over it, depending upon whether the particular identification is needed (e.g., a listing of well-known cities may not need a heading; a listing of small towns in a particular state may require a heading).

Total Income	If Income All "Earned" Income		If Income All "Investment" Income	
	Income Tax (Including Supertax, If Any)	Effective Rate	Income Tax (Including Supertax, If Any)	Effective Rate
00,000,000	00,000	00	00,000	00

If columns have primary and secondary, or *decked*, headings, as in the table above, they may or may not all be set in the same type. In an unruled table, however, different type should be used for primary and secondary headings, or else braces should be used.

It is sometimes necessary to set boxheads sideways—reading bottom to top—in order to get them into the measure of a narrow column. For headings turned sideways, as in the table below, normal spacing is used and no attempt is made to justify the right-hand side of the headings. Run-over lines may be set without indention. The heading over a stub is set horizontally if possible, even if other headings are vertical.

| Cumulative total of words arranged in an order of decreasing common usage | Number of words actually found in contest copy | Errors | | Average Errors | | Average decrease in number of errors made by second-year students |
		By 79 first-year students	By 84 second-year students	Per first-year student	Per second-year student	
5	5	149	165	1.8	1.9	−.08
10	10	201	235	2.5	2.8	−.25
25	23	292	326	3.6	3.8	−.20
50	47	438	453	5.5	5.3	+.15
100	84	549	573	6.9	6.8	+.12

When tables are set broadside of the page—that is, reading across from the bottom to the top of the page, the stub of the table being at the bottom of the page—vertical column headings will be upside down when the book is in its normal position.

Rules. The two main points to be considered when deciding whether or not to use rules in a table are the complexity of the data involved and the kind of type composition to be used. With Monotype and Linotype, intersection horizontal and vertical rules require the cutting of slugs and are expensive to set; with cold type or photocomposition, intersecting rules do not create this problem. However, the more recent practice is to avoid using rules in tables whenever possible. When the complexity of the data requires the use of some rules in the table, an attempt should be made to use only one kind, as with the following, which uses only horizontal:

TABLE 1—SECTION AND BRANCH STATISTICS

| | For Fiscal Year Ending | | | |
	April 30, 1967	April 30, 1969	April 30, 1971	April 30, 1973
Sections				
Number of sections	54	59	60	61
Number of section meetings held	460	491	498	521
Total attendance	73,254	108,523	73,806	73,381
Branches				
Number of branches	100	109	111	117
Number of branch meetings held	940	1,137	1,036	986
Total attendance	47,408	51,807	59,439	36,629

At times the complexity of the table warrants full ruling. In a ruled table, a double horizontal rule should follow the table caption, and single horizontal rules separate column heads from the data beneath and decked column heads from each other. A cut-in head—one which cuts across the columns, as in the following table—is preceded and followed by a single rule. Vertical rules separate the columns and column heads. The end of the table may be marked by either a single or a double rule. A double vertical rule should be used to separate the parts of a table that is doubled up.

TABLE 11. APPROXIMATE FEED REQUIREMENTS FOR
DAIRY COWS — QUANTITIES PER ANIMAL PER YEAR

PRODUCTION OF MILK PER COW, POUNDS	PASTURE, DAYS	CORN, BUSHELS	OATS, BUSHELS	CONCENTRATE, PROTEIN, POUNDS	SILAGE, TONS
I. FARMS WITH SILAGE					
Under 7000 7000–8000 8000–9000 9000–10,000 Over 10,000	175	14	14	265	3½
II. FARMS WITH NO SILAGE					
Under 7000 7000–8000 8000–9000 9000–10,000 Over 10,000	175	18	13	260	. . .

In an open table in which a column of numbers are totaled at the end, the totaling rule, or line separating the total from the rest of the column, should be the length of the total, including the dollar sign if there is one.

A double rule marks the end of a particular part. A total that is to appear in the column that follows the sums added, and used in a second addition, may be set on the line with the rule or on the line with the last figure.

```
           675                                          675
     500                              or           500
   10,650                                        10,650   11,150   11,825
   ───────   11,150                              ──────   ──────   ──────
              ───────   11,825
                         ──────
```

Braces. A table may be constructed with braces instead of headings and rules.

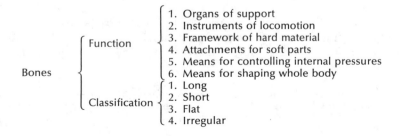

```
                     ⎧ 1. Organs of support
                     ⎪ 2. Instruments of locomotion
            ⎧ Function⎨ 3. Framework of hard material
            ⎪        ⎪ 4. Attachments for soft parts
            ⎪        ⎪ 5. Means for controlling internal pressures
Bones ⎨        ⎩ 6. Means for shaping whole body
            ⎪                ⎧ 1. Long
            ⎪ Classification ⎨ 2. Short
            ⎩                ⎪ 3. Flat
                             ⎩ 4. Irregular
```

Braces present no problems in tables set by Monotype or photocompo-sition, but they are best avoided in Linotype, where they are difficult and expensive to set.

"Continued" tables. If a table is more than one page in length, the table number—the table caption if there is no number—should be repeated at the top of each page, followed by an em dash and the word *Continued*. It is not necessary to repeat both number and caption. The column heads should be repeated, separated from the caption by a single rule. Double rules, if used, should be placed only at the beginning and at the end of the table, single rules at the bottom of pages.

The table number and the column heads of a wide table set broadside of the page need be repeated only on left-hand pages.

Whenever a page begins with an indented line, a catchline should be inserted, made up of the preceding flush line, followed by "*—continued,*" condensed into one line. Whenever in any column there is an addition continuing from the preceding page, there should be inserted a line at the top: *Brought forward . . .*, corresponding to a line at the bottom of the preceding page, *Carried forward . . .*

Footnotes. Reference marks, or *references,* for footnotes on tables can take the form of superior numbers for tables giving data in words, superior letters for tables with number data, or superior symbols, when either of the others could cause confusion or be mistaken for part of the data. When the text of the book is also footnoted, the references for the table footnotes should differ in form from those used for text footnotes. The reference should, if possible, be placed after the heading or entry explained in the footnote. If the reference cannot be inserted without breaking the align-ment, it may be placed before the entry. (See p. 33.) If a reference stands alone, it should be enclosed in parentheses and centered in the column.

The notes are best placed immediately below the table; they may be further distinguished from footnotes to text by setting them in a smaller size of type. A general note on the table should appear on the first page only, and is differentiated from the other footnotes by having no reference; it is introduced by the word NOTE and a colon.

For a table more than a page in length, footnote references on column headings and the notes to which they refer should appear on the first page and need not be repeated, though in some instances repetition on left-hand pages may be advisable. A footnote on a specific item in a table can appear on the same page as the item or can be grouped with other specific footnotes on the last page of the table.

Credits. The source of a table or of the data for a table is generally noted below the table along with other general notes. If the whole table is reprinted from another source, the reprint permission is given in the form of a general note.

NOTE: Reprinted, by permission, from *Readings in Business Planning and Policy Formulation,* ed. Robert J. Mockler (New York: Appleton-Century-Crofts, 1972), p. 234. © 1972 by Robert J. Mockler.

If the data for the table is taken from another source, a source note is given; it is generally placed below the table along with the other general notes.

SOURCE: Bureau of the Census, Department of Commerce.

CAPTIONS AND LEGENDS

Figures, illustrations, plates, and tables in a book or article (see pp. 44–45) are often accompanied by a short identification in the form of captions or legends. Although the terms caption and legend are often used interchangeably, technically *caption* refers to the title or headline for a figure or illustration, and *legend* refers to an explanatory or descriptive statement about the figure or illustration. Traditionally, the caption was placed above the illustration, but today it is often placed beneath. The legend is usually placed beneath the illustration; if there is a caption, the legend may be set either run on with it or on a separate line beneath it.

Gainsborough: *The Morning Walk*
Note the interplay between the formalized, stylized clothing and the undisciplined natural surroundings

Altamira Cave (Santander, Spain). A number of excellent examples of Old Stone Age animal paintings are preserved on the roof and walls of this cave.

The type used for captions and legends is almost always of a font smaller than the body of the text, but larger than the type used for footnotes. The caption or legend should be no wider than the cut, except that under a cut slightly narrower than the width of the page, the caption or legend may be the width of the page.

Capitalization for captions is generally headline-style, while legends ordinarily use sentence-style capitalization. However, when the legend is only a short identifying remark, it can be set caption-style.

YOUTH OF THE APOLLO TYPE

No period is used after a caption unless it is run in on the same line with a legend. The period may be omitted from the end of a short legend that resembles a caption. If a legend consists of two or more sentences, however, it must have sentence-style punctuation.

The Empire State Building
Once the tallest building in the world, the Empire State Building is still a sentimental favorite. After all, where would King Kong find a handhold on the World Trade Center or the Sears Tower?

Quotations from the text that are used as legends may be given with or without quotes.

"It beareth the name of Vanity Fair"

You'll find us rough, Sir, but you'll find us ready.

Any words (e.g., *top left; left to right; foreground*), letters, or symbols used in a legend to indicate certain parts of an illustration should be italicized and placed before the identifying name or descriptive phrase in the legend.

> The Bloomsbury Group. *Above left,* Virginia Woolf; *center,* E.M. Forster; *right,* Lytton Strachey; *below left,* Vanessa Bell; *center,* G.E. Moore; *right,* Clive Bell.

CASABLANCA
> *Left to right:* Dooley Wilson (Sam), Humphrey Bogart (Rick Blaine), Ingrid Bergman (Ilsa Lund), and Claude Rains (Captain Renault).

> FIG. 104. *A,* diagrammatic representation of staminate inflorescence; *B,* diagram of branch of staminate inflorescence showing three flowers; . . . *stp,* stipule. (Redrawn and adapted from Briosi and Tognini, *Istituto Botanico di Pavia*.)

Plates, illustrations, and figures need to be numbered only if they are referred to in the text and they would not be easily identifiable in any other way. If possible, all illustrations and figures printed in the text that are to be identified should be numbered consecutively, either throughout the book or with double numeration (i.e., chapter number and figure number) consecutive within each chapter. It is preferable to use arabic numerals for all numeration, even for a separate section of plates. The figure number generally precedes all other elements in the caption or legend. It may be set on a separate line or run in with the caption, or with the legend if there is no caption.

Credits. Acknowledgments should always be made for any illustrations or tables except those by the author himself. (For credits on tables, see p. 262.) The simplest credit is a line in 4-point to 6-point italic, set either close under the cut, flush left or right, or else along one of the vertical edges of the cut.

Courtesy U. S. Geological Survey   *W. H. Jackson*

FREMONT PEAK FROM SENECA LAKE

Ewing Galloway

If the illustration requires a longer credit line, this may be set on a separate line beneath the legend, placed in parentheses and run on with the legend, or given as part of the legend itself. Occasionally, the credit line is given in a different style of type to differentiate it from the legend. When the credit is set on a line by itself, no end punctuation is necessary.

If an illustration is protected by a copyright, more than a simple credit line such as those above may be required. Very often the copyright holder will specify the form for the credit line. When the form has not been specified, a consistent form should be established and followed throughout a work.

If an illustration is reproduced from a previously published source, the credit line should note title and author of the source, page or figure number, and permission of the copyright holder.

> The stripped, linear look of construction scaffolding can make an effective stage setting when other elements, such as costumes, provide color and interest. (Reprinted, with permission, from Hubert C. Heffner, Samuel Selden, and Hunton D. Sellman, *Modern Theatre Practice*, p. 163. © 1973 by Meredith Corporation.)

A shortened form for this information may be used if the work is also listed in the bibliography.

> A cardboard model of the stage set may be used for preliminary planning of the production.
> Reprinted, with permission, from Selden and Rezzuto, *Essentials of Stage Scenery*, fig. 4–9.

If much of the illustration for a work is from a single source, this may be acknowledged in the front matter.

When an agency provides illustration material, the expected credit line is usually specified. This is often set as a simple italicized credit line adjacent to the illustration rather than as part of the legend.

> *Alinari-Art Reference Bureau*

If free illustration material has been obtained from a source, acknowledgment of this is usually made by including "Courtesy" or "Courtesy of" in the credit line.

> Setting for *Macbeth*, designed by Robert Edmond Jones. Courtesy of The Theatre Collection, The New York Public Library at Lincoln Center; Astor, Lenox, and Tilden Foundations.

When the source of an illustration is in the public domain, no permission is needed to reproduce it, but a credit line should be given for the source, if only to identify it for the reader.

> FIG. 103. Habit of carpellate flowers of hemp.
> (Reproduced from *Yearbook of U. S. Dept. of Agriculture*, 1913.)

If the illustration material has been commissioned for the work, professional courtesy requires that acknowledgment be given. If the illustration is signed and the signature is reproduced, that is sufficient. Otherwise a short credit line, such as "Map by Molly Marsh," "Drawing by Carolyn Gastright," or "Photo by Alfred Gescheidt," should be given.

BIBLIOGRAPHIES

Bibliographies require careful editing. The first decision to be made by the editor is whether the bibliography is in the form most appropriate for the text. If bibliographical lists occur at intervals throughout a work, he should note whether the same kind of information is given in each list. For instance, if an annotated list is given for some chapters, all the lists should be annotated; if the book-list form is chosen for some, all others should have manufacturing and price data.

Authors have been instructed in Part I in the makeup and typing of bibliographies (see pp. 37–43). As there presented, the examples of bibliographical form do not show the possible type variations. These variations will be considered here.

Names of authors. Often the author's name is set in the same type style as the rest of the entry; other possible forms include setting the name in cap-and-small, small caps, bold face, or setting the last name in caps. A period usually follows the name part of the entry, although a comma may be used. A variety of possible forms are given below.

> Anouilh, Jean.
> **Salinger, J.D.**
> BROOKS, CLEANTH, and WARREN, ROBERT PENN.
> PETTY, WALTER T., and MARY BOWEN.
> PASAMANICK, Benjamin, and others.
> Evans, M. Blakemore; Röseler, Robert O.; and Reichard, Joseph R.
> Prybyla, Jan S., ed.

If two or more books or articles by the same author are listed, a 3-em dash is usually substituted for the name in the second and following entries. Since the dash is understood to replace the name given in the preceding entry, it is followed by a period and a space before the title.

> Skinner, Burrhus Frederic. *Contingencies of Reinforcement: A Theoretical Analysis.* Century Psychology Series. New York: Appleton-Century-Crofts, 1969.
> ———. *Cumulative Record: A Selection of Papers.* 3rd ed. Century Psychology Series. New York: Appleton-Century-Crofts, 1972.
> ———. "Autoshaping." *Science* 173 (1971):752–3.

Titles. Book titles and titles of periodicals are usually italicized and titles of articles roman quoted. Subtitles should be retained and set like the rest of the title; they should be separated from the rest of the title by a colon. The title of another full-length work included in an italicized title should be quoted.

The Middle Ages: 395–1500.
The Madness in Sports: Psychosocial Observations on Sports.
Mensch und Zeit: Anthology of German Radio Plays.
Critical Considerations of "Pale Fire."

"A Re-evaluation of Plato's *Republic*."
"The Nuer: An African Tribe."
"How to Avoid Those Land-Sale Frauds."

For bibliographies in scientific works, the titles are often given with sentence-style rather than headline-style capitalization. In this style entry, book titles are italicized; titles of articles are set roman and are not quoted.

Diagnosis and treatment of cardiac arrhythmias.
Contributions to behavior-genetic analysis.

Foreign gas broadening in cool white dwarfs.
Four-hundredth-order harmonic mixing of microwave and infrared laser radiation using a Josephson junction and a maser.

As titles of scientific articles are often long, it is acceptable form to drop the article title from a bibliographical entry when the author, journal, volume number, and page numbers are all included.

The capitalization of books or articles in a foreign language should follow the grammatical rules of that language.

Précis de civilisation française.
La familia de Pascual Duarte.
El gesticulador: Pieza para demagogos en tres actos.
Die Energie: Von der Tretmühle zum Kernreaktor.
La luna e i faló.
Teatro Brasileiro contemporâneo.

With unpublished material, such as a dissertation, thesis, or paper presented at a symposium, the title, even for a book-length work, is given in quotes, except in scientific bibliographies where titles of articles are also not being quoted.

"The Status of Working Mothers in Middle-Class Society."
"*My Mortal Enemy:* The Use of Allusion in the Novel Démeublé."

For works issued in microfilm, titles are italicized.

Investigations of Hydroelectric Power Sources in Thailand.
Official Records of the Union and Confederate Navies in the War of the Rebellion.

Publication data. For most general bibliographical entries, the place of publication, the name of the publisher, and the date of publication are grouped together. Very often a colon is used after the place, and the publisher and date are separated by a comma, but sometimes all commas are used.

New York: Appleton-Century-Crofts, 1973.
New York, Walker and Co., 1973.

When books are issued by a publisher under a special division or imprint, this should be noted.

New York: New American Library, Signet Books, 1970.
New York: Appleton-Century-Crofts, Goldentree Bibliographies, 1969.

If a publisher's name has changed since the publication of the book, the name on the title page should be used.

Bibliographical entries for scientific works often give the date of publication immediately following the author's name and group the other publication data after the title of the work.

Kelman, G.R. 1971. *Applied cardiovascular physiology.* New York: Appleton-Century-Crofts.

For an article from a magazine or journal, the entry should include the title of the periodical, in italics, the volume or issue number in arabic numerals followed by a colon, and the page numbers. For the reader's convenience, the year of publication may be given in parentheses after the volume number. For popular magazines and others whose page numbers are not consecutive throughout a volume, the date of the publication should be used instead of the volume number.

For daily or weekly newspapers and other such publications, the date of the issue—day, month, and year—should be given. For newspapers, page numbers should be omitted unless the specific edition that was used is also noted. For daily newspapers in the United States, the city of publication may be italicized along with the title, whether or not it is part of the official title: *Boston Globe, Cincinnati Enquirer, Honolulu Advertiser.* (If meticulous research has been done to ascertain the official titles of all newspapers referred to in the bibliography, then the city should be italicized only if it is part of the official title.) If a city is not well known or if it may be confused with another city of the same name, the abbreviation of the state it is in may be given, roman and in parentheses, after the name of the city: *Camden* (N.J.) *Courier-Post, Springfield* (Mass.) *Union.* A city name should not be given with newspapers such as *Wall Street Journal* or *Christian Science Monitor* or with those whose titles already have a geographical designation, such as *Rocky Mountain News, Florida Times-Union,* or *Daily Oklahoman.* For foreign daily newspapers, the city's name can be given in parentheses after the name of the paper, unless it forms part of the official title.

Annotations. A variety of styles for annotations are in use. They may be run in after the entry, or set as a separate paragraph in the same type, full measure or indented, or set in smaller type, full measure or indented.

Rose, Barbara, *American Art Since 1900: A Critical History* (New York: Frederick A. Praeger, 1967), 320 pp. *Discusses individual artists and works within the broader framework of the evolution of the various movements within 20th-century American art.*

Hansen, James C., and Cramer, Stanley H., eds., *Group Guidance and Counseling in the Schools.* New York: Appleton-Century-Crofts, 1971, 396 pp.
Readings from a number of authorities on the theory and practice of the use of the group in school counseling and guidance.

Phillips, B.J., "On Location with the WACS." *Ms.*, November, 1972, pp. 53–63.

A discussion of the present-day Women's Army Corps, what is and is not modernized about it, using basic training at Fort McClellan as a perspective.

Book lists. Here, too, a variety of possible forms can be used to present the material. Some are businesslike, or resemble a catalogue; others may use a more open, imaginative arrangement.

Kronenberger, Louis, ed., *Atlantic Brief Lives: A Biographical Companion to the Arts.* Boston: Atlantic-Little, Brown, 1965. pp. xxii, 900. Cloth. $15.00

Selden, Samuel, and Rezzuto, Tom. Essentials of Stage Scenery. Appleton-Century-Crofts, 1972. 263 pp. illus. $10.95.

BRAUTIGAN, RICHARD
REVENGE OF THE LAWN: STORIES 1962–1970 SIMON & SCHUSTER $1.95
Stories from *Rolling Stone, New American Review, Esquire,* and others, including "Revenge of the Lawn," "The Lost Chapters of *Trout Fishing in America*," "Ernest Hemingway's Typist," and "Greyhound Tragedy."
IN WATERMELON SUGAR DELTA $1.95

DETAILS OF PAGE MAKEUP

The revision of pages, often called makeup, requires careful attention to a great many details. Aside from verifying corrections, many things incidental to the changing of strips of type or pieces of film into pages are to be noticed. In the compositor's office, this work is sometimes done by the proofreader, oftentimes by a *reviser.* The editor should check carefully the accuracy of the reviser's work, but should direct his attention particularly to the position of the cuts and the addition or deletion of words to avoid short or long pages.

Folios. As a general rule, all pages that are part of the printed signature, or are of the same paper as the text paper, are included in the pagination of the book. This would exclude the endpapers and any inserts, such as a frontispiece or a section of photographs, that are on different paper and are added after the book is printed up. Flyleaves are included in pagination. A book may be paginated in one of two ways:

(a) The first page of front matter may be counted as page 1, the paging to follow consecutively throughout the entire book with the use of arabic numerals.

(b) The more common method is to use *lowercase* roman numerals for the front matter. The text then begins with 1 in arabic numerals. Occasionally there is some difficulty in determining what is the first page of text. Where there is a second half-title, it should be considered as the first text page. In practice, however, the first page of the printed text proper is sometimes numbered as page 1, and the half-title is considered as the end of the front matter. If the text proper is preceded by a part title, the part title must be considered as page 1. When an introduction, especially one written by the author, is directly related to the text and should be seen before the text is read, it is considered part of the text and should be paginated with it.

The folios may be placed at the top of the page, usually flush outside, or at the bottom of the page (*drop-folios*) centered, flush outside, or slightly indented. For alphabetically arranged books (dictionaries, encyclopedias) where folios are unimportant, they may be flush inside, top or bottom. In the front matter the folios are printed only on the table of contents and the following parts, which in many books means the list of illustrations, preface, foreword, or introduction when it is not part of the text. They may be omitted on the first page of each of these sections, but most publishers use drop-folios. Drop-folios are also used on the first page of each chapter or other pages with sinkage. Under full-page tables or cuts folios are generally not used, except in cases where there is a long series of tables or cuts. Folios should not be used on half-titles or part titles in the body of the book.

Front matter. The type and design of the half-title should relate it directly to the title. It should be placed in about the same position on the page as is the title on the following page.

The frontispiece is usually an insert "tipped in" to face the title page; that is, a page printed separately because it requires a different kind of paper, and pasted in when the book is bound. If it is not an insert but is printed on the same kind of paper as the text, it is often included in the pagination of the front matter.

The table of contents usually begins on the first right-hand page following the dedication, or if there is no dedication, on the page facing the copyright notice. However, when the table of contents is two pages long, it may be more helpful to run it on facing pages, starting on a left-hand page, so that all the contents may be seen at a glance. On the contents page it is unnecessary to use headings such as *chapter* and *page,* except in those rare cases where the elements must be identified to avoid confusion.

A list of illustrations or acknowledgments follows the table of contents, on the first right-hand page. Other preliminary matter, each section of which begins on a right-hand page, follows. (See p. 62 for the order of the parts.)

The card page (list of author's other books), is often a left-hand page facing the title, but may also be on a right-hand page preceding the title. The copyright and the frontispiece are left-hand pages, and the epigraph may be. Otherwise each new subject in the front matter and in the appendix must start on a right-hand page; if the section ends on a right-hand page, it must therefore be followed by a blank page.

Running heads. The form and wording of running heads, if they are used, should be a part of the instructions sent to the printer by the publisher. The commonest practice is to use the book title on left-hand pages and the chapter title on right-hand pages, but other arrangements are common. (See pp. 249–250.)

The line carrying running head and folio does not ordinarily extend over marginal notes or line numbers.

Widows. A widow is a line of less than full measure occurring as the first line of a page. Widows must be eliminated. A line may be taken from the preceding page, making it short, or the widow may be run on one line long at the bottom of a page instead of being carried over (see the following section on long and short pages.) Care must be taken, however, that these adjustments to eliminate the widows do not create new widows. A potentially less complicated way to eliminate widows is for the editor or author to add enough words to fill out the widow line or else to delete enough words to eliminate it. These additions or deletions should be made near the end of the paragraph in order to keep the number of lines to be reset to the minimum.

Pages that are not full length, such as at the ends of chapters, should have at least four to five lines of text to keep them from seeming blank.

Long and short pages. If the exigencies of the makeup result in long or short pages, facing pages should be the same length. Often a page can be lengthened or shortened to match the facing page by adding or deleting words from one of the paragraphs ending on the page, in order to gain or lose a line. Occasionally a line can be spaced out or spaced tighter at the end of a paragraph in order to gain or lose a line. Where there are headings, subheads, illustrations, or other breaks in the page, it is sometimes possible to adjust the space around these to correct the length of the page. When none of these methods is possible, it is permissible, though not generally desirable, to *card* a short page — insert very thin strips of cardboard between the lines of type to gain the desired length — or in photocomposition, to advance the film a fraction more than the established "leading."

Headings. A heading should not be the last line on a page, but should have at least two lines below it or else be carried over. The page preceding a heading may run short. Cut-in heads often cause the makeup man much trouble, for they too should have a line — preferably two lines — between them and the bottom of the page.

Spacebreaks. If spacebreaks (see p. 244) are used to indicate a lapse of time or a break in the continuity, the break should be within the page with at least two lines of type between it and the top or the bottom. If the page ends at the break, the break can be indicated by letting the bottom of the page run short, or sinking the top of the next page, or both. If asterisks are used instead of a spacebreak, they may stand at the bottom of the page but not at the top.

Poetry. If poetry must be divided elsewhere than between stanzas, it should not be divided after the first line or before the last line of a stanza. It is best not to divide between rhymed lines if such a division can be avoided. If a page breaks within a group of poetical extracts centered by the longest line of the group, the position of each part should be carefully checked, for the part which does not contain the longest line may be off center. Ordinarily such a part should be centered.

Columns. If matter appearing on the galley in two or more columns has to be broken, the page should not be cut straight across the columns unless they are to be read across the page, not down each column separately. If a page breaks within matter set in columns with headings over the columns, the headings should be repeated at the top of a left-hand page but not necessarily on a right-hand page.

Ditto marks. When matter containing ditto marks is made into pages, the reviser must see that no ditto marks are allowed to stand in the first line of a column. On the other hand, in beginning to revise from a new galley the reviser should note whether a change to ditto marks is possible. (Errors in these details occur often when galleys are set from reprint page copy and the pages break differently in the new makeup.) The same caution applies to dashes used for the author's name in bibliographies.

"Continued." When an entry-a-line index or any similar matter consisting of short lines with one or more indentions is made up, an indented line should not be allowed to stand as the first line on a left-hand page. A "continued" line should be inserted, composed of the key word of the preceding flush line, followed by *continued*, in parentheses. For example: Education *(continued)*. Such a line is often used to avoid beginning a second column with an indented line, but usually this is not considered necessary.

See also *"Continued" tables*, page 262.

Illustrations. The placement of one or more illustrations on a text page involves consideration of the effect on the two-page spread and of the general unity of design throughout the book. Another factor that must be considered is how closely the illustrations and text are related — if the text discusses the illustration very specifically, the illustration may have to be within the two-page spread that contains the discussion, regardless of the effect this has on the design unity of the book. Where the text discusses a number of illustrations, design problems can be simplified by giving each illustration a plate number, or figure number, by which it can be referred to in the text. This is especially helpful when insert illustrations are discussed in the text. Plate numbers can be consecutive throughout the book or for each chapter (e.g., the third illustration in Chapter 9 would be Plate 9–3). Plate numbers are treated as part of the legend or caption (see p. 264). Where there are a number of illustrations involved, they can best be placed by constructing a dummy (see p. 524) of the complete book.

Broadside cuts. Illustrations that are wider than the page and must therefore be placed broadside (turned sideways) should have the left-hand side of the cut at the bottom of the page; that is, the legend should be on the right-hand side of the page. Whenever possible, such illustrations should be on a left-hand page, because in this position they are easier to read when the book is turned sidewise. Running heads and folios are usually omitted; drop-folios may be used if there is a long series of such illustrations. Tipped-in cuts never carry either headings or folios.

Page-width cuts. A page-width illustration set within the text on a page must not be followed by less than a full line of type. (This would be similar to the problem of a widow at the top of a page.) If an illustration is placed near, but not at, the top or bottom of the page, it should be preceded or followed by enough lines of type to differentiate the text from the captions, or legends, accompanying the illustration. When an illustration is at the top of the page, the running head and folio may be omitted, as they tend to detract from the appearance of the cut. They are usually retained when there are a number of such cuts.

Narrow and small cuts. Generally a cut narrower than the page should be centered in the width of the page, leaving a space on each side blank. If there are several narrow cuts, they may be centered as a group, thereby cutting down on wasted white space. (Care must be taken that illustrations grouped together in this way do not conflict with each other, either visually or in terms of the text.) When illustrations are set as a group, a caption, or legend, applying to all of them may be set the width of the group. Occasionally runarounds (small or narrow cuts with text on three sides) may be included in a book. These are now seldom used, both because they are expensive to set, requiring resetting when the pages are made up, and because if the type lines alongside are very short they may require much letterspacing to be justified. An alternative to runarounds, when there are a large number of narrow cuts in a book, is to have wide outside margins and set the cuts in the margins alongside the text. In any case, the desired treatment of any cut should be clearly indicated to the compositor.

Footnotes. Text footnotes may be placed at the bottom of the page on which the reference occurs, or they may be grouped at the end of an article or book or, less desirably, at the end of each chapter. Placement depends on the kind of book in which the footnotes occur—in a scholarly work, footnotes may be better at the bottom of the page, while in a general book, they may prove distracting there—and on the predominant kind of footnotes being used—reference notes may be grouped in one place, while explanatory notes should be on the page to which they refer. When footnotes are placed at the bottom of the page, they are set apart from the text by a blank line, a short rule, or a rule the full width of the page. On a short page at the end of a chapter the note follows the text with the same space intervening; in other words, it is not dropped to the foot of the page.

Turnovers. It is sometimes necessary to break a page in such a way that part of a footnote has to be run over to the following page. If this happens, care should be taken, especially in running over to an even-numbered page, to see that the note is broken in the middle of a sentence, so that the reader will know that the footnote is not complete on that page. The run-over must, of course, be more than one line.

Position in line. Short footnotes may be indented like longer notes or they may be centered; if they are very short and other short notes occur on the same page, they may all be set in one line. For instance:

   [1]Hamilton's *History of the Republic,* Vol. IV, p. 25.
   [2]See page 333.

or

   [1]See page 333.   [2]*The Federalist.*

A series of many short footnotes on one page may be run in or set in columns. Whichever style is indicated as the preference of the author or publisher should be followed consistently throughout a work.

For the meaning and use of *ibid., op. cit.,* and *loc. cit.* see pages 29–30.

TYPOGRAPHY OF VARIOUS FORMS

In the publishing field, books fall into two general divisions, *trade books* and *technical books.* Trade books—books published for general sale through bookstores—include books such as fiction, biographies, poetry, plays, art books, and general nonfiction. The typography of trade books generally presents few difficulties and often provides an opportunity for the use of more interesting and decorative typefaces and imaginative type page designs. Technical books, which include textbooks, reference books, and other highly specialized books, have clarity as their central typographical consideration and often require specific typographical forms, as in legal citations or scientific writings. Typographical considerations for various kinds of trade and technical books are discussed below.

FICTION

In fiction, the typographical design should reflect the spirit of the book, and as far as is possible, the wishes of the author about typographical format should be respected.

Paragraphing. In dialogue a change of speaker is traditionally a signal for a new paragraph. Contemporary writers often dispense with this convention, especially when writing interacting conversation between two persons.

Quotation marks should be used to open and close each speaker's dialogue, so that it will be clear when there is a change in speaker. At times, especially in a book written in the first person, the narrator's remarks will be given without quotes, the other speaker's remarks with quotes.

Nicknames. It is usually undesirable to enclose nicknames in quotation marks, though some offices follow the rule of quoting them the first time they are used. There are two reasons for this policy: (1) It avoids the spotty or bristling effect of a page sprinkled or thickly sown with quotes; (2) it assumes that the reader can guess that "Dusty" is a person without being nudged in the ribs by quotes. It is better to overrate than to underrate the intelligence of a reader. (See also pp. 220–221.)

Slang, cant, specialized words. Slang should not be enclosed in quotes, unless it is not normally part of the speaker's vocabulary. Words used in a localized setting should not be quoted merely because they may be new to the reader. For instance, in a Western cowboy story the single quotation marks are unnecessary in sentences like "He went out to 'bulldog' a steer"; "It's just a 'charco.'" War, aviation, nautical, racketeer, northern and other localized stories all use words or terms that might be quoted if used out of their setting but for which quotation marks are unnecessary when they are in their proper environment.

Abbreviations. No abbreviations should be used in dialogue except Mr., Ms., and Mrs. and abbreviations the speaker himself would use, such, perhaps, as TV, UFO, UN, OK, USSR, USA.

In dialogue, numbers are usually spelled out, as this is the way they would be spoken, but numbers used for dates, years, page numbers, and large uneven numbers are often given in numerals.

Dialect. No correction of grammar should be attempted in quoted speech. Such errors are usually intentional, to indicate a particular characteristic, such as education, origins, or life-style, of the character speaking. No distortion of spelling should be used if the distortion would be pronounced exactly or approximately as the correct spelling is pronounced. Thus, "could uv" is a needless hardship because it is a common way of pronouncing "could of."

POETRY

Position. Poetry can be set with a fixed indention throughout a book or it can be centered. The use of fixed indention can lend stylistic unity to a book of poetry, especially when the poems have greatly varying line widths. When the indention style is used, an average should be taken of the long lines and this should be used to help compute the width of the text page, so that most, if not all, of the long lines can be accommodated without necessitating turnovers.

Centered poetry is more complicated, and more expensive to set, than fixed indention, as it involves centering each poem and, if a poem is carried over on another page, centering that part of the poem on the page. It is often recommended that centered poems be centered optically to avoid an uncentered appearance for a poem with only a few long lines. However, optical centering can be destructive to the shape of the poem when it necessitates turnovers for many of the long lines. An alternative is to center according to the longest line in the poem, or on that page of the poem, but this often results in the appearance of a fixed indention without the convenience of that form. Centering can also become difficult with the modern forms of verse where unusual typographical arrangements, spacing, and indentions within the poem must be preserved. Centering is usually successful for formal poetry with uniform line lengths.

Indention. Poetry with rhyming lines is sometimes set with all lines aligned on the left, sometimes with rhyming lines indented alike. The indentions of the original must be scrupulously followed. The indention of runover lines should be different from any regular line if possible, usually one em more; but this is often impossible in forms that have some lines deeply indented.

Scansion. If the author has placed an accent over a vowel to indicate that a word should be pronounced with an extra syllable, the grave accent is the one to use.

> She saw the horseshoe's curvèd charm
> To guard against her mother's harm.

For details of typographical style concerning poetical excerpts, see pages 17–22.

PLAYS

The major consideration in designing typography for plays is whether the book is to be used primarily for reading or for performance. A play printed for performance requires a high degree of typographical clarity, including large enough type to allow for ease of reading in audition or rehearsal, where the actor will be concentrating as much on his performance and stage business as on the words. Specific typographical problems for performance editions will be considered below.

Dramatis personae. The list of the characters in a play is ordinarily centered on the page facing the beginning of the play, under the caption *Dramatis Personae* or simply *Characters*. The list may be given alphabetically or in order of appearance or importance. Any short identification of a character is given after the name, separated by a comma; a long identification is treated as a separate sentence. Often, the characters' names are set in roman, the identifications in italic.

CHARACTERS
Armstrong St. Mark, *popular society poet*
Patricia Gregory, *fading Fifties film star*
Dr. Josephine Adrienne, *head of the sanatarium*
Louise Hiltz, *society matron*
Mr. Dale, *gardener and part-time sage*

Speakers' names. The names of the speakers in a play must be clearly distinguished typographically from the text and stage directions. They may be set in caps, small caps, cap-and-small, or boldface. (Names set in italic may become confused with stage directions.) The names may be set on a separate line, flush left or centered over the speech—centering should not be used when there are many short speeches, as this may often result in the name being centered over a blank space. Where space is a consideration, or where there are a number of short speeches, names may be set in the left margin next to the speech, followed by a period for reading editions or a colon, dash, or space for performance editions. Turned-over lines may be indented for further clarity. Abbreviations may be used for names if they are used consistently. For performance editions, the typography of the names must make them easily recognizable, so the actors may readily locate their lines.

Stage directions. To differentiate them from the text, settings and stage directions are usually set italic, and stage directions are given in brackets, or occasionally parentheses. Names of characters appearing in settings or stage directions are often set roman. Settings are generally one or more full sentences and should be punctuated as such. They are most often set in paragraph form at the opening of an act or scene.

Capitalization and punctuation for stage directions should be consistent; a variety of styles are possible. Sentence-style capitalization (the first word and all proper names) may be used for all directions or only for those that are sentences, or virtual sentences like *Exit*. When all directions have sentence style punctuation, they may all be followed by a period; when only full sentences are so capitalized, only they would be followed by a period. However, the period may be omitted from the end of all directions, and only retained internally if the direction is composed of several sentences.

Often, short specific directions are inserted in the character's speech; in a performance edition these must be sufficiently differentiated from the text to allow the actor to react to them while reading the speech. Longer directions are usually set on a separate line, either with a fixed indention throughout the book, or with paragraph indention for the first line. Several possible arrangements are shown below.

ACT I

A cavernous room, the edges and ceiling lost in darkness. The only light is centered on a heavy, ornate desk, one missing leg replaced by a stack of books. More books are stacked randomly around the desk. At both ends of the eight-foot-long desk are attached candelabra, on one of which (stage right) a frock coat is hung. At the desk is VICTOR, *assiduously signing pieces of newspaper and handing them to the* ASSISTANT, *wearing a frock coat and standing immediately behind him at the left.*

VICTOR [*His head bent over the papers*].
    This to Lansdowne, these two to Frammedge, and that to Borum [*Signing and handing back the papers as he speaks*].

    [*The* ASSISTANT, *his face immobile, takes each paper, crumples it, and stuffs it into a brown paper shopping bag on his left arm. During this, a* POLICEMAN *enters from stage right and stalks into the circle of light.*]

POLICEMAN.
    Harumpf-rumpf [*Clearing throat*]. Beg pardon, sir [*Taking off his hat*] but you're under arrest.
ASSISTANT [*Still looking straight ahead*].
    No, no—that's later, after the shooting.
POLICEMAN [*Exiting left*].
    Don't see why I always have to be in Act Two.

AUNT NANCY. We'll wait, Jim might come.
    (*There is another knock on the door.*)
SILAS (*rising*). I wonder what they want now?
    (*He goes out again.*)
AUNT NANCY. That might be Jim.
LIZA. Hope it is.

    (*Silas returns, followed by* JIM. *When* AUNT NANCY *sees who it is she flies to him.* JIM *opens his arms and draws her to him.*)

AUNT NANCY (*holding him off and looking at him*). Is it you sure enough, Jim?

Mark

Well, then, it would have to have been Hollister. He was the only one with a key to the room.

    (**Hollister** *half-rises from his chair.*)

Frayer
    (*Turning to face the group.*)

That's what we were supposed to think. But actually someone else *did* have a key to the room—and much more reason for wanting Wilson dead.

    (*He pauses, then steps toward the sofa.*)

Didn't you——

Dedria
    (*Sobbing, her hands to her face.*)

No! no! It wasn't me! It was——

The lights suddenly go out. Confusion and running. Two shots are fired. **Dedria** *screams, the sound spiraling upward, then suddenly cut off.*

Metrical lines. Metrical lines are frequently broken in plays, two or more characters each speaking part of a line. When the page is wide enough to allow it, the second part should begin one em beyond the end of the first part. If the page is too narrow to allow this, the second part may be set flush to the right margin. If there would be no extra space at the left end if the part were set in this manner, it should begin one em beyond the preceding line and the runover indented like other turned-over lines. The following examples show the forms recommended.

> GIRL.   You must not talk of Death.
> (*She comes hastily to him and holds*
> *him tightly in her arms.*)
> He is abroad this night. He rides on horses.
> He walks the highways and he creeps through lanes
> And knocks at many doors.
> BOY.   He cannot break
> The bolts upon our doors.
> GIRL.   Poor fool, he breathes
> With his thin icy breath out of the night
> And our stout doors are thinner than thin mist.

For verse plays, especially for reading editions in which there may be notes for various lines, line numbers may be given, in the margin, for every fifth or tenth line. The line numbers are counted by the full metrical line, whether it is broken between several speakers or has a turned-over line.

WRITINGS ON RELIGION

References. The form in which references to Scripture are set is largely a matter of the personal preference of the author and the editor, but clearness is essential. Book names containing five or more letters are usually abbreviated. (See forms listed on pp. 101–104.) The following examples illustrate a style that is preferred by many, using a centered colon (equal spacing on both sides) between the chapter and verse, an en dash when a reference extends for several verses or chapters, and a semicolon when the reference is to two or more separate passages.

Luke 4:16	II Cor. 1:16–20; 2:5
Matt. 2:5–13	Luke 3:6–5:2

An em dash is often used when the reference extends from one chapter to another, Luke 3:6—5:2, and an en dash when it extends from one verse to another within a chapter, Matt. 2:5–13.

Proper names. There are a number of different versions of the Bible in use, including the King James, the Jerusalem Bible, the Revised Standard, the New English Bible, the New American Bible, and the older Douay Version. As the spelling of various proper names differs in these versions, care must be taken that all spellings are taken consistently from whichever version is used.

Type. The words LORD and GOD are sometimes set cap-and-small in the Bible and hymnbooks, but this style is by no means universal.

> The LORD will perfect *that which* concerneth me: thy mercy, O LORD, *endureth* for ever.—Ps. 138.
> [The italic indicates that these words were supplied by the translator.]

Symbols. The following symbols may be used in religious work:

℞ Response—indicating the part repeated by the congregation in a responsive religious service.

℣ Versicle—indicating the part recited or sung by the priest.

✠ or + A sign of the cross used by the Pope, and by the Roman Catholic bishops and archbishops, immediately before the subscription of their names. In some service books it is used in those places where the sign of the cross is to be made.

LETTERS

Extract. When a letter is being quoted as extract, the first consideration is to follow the form and style of the letter as it was written. If a number of letters are being quoted, some styling may be done, as setting all parts such as headings (sender's address and the date), complimentary closes, and signatures either left or right on the page, and aligning the lines in these parts either flush left or flush right. Often, if information in the heading, salutation, or signature is already known from the context in which the quote is given, part or all of these may be omitted (e.g., if it is known that a series of letters was written from Philadelphia by Benjamin Franklin, the heading address and the signature could be omitted). Punctuation should follow that of the original letter, and the styling of the typography should reflect the style of the letter as much as possible.

Use of quotation marks. Letters that are excerpts from another work are usually set in type smaller than the text. It is then unnecessary to enclose them in quotation marks. If they are presented complete with heading and signature in a book in which excerpts are enclosed in quotes, it is correct to use opening quotes before the first line of the heading or salutation and closing quotes at the end of the signature (see pp. 18–19).

Printed letters. Letters, such as form letters, that are printed are usually styled in a standard business letter style. The letter should be centered vertically according to the optical center of the page (one-eighth of the page length above the actual center).

Heading. If the letter is printed on letterhead paper, the date, if one is required, can be printed on the right, with the end of the line one em from the right margin. If a full heading (address and date) is required, it should be printed block style—the lines aligned on the left—with the longest line being one em from the right margin. There is no punctuation at the ends of the lines; in business and informal letters abbreviations may be used.

> 687 Lexington Avenue
> New York, N.Y. 10022
> February 2, 1973

Salutation. The salutation should be set block style, flush with the left margin. No punctuation is used at the ends of the identification lines (name and address of recipient), but a comma, or in business letters a colon, is used at the end of the direct salutation line.

Mr. L. Royce Thomas, President
K & K Enterprises, Inc.
Chadwell Building
Columbia, Missouri 65201

Dear Mr. Thomas:

When printed form letters are to be sent to a number of recipients, they generally do not have any identification lines in the salutation and often have a general direct salutation such as "Dear Fellow Citizen."

Complimentary close and signature. The complimentary close and signature should be set block style, with the end of the longest line one em from the right margin. A notation of official position follows on the same line as the signature, or if it is very long, on the line beneath, aligned on the left.

LEGAL WRITING

Publishers of legal books and journals have special style sheets of forms which are technical and largely inflexible. The following pages cover only such problems as would normally confront an editor or designer in a general publishing house. Precedents could be found for styles other than those here shown, especially in the use of italics.

Names of cases. In a complete citation[1] italics are unnecessary.

Davison v. Johonnot, 7 Met. (Mass.) 388

The name of a legal case appearing without citation numbers is usually set in italic, with the *v.* (versus) in roman.

McKeeman v. *Goodridge*   *Hogan* v. *French*

---

[1] A useful guide for the styling of citations is *A Uniform System of Citation* by the Harvard Law Review Association (Cambridge, Mass.), 11th ed., 1967. For a general guide to legal writing, see Elliott Biskind, *Legal Writing Simplified* (New York: Clark, Clark, Boardman, 1971).

The phrases *in re, et al., ex rel., ex parte* are generally italic in a roman title, roman in an italic one.

> *People* ex rel. *Hogan* v. *French*
> *Ex parte* Harold Elliston *et al.*, 36 Atl. 312

The first word in the name of a case, proper names, and words replacing proper names should be capitalized.

> In the matter of Troutman
> Hartwell, appellant, v. Carter, respondent
> United States v. Nineteen Quarts of M-E Chlorine Solution

References. The name of the case should be separated from the reference by a comma. Within the reference, punctuation is reduced to a minimum, avoiding double punctuation.

> Raher v. Raher, 150 Iowa 521, 35 L. R. A. (N.S.) 292
> Rev. Stat. 1915 (Civ. Code) p. 1234

If the name of a state or of a reporter is given in full, no comma is needed between it and the page number or date.

> 140 Iowa 410   2 Black 635, 668

The copy editor should be cautious about making changes in the form of a citation for the sake of apparent consistency. For instance, in citations of Acts of Parliament, public Acts are cited thus:

> 3 & 4 Geo. V, c. 12, ss. 18, 19

Private Acts are cited:

> 3 & 4 Geo. V, c. xxii, ss. 18, 19

Parentheses and brackets are not interchangeable. In references to Law Reports published by the Incorporated Council of Law, dates up to 1890 are enclosed in parentheses; 1891 and later dates are in brackets. Note also the position of the comma:[2]

> Cooper v. Whittingham (1880), 15 Ch. D. 501
> Thompson v. City Glass Bottle Co., [1902] 1 K. B. 233

Abbreviations are used freely in references. Besides the usual abbreviations like bk., ed., pt., 2nd, 3rd, n. (note), and c. (chapter), names of

---

[2]For more detailed information on this, see Horace Hart, *Rules for Compositors and Readers at the Oxford University Press,* ed. J.A. Murray and H. Bradley, 37th ed. (London: Oxford University Press, 1967).

persons, places, courts, books, and periodicals are all shortened whenever possible, not according to the whim of the writer but in accordance with established forms.[3]

Book titles. In citations the names of books, government publications, and periodicals are often set cap-and-small.

ANN. CAS.: *American Annotated Cases*
L. R. A. (N.S.): *Lawyers Reports Annotated,* New Series
See 2 WIGMORE, EVIDENCE (2nd ed. 1923) §856.
47 HARV. L. REV. at 618

MATHEMATICAL AND SCIENTIFIC WRITING

Most writings using equations, formulas, or other mathematical expressions require a number of specialized mathematical signs and symbols. When a manuscript is first submitted for editing, the author should include a list of all special signs, indicating their frequency of use. This list can immediately be given to the compositor, who can then indicate which signs he has in stock, and which, if any, of the others he is able to order for the job; compositors do not commonly have even a majority of these special signs in their usual type supply. (The largest number of these special signs are generally found in 10-point type, since this is the size most commonly used for mathematics.)

If the compositor is given a list of the special signs while the manuscript is still being edited, it allows him time to order any new type required before production is begun and also gives the author or editor time to select substitute signs or symbols for those which are unavailable.

Type. Letters in mathematical equations and letters referring to geometrical figures should be set in italics.

$$x + y = 10$$
$$\angle ABC = \angle ABD$$

These letters should not be changed to roman when they are used in an italic context.

It appears from (2) that *lines parallel to the x-axis* (including the x-axis itself) *have the slope O.* For in that case $y_2 = y_1$.

Vectors are often given in boldface roman, to differentiate them from their components, which are italicized.

In scientific writing many shortened forms are written without an abbreviating point as in the examples on page 284.

[3]For authoritative lists of abbreviations as well as spellings and definitions of legal terms, ancient and modern, see Henry Campbell Black, *Black's Law Dictionary,* 4th ed. rev. (St. Paul, Minn.: West, 1968) and James Arthur Ballentine, *Ballentine's Law Dictionary,* ed. William S. Anderson, 3rd ed. (Rochester, N.Y.: Lawyer's Co-operative Publishing Co., 1969).

Chemical symbols are always roman and without a point.

   Na   O   NaCl HNO_3 P CO_2

The same is true of trigonometric terms.

   cos   cosec   cot   log   mod   sin   tan
   cosine   cosecant   cotangent   logarithm   modulus   sine   tangent

Opening and closing parentheses, brackets, and braces used in mathematical equations should be of sufficient height to enclose all elements within them. Except where mathematical conventions dictate otherwise, the usual order in which they are used for enclosing is parentheses, brackets, braces, large parentheses, large brackets, and then large braces.

Superior figures. When both inferior (subscript) and superior (superscript) figures follow a number, the inferior should be set first: 5_2^2. A prime is always set close to the figure or letter. Exponents precede all punctuation except the period of an abbreviation.

$$A' > B', a^2 > b^2$$

cm^2, square centimeter, $in.^3$, cubic inch

Spacing. Signs—plus, minus, equality, arrow—are often hair-spaced or thin-spaced, but they may be set closed with accompanying numerals or symbols. (An em dash should not be used for a minus sign.) It may be advisable to send the compositor a sample showing the amount of spacing preferred, because some compositors tend to use more than necessary.

$$x + 3y - z = O$$
Find the sum of 8 terms, of the series 1, 2, 4, $\cdots$.
Use $x = -3, -2, \cdots, 6$.

If the three points are set on the line instead of being centered as above, a justifier or spaceband should precede them: "The notation $B_1, B_2, B_3, \ldots$ is."

A hair space or a thin space is often used between numerals and letters in formulas and equations.

$$2KClO_3 \rightarrow 2KCl + 3O_2$$

A thin space should be used before N and M in expressing normality or molarity.

$$2.5N \qquad 5M \qquad 0.1M$$

Position. Short equations and formulas can often be run in the text to save space, but important equations or formulas should be *displayed* (placed on a separate line). Equations and formulas that are displayed may be centered or they may have a standard indention throughout the book or article.

A series of displayed equations should be aligned by the equality signs $(=)$, proportion signs $(::)$, or arrows $(\rightarrow)$. If standard indention is being used, the series should be indented the standard amount according to the equation that is longest on the left of the sign.

$$SO_2 + Cl = SO_2Cl$$
$$SO_2Cl_2 + 2H_2O = H_2SO_4 + 2HCl$$

If centering is being used, the series should be centered optically (set so as to give the appearance of being centered as a whole).

$$x + y + z = 21{,}000$$
$$x = 2z$$
$$.05y = .06z$$

Equation numbers. When equations, formulas, or other mathematical expressions are discussed several places in the text or where cross-references to them are required, they may be given equation numbers. The usual system is that of double numeration, using the chapter number and a sequence of equation numbers beginning with 1 for each new chapter. All numbered equations must be displayed. Equation numbers are given in parentheses and are usually set flush right, but occasionally flush left.

$$b\left[7a + b\left(6 - \frac{b}{a}\right)\right] \tag{1.1}$$

Division. If an algebraic expression must be divided, it is best to break it before an operational sign.

$$\left|\begin{array}{l} 15r + 6s - 11t + r - 9s + t - 2s \\ + 5r - 2t - 10r \end{array}\right|$$

An equation may be divided, but it is much better to set it on a line by itself, displayed, in preference to breaking it. When an equation must be divided, the first line is often set flush left; the turned-over line may be set flush right, indented from the right to allow for an equation number, or aligned by an operation sign.

$$\left|\begin{array}{l} (42)^3 = (40 + 2)^3 = 40^3 + 3 \times 40^2 \times 2 + 3 \times 40 \times 2^2 + 2^3 \\ = 64{,}000 + 9600 + 480 + 8 \\ = 74{,}088 \end{array}\right|$$

$$\left|\begin{array}{l} CH_3 \cdot CH(NH_2) \cdot COOH + H_2O \\ \rightarrow NH_3 + CH_3 \cdot CHOH \cdot COOH \end{array}\right|$$

Note: The vertical lines show width of page.

An equation like the following that is too wide to set in one line is best set with the turnover indented to clear the equality sign.

$$S_1 = \text{the stress resulting from the direct tension in the cable.}$$

A formula written without operational signs is indivisible.

$$\text{Ferric ammonium alum,}$$
$$[(NH_4)_2SO_4]\ [Fe_2(SO_4)_3]\ [H_2O]_{24}$$

When an expression that occupies several lines cannot be divided (part at the bottom of one page, the rest at the top of the next), it should be given a figure or equation number and be set as a figure in the nearest convenient place in the text.

Capitalization. Equations are essentially sentences; therefore capitalization in the first half of an equation (the subject) is unnecessary for all except the first word:

$$\text{Life in hours} = 1{,}500 \div (\text{P.F.})^2$$
$$\text{Radiant efficiency} = \frac{2 \times 25.1}{A \times G \times H}$$

Capitalization following the equality sign is largely a matter of the author's preference. Practice varies widely, but consistency of form should be observed throughout a work.

$$\text{Power} = \text{force} \times \text{velocity}$$
$$\text{Percent of error} = \frac{\text{Possible error}}{\text{Measurement}} \times 100\%$$
$$\frac{\text{Length of the arc}}{\text{Circumference}} = \frac{\angle \text{ of sector}}{360°}$$

Punctuation. No punctuation is necessary after displayed equations.

$$\text{The equation for the above reaction is}$$
$$2HCl + Zn \rightarrow H_2 + ZnCl_2$$
$$\text{It gives us much more information} \ldots$$

Where there are a large number of signs and symbols used, however, it may be desirable to use punctuation after the displayed equations in order to make it clear where each sentence ends.

Notes on the use of mathematical signs. In order to cut down on printing costs and the amount of handsetting involved, all mathematical expressions should be given in a simplified manner, using as few special symbols and setting as much on a regular type line as possible. When mathematical expressions are set within the text rather than displayed, it is especially important to avoid items that will extend above or below the type line to the extent that extra leading would be required. Fractions can be set with a solidus (/) so that they may be accommodated on one line. For example, $\dfrac{a+b}{x-y}$ can be set as $(a + b)/(x - 6)$. Whenever possible, the partial radical sign should be used in place of a full radical sign, which requires handsetting. For example, $\sqrt{a^2 + ab + b^2}$ could be set $\sqrt{}(a^2 + ab + b^2)$.

In mathematical expressions, the signs $<$ and $>$ mean "is less than" and "is greater than," respectively. It is therefore unnecessary to use a verb with them, or with similar signs.

Wrong: . . . where A is $< B$. $C \parallel$ to AB
Right: . . . where $A < B$. $C \parallel AB$

In mathematical or scientific texts it is permissible to use $<$ and $>$ and similar signs within a sentence dealing with non-variables, but in this case the verb should be used with the sign: e.g., "In this case the gravity is $<$ the force . . ." Usually this is simply written out: "is less than" or "is greater than." (In etymological text, especially in dictionaries, $<$ is used to mean "derived from"; the etymological and mathematical meanings should not be confused.)

The verb used with the sign for equality should be consistent throughout a work. In the following *is* is preferable.

Possible: Does $\angle\,C = \angle\,Z$?
Preferable: Is $\angle\,C = \angle\,Z$?

The word following the sign $\therefore$ ("therefore") or $\because$ ("because") is usually lowercase:

$\therefore$ the distance is greater than 230 kilometers.

Mathematical signs in a reading text: The percent sign should be used only when the nature of the text makes the use of signs appropriate. If a phrase like "44 to 50 percent" is expressed with the sign, both numbers should be followed by the sign:

This contains 44% to 50% of ethyl alcohol by weight, and about 50% to 56% by volume.

There is no justification for the use in straight reading matter of such forms as: + sign, "+" sign. If the sign itself is used, the word *sign* is redundant. If the word *sign* must be used the preferable form is *plus sign*.

Similarly incorrect are the forms "caret sign," "a dagger sign," "a $ mark." These should be written "a caret," "a dagger," "a dollar mark."

A superior plus sign following a letter symbol signifies a unit charge of positive electricity, a superior minus indicates a negative charge. If two or more of these superiors follow a symbol, they are best placed side by side. Setting them one below another is difficult and expensive. CA^{++} signifies a calcium ion carrying two positive charges.

A plus sign placed above the initial letter of a substance indicates that it is a base of alkaloid: $\overset{+}{M}$, morphine; $\overset{+}{Q}$, quinine. A minus sign over a letter indicates an acid.

The sign of multiplication (not an x) should be used to indicate magnification: $\times$ 1000 indicates magnification of 1000 diameters. Less-than-life-size would be indicated: $\times$ ½.

For the expression of dimensions, see page 130.

COMPOSITION OF FOREIGN LANGUAGES

Copy editors and proofreaders find a knowledge of languages frequently useful in their work. They would, for instance, be seriously handicapped if they could not recognize and call by name the Greek letters so often used as signs and symbols in technical books of many kinds. Even a smattering of foreign languages may help them to read correctly an indistinctly written foreign word. They often have to answer questions of how to divide or to cap in citations of works in foreign languages. The following sections may be helpful for reference when questions of form arise concerning foreign words, phrases, and titles used in an English text.

For additional information on the composition of foreign languages, see the *Government Printing Office Style Manual* (1973) and the *Manual of Foreign Languages* by George F. von Ostermann (4th ed.; New York: Central Book Company, 1952). Specific information on capitalization of foreign languages is given in *Anglo-American Cataloging Rules* (Chicago: American Library Association, 1967), and general instructions for capitalization, use of accents, and transliteration for a number of languages are covered by *The MLA Style Sheet* (2nd ed.; New York: Modern Language Assn., 1970). For instructions on transliterating languages from nonroman alphabets, see *A.L.A. Cataloging Rules for Author and Title Entries* (Chicago: American Library Association, 1949) and *Cataloging Rules of the American Library Association and the Library of Congress: Additions and Changes, 1949–1958* (Washington: Library of Congress, 1959).

DANISH

The alphabet. The Danish language uses twenty-nine letters, our alphabet with the addition of æ, ø, and å. In arranging Danish words alphabetically, æ, ø, and å usually follow z, although å may be placed before a. The letter ø may be replaced by ö, but not by œ. The letters c, q, w, x, and z are used only in foreign words and proper names.

Capitalization. Danish nouns are capped. The pronouns De, Deres, Dem are capped when they mean you (polite form), lowercase when they mean he, she, or it.

Divisions. When a word is divided at the end of a line, at least three letters should be carried over unless the lines are very short. Compounds should be divided according to their construction, prefixes and suffixes being kept intact. A word may be divided before a consonant standing between vowels. The combinations sk, sp, st, and str should not be separated, and belong with a following vowel. Other combinations of consonants may be divided.

Punctuation. The clausal construction of the sentences is closely followed.

DUTCH

The alphabet. The Dutch language uses the English alphabet, but the letters q, x, and y appear only in foreign words. The following accents may be used: ā, ă, ē, ĕ, ī, ĭ, ō, ŏ. The vowel ij, peculiar to this language, is equivalent to our y: Bilderdijk, Huijgens. Both letters are capped if they begin a sentence or a proper name: IJssel Lake, IJmuiden.

Capitalization. Caps are used in Dutch much the same as in English, except that if a sentence begins with a one-letter word, the second word is capped and not the first.

The pronoun ik (I) is not capped, but U (you) and Uw (your) are.

In family names ten and van are lowercase: van 't Hoff, van Leeuwenhoek. (See also pp. 167–168.)

Divisions. When a word is divided at the end of a line, at least three letters should be carried over unless the lines are very short. The component parts of a compound word should be kept intact. A word may be divided before a consonant standing between vowels. The combination ch is inseparable and belongs with a following vowel. A division may usually be made between two consonants standing between vowels. A combination of three consonants is divided phonetically.

FRENCH

A useful reference for punctuation, abbreviations, capitalization, and other technical problems of French is Nouveau dictionnaire des difficultés du français, by Jean-Paul Colin (Paris: Hachette-Tchou, 1971).

The alphabet. All the letters of our alphabet are used in French except the w, which appears only in words foreign to French. The following accented letters are used: à, â, ç, é, è, ê, î, ô, ù, û. The cedilla is placed under ç preceding a, o, or u whenever c is soft and should be pronounced like s in *sin* (*façade*). The dieresis is placed over the second of two vowels to show that the vowels do not form a diphthong: *naïf*. Accents may be omitted on the capital A, but they are always used on small caps.

La Société provençale à la fin du moyen âge
LA SOCIÉTÉ PROVENÇALE À LA FIN DU MOYEN ÂGE
LA SOCIÉTÉ PROVENÇALE A LA FIN DU MOYEN AGE

Capitalization. Caps are not used so freely in French as in English. The following practice should be observed.

1. Proper names: In names of persons *La* or *Le* is usually capped, unless the name is taken from the Italian.

La Fayette	le Dante
La Rochefoucauld	le Tasse

In names of places an article is lowercase; likewise common-noun elements like *golfe, mer, mont*.

les États-Unis	la mer Rouge	le mont Blanc

2. Names of the Deity:

le Saint-Esprit	Notre-Seigneur

3. Names of religious festivals and holidays:

la nuit de Noël	la semaine de la Passion

4. Names of historical events and periods:

la Grande Guerre	les croisades
la guerre de Cent ans	le Moyen Age
la Conférence de la paix	

5. Names of streets, squares, buildings, and the like:

la rue Saint-Jacques	le lycée Louis-le-Grand
la place de la Concorde	l'église de Saint-Étienne-du-Mont

6. Names of institutions and orders and their members:

la Légion d'honneur	un jésuite
l'ordre des Templiers	un templier
l'État, l'Église (as institutions)	le protestantisme

7. Titles:

Sa Majesté   M. le Prince
Votre Majesté   le général Gamelin

8. Nationals:

un Anglais   un Irlandais

9. Roman numerals:

Louis XV   tome IX
le XIVe arrondissement   l'acte V
l'an IV

Numerals denoting centuries are usually set in small caps:

XIVe siècle

Caps are not used for the following:

1. Names of the months and the days of the week:

avril 1974   le samedi passé

2. Nouns or adjectives derived from proper names:

la Guerre franco-allemande   l'Académie française
le calendrier julien   Parlez-vous français?

3. The word *saint* in naming the saints themselves (but see 5 above):

saint Paul, saint Louis   l'histoire de sainte Geneviève
L'autre est consacré à saint Maur.

4. The titles *monsieur, madame,* and the like, in direct address (see also page 119):

Oui, monsieur.

5. Salutations of letters:

Mes chers parents:

Titles and captions. The nature of the work in which a French title is used should be considered in deciding upon the best form of capitalization to use. In the text of a book or article in English the title of a French book, magazine, story, poem, or the like is best capped in the same manner as similar titles in English.

> This was Jean Baptiste Lamarck, whose book *La Philosophie Zoologique* appeared in 1809.

In French texts, textbooks in the French language, catalogs of French books, footnotes in English texts when citations of French works are frequent, and similar context, French book titles, headings, captions, and the like should follow a style more in accord with French practice.
The first word and proper nouns should be capped.

> Vue générale de la littérature française
> En Champagne

Also cap the first substantive and a preceding adjective.

> Les Mois et les saisons
> Le Tour du monde en quatre-vingt jours
> Des Hommes exceptionnels
> Les Pauvres Gens
> Les Deux Sourds
> La Vieille Maison retrouvée

> *Note.* The above rules are those most commonly followed by American printers, but the proofreader should not without specific instruction presume to change copy prepared in accordance with the practice indicated by the following:
> Cap proper nouns and cap the first word unless it is *le, la, les,* in which case the next word should be capped.

> Petite pluie   le Monde où l'on s'ennuie
> Une ténébreuse affaire   les Fauts ménages

Divisions. When a word is divided at the end of a line, at least three letters should be carried over unless the lines are very short. Division of abbreviated words should be avoided. Compound words should be divided according to their construction, prefixes and suffixes being kept intact.

> extra-ordinaire   pre-avertir

Words or phrases containing an apostrophe may not be divided at the apostrophe.

> aujour-d'hui   lors-qu'il

A division may usually be made before a consonant standing between vowels.

lé-ga-taire   ba-lan-cer
che-va-lier   jeu-nesse

No division should be made before *x* or *y*, or after *x* or *y* before a vowel or *h*.

roya-liste   soixan-tième   Alexan-dre

When hyphenated phrases like *viendra-t-il* must be divided, the *t* should be carried over.

The combinations *bl, br, ch, cl, cr, dr, fl, fr, gl, gr, gh, ph, pl, pr, th, tr,* and *vr* should not be separated.

per-dre   qua-drille   pro-phé-ti-ser
cé-lé-brer   dé-pê-cher   Vau-ghan
ré-pu-bli-que   ca-tho-li-que   ou-vrier
dé-peu-ple-ment   qua-tre   au-tre-ment
dé-cret   pa-thé-ti-que   li-vrai-son

The combination *gn* should not be divided when it has the sound of *ni* in *onion*, but where the *g* and *n* have separate sounds they may be divided.

com-pa-gnon   in-di-gni-té   diag-nos-ti-que

Any other two consonants may be divided, including doubled consonants.

par-tir   artis-tique   ma-nus-crit   mons-trueux   mil-lion

As a rule, two vowels are not separated.

oeu-vre   théâ-tre   joue-rai   étions   fouet-ter

Mute syllables may be turned over to the next line, but such divisions should be avoided if possible.

élé-gan-ces   mar-che   mar-quent   comp-tes

Punctuation. Commas are not used so much in French as in English. No comma is used before a conjunction linking the last two members of a series of coordinate words, phrases, or clauses. Nonrestrictive relative clauses are often not set off. In dates the month and year are not, as a rule, separated by a comma.

Quotation marks. English quotation marks are often used in French texts and ordinarily in a text of which any part is in English. The French have, however, quotation marks, called guillemets, of a different form—small angle marks resting on the line. They should be set with a thin space between them and the text.

«J'ai, nous a dit le général Gouraud, fait mon livre pour les jeunes.»

Note that fewer quotation marks are considered necessary than would be used in English, since none are used, for instance, after *J'ai* or before *fait* in the preceding example.

If a quotation is two or more paragraphs long, opening guillemets should be used before the first paragraph, closing guillemets before each following paragraph and at the end of the last paragraph. If a quotation occurs within a quotation, the beginning of this subquotation is marked by opening guillemets, and closing guillemets are used before each line of the subquotation and at the end.

> Le reliquat d'*Actes et paroles* donne cette variante inédite:
> «Le moment est venu de jeter sur l'ocean qui nous sépare ce pont immense, la solidarité.
> »Les peuples se parlent par-dessus les gouvernements, les peuples sont de grandes âmes qui s'entendent à travers l'obstacle, et qui proclament les principes pendant que la politique cherche les expédients. En ce moment suprême, le cœur de la France parle au cœur de l'Amérique, et voici ce que la France vous dit:
> «Vous êtes l'admirable Amérique, aucune nation n'est plus vénérable que »vous. Vous êtes colonisateurs et civilisateurs. Vos grands hommes font votre »terre illustre comme la Grèce et l'Italie. L'exemple de la grande vie a été donné »chez vous par Washington et de la grande mort par Lincoln; vous avez eu »Franklin qui a dompté la foudre, vous avez eu Fulton qui a dompté la mer; vous »avez eu John Brown et Peabody qui tous deux ont imité le Christ, l'un du côté »libérateur, l'autre du côté secourable. Peabody la main qui donne, John Brown, »la main qui délivre.» [One pair of guillemets dropped.]
> Six mois avant sa mort, le 24 novembre 1884, la dernière visite officielle que fit Victor Hugo fut d'aller voir la statue de la Liberté, par Bartholdi.

If guillemets and punctuation come together, the following combinations are allowable.

Within a sentence: », !», ?»,

At the end of a sentence: ». »! »? .» !» ?» ?»! !»? !»! ?»?

A period at the end of a quotation is dropped before a comma, an exclamation point, or a question mark; and a sentence period is dropped after a period, an exclamation point, or a question mark.

> Il ajouta: «Oui, cette belle œuvre tend à ce que j'ai toujours aimé, appelé: la paix.» [The sentence period is dropped.]
> Le deuxième jour, il dit: «Que la lumière électrique soit!»

If the quotation is a word or phrase, the sentence punctuation should be outside the guillemets.

> Vous envoyer «l'Illustration».
> Mais chacun ici sait que, si elle a lieu, les agresseurs seront la France, que l'on estime tant, et l'Angleterre, «poussées toutes deux par la juiverie et la finance internationale»!

The dash. In direct discourse, opening quotation marks are often replaced by em dashes (always spaced at least a hair), no closing quotation marks being used.

> —Moi, dit la jeune fille.
> —Et moi, s'écria l'homme âgé.
>
> Le professeur, impatient, s'ecrie malencontreusement:
> —Vous n'avez donc jamais vu la lune?
> Dans le silence, la réponse tombe, inattendue:
> —Non, monsieur!
> N'en pouvant croire ses oreilles, il insiste:
> —Comment! Vous n'avez jamais vu la lune?

The beginning and the end of such conversational matter is sometimes marked by guillemets, with each intervening change of speaker indicated by an em dash; but the two marks should be used together only when the dialogue constitutes a quotation.

Usually a new paragraph marks every change of speaker, but sometimes the words of different speakers are run into one paragraph.

> Un jeune Américain, qui visitait Paris pour la première fois, demanda à un de ses amis où se trouvait «Complet». —Toutes les fois le même mot «Complet». Dites-moi donc où est cet endroit. —Comment! dit l'autre, vous n'avez pas encore vu «Complet»? Il vous faut voir «Complet».

English quotation marks may be preferred, placed at the beginning and at the end of a quoted speech.

> "Partage," dit le lion au loup. "Sire, répondit celui-ci, à vous le morceau de choix, le bœuf." "Tu crois!" s'écria le lion. "Allons à toi, dit-il au renard, partage, et sois juste."

Points of suspension. Points of suspension are three points, set close together (a single type in some fonts), without any space at the left and always followed by a space. They are used in French text much as dashes are used in English—to mark an interruption or a pause or an abrupt change in thought or construction.

—Pierre, François, Jean-Paul, prenez vos bicyclettes, on descend au jardin...
Anne-Marie, Hélène, vous vous occuperez de Gérard.

Le grand-père fait l'appel des petits qui sortent en gai tumulte de la chambre
des jeux. Trois garçonnets, deux fillettes, et une poupée... un poupon, Gérard,
dans les bras de sa grand-mère.

Les céréales, les animaux domestiques, la maison, la cité, la vie sédentaire... «La
plus grande revolution de l'histoire... » Cette parole d'un professeur me
revient à l'esprit.

«L'enseigne fait la chalandise», disait La Fontaine... et l'on voit au XVIIᵉ siècle.
Dès maintenant vous pouvez le faire... et facilement.

Numbers. In numbers of five or more digits, groups of three digits may
be separated by either a point or a thin space. A decimal comma is used instead of a decimal point.

45.600	45 600	45,7

Note the following forms of expressing sums of money or measures of
length or weight:

60 francs	60 mètres	20 kilogrammes
60 fr. 25	60 m. 50	20 kg. 5
60^{fr},25	60^{m},50	20^{kg},5
à 23 h. 15		

Percent is represented by *0/0, pour 100, p. 100, pour 0/0,* or *pour cent.*

Abbreviations. Abbreviations are used freely in French. Some of them
are listed below. Note that no period is used after an abbreviation when the
last letter is the same as the last letter of the word for which it stands.

a., *acte,* act
A.J.C., *avant Jésus-Christ* (used in dates: 41 A.J.C.)
à/l., *à livraison,* on delivery (of goods)
Alt., *Altesse,* Highness
art., *article,* article
a/s. de, *aux soins de,* c/o
av., *avec,* with; *avril,* April
av. J.-C., *avant Jésus-Christ,* before Christ
b. à p., *billets à payer,* bills payable
b. à r., *billets à recevoir,* bills receivable
B. ès L., *Bachelier ès lettres,* Bachelor of Letters
B. ès S., *Bachelier ès sciences,* Bachelor of Science
b.p.f., *bon pour francs,* value in francs
c., *centime(s),* centime(s)
ch., *chapitre,* chapter
cie, *compagnie,* company
c.-à-d., *c'est-à-dire,* that is

Cte, *comte,* Count
déc., *décédé(e),* deceased; *décembre,* December
Dr, *docteur,* Doctor
dzne, *douzaine,* dozen
e.o.o.e., *erreur ou omission exceptée,* errors or omissions excepted
esc., *escompte,* discount
etc., *et cætera,* et cetera
fév., *février,* February
fig., *figure,* figure
fr., f., *franc(s),* franc(s)
h., *heure,* hour
in-f., *in-folio,* folio
in-8°, *in-octavo,* octavo
in-4°, *in-quarto,* quarto
1er, *premier (m.),* first
1ère, *première (f.),* first
IIe, 2e, *deuxième,* second
j., *journal,* newspaper
janv., *janvier,* January
juil., *juillet,* July
l/a., *lettre d'avis,* letter of advice
l/cr., *lettre de crédit,* letter of credit
liv., *livre,* book
LL. MM.,[4] *Leurs Majestés,* Their Majesties
M.,[5] *Monsieur,* Mr. (*pl.* MM.)
m. à m., *mot à mot,* word for word
Me, *maître,* lawyer
Mgr, *monseigneur,* my lord
Mis, *marquis,* marquis
Mise, *marquise,* marchioness
Mlle, *mademoiselle,* Miss
Mme, *madame,* Mrs.
Mn, *maison,* house
MS. or ms., *manuscrit,* manuscript (*pl.* MSS.)
N.-D., *Notre-Dame,* Our Lady
No, *numero,* number (*pl.* Nos)
nov., *novembre,* November
oct., *octobre,* October
p.d.a., *pour dire adieu,* to say good-by
p.ex., *par exemple,* for instance
p.f., *pour féliciter,* to congratulate
p.f.s.a., *pour faire ses adieux,* to say good-by
p.f.v., *pour faire visite,* to make a call
p.p., *publié par,* published by
p.p.c., *pour prendre congé,* to take leave
p.r.v., *pour rendre visite,* to return a call
P.S., *post-scriptum,* postcript
p.v.t., *par voie télégraphique,* by telegraph
R.S.V.P., *Répondez s'il vous plait,* An answer is requested
s., *sur,* on; e.g., Boulogne s/M (= *sur mer*)

---

[4]LL. MM., S. M., and so on, are used only before another title: S. M. l'Empéreur.
[5]The abbreviations M., Mlle, Mme, Mgr, are not used in reproducing direct discourse (see p. 119).

S., Ste, *Saint, Sainte* (for names of saints)
S.A.I., *Son Altesse Imperiale,* His (Her) Imperial Highness (*pl.* SS.AA.II.)
S.A.R., *Son Altesse Royale,* His Royal Highness
sc., *scène,* scene
S.E., *Son Éminence,* His Eminence
s.-ent., *sous-entendu,* understood
s.e.o.o., *sauf erreur ou omission,* errors or omissions excepted
sept., *septembre,* September
S.Exc., *Son Excellence,* His Excellency
s.g.d.g., *sans garantie du gouvernement,* without government guarantee
S.M., *Sa Majesté,* His (Her) Majesty
S.M.I., *Sa Majesté Imperiale,* His (Her) Imperial Majesty
S.S., *Sa Sainteté,* His Holiness
St-, Ste-, *Saint, Sainte* (for names of places)
s.v.p., *s'il vous plait,* if you please
t., *tome,* volume
tit., *titre,* title
t.p., *timbre-poste,* postage stamp
t.s.v.p., *tournez s'il vous plait,* please turn over
v., vol., *volume,* volume
voy., v., *voyez, voir,* see
vve, *veuve,* widow

Phonetics. The following symbols are those commonly used in the phonetic transcription of French.

Letter	English Equivalent	Symbol	Example	
Vowels				
a, â	father	[ɑ]	bas	[bɑ]
a, à	ask	[a]	la	[la]
e	sofa	[ə]	le	[lə]
e, é, ai	made	[e]	les	[le]
e, è ê, ë ai, ei	let	[ɛ]	très	[trɛ]
i, î, ï, ie, y	mach*i*ne	[i]	midi	[midi]
i = y		[j]	vieux	[vjø]
o, au *final or followed by a consonant*	cloth	[ɔ]	mol	[mɔl]
o, ô, au, eau	*go*	[o]	vos	[vo]
u, û	(None)	[y]	une	[yn]
eu, œu	(None)	[ø]	deux	[dø]
eu, œu, ue	(None)	[œ̃]	seul	[sœ̃l]
oi	*wa*sh	[wɑ]	bois	[bwɑ]
oi	*wa*ft	[wa]	voici	[vwasi]
ou	s*ou*p	[u]	sous	[su]
ui	(None)	[ɥ]	suis	[sɥi]
Nasal Vowels				
an, am, en, em	(None)	[ɑ̃]	dans	[dɑ̃]
			encore	[ɑ̃kɔːr]
in, im, ain, aim, ein, eim, yn, ym	(None)	[ɛ̃]	fin	[fɛ̃]
			main	[mɛ̃]
			faim	[fɛ̃]
on, om	(None)	[ɔ̃]	bon	[bɔ̃]
un, um	(None)	[œ̃]	lundi	[lœ̃di]

LETTER	ENGLISH EQUIVALENT	SYMBOL	EXAMPLE	
Consonants				
b	*bat*	[b]	balle	[bal]
c (*before* a, o, u)	*coat*	[k]	car	[kar]
c (*before* e, i, y)	*cent*	[s]	ici	[isi]
ch	ma*ch*ine	[ʃ]	chaise	[ʃɛːz]
d	*d*are	[d]	des	[dɛ]
f	*f*eet	[f]	face	[fas]
g (*before* a, o, u)	*g*o	[g]	gare	[gaːr]
g (*before* e, i, γ)	plea*s*ure	[ʒ]	geste	[ʒɛst]
gn	o*ni*on	[ɲ]	digne	[diɲ]
h	(Silent)		homme	[ɔm]
j	plea*s*ure	[ʒ]	joli	[ʒɔli]
k	*k*ick	[k]	képi	[kepi]
l	*l*ate	[l]	lait	[lɛ]
ll (*after* i) but ville-[vil]	*y*et	[j]	famille	[famiːj]
m	*m*an	[m]	midi	[midi]
n	*n*eed	[n]	net	[nɛt]
p	*p*age	[p]	paix	[pɛ]
q (qu)	*k*ick	[k]	qui	[ki]
r	*r*ay	[r]	race	[ras]
s	mi*ss*	[s]	salle	[sal]
s (*between vowels*)	ga*z*e	[z]	église	[egliːz]
t	*t*able	[t]	table	[tabl]
v	*v*ery	[v]	votre	[vɔtr]
w	*v*ery	[v]	wagon	[vagɔ̃]
x	bo*x*	[ks]	luxe	[lyks]
x	ba*gs*	[gz]	exemple	[egzɑ̃ːpl]
z	ga*z*e	[z]	douze	[duːz]

: is a sign of length

GERMAN

The alphabet. The German alphabet contains the following letters:

A	𝔄		a	𝔞		N	𝔑		n	𝔫
B	𝔅		b	𝔟		O	𝔒		o	𝔬
C	ℭ		c	𝔠		P	𝔓		p	𝔭
D	𝔇		d	𝔡		Q	𝔔		q	𝔮
E	𝔈		e	𝔢		R	𝔑		r	𝔯
F	𝔉		f	𝔣		S	𝔖		s	ß ſ
G	𝔊		g	𝔤		T	𝔗		t	𝔱
H	𝔥		h	𝔥		U	𝔘		u	𝔲
I	ℑ		i	𝔦		V	𝔙		v	𝔳
J	ℑ		j	𝔧		W	𝔚		w	𝔴
K	𝔎		k	𝔨		X	𝔛		x	𝔵
L	𝔏		l	𝔩		Y	𝔜		y	𝔶
M	𝔐		m	𝔪		Z	𝔷		z	𝔷

Three umlauted letters are used: ä, ö, ü and the diphthong äu.

Fonts of German type, technically known as *Fraktur*, contain no small caps or italic.

Several of the letters in *Fraktur* resemble other letters so closely that they are likely to be confused. Such letters are shown together and the difference is pointed out.

𝕭 (B) and 𝖁 (V). *V* is open in the middle; *B* is joined across.

ℭ (C) and 𝕰 (E). *E* has a little stroke in the middle, projecting to the right; *C* does not.

𝕲 (G) and 𝕾 (S). *S* has an opening at the upper right; *G* is closed, and has, besides, a perpendicular stroke within.

ℑ (I) and ℑ (J) have the same form—ℑ. If a consonant follows, the letter is *I*; if a vowel, it is *J*. (Sometimes the two caps differ slightly in form.)

𝕶 (K), 𝕹 (N), 𝕽 (R). *K* is rounded at the top; *N* is open in the middle; *R*, like *K*, is united about the middle.

𝕸 (M) and 𝖂 (W). *M* is open at the bottom; *W* is closed.

𝔟 (b) and 𝔥 (h). *b* is entirely closed below; *h* is somewhat open, and ends at the bottom, on one side, with a projecting hair-stroke.

𝔣 (f) and ſ (s). *f* has a horizontal line *through* it; *s*, on the left side only.

𝔪 (m) and 𝔴 (w). *m* is entirely open at the bottom; *w* is partly closed.

𝔯 (r) and 𝔵 (x). *x* has a little hair-stroke below, on the left.

𝔳 (v) and 𝔶 (y). *v* is closed; *y* is somewhat open below, and ends with a hair-stroke. Note also that the German *y* looks much like an English *n*.

Ligatures. Fonts of German type contain the following ligatures: ch, ck, ff, fi, fl, ll, ſi, ſſ, ſt, ß, tz. These letters are not united, however, if they are brought together by the combination of prefixes or suffixes with roots. For instance:

auffahren	vielleicht
Auflage	dasselbe
schlaflos	achtzig
Wasserstoffionen	entzweien
verwerflich	heutzutage

There are no triple letters corresponding to English ffi and ffl. These combinations are set ſtoffig, trefflich, and so on.

When German is printed in roman type (called by the Germans *Latein-isch*), as much of it is nowadays, a special character, ß, is used for ß. But many prefer the simple roman ss. (The substitution of sz for this combination is erroneous, yet unfortunately it has been used.)

Use of ß and ſ. The letter ß is used at the end of words, prefixes, and roots; the long ſ, in all other positions except in some words of foreign origin.

Haus, Häuschen   bis, bisweilen   ſagen   leſen   Schickſal

The following prefixes end with an $:

aus=   bis=   bis=   bys=   ins=   plus=   fus=   trans=

The long ſ is used at the beginning of a root following the prefixes bi=, ſu=, or tran=.

The round $ is used at the end of a root before the following suffixes:

=bar	=ka	=ler	=los	=tum
=chen	=ke	=lich	=ma	=ver
=haft	=ker	=ling	=mus	=voll
=heit	=kus			

Capitalization. German practices in capitalization vary widely from English. Caps are used for the following:

1. The first word of a sentence.
2. The first word of a line of poetry.
3. The first word of a formal quotation.
4. Proper names. The preposition von preceding a name is always set lowercase in German, regardless of its position in the sentence or of whether a title or name precedes. A thin space between von and the name is sufficient.

Es war von Bismarck.

5. Nouns and words used as nouns.

der Mann   die Armen

Certain nouns are set lowercase when used to form part of an adverbial phrase:

heute morgen   gestern abend

Neuter adjectives used as nouns after etwas, viel, nichts, alles, and allerlei should be capped.

etwas Gutes für seine Arbeit

6. The pronoun Sie, "you," and its possessive form Ihr, "yours."
7. Adjectives and pronouns in titles.

Eure Königliche Hoheit   Seine Majestät

In contrast to English usage, caps are not used for the following:

1. Proper adjectives, except those derived from the names of persons and those used in geographical proper names.

die Grimmſchen Märchen   in den deutſchöſterreichiſchen
der Kölner Dom   Provinzen

2. Proper adjectives denoting nationality, unless they are used in titles.

das deutſche Volk   das Deutſche Reich

Note auf deutſch, "in German," but Studieren Sie Deutſch? "Are you studying German?"

Capitalization in titles, headings, captions, and the like is the same as in text.

Division. When a word is divided at the end of a line, at least three letters should be carried over unless the lines are very short.

Compound words, as a rule, should be divided according to their elements—prefixes and suffixes should be kept intact.

dar-auf   hin-auf   Auf-lage
Haus-tür   dort-hin   aus-ſehen

In simple words a division may be made before a single consonant between two vowels.

tra-gen   Lie-fe-rung   Ver-tei-lung
Träu-me-rei   Phy-ſio-lo-gie   Li-te-ra-tur

The combinations ch, ſch, ſt, ph and th are never separated, unless they belong to different elements of a compound.

brau-chen   ger-ma-ni-ſchen   Fen-ſter
wa-ſchen   Be-ſte   Ko-ſten

When two or more consonants other than the combinations just noted occur between vowels, usually only the last goes with the following vowel; but in words of foreign origin-b, p, d, t, g, k, when followed by l or r, go with the next syllable.

Fort-bil-dung   kämp-fen   Re-pu-blik
Waſ-ſer   Bi-bli-o-thek   Qua-drat
ſel-ten   Fe-bru-ar

If a word is divided at ck the c should be changed to k.

ſchrecken, ſchrek-ken   Glocken, Glok-ken
Flicken, Flik-ken   rucken, ruk-ken

Three identical letters should not come together; therefore the form Schiffahrt is correct, not Schifffahrt. But if such a word as this is divided, the third letter should be restored:

Schiffahrt, Schiff-fahrt   Schalloch, Schall-loch

Punctuation. Dependent clauses are always set off in German, no distinction being made between restrictive and nonrestrictive as in English. Therefore a comma is used before daß, der, die, das, welcher, womit, and other subordinate conjunctions, relative pronouns, and the like.

Die Dame, die uns jetzt besucht, ist sehr reich.

A comma is used before ohne . . . zu, um . . . zu, and anstatt . . . zu (all of which are followed by an infinitive).

Er ging an mir vorbei, ohne etwas zu sagen.

A comma is ordinarily used before the coordinate conjunctions und and oder, provided the following clause contains both subject and verb.

Die Luft ist blau, und die Felder sind grün.
Er legte sich hin und schlief sogleich ein.

The word aber, meaning "however," is not set off by commas; no comma is used before und or oder in a series of words or phrases; or to separate month and year in dates.

Der alte Mann aber verlor den Mut nicht.
Karl, Fritz und Johann sind meine besten Freunde.
6. August 1936.

No comma is used before usw.
Usually exclamation points are used after the salutations of letters and after commands.

Lieber Vater!   Folgen Sie mir!

Quotation marks („   ") are used as in English, to enclose a direct quotation. A colon regularly precedes a quotation.

Der Fuchs sprach: „Die Trauben sind mir zu sauer."

If a word or phrase is enclosed in quotation marks, the closing quotation marks should precede a period, a comma or a semicolon that is required by the sentence at that point.

Der „Gebirgspfarrer", der allerdings ähnlich wie „Germelshausen" . . . das Trauerspiel „Eines Bürgers Recht"; das Volksdrama „Theophrastus Paracelsus".

A quotation within a quotation is marked by single quotation marks of the same font as the double quotes.

„Der ‚Gebirgspfarrer', der allerdings ähnlich wie ‚Germelshausen' . . ."

The apostrophe is used to denote the possessive only after proper names ending in ß, ß, or ʒ. In other cases ß only is used.

Demosthenes' Reden   Schillers Gedichte
Horaʒ' Oden

The apostrophe is used to denote the elision of letters in ist's, geht's, hab' ich, and the like, but not when a preposition and das are merged or when the final e of the imperative singular of a verb is suppressed: ans, ins, durchs, fürs; komm! sag!

Italics. Italics are represented in German type by letterspacing. In spaced words the following letters should never be separated unless they belong to separate elements of a compound: ch, ck, ft, ß, ß; but ff, fi, fl, ll, fi, and ff may be spaced. If the composition is in roman type, all but ß, may be separated.

Denn alle, mit alleiniger Ausnahme der im deutschen und im österreichisch-ungarischen Heere noch heute vorgeschriebenen Schreibung K o m p a g n i e neben Kompanie, beseitigte n u r der „Buchdrucker-Duden".

Numbers. In numbers of four or more figures groups of three digits are set off from those on the left by either a thin space or a period. Instead of the English decimal point, German usage requires a comma.

200 000 or 200.000   2,50 rm.   76,5%

Superior figures referring to footnotes precede punctuation.

. . . eingegeben[1].

Ordinal numerals are indicated by using a period after the cardinal number.

1, ein, one   1., erste, first   2. Band, second volume

Dates. Several forms of writing dates are acceptable practice in German. For instance:

den 3. Mai 1930   d. 3. Mai 1930   3. M. 1930

Hyphen. In a series of compounds in which the common element is expressed only in the last of the series, the suppressed element is represented by a hyphen.

Käse-, Butter- und Milchvorräte, cheese, butter, and milk stores.

Accented letters. Roman accented letters are sometimes used with German type in such foreign words as *Café*.

Abbreviations. Abbreviations are used much more freely in German than in English. Following are some of the standard German abbreviations.

a., *akzeptiert,* accepted
A., *Acker,* acre
a.a.O., *am angeführten Orte,* in the place cited
Abh., *Abhandlungen,* transactions (of a society)
Abschn., *Abschnitt,* section
Abt., *Abteilung,* division
a.c., *anni currentis,* current year
A.G., *Aktiengesellschaft,* joint stock company
Anm., *Anmerkung,* note
Art., *Artikel,* article
A.T., *Altes Testament,* Old Testament
Aufl., *Auflage,* impression
Ausg., *Ausgabe,* edition
Bd., *Band,* volume; Bde., *Bände,* vols.
bearb., *bearbeitet,* edited, compiled, adapted
beif, beiflgd., *beifolgend,* (sent) herewith
beigeb., *beigebunden,* bound (in with something else)
bes., *besonders,* especially
bez., *bezüglich,* respecting
Bl., *Blatt,* newspaper
bzw., *beziehungsweise,* respectively
ca., *circa,* about
d.G., *durch Güte,* kindness of
dgl., *dergleichen,* such like, similar case
d.h., *das heisst,* that is to say (i.e.)
d.i., *das ist,* that is
d.l.M., *des laufenden Monats,* of the current month
Dr., *Doktor,* Dr.
Dr.u.Vrl., *Druck und Verlag,* publisher, printed and published by
D-Zug., *Durchgangs-Zug,* a through train
ebd., *ebenda* (ibid.)
E.M.K., *electromotorische Kraft,* electromotive force
eng., *englisch,* English
entspr., *entsprechend,* corresponding
Ew., *Euer,* your
Exz., *Exzellenz,* Excellency
f., *folgende Seite,* next page

ff., *folgende*, following
F.f., *Fortsetzung folgt*, to be continued
Forts., *Fortsetzung*, continuation
fr., *franko*, postpaid; *frei*, free
Fr., *Frau*, Mrs.
Frhr., *Freiherr*, Baron
Frl., *Fräulein*, Miss
F.u.S.f., *Fortsetzung und Schluss folgen*, to be continued and concluded
geb., *geboren*, born; *gebunden*, bound
Geb., *Gebrüder*, brothers
ges., *gesammelte*, collected, compiled; *gesamt*, complete
Ges., *Gesellschaft*, company or society
Gesch., *Geschichte*, history
gest., *gestorben*, deceased
G.m.b.H., *Gesellschaft mit beschränkter Haftung*, corporation with limited
   liability
Gr. 8°, *Grossoktav*, large 8vo
g.u.v., *gerecht und vollkommen*, correct and complete
H., *Heft*, number or part (of a publication)
hl., *heilig*, holy
hrsg., *herausgegeben*, published
Hs., *Handschrift*, manuscript (*pl.* Hss.)
i.a., *im allgemeinen*, in general
i.A., *im Auftrage*, by order of
I.G., *Interessengemeinschaft*, amalgamation, trust
i.J., *im Jahre*, in the year
i.J.d.W., *im Jahre der Welt*, in the year of the world
I.K.H., *Ihre königliche Hoheit*, Her Royal Highness
Ing., *Ingenieur*, engineer
J., *Jahr*, year
Jahrg., *Jahrgang*, annual set
Kap., *Kapitel*, chapter
kgl., *königlich*, royal
Kl., *Klasse*, class
Lief., *Lieferung*, number
M., *Mark*, mark (coin)
m.E., *meines Erachtens*, in my opinion
M.E.Z., *Mitteleuropäische Zeit*, Middle European time
näml., *nämlich*, namely
n.Chr., *nach Christo*, anno Domini
n.F., *neue Folge*, new series
no., *netto*, net
Nr., *Nummer*, number
n.S., *nächste Seite*, next page
od., *oder*, or
O.W., *österreichische Währung*, Austrian currency
p.c., *pro centum*, percent
Pf., *Pfennig*, penny
Pfd., *Pfund*, pound
Q., *Quadrat*, square
Rab., *Rabatt*, discount
resp., *respektiv*, respectively
Rm., *Reichsmark*, reichsmark (coin)
s., *siehe*, see
S., *Seite*, page
sämt., *sämtliche*, complete
sel., *selig*, deceased, late

Ser., *Serie,* series
S.M., *Seine Majestät,* His Majesty
s.o., *siehe oben,* see above
sog., *sogenannt,* so-called
Sp., *Spalte,* column
st., *statt,* instead of
St., *Stück,* each; *Sankt,* Saint
s.u., *siehe unten,* see below
Thlr., *Thaler,* dollar
u., *und,* and; *unter,* among
U., *Uhr,* clock, o'clock
u.a., *unter anderen,* among others; *und andere,* and others
u.a.m., *und andere mehr,* and others
u.A.w.g., *um Antwort wird gebeten,* an answer is requested
übers., *übersetzt,* translated
u.dgl., *und dergleichen,* and the like
unbest., *unbestimmt,* indefinite
usf., *und so fort,* and so on
usw., *und so weiter,* etc., and so forth
u.v.a., *und viele andere,* and many others
v.Chr., *vor Christus,* before Christ
v.d., *von der* (in names, as *J. v.d. Traum,* written in full *Julius von der Traum*)
verb., *verbessert,* improved, revised
verm., *vermehrt,* augmented
Ver.St., *Vereinigte Staaten,* U.S.A.
vgl., *vergleichen,* compare
v.H., *vom Hundert,* of the hundred
v.M., *vorigen Monats,* last month
v.o., *von oben,* from above
vorm., *vormittags,* in the forenoon, a.m.; *vormals,* formerly
v.u., *von unten,* from below
Wwe, *Witwe,* widow
Xr., *Kreuzer,* cruiser; *Kreutzer* (a coin)
z., *zur,* to the
Z., *Zeile,* line; *Zoll,* inch, toll
z.B., *zum Beispiel,* for example (e.g.)
Zf., *Zeitschrift,* periodical
z.T., *zum Teil,* partly, in part
Ztr., *Zentner,* hundredweight
zw., *zwischen,* between
z.Z., *zur Zeit,* at present, now

CLASSIC GREEK

The alphabet. The Greek language used twenty-four letters.

FORMS			NAMES	SOUNDS
	Italic	Roman		
A	*α*	α	alpha	a
B	*б β*	β	beta	b
Γ	*γ*	γ	gamma	g
Δ	*δ*	δ	delta	d
E	*ε*	ε	epsilon	e short
Z	*ζ*	ζ	zeta	z
H	*η*	η	eta	e long

Θ	θ	θ	theta	th
I	ι	ι	iota	i
K	κ	κ	kappa	k, c
Λ	λ	λ	lambda	l
M	μ	μ	mu	m
N	ν	ν	nu	n
Ξ	ξ	ξ	xi	x
O	ο	ο	omicron	o short
Π	π	π	pi	p
P	ρ	ρ	rho	r
Σ	σ	σ ς	sigma	s
T	τ	τ	tau	t
Y	υ	υ	upsilon	u
Φ	φ	φ	phi	ph
X	χ	χ	chi	ch
Ψ	ψ	ψ	psi	ps
Ω	ω	ω	omega	o long

The letters ϐ and ϑ are rare forms of the delta, and should not be used as symbols.

The letter σ is the sigma used at the beginning or in the middle of a word; ς is used at the end of a word.

Accents and aspirates. The accents and aspirates used in the Greek are as follows:

- ᾽ lenis (smooth breathing)
- ῾ asper (rough breathing)
- ´ acute
- ` grave
- ῀ circumflex

- ῎ lenis acute
- ῍ lenis grave
- ῞ asper acute
- ῝ asper grave
- ῟ circumflex lenis

- ῏ circumflex asper
- ·· dieresis
- ∴ dieresis acute
- ∵ dieresis grave

These are placed over small letters, in front of caps. In diphthongs the breathing is placed over the second vowel. Care should be taken not to misread a lenis for an apostrophe, or vice versa.

The acute accent is used only upon one of the last three syllables of a word. An acute accent on the last syllable is changed to a grave accent when another accented word immediately follows it in the same clause. The circumflex accent may be used only on one of the last two syllables of a word. The grave accent may be used only on the last syllable. A dieresis is used over ι or υ to show that it is a separate letter, not part of a diphthong.

Subscript. The tiny ι sometimes appearing under vowels is called the "subscript iota": καιρῷ

Diastole. The diastole, which looks like a comma, is placed between two letters, without spacing on either side. Its purpose is to distinguish certain words from other words of the same form: ὅ,τε, neuter of ὅστε, distinguished from ὅτε, conjunction.

Capitalization. The ancient Greeks used only capital letters; so when caps and lowercase letters are used, the capitalization follows a practice established by modern editors, among whom there is some difference of opinion. Caps are used for the following:

1. The first word of a paragraph
2. The first word of a stanza of poetry, but not for the first word of each line
3. The first word of a sentence, as a rule
4. The first word of a direct quotation
5. Proper names

Punctuation. The sign of interrogation is a semicolon. The colon and semicolon are represented by a raised period (·). The apostrophe is used to mark the omission of a vowel or diphthong; when this occurs at the end of a word, the ordinary spacing of the rest of the line is used before the word which immediately follows. As punctuation was not used by the ancient Greeks, modern editors generally follow English style (except for the interrogation mark, colon, and semicolon) when they insert it.

Divisions. When a word is divided at the end of a line, at least two letters should remain on the first line and no less than three letters should be carried over unless the lines are very short. Compounds should be divided according to their construction, prefixes and suffixes being kept intact. Great care must be used not to carry over a final consonant belonging to the first element, as, for example, the prefixes ἀμφ-, εἰσ-, κατ-, καθ-, προσ-, συγ-.

A Greek word has as many syllables as it has vowels or diphthongs. The following are the principal diphthongs in Greek:

αι	υι	ου
ει	αυ	ηυ
οι	ευ	

In a simple word a single consonant between vowels belongs with the vowel following it.

χαλε-πῶν   τού-τοις

A combination of consonants that can begin a word should not be separated. The following forty-one combinations should not be separated:

βδ,	θλ,	μν,	σμ,	φθ,
βλ,	θν,	πλ,	σπ,	φλ,
βρ,	θρ,	πν,	στ,	φν,
γλ,	κλ,	πρ,	σφ,	φρ,
γν,	κμ,	πτ,	σχ,	χθ,
γρ,	κν,	σβ,	τλ,	χλ,
δμ,	κρ,	σθ,	τμ,	χν,
δν,	κτ,	σκ,	τρ,	χρ,
δρ,				

Other combinations of consonants may be divided.

$$\pi o\lambda\text{-}\lambda\acute{\alpha}\kappa\iota\varsigma$$

Numerals. Numerals may be represented in Greek by the letters of the alphabet together with an accent mark placed after the letter or inverted before the letter. Three special characters are used; ς', called *stigma*, for 6; Ϙ', called *koppa*, for 90; and ϡ, called *sampi*, for 900. (Another symbol for 6 is the digamma F.)

CARDINAL NUMBERS

1 α'	19 ιθ'	700 ψ'
2 β'	20 κ'	800 ω'
3 γ'	21 κα'	900 ϡ'
4 δ'	22 κβ'	1000 ͵α
5 ε'	23 κγ'	1111 ͵αρια'
6 ς'	30 λ'	2000 ͵β
7 ζ'	40 μ'	2222 ͵βσκβ'
8 η'	50 ν'	3000 ͵γ
9 θ'	60 ξ'	4000 ͵δ
10 ι'	70 ο'	5000 ͵ε
11 ια'	80 π'	6000 ͵ς
12 ιβ'	90 Ϙ'	7000 ͵ζ
13 ιγ'	100 ρ'	8000 ͵η
14 ιδ'	200 σ'	9000 ͵θ
15 ιε'	300 τ'	10,000 ͵ι
16 ις'	400 υ'	20,000 ͵κ
17 ιζ'	500 φ'	100,000 ͵ρ
18 ιη'	600 χ'	

HEBREW

(The material on the composition of Hebrew is reprinted from the *Government Printing Office Style Manual*, 1973.)

Alphabet, transliteration, and pronunciation

	Name	*Translitera-tion*	*Phonetic value*	*Numeral value*
א	'Alef	' or omit	originally a glottal stop; now silent	1
ב	Bēth	*b, v*	*b, v*	2
ג	Gīmel	*g*	*g* in go	3

		Name	Transliteration	Phonetic value	Numeral value
ד		Daleth	d	d	4
ה		Hē	h	h; silent at end of word	5
ו		Wāw	w	originally w; now v	6
ז		Zayin	z	z	7
ח		Ḥēth	ḥ	a strong h	8
ט		Ṭēth	ṭ	originally emphatic t; now t	9
י		Yōd	y	y in yes	10
כ	ך	Kaf	k, kh	k, kh as German ch	20
ל		Lamed	l	l	30
מ	ם	Mēm	m	m	40
נ	ן	Nūn	n	n	50
ס		Samekh	s	s in so	60
ע		ʿAyin	ʿ	originally a laryngal voiced spirant; now silent	70
פ	ף	Pē	p, f	p, f	80
צ	ץ	Ṣadē	ṣ	originally emphatic s; now ts in pets	90
ק		Qōf	q	originally velar k; now k	100
ר		Rēsh	r	r, as in French uvular or Italian trilled	200
ש		Śīn, Shīn	ś, sh	ś; originally palatal; now s in so; sh as in shoe	300
ת		Tāw	t	t; originally also like th in thin	400

Hebrew uses no capitals at beginning of words, such as proper names.

Hebrew follows English and American usage with regard to quotation marks and italics.

In transliteration, especially of names, the macrons over vowels and the dots under consonants, as well as ʾ and ʿ, are often omitted; ʿ is also printed as ʾ. For f, ph is often used. For ś, an ordinary s is often found, and then samekh is sometimes represented by ś. For sh, š is sometimes used, especially in scholarly works. There are other special transliteration practices to be found in scholarly works.

Hebrew is read from right to left. Its alphabet consists of 22 letters, all consonants; the vowels are represented by vowel signs or points, as explained under Vowels below.

Special characters

Five of the letters (*kaf, mēm, nūn, pē,* and *ṣadē*) have a so-called final form, shown immediately to the right of its respective regular form. This final form is used as the final letter of a word.

Eight of the letters represent two sounds each, distinguished by means of a dot, as follows:

ב	as *b* or *v*	בּ	as *b* or *bb*
ג	as *g;* also like Dutch *g*	גּ	as *g* in big, *gg*
ד	as *d;* and like *th* in then	דּ	as *d, dd*
ה	as *h* or silent	הּ	as *hh* (stronger aspiration)
כ	as *k* or German *ch*	כּ	as *k, kk*
פ	as *p* or *f*	פּ	as *p, pp*
שׁ	as *sh*	שׂ	as *s* in sin
ת	as *t* or *th*	תּ	as *t, tt*

Some of the letters seem to be more or less similar. These are grouped, for the convenience of identification, within brackets below:

Vowels

The vowels are represented by marks called vowel points. These are placed above or below the consonant and, with the exception of the furtive pataḥ, have the effect of a vowel following the consonant; e.g., בַ (*ba*), בֵ (*bē*). The forms, names, and sounds of the vowels are as follows:

Long Vowels		Short Vowels	
֫ Qameṣ *ā*	*a* as in palm	֫ Pataḥ *a*	*a* as in part (short)
֑ Ṣere *ē*	*ei* as in vein	֫ Segol *e*	*e* as in bed
֑ Hirik gadol *ī*	*i* as in machine	֫ Hirik katon *i*	*i* as in big
֑ Holam *ō*	*o* as in no	֫ Qameṣ katon *o*	*o* as in soft
֑ Shuruk *ū*	*oo* as in moon	֫ Kubbuts *u*	*u* as in full

The furtive pataḥ

All vowels are pronounced as if they follow the consonant to which they are ascribed, with the exception of final ֻ, which is pronounced not *ḥa*, but *ah*. This pataḥ is termed "furtive pataḥ."

The shwa

Sometimes shwa represents the sound of the first *e* in believe; e.g., שְׁמַע (shema); it may be transliterated *ᵉ*. At other times it is not pronounced, as in אברם (avrom), so that a consonant cluster results. Also, shwa is written, according to certain rules for writing Hebrew, before the points for *a, e,* and *o* to represent a very short vowel; e.g., חֲלִי,אֱמֶת,אֲנִי. These vowel point combinations, ֲ, ֱ, and ֳ are transliterated *ă, ĕ,* and *ŏ,* respectively.

Punctuation and accentuation

Although the principles and marks of punctuation in modern Hebrew are, in the main, as in English, Scriptural Hebrew employs, in addition to the vowel points, 21 accent marks, which are placed either singly or in various combinations above or below the consonantal characters they modify. These have a threefold object: (*a*) to indicate stress; (*b*) to direct cantillation—the chanting in which the Scriptures are intoned; and (*c*) to indicate distinctions in the meanings of words, e.g., בָּנוּ they build, but בָּנוּ in us.

As marks of cantillation, accent marks are divided into two classes: disjunctives and conjunctives, the former corresponding to marks of separation in English—the period, semicolon, comma, etc., the latter indicating that the word bearing them is connected in sense with that which follows. The table presents the forms, names, and classifications of these accents:

Disjunctives

Form	EMPERORS (קֵסָרִים)	Name	Form	PRINCES (מִשְׁנִים)	Name
			בֿ Zarqā'		זַרְקָא
בֿ Silluq		סִלּוּק	בֿ Paštā'		פַּשְׁטָא
בֿ 'Ethnāh		אֶתְנָח	בֿ Yᵉthīv		יְתִיב
			בֿ Tᵉvīr		תְּבִיר
			בֿ 'Azlā'		אַזְלָא
	KINGS (מְלָכִים)		בֿ Gērēš		גֶּרֶשׁ
			בֿ Gēršayīm		גֵּרְשַׁיִם
בֿ Sᵉgōltā'		סְגוֹלְתָּא	COUNTS (שְׁלִישִׁים)		
בֿ Zāqēf Qāṭōn	זָקֵף קָטֹן		בֿ Pāzēr		פָּזֵר
בֿ Zāqēf Gādōl	זָקֵף גָּדוֹל		בֿ Qarnēy Fārāh		קַרְנֵי פָרָה
בֿ Ṭippᵉḥā'		טִפְּחָא	בֿ Tᵉlīšāh Gᵉdōlāh		תְּלִישָׁה גְדוֹלָה
בֿ Rᵉvīaʿ		רְבִיעַ	בֿ Tᵉlīšāh Qᵉṭannāh		תְּלִישָׁה קְטַנָּה
בֿ Šalšeleth		שַׁלְשֶׁלֶת	בֿ | Pᵉsīq		פְּסִיק

Conjunctives

Form	Name		Form	Name	
בֿ Mūnaḥ	מוּנַח	בֿ Dargā'		דַּרְגָּא
בֿ Mahpakh	מַהְפָּךְ	בֿ Merkā'		מֵרְכָא
בֿ Qadmā'	קַדְמָא	בֿ Merkā' Kᵉfūlāh		מֵרְכָא כְּפוּלָה

There are also three supplementary marks of interpunction: The *soph-pasuk* (:), terminal mark of a verse; the *pesik* (|), for a pause within the verse; and *makkeph* (-), the elevated hyphen between words.

Syllabification
Words in modern Hebrew may be divided between syllables of three or more letters.

The calendar
The Hebrew calendar was given its present fixed form by Hillel II about A.D. 360. It is based on a year of 12 months, alternating 30 and 29 days, with an intercalary month of 29 days in leap year. These months, with their corresponding periods in the Gregorian calendar, are as follows:

Tishri	תשרי	September–October
Heshvan	חשון	October–November
Kislev	כסלו	November–December
Tevet	טבת	December–January
Shevat	שבט	January–February
Adar	אדר	February–March
Veadar	ואדר	Intercalary month
Nisan	ניסן	March–April
Iyar	איר	April–May
Sivan	סיון	May–June
Tammuz	תמוז	June–July
Av	אב	July–August
Elul	אלול	August–September

The year begins on the first day of the month of Tishri, which is the day of the Molad, or appearance of the new moon, nearest the autumnal equinox. The actual date is, however, sometimes shifted 1 or 2 days, according to specific regulations; thus, New Year may not fall on either a Friday or a Sunday, since that would conflict with the observance of the Sabbath; nor, for a like reason, may it come on a Wednesday, since that would cause Atonement Day to come on a Friday.

To convert a given year (anno Domini) into its corresponding Hebrew year (anno mundi), add 3,760 to the former, bearing in mind, however, that the year begins in September. As the Hebrew calendar omits the thousands, the year 5705, corresponding to the Christian year 1945, is represented in Hebrew characters by תשה, 705, these characters, as already explained, denoting 400, 300, and 5, respectively.

The days of the week are referred to as first day, second day, etc., the seventh being called Sabbath (שבת). The holidays, festivals, and fasts, with their dates, are as follows:

Rosh Hashana (New Year, Tishri 1)	ראש השנה
Tsom Gedaliah (Fast of Gedaliah, Tishri 3)	צום גדליה
Yom Kippur (Day of Atonement, Tishri 10)	יום כפור
Sukkoth (Feast of Tabernacles, Tishri 15–22)	סכות
Simhath Torah (Rejoicing Over the Law, Tishri 23)	שמחת תורה
Hanukkah (Feast of Dedication, Kislev 25)	חנכה
Asarah be-Tevet (Fast of Tevet, Tevet 10)	עשרה בטבת
Purim (Feast of Lots, Adar 14)	פורים
Pesah (Passover, Nisan 15–21)	פסח
Shabuoth (Feast of Weeks, Sivan 6)	שבועות
Tishah be-Av (Fast of Av, Av 9)	תשעה באב

Abbreviations

In Hebrew, abbreviations are set as follows: If of one letter, one prime mark (') is used after the letter; if of more than one letter, a double prime ('') is used just before the last letter. Vowel points are always omitted. The abbreviations most frequently used are as follows:

Sir, Master, Mr.; thousand	א', אדון; אלף
Aleph Beth (the alphabet)	א״ב, אלף בית
Said our learned ones of blessed memory	אחז״ל, אמרו חכמינו זכרונם לברכה
The Land of Israel (Palestine)	א״י, ארץ ישראל
God willing	אי״ה, אם ירצה השם
Synagogue	בהכנ״ס, בית הכנסת
Sons of Israel, the Jews	ב״י, בני ישראל
In these words, viz	בזה״ל, בזה הלשון
The author	בע״מ, בעל מחבר
Gaon (title of Jewish princes in the Babylonian exile), His Highness, His Majesty.	ג', גאון

The laws of Israel	ד"י, דיני ישראל
The Holy One, Blessed be He (the Lord)	הקב"ה, הקדוש ברוך הוא
Destruction of the First Temple	חב"ר, חרבן בית ראשון
Destruction of the Second Temple	חב"ש, חרבן בית שני
Exodus from Egypt	יצ"מ, יציאת מצרים
As it was said; as it was written	כמ"ש, כמו שנאמר; כמו שכתב
A.M. (anno mundi)	לב"ע, לבריאת עולם
The Holy Language (Hebrew)	לה"ק, לשון הקדש
Good luck; I congratulate you	מז"ט, מזל טוב
The Sacred Books	סה ק, ספרים הקדושים
The Holy Scroll	ס"ת, ספר תורה
May he rest in peace	ע"ה, עליו השלום
In the Hereafter	עוה"ב, עולם הבא
New Year's Eve	ער"ה, ערב ראש השנה
Sabbath Eve	ע"ש, ערב שבת
Verse; chapter	פ', פסוק; פרק
The judgment of the court	פב"ד, פסק בית דין
Saint (St.); Zion	צ', צדיק; ציון
Recognition of God's justice	צה"ד, צדוק הדין
The reading of the Holy Scroll	קה"ת, קריאת התורה
First of all	קכ"ד, קדם כל דבר
Our Rabbis of Blessed Memory	רו"ל, רבותינו זכרונם לברכה
Rabbi Moses, son of Maimon (Maimonides)	רמב"ם, ר' משה בן מימון
Catalog	רש"ס, רשימת ספרים
Year; line; hour	ש , שנה; שורה; שעה
Sabbath days and holidays	שוי"ט, שבתות וימים טובים
As stated	שנ', שנאמר
Babylonian Talmud	ת"ב, תלמוד בבלי
The Books of the Law, the Prophets, and Hagiographa (Old Testament)	תנ"ך, תורה, נביאים, כתובים

Cardinal numbers

one	אחד, אחת	twenty	עשרים
two	שנים, שתים	thirty	שלשים
three	שלשה, שלש	forty	ארבעים
four	ארבעה, ארבע	fifty	חמשים
five	חמשה, חמש	sixty	ששים
six	ששה, שש	seventy	שבעים
seven	שבעה, שבע	eighty	שמנים
eight	שמנה	ninety	תשעים
nine	תשעה, תשע	hundred	מאה
ten	עשרה, עשר	thousand	אלף

In forming the numbers from 11 to 19, the terms עשרה in the feminine and עשר in the masculine are used, preceded by the proper unit number; for 21 and upward, the term corresponding to the proper tenth digit is followed by the proper unit term preceded by the conjunction ו, and; e.g., twelve שנים עשר, twenty-four עשרים וארבע, etc.

Ordinal numbers

first	ראשן	sixth	ששי
second	שני	seventh	שביעי
third	שלישי	eighth	שמיני
fourth	רביעי	ninth	תשיעי
fifth	חמישי	tenth	עשירי

After 10 the ordinals are similar in form to the cardinals with the addition of the definite article ה; e.g., העשרים, the twentieth.

Seasons

spring	אביב	autumn	סתיו
summer	קיץ	winter	חרף

Time

hour	שעה	month	חדש
day	יום	season	מועד
week	שבוע	year	שנה

REFERENCES.—J. Philips and A. Hyman, Complete Instructor in Hebrew (1919); J. Weingreen, A Practical Grammar for Classical Hebrew (1939); A. S. Waldstein, English Hebrew and Hebrew English Dictionary (1936); P. Arnold-Kellner and M. D. Gross, Complete Hebrew-English Dictionary (1923).

ITALIAN

The alphabet. The Italian alphabet is the same as the English, though *k, w,* and *x* are used only in words of foreign origin. The grave accent is used on monosyllabic words to distinguish them from similar words: è, *is;* e, *and;* nè, *nor;* ne, *thence.* The grave or the acute accent may be used to distinguish *-io* and *-ia* from unstressed *-io* and *-ia* at the end of a word.

Capitalization. Caps are not used so freely in Italian as in English. When in doubt, do not cap.

Proper names are capped, but not adjectives derived from proper names.

gli Stati Uniti	mare Mediterraneo
nodo gordiano	gotico

Names of the Deity are capped:

Nostro Signore	Messia	Spirito Santo

But note that the following are not capped as they would be in English:

papista	religione cattolica
capo d'anno	cristiano
giorno di Pasqua	gesuita

Caps are not used for the names of the months and the days of the week.

marzo, aprile	sabato, domenica

(In the date line of letters the name of the month is sometimes capped: Napoli, 23 Ottobre, 1967.)

In the salutation of a letter only the first word and proper nouns are capped, in the complimentary close none are.

Nonna carissima	dal tuo affezionatissimo nipote

Headings, titles, captions. In headings, book titles, and other matter set in caps and lowercase, ordinarily only the first word and proper names are capped.

Storia della letteratura italiana
Le più belle pagine di Giacomo Leopardi
Fondamenti di grafia fonetica secondo il sistema dell'Associazione fonetica
    internazionale

Names of periodicals may, however, be capped as in English.

Divisions. If it is necessary to divide a word, at least three letters should be carried over. A single consonant between two vowels belongs to the following vowel.

ta-vo-lino   pre-ci-pi-tare   pre-zi-o-sa-mente

Likewise *ch, gh, cl, fl, gl, pl, gn,* combinations of *s* plus consonant, and consonant plus *r.*

ca-sti-ghi   re-flet-tere   mi-gliore
in-chio-stro   bi-so-gno   re-pli-ca-zi-one

Exception: An *s* belonging to a prefix should not be separated from it.

dis-gra-zia   tras-porto

Double consonants may be divided.

frat-tanto   nar-rare   mag-gio
af-flitto   cac-cia   vec-chio
ac-qua   nac-que

If the first of two or more consonants in the middle of a word is *l, m, n,* or *r,* it belongs with the preceding vowel, and the other consonant or consonants go with the following.

tem-pra   ar-ti-gi-ano   esul-tanza

Two adjacent vowels should not be separated unless the second is accented.

geo-gra-fia   ma-e-stro

A division should not be made after an apostrophe.

al-l'aura   del-l'acqua

Punctuation. Italian punctuation usage resembles French more than English. The following may be noted.

Conversational matter is indicated by dashes (spaced); and in contrast to French practice, a dash follows a quoted speech if other matter follows in the same paragraph.

— Gelata. Il cavaliere non vorrebbe che bevessi acqua gelata, perchè mi disturba, ma non ci resisto! — e, ripresa la padronanza di sè stessa, con aria di lieve canzonatura:

— Ammira la mia fretta di tornare qua? Ebbene, sapevo che Lei domani ci lascia e non ho voluto mancare, mi perdona? al mio dovere.

— Quale?

— Quello di salutarla.

Aldo fece un gesto di sorpresa:

— Lo chiama dovere?

— Se non Le va, metta un altro vocabolo, per esempio . . . piacere.

Points of suspension (. . .) are used to indicate a pause or interruption; if the interruption ends the sentence, even if the sentence is not complete, a period is added to the points of suspension unless other punctuation immediately precedes.

Abbreviations. Following are a few of the abbreviations commonly used in Italian.

a.C., *avanti Christo,* B.C.

a.D., *anno Domini,* A.D.

a.f., *anno futuro,* next year

affmo, *affezionatissimo,* most affectionately

a.p., *anno passato,* last year

b.p., *buono per,* good for

c^{ia}, *compagnia,* company

c^{o}, *compagno,* partner

d.C., *dopo Cristo,* after Christ, A.D.

d.c., *da capo,* again

dic., *dicembre,* December

ecc., *eccetera,* etc.

ediz., *edizione,* edition

es., *esempio,* example

febb., *febbraio,* February

ferr., *ferrovia,* railroad

frat., *fratello,* brother

genn., *gennaio,* January

jun., *juniore,* junior

L.it., *lire italiane,* Italian lire

magg., *maggiore,* major; *maggio,* May

Med., *Medico,* Dr.

N^{o}, *numero,* number

nov., *novembre,* November

N.S., *Nostro Signore,* Our Lord

On., *Onorevole,* Honorable

ott., *ottobre,* October

p., *per,* for

pag., *pagina,* page

p.es., *per esempio,* e.g.

S.A., *Sua Altezza,* His (Her) Highness

scel., *scellino,* shilling

S.E., *Sua Eccellenza,* His Excellency

sett., *settembre,* September

sez., *sezione,* section

Sigg., *Signori,* Messrs.

Sign., *Signor,* Sir, Mr.

St., *San, Santo,* Saint

v/, *vostra,* your

v., *vedi,* see

V.S., *Vostra Signoria,* Your Honor

Ordinal numerals are represented by the cardinal numeral followed by a superior *o*: 1°, *primo,* first; 2°, *secondo,* second.

LATIN

Capitalization. The Romans used only one style of letters; therefore when caps and lowercase are used in Latin, the capitalization follows a practice established by modern editors. Caps are used for the following:

1. The first word of a paragraph

2. The first word of a stanza of poetry, but not for the first word of each line

3. The first word of a sentence (usually)

4. Proper nouns
5. Proper adjectives

In headings and titles the usual practice is to cap the first word and proper nouns and proper adjectives.

Aristotle: *De generatione et corruptione*

Capitalization in the style of titles in English may sometimes be preferable.

Philosophiae Naturalis Principia Mathematica. Newton (London, 1686).

Divisions. When a word is divided at the end of a line, at least three letters should be carried over unless the lines are very short.

Compounds should be divided according to their construction, prefixes and suffixes being kept intact.

prod-est	con-ci-pio
ad-est	eius-modi

There are as many syllables in a Latin word as there are separate vowels and diphthongs. Diphthongs are *ae, au, oe, ei, eu, ui.*

In simple words a division may usually be made before a consonant standing between vowels.

vo-lat	pa-ter
ge-rit	di-li-gen-ter
me-ri-dies	tre-pi-da

Unless unavoidable, a division should not be made either before or after *x*; if necessary the *x* may be carried over to the next line.

dixe-rat	di-xe-rat	maxi-mus	ma-xi-mus

The combinations *qu* and *ch, ph,* and *th* are treated as single consonants.

Doubled consonants like *tt* and *ss* may be divided.

mis-sus	ne-ces-sa-ri-is

A division may be made before the combinations *bl, br, ch, cl, cr, dl, dr, fl, fr, gh, gl, gr, pl, pr, th, tl, tr,* unless the second letter introduces the second part of a compound word.

vo-lu-cris	ab-rum-po
pa-tris	ad-la-tus

Usage varies when the following combinations of consonants are involved: *bd, ct, gn, mn, ps, pt, phth, st, str, thl, zm.* Some authorities divide before them; others separate them. The proofreader should therefore be informed of the writer's preference.

Other combinations of two or more consonants may be separated after the first consonant.

mon-strum	quan-tum	ad-ven-tant
nar-cis-sus	re-fe-ren-dae	for-tu-na

NORWEGIAN

The alphabet. The Norwegian language—or languages, since Norway has two official languages, *riksmål* and *landsmål*—uses twenty-nine letters, the English alphabet with the addition of æ, ø or ö, and å or aa. In alphabetizing, these letters usually follow *z*. The letters *c, q, w, x,* and *z* are used only in foreign words and proper nouns.

Capitalization. Proper nouns are capped, but not adjectives derived from them. The pronouns *De, Dem,* and *Deres* are capped. Names of the months and of the days of the week are not capped. In other respects English usage is followed.

Divisions. When a word is divided at the end of a line, at least three letters should be carried over unless the lines are very short. Compounds should be divided according to their construction, prefixes and suffixes being kept intact. A simple word may be divided before a consonant standing between vowels. Two or more consonants between vowels may be divided before the last consonant, except *sk, sp, st,* and *str,* which are inseparable and belong with a following vowel.

POLISH

The alphabet. The Polish language uses twenty-three letters of the English alphabet, *q, v,* and *x* being omitted. The following accented letters are used: ą, ć, ę, ł, ń, ó, ś, ź, ż.

There are six words of one letter each in Polish: *w,* in; *ż,* with; *i,* and, also; *a,* and; *o,* about; *u,* by, at, among, with, near, in.

Divisions. When a word is divided at the end of a line, at least three letters should be carried over unless the lines are very short. Compounds should be divided according to their construction, prefixes and suffixes being kept intact. A word may be divided before a consonant standing between vowels. The following combinations of consonants are treated as single consonants: *ch, cz, dz, dź, dż, rz, sz,* and *szcz, zd, dg.* Other combinations of consonants may be divided. The following combinations of consonants and vowels may not be divided: *bi, fi, gi, gie, ki, kie, mi, ni, pi,* and *wi.*

Punctuation. Punctuation is practically the same as in English.

PORTUGUESE

The alphabet. The Portuguese language uses the English alphabet, but *k* appears only in foreign words, and *w* and *y* are rarely used. The tilde or til in Portuguese represents a nasalized vowel and is written over the first vowel of a diphthong, as in São Paulo, pronounced *souɴ powlŏŏ*.

Divisions. When a word is divided at the end of a line, at least three letters should be carried over unless the lines are very short. A word may be divided before a consonant standing between vowels. The following combinations of consonants standing between vowels belong with the following vowel: *bl, br, ch, cl, cr, ct, dl, dr, fl, fr, gl, gn, gr, gu, lh, nh, ph, pl, pr, ps, pt, qu, sc, sp, st, th, tl, tr, vl, vr.*

Following a consonant *st* may be separated.

de-mons-tra-ção   cons-ti-tui-ção

Other combinations than the above may be separated. No division should be made between two vowels. The triphthongs *eia, éia, eão, ião, oei* should not be divided.

RUSSIAN

(The material on the composition of Russian is reprinted from the *Government Printing Office Style Manual*, 1973.)

Alphabet, transliteration,[1] and pronunciation

А	а	a	*a* in far [2]
Б	б	b	*b*
В	в	v	*v*
Г	г	g	*g* in go [3]
Д	д	d	*d*
Е	е	ye, e [4]	*ye* in yell, *e* in fell [5]
Ё	ё	yë, ë [6]	*yo* in yore, *o* in order [7]
Ж	ж	zh	*z* in azure
З	з	z	*z* in zeal
И	и	i	*i* in machine [8]
Й	й	y	*y* in boy
К	к	k	*k*
Л	л	l	*l*

---

[1] U. S. Board on Geographic Names transliteration, 1944.
[2] When stressed; when unstressed, like *a* in sofa.
[3] Also pronounced as *v* in the genitive ending -го; often used for original *h* in non-Russian words, but is pronounced as *g* by Russians.
[4] *Ye* initially, after vowels, and after ъ, ь.
[5] Pronounced as *i* in habit, or the same sound with preceding *y*, when unstressed.
[6] *Yё* as for *ye*. The sign ё is not considered a separate letter of the alphabet, and the ¨ is often omitted. Transliterate as *ё, yё* when printed in Russian as ё; otherwise use *e, ye*.
[7] Only stressed.
[8] Like *i* in habit when unstressed; like *yie* in yield after a vowel and after ь.

М	м	m	*m*
Н	н	n	*n*
О	о	o	*o* in order [9]
П	п	p	*p*
Р	р	r	*r*
С	с	s	*s* in so
Т	т	t	*t*
У	у	u	*u* like the *oo* in Moon.
Ф	ф	f	*f*
Х	х	kh	*h* in how, but stronger, or *ch* in Scottish loch
Ц	ц	ts	*ts* in hats
Ч	ч	ch	*ch* in church
Ш	ш	sh	*sh* in shoe
Щ	щ	shch	*sh* plus *ch*, somewhat like *sti* in question
Ъ	ъ	" [10]	([11])
Ы	ы	y	*y* in rhythm
Ь	ь	' [12]	([13])
Э	э	e	*e* in elder
Ю	ю	yu	*u* in union
Я	я	ya	*ya* in yard

[9] Like *o* in abbot when unstressed.
[10] The symbol " (double apostrophe), not a repetition of the line above.
[11] No sound; used only after certain prefixes before the vowel letters e, ё, я, ю. Formerly used also at the end of all words now ending in a consonant letter.
[12] ' (apostrophe).
[13] Palatalizes a preceding consonant, giving a sound resembling the consonant plus *y*, somewhat as in English meet you, did you.

Special characters

Russian uses the Cyrillic alphabet. Many of the characters are the same as in Latin, with the following special ones: Б б, Г г, Д д, Ж ж, Й й, Л л, П п, Ф ф, Ц ц, Ш ш, Щ щ, Ъ ъ, Ы ы, Э э, Ю ю, and Я я. Note the following somewhat similar characters: З Э, Л П, У Ч, Ш Щ, з э, л п, ш щ. The Ы is a separate character and not a combination of Ь and I.

Transliteration

This is a mechanical process of substituting the transliteration letter or combination of letters for each Russian letter: Москва = *Moskva*, Киев = *Kiyev*, Русский = *Russkiy*, etc.

Vowels and consonants

The vowel letters are а, е, ё, и, о, у, ы, э, ю, and я, represented, respectively, by *a*, *e* or *ye*, *ё* or *yё*, *i*, *o*, *u*, *y*, *e*, *yu*, *ya*. The letters й, ъ, and ь are not called either vowels or consonants.   All other letters are consonants.

Diphthongs

The sequences of a vowel followed by й are often called diphthongs. Their sounds are:

ай (*ay*) *ai* in aisle
ей (*ey*, *yey*) *ey* in they, or as *yea* (=yes)
ий (*iy*) like prolonged English *ee*
ой (*oy*) *oy*
уй (*uy*) *uoy* in buoy as pronounced by some ($\breve{oo}$ plus *y*)

ый (*yy*) *y* in rhythm plus *y* in yield
эй (*ey*) *ey* in they
юй (*yuy*) *you* plus *y* in yield
яй (*yay*) *ya* in yard plus *y* in yield

Digraphs

The transliterations *ye, zh, kh, ts, ch, sh, shch, yu, ya* represent single Russian letters and should not be divided in syllabification.

Consonantal units

The following combinations of consonants should be treated, for syllabification purposes, as indivisible units:

бл, бр (*bl, br*)	мл (*ml*)
вл, вр (*vl, vr*)	пл, пр (*pl, pr*)
гл, гр (*gl, gr*)	ск, скв, скр, ст, ств, стр (*sk, skv, skr,*
дв, др (*dv, dr*)	*st, stv, str*)
жд (*zhd*)	тв, тр (*tv, tr*)
кл, кр (*kl, kr*)	фл, фр (*fl, fr*)

These simplified rules have been followed for the past 2 years by the Library of Congress Card Division. (Based on practice in Bol'shaia sovetskaia entsiklopediia, v. 36.)

General:

1. A single letter is not separated from the rest of the word.
2. A soft or hard sign is not separated from the preceding consonant.
3. Division is made at the end of the prefix (a fill-vowel is considered part of the prefix): со-глас-но воз-дух по-треб-ле-ние объ-ем пре-до-ста-вить.
4. In compound words, letters are not separated from the component parts of the word, and a fill-vowel goes with the preceding syllable:
сов-хоз зем-ле-вла-де-лец

Two vowels together:

1. Division is made between the vowels: сто-ит (*but:* рос-сий-ский).

One consonant between two vowels:

1. The consonant goes with the following vowel:
ма-не-ры по-вы-ше-ни-ем ста-тья-ми.

Two consonants between two vowels:

1. Division is made between the consonants. (*Exception:* ст goes with the following vowel): топ-ли-во управ-ле-ние ре-ак-тив-ный биб-ли-о-те-ка Поль-ша (*but:* пу-скает ча-сти).

Three or more consonants between two vowels:

1. If a consonant is doubled, division is made between the two:
искус-ство диф-фрак-ция.
2. ст is never separated.
3. Division is not made before the first nor after the last consonant. (*Exception:* When ст begins the consonant group, it may be separated from the preceding vowel): мест-ность *or* ме-стность
4. Otherwise, division is optional: элек-три-че-ство *or* элект-ри-че-ство. Ан-глия *or* Анг-лия цент-раль-ный *or* цен-траль-ный
Exception: The following are consistently divided as shown: марк-сизм Мо-сква

Rules for syllabification [1]

1. Diphthongs, digraphs, and consonantal units may not be divided.
2. Division is made on a vowel or on a diphthong before a single consonant, a digraph, or a consonantal unit: ба-гаж (*ba-gazh*), Бай-кал (*Bay-kal*), му-ха (*mu-kha*), рё-бра (*rë-bra*), каче-ство (*kache-stvo*), свой-ство (*svoy-stvo*).
3. In a group of two or more consonants, division is made before the last consonant, digraph, or consonantal unit: мас-ло (*mas-lo*), мас-са (*mas-sa*), мар-шал (*mar-shal*), точ-ка (*toch-ka*), долж-ность (*dolzh-nost'*), сред-ство (*sred-stvo*).

---

[1] Since the orthographic reform of 1918, the rules for syllabification have been considerably liberalized. It is generally permitted now to divide according to convenience, provided that phonetics and etymology are not severely overridden. These rules, designed as a guide for workers who might not be thoroughly familiar with the Russian language, are of necessity somewhat restrictive, but they insure invariably correct word division in conformity with generally approved usage.

As a great deal of Russian matter, especially bibliography, is printed in transliterated form, these rules have been formulated so as to apply with equal accuracy whether matter is in Russian characters or in transliteration.

4. Division may be made between vowels not constituting a diphthong or between a diphthong and another vowel: оке-ан (*oke-an*), ма-як (*ma-yak*).

5. Certain adverbial prefixes are kept intact, except before ы. These are: без (бес), во, воз (вос), вы, до, за, из (ис), на, над, не, ни, низ (нис), о, об обо, от, ото, пере, по, под, пред(и), пред(о), при, про, раз (рас), с(о), and у. In transliteration these prefixes are respectively *bez* (*bes*), *vo*, *voz* (*vos*), *vy*, *do*, *za*, *iz* (*is*), *na*, *nad*, *ne*, *ni*, *niz* (*nis*), *o*, *ob*, *obo*, *ot*, *oto*, *pere*, *po*, *pod*, *pred*(*i*), *pred*(*o*), *pri*, *pro*, *raz* (*ras*), *s*(*o*), and *u*: без-вкусныЙ (*bez-vkusnyy*), бес-связь (*bes-svyaz'*), во-круг (*vo-krug*), but раз-ыскать (*ra-zyskat'*), etc.

6. Compound words are divided according to their component parts (and each part according to rules 1 to 5): радио-связь (*radio-svyaz'*), фото-снимка (*foto-snimka*).

7. It is to be noted that the й (*ĭ*) always terminates a syllable: бой-кий (*boy-kiy*), рай-он (*ray-on*); the ъ (") terminates a syllable except in words beginning with въ (*v"*), взъ (*vz"*), and съ (*s"*): отъ-ехать (*ot"-yekhat'*) but съём-ка (*s"yëm-ka*), съест-ной (*s"yest-noy*); the ь (') terminates a syllable except before the soft vowels е (*e*), и (*i*), ю (*yu*), and я (*ya*): маль-чик (*mal'-chik*), but соло-вьев (*solo-v'yev*), бри-льянт (*bri-l'yant*). се-мья (*se-m'ya*).

8. Foreign words and components of foreign words (not naturalized) follow the conventions of the language of origin: Шек-спир (*Shek-spir*), мас-штаб (*mas-shtab*), Лоа-ра (*Loa-ra*) [not Ло-ара (*Lo-ara*) (from the French *Loire*)], се-ньор (*se-n'or*).

Illustrative word divisions

[The numbers in parentheses refer to the syllabification rules]

аме-ри-кан-ский *ame-ri-kan-skiy*	(2, 2, 3)	не-сго-ра-е-мый *ne-sgo-ra-e-myy*	(5, 2, 4, 2)
ан-глий-ская *an-gliy-skaya*	(3, 2)	неф-те-хра-ни-ли-ще *nef-te-khra-ni-li-shche*	(3, 6, 2, 2, 2)
без-ал-ко-голь-ный *bez-al-ko-gol'-nyy*	(5, 3, 2, 7)	ни-сколь-ко *ni-skol'-ko*	(5, 7)
бес-сроч-ный *bes-sroch-nyy*	(5, 3)	об-ло-же-ние *ob-lo-zhe-niye*	(5, 2, 2)
ва-ку-ум *va-ku-um*	(2, 4)	обо-зна-че-ние *obo-zna-che-niye*	(5, 2, 2)
во-гну-тость *vo-gnu-tost'*	(5, 2)	объ-яс-ни-тель-ный *ob"-yas-ni-tel'-nyy*	(7, 3, 2, 7)
во-до-вме-сти-ли-ще *vo-do-vme-sti-li-shche*	(2, 6, 2, 2, 2)	од-но-звуч-ный *od-no-zvuch-nyy*	(3, 6, 3)
воз-зре-ние *vo z-zre-niye*	(5, 2)	от-зву-чать *ot-zvu-chat'*	(5, 2)
вос-хва-ле-ние *vos-khva-le-niye*	(5, 2, 2)	ото-зва-ние *oto-zva-niye*	(5, 2)
вы-здо-ро-веть *vy-zdo-ro-vet'*	(5, 2, 2)	отъ-ез-жа-ю-щий *ot"-yez-zha-yu-shchiy*	(7, 3, 4, 2)
вы-со-ко-нрав-ство *vy-so-ko-nrav-stvo*	(2, 2, 6, 3)	Па-ра-гвай *Pa-ra-gvay*	(2, 8)
го-су-дар-ствен-ный *go-su-dar-stven-nyy*	(2, 2, 3, 3)	пе-ре-гнать *pe-re-gnat'*	(2, 5)
до-школь-ное *do-shkol'-noe*	(5, 7)	пер-спек-ти-ва *per-spek-ti-va*	(8, 3, 2)
зав-траш-ний *zav-trash-niy*	(3, 3)	пи-о-нер-ский *pi-o-ner-skiy*	(4, 2, 3)
изъ-яс-не-ние *iz"-yas-ne-niye*	(7, 3, 2)	по-глуб-же *po-glub-zhe*	(5, 3)
ис-сле-до-ва-тель-ский *is-sle-do-va-tel'-skiy*	(5, 2, 2, 2, 7)	по-гля-ды-вать *po-glya-dy-vat'*	(5, 2, 2)
Крон-штадт-ский *Kron-shtadt-skiy*	(8, 3)	по-да-вать-ся *po-da-vat'-sya*	(5, 2, 7)
на-всег-да *na-vseg-da*	(5, 3)	под-жи-да-ние *pod-zhi-da-niye*	(5, 2, 2)
на-дви-га-ю-щий-ся *na-dvi-ga-yu-shchiy-sya*	(5, 2, 4, 2, 7)	пред-ва-ри-тель-ный *pred-va-ri-tel'-nyy*	(5, 2, 2, 7)
над-вя-зать *nad-vya-zat'*	(5, 2)	пре-ди-сло-вие *pre-di-slo-viye*	(2, 5, 2)

Illustrative word divisions—Continued

пре-до-хра-нять *pre-do-khra-nyat'*	(2, 5, 2)	рас-ска-зы-вать *ras-ska-zy-vat'*	(5, 2, 2)
при-вхо-дя-щий *pri-vkho-dya-shchiy*	(5, 2, 2)	соб-ствен-ный *sob-stven-nyy*	(3, 3)
про-све-ще-ние *pro-sve-shche-niye*	(5, 2, 2)	со-дей-ство-вать *so-dey-stvo-vat'*	(5, 7, 2)
про-те-стант-ство *pro-te-stant-stvo*	(2, 2, 3)	со-е-ди-нён-ные *so-ye-di-nën-nyye*	(5, 2, 2, 3)
про-хва-тить *pro-khva-tit'*	(5, 2)	сол-неч-ный *sol-nech-nyy*	(3, 3)
раз-вью-чи-вать *raz-v'yu-chi-vat'*	(5, 2, 2)	солн-це-сто-я-ние *soln-tse-sto-ya-niye*	(3, 6, 4, 2)
раз-мно-жать *raz-mno-zhat'*	(5, 2)	удоб-ней-ше *udob-ney-she*	(3, 7)

Stress and diacritics

No simple set of rules for syllabic stress can be formulated. The only dependable guide is a native, or a dictionary in the case of basic forms and a grammar for their inflectional shiftings.

The only diacritics are the dieresis and the breve. These do not indicate stress but modification of sound. Note alphabet.

Capitalization

Capitalization is practically as in English, except that proper adjectives, names of the months (except when abbreviated), and days of the week are lowercased.

Punctuation

Punctuation is very similar to that of English, but the comma is used for restrictive as well as nonrestrictive clauses. The dash is used between a subject and a complement when there is no verb *is* or *are*, and sometimes before a clause where the equivalent of the conjunction *that* has been omitted. Dialog is usually shown by dashes rather than quotation marks. Cited material is enclosed in quotation marks, which are usually in the French form—« », though sometimes in the German form—,, ", and rarely as in English.

Abbreviations

амер.	американский, American
АН	Академия наук, Academy of Sciences
б.г.	без года, no date
б.м.	без места, no place
ВКП (б)	Всесоюзная Коммунистическая Партия (большевиков) All-Union Communist Party (Bolshevik)
г.	год, year; город, city; господин, Mr.
г-жа	госпожа, Mrs.
гл.	глава, chapter
гр.	гражданин, citizen; гражданка, citizen (female)
до н. э.	до нашей эры, B.C.
ж. д.	железная дорога, railroad
и т. д.	и так далее etc.
км.	километр, kilometer
КПСС	Коммунистическая партия Советского, Союза, Communist Party of the Soviet Union
м.	метр, meter
мм.	миллиметр, millimeter
н. ст.	новый стиль, new style
н. э.	нашей эры, A.D.
обл.	область, oblast
отд.	отделение, section
по Р. Х.	по Рождестве Христове, anno Domini
см.	сантиметр, centimeter; смотри, see, cf.
СССР	Союз Советских Социалистических Республик, Union of Soviet Socialist Republics
с. ст.	старый стиль, old style
США	Соединенные Штаты Америки, United States of America
ст.	статья, article; столбец, column
стр.	страница, page
т.	том, volume; товарищ, comrade
т.е.	то есть, that is
ЦК	Центральный Комитет, Central Committee
ч.	часть, part

Cardinal numbers

один, одна, одно *m., f., n.*	one	семнадцать	seventeen
		восемнадцать	eighteen
два, две *m. & n., f.*	two	девятнадцать	nineteen
три	three	двадцать	twenty
четыре	four	двадцать один, etc.	twenty-one, etc.
пять	five	тридцать	thirty
шесть	six	сорок	forty
семь	seven	пятьдесят, etc.	fifty, etc.
восемь	eight	девяносто	ninety
девять	nine	сто	hundred
десять	ten	сто один, etc.	one hundred and one, etc.
одиннадцать	eleven		
двенадцать	twelve	двести	two hundred
тринадцать	thirteen	триста, etc.	three hundred, etc.
четырнадцать	fourteen		
пятнадцать	fifteen	пятьсот, etc.	five hundred, etc.
шестнадцать	sixteen	тысяча	thousand

Ordinal numbers [2]

первый	first	шестнадцатый	sixteenth
второй	second	семнадцатый	seventeenth
третий	third	восемнадцатый	eighteenth
четвёртый	fourth	девятнадцатый	nineteenth
пятый	fifth	двадцатый	twentieth
шестой	sixth	двадцать первый	twenty-first
седьмой	seventh	сотый	hundredth
восьмой	eighth	сто первый, etc.	one hundred and first, etc.
девятый	ninth		
десятый	tenth	двухсотый	two hundredth
одиннадцатый	eleventh	трехсотый	three hundredth
двенадцатый	twelfth	четырехсотый	four hundredth
тринадцатый	thirteenth	пятьсотый, etc.	five hundredth, etc.
четырнадцатый	fourteenth		
пятнадцатый	fifteenth	тысячный	thousandth

Months

январь (Янв.)	January	июль	July
февраль (Февр.)	February	август (Авг.)	August
март	March	сентябрь (Сент.)	September
апрель (Апр.)	April	октябрь (Окт.)	October
май	May	ноябрь	November
июнь	June	декабрь (Дек.)	December

Days

воскресенье	Sunday	четверг	Thursday
понедельник	Monday	пятница	Friday
вторник	Tuesday	суббота	Saturday
среда	Wednesday		

Seasons

весна	spring	осень	autumn
лето	summer	зима	winter

Time

час	hour	месяц	month
день	day	год	year
неделя	week		

---

[2] The ordinal numbers here given are of the masculine gender. To convert them to feminine or neuter, it is only necessary to effect the proper gender changes: For the feminine, change ый to ая, ий to ья, ой to ая. For the neuter, change ый to ое, ий to ье, and ой to ое.

NOTE ON OLD SPELLING

On October 10, 1918, the Council of People's Commissars decreed the introduction of a spelling reform that had been proposed many years before but never adopted. The spelling used from that time in all official publications, except those of the Academy of Sciences (Akademiya Nauk), was this new spelling. The academy adopted the new spelling in 1924. All Russian publications, except for a few printed outside the Soviet Union, have used the new spelling since the institution of the reform.

The old spelling, found in books printed before the dates mentioned, differed in the following ways:

1. There were used the additional *i* (in the alphabet, after и and before к, as й was not considered a separate letter), ѣ (after ь), ѳ (after я), and ѵ (after ѳ).

2. *I* was used only before another vowel letter and in the word мiръ, world. It is now replaced by и (мiръ became мир).

3. Ѣ occurred in certain words and in some grammatical endings. It represented the same sound as e and is now replaced by e everywhere. In a few cases ѣ was pronounced like ë, and where e is now printed with dieresis (¨), the replacement of ѣ is, of course, ë.

4. Ѳ was used in words of Greek origin, for Greek θ (th). It was pronounced *f*, and is now replaced by *f*.

5. Ѵ was used in a few ecclesiastical words, for Greek υ (*u, y*). It was pronounced like и, and is replaced by that letter.

6. Ъ was used at the end of all words after a consonant not followed by ь. In this position ъ has simply been omitted since the reform. For some years after 1918, some publishers omitted ъ altogether, using an apostrophe for it after prefixes, but the use of the apostrophe is now discouraged, and ъ is used.

7. The prefixes из, воз, вз, раз, низ, без, чрез, через were written with final з everywhere, whereas now they are written ис, вос, etc., before к, п, с, т, х, ц, ч, ш, ф, щ.

8. Some adjective endings in the genitive singular were written -аго, -яго; these were replaced by -ого, -его.

9. The plural nominative of adjectives agreeing with feminine and neuter nouns was written -ыя, -iя; these endings were replaced by -ые, -ие, which had formerly been used only for adjectives agreeing with masculine nouns.

10. The pronoun "they" in referring to the feminine gender was written онѣ; this was replaced by они, previously used only for masculine reference.

11. Similarly, однѣ, однѣх, однѣми were replaced by одни, одних, одними.

12. The genitive pronoun "her" was written ея; this was replaced by её, formerly used only as accusative.

13. Ё was printed only in schoolbooks.

REFERENCES.—R. I. Avanesov and V. N. Simonov, Ocherk Grammatiki Russkogo Literaturnogo Yazyka (1945); S. C. Boyanus, A Manual of Russian Pronunciation (1935); V. K. Müller, Russian-English and English-Russian Dictionary (1944); Pravila russkoi orfografii i punktuatsii (1957); A. B. Shapiro, Russkoe pravopisanie (1961).

SPANISH

The alphabet. The Spanish language uses all the letters of the English alphabet. In addition it has the double letters *ch, ll,* and *rr,* and the *ñ,* which in alphabetizing follow *c, l, r,* and *n* respectively.

Accents. The acute accent is used to indicate a stressed syllable and to distinguish words otherwise identical in form but of different meaning or use.

mas	*but*	más	*more*	solo	*alone*	sólo	*only*
este	*this*	éste	*this one*	se	*himself*	sé	*I know*
tu	*thy*	tú	*thou*	el	*the*	él	*he*
te	*thee*	té	*tea*	de	*of*	dé	*give*
mi	*my*	mí	*me*	si	*if*	sí	*yes*

The following are written with an acute accent whenever they are used in an exclamation or a question.

¡cómo!	¿cuándo?	¿dónde?	¡qué!
¡cuál!	¡cuánto!	¿por qué?	¿quién?

The dieresis is used over *u* in the syllables *gue* and *gui* when the *u* is not silent but pronounced.

Capitalization. Caps are used somewhat less than in English. They are used for the following:

1. Proper names

el Mar Cantábrico	el Océano Atlantico	el Nuevo Mundo
los Estados Unidos	la España	la Sudamérica

2. Names of the Deity
3. Sobriquets

el Gran Capitán	el Caballero de la Triste Figura

4. Titles of honor

Vuestra Alteza	Vuestra Majestad
Vuestra Excelencia	Vuestra Merced

5. The first word and proper names in salutations of letters

Muy señor mío:	Muy distinguida señora:

Note than an abbreviation used here is capped:

Muy Sres. míos:

Caps are not used for the following:

1. Names of the months and of the days of the week

marzo, abril	25 de junio de 1964
lunes, martes	el miercoles pasado

2. Titles, unless they are abbreviated or form part of a place name

Sí, señor   el señor don Enrique Palava
Sr. Palava   la avenida del Conde de Peñalver

3. Proper adjectives or nouns derived from proper names

el idioma castellana   los jóvenes americanos
Los chinos hablan y leen el chino.

Headings and captions. In headings, captions, titles of books, and the like, only the first word and proper nouns are capped.

El caballo del moro   *Antología de poesía española*
Las novelas ejemplares   Vista general de la ciudad de Valparaiso

La Alhambra: Interior de la Torre de la Cautiva
Secure the phonograph records of "La golondrina" and "La paloma."

Names of periodicals are usually capped.

El señor Norton[6] escribe artículos para "La Revista de Filología Española."

Punctuation. Punctuation marks are used in Spanish very much as in English, but the following practices should be noted.

A question in Spanish is enclosed in question marks, the first one inverted and placed before the interrogative word beginning the question. Exclamations are punctuated similarly.

Si no escribe a su padre ¿qué ocurre?
Pero, ¿cómo podremos entrar aquí?
El sereno canta: ¡Las once en punto y sereno!

Note that quotation marks are considered unnecessary in the third example.

If a sentence is both exclamatory and interrogative, the inverted exclamation point should be used at the beginning and the question mark at the end.

¡Usted se atreve a hablarme así?   (Do you dare to speak to me in that way?)

Es engaño.
¡El Rey procurar mi daño,
Solo, embozado y sin gente?

[6]Note that the definite article is always used before titles except in direct address, unless the title is *don, doña, san, santo,* or *santa.*

Spanish quotation marks differ from English quotes: « » They are used to mark quotations but not to enclose dialogue.

«Cortesía» significa «urbanidad».
...y le dió permiso «para explorar, conquistar, pacificar y colonizar» el oeste de la costa sudamericana...

Dialogue is indicated by dashes, not by quotation marks.

—¿Qué tienes tú ahora?—rugió el maestro.   —A ti no te toca todavía leer. ¡Silencio!
—Pero, profesor—gimoteó el niño; —mire Vd. ¡Aquí vienen en el párrafo que tendré que leer esos mismos tres hombres terribles!

Note that the end of the spoken words is marked by a dash when other words follow in the same paragraph. Each change of speaker should be indicated by a new paragraph.

English quotation marks are sometimes used:

Comedias de cautivos: "El trato de Argel," "El gallardo español," "Los baños de Argel," "La gran sultana."

Points of suspension. Points of suspension are three points, set close together (a single type in some fonts), without any space at the left. They are used in Spanish text much as dashes are used in English—to mark an interruption or a pause or an abrupt change in thought or construction.

¿Qué espero?   ¡Oh, amor gigante!...   ¿En qué dudo?...   Honor ¿qué es esto?...

Divisions. Whenever it is necessary to divide a Spanish word at the end of a line, carrying over a part, the following principles should be observed.

No fewer than three letters should be carried over unless the measure is very narrow.

In simple words a single consonant between two vowels belongs with the following vowel. The double letters *ch, ll,* and *rr* are considered as simple consonants. The letter *y* standing between vowels is treated as a consonant.

ho-nor	mu-cha-cho	ca-rro
ca-ñón	ga-llina	re-yes

Prefixes and suffixes form separate syllables.

des-ampara   sub-inspección   super-abundante
Similarly: es-otros, nos-otros, vos-otros.

The letters *l* and *r* preceded by a consonant, except *rl, sl, tl,* and *sr,* must not be separated from it, except to preserve a prefix intact.

ta-bla	sa-cro	con-tra-po-ner
bi-blio-teca	pu-drir	sub-lu-nar
si-glo	pa-la-bra	sub-ra-yar

Other combinations of consonants may be divided, including cc and *nn,* the only two consonants besides *ll* and *rr* that are ever doubled.

ex-cepto	de-sig-nar	sec-ci-ón
ac-tual-mente	sig-ni-fi-ca-ción	frac-ci-ón
at-lán-tico	avan-zar	in-no-var

The letter *s* does not combine with a following consonant.

cons-pi-ra-dor	ins-pi-rar	trans-fi-gu-rar
es-ca-lera	cons-tante	abs-ti-nen-cia
es-tar	pers-pi-ca-cia	ads-cri-to

Exception: ist-mo

Division between vowels should be avoided, though if necessary two strong vowels (*a, e, o*), or an accented vowel and a strong vowel, may be separated. Other combinations of vowels form inseparable diphthongs or triphthongs.

Abbreviations. Abbreviations are used very freely in Spanish composition. In addition to the usual abbreviations of English—such as those of months, days, titles, weights and measures, coins, street, avenue, chapter, article, page—abbreviations for phrases used often in letters are in common use. For instance, note the following abbreviations used to conclude letters:

B.S.M., *beso, sus manos,* with great respect
C.M.B., *cuyas manos besos,* very respectfully
S.S.S., *su seguro servidor,* your faithful servant
S.A.S.S., *su atento y seguro servidor*
Su atto., afmo. y s. s., *su atento, afectísimo y seguro servidor*

Similarly, in the salutations of letters:

Muy Sres. n/s, *or* Muy Sres. nuestros, *Muy señores nuestros*

The custom has been to use a great many superior letters in abbreviations, just as English did in Colonial days and before. These, of course, are hard to set in type and are being used much less than formerly. Instead of *Ag^{to}* we now see *Agto; afmo.* has replaced *af^{mo}*, and so on.

The following abbreviations are commonly used in Spanish:

@, *arroba;* @@, *arrobas* (Spanish weight—about 25 lb; Spanish measure—about 4 gal)

A., *autor,* author (*pl.* AA.)

ab., *abad,* abbot

a/c., *a cuenta,* on account

A.C. *or* A. de C., *Año de Cristo,* A.D.

a/f., *a favor de,* in behalf of

afmo., *afectísimo,* most affectionate

ap., *aparte,* aside; *apóstol,* apostle

apda., *apreciada,* valued

art., *articulo,* article

Avda., *avenida,* avenue

B., *beato,* blessed

B.L.M., *beso la mano* (I kiss your hand), respectfully

B.p., *Bendición papal,* papal benediction

B.S.M., *beso sus manos,* with great respect

Br. *or* br., *bachiller,* Bachelor (academic degree)

c. *or* cap., *capítulo,* chapter

c^a, c^{ia}, *or* compa., *companía,* company

C.A., *Centroamérica,* Central America

C. de J., *Compania de Jesús,* Society of Jesus (S.J.)

c/l. *curso legal,* legal procedure

cllo., *cuartillo* (a unit of measure)

C.M.B., *cuyas manos beso,* very respectfully

col., *columna,* column

C.P.B., *cuyos pies beso,* very respectfully

cps., *compañeros,* partners

cta., *cuenta,* account

ctmo., *céntimo,* centime; *centésimo,* hundredth

ctvo., *centavo,* cent

c/u., *cada uno,* every one, each

D., *don,* Mr.

DD., *doctores,* Drs.

D.F., *Distrito Federal,* Federal District

EE. UU., *or* E.U., *Estados Unidos,* United States

E.P.M., *en propia mano,* in his own hand

E.S.M., *estrecho su mano,* I press your hand

est., *estimada,* respected, esteemed

E.U.A., *Estados Unidos de America,* U.S.A.

F.C., *ferrocarril,* railroad (*pl.,* FF. CC.)

fha., *fecha,* enacted

fra., *factura,* invoice of merchandise

fvda., *favorecida,* esteemed

G., *gracia,* favor

Gral., *general,* general

hh., *hojas,* leaves

Hnos., *hermanos,* brothers

L., Lic., Lcdo., *licenciado,* licensed

l., *ley,* law; *libro,* book

lin., *línea,* line

L.S., *lugar del sello,* place of the seal

Mons., *Monseñor,* Monsignor

m/m., *más o menos,* more or less
m/n., *moneda nacional,* national currency
n., *noche,* night
n.a., *nota del autor,* author's note
No. *or* núm., *número,* number
P., peso; *padre,* Father
pág., *página,* page (*pl.,* págs.)
pár., *párrafo,* paragraph
p/cta., *por cuenta,* on account
P.D., *posdata,* postscript
pdo., *pasado,* past
p.ej., *per ejemplo,* for example
P.O., *por orden,* by order
p.pdo., *próximo pasado,* preceding
P.R., Puerto Rico
P.S.M., *por su mandato,* by his orders
pta., *peseta* (a Spanish coin) (*pl.,* ptas.)
pte., *parte,* part
Q.E.G.E., *que en gloria esté,* deceased
Q.E.P.D., *que en paz descanse,* deceased
r., real (coin) (*pl.,* rs.)
R.A., *República Argentina,* Argentine Republic
S.A., *Sudamérica,* South America; *Su Alteza,* His Highness; *Sociedad Anónima,*
  stock company
S.E., *Su Excelencia,* His Excellency
sec., *sección,* section
S.M., *Su Majestad,* His Majesty
Sr., *señor,* sir
Sra., *señora,* lady
Sres., *señores,* sirs
Srio, *secretario,* secretary
Srta., *señorita,* young lady
Sto., *Santo,* saint
tom., *tomo,* volume
tip., *tipografía,* printing office
tit., *título,* title
U. *or* Ud., *usted,* you (*pl.,* UU. *or* Uds.)
últ., *último,* last
v., *véase,* see
V., *usted,* you; *vale,* bond, promissory note, IOU
v. *or* verso., *versículo,* verse
V.A., *Vuestra Alteza,* Your Highness
V.B., *visto bueno.* (The preceding document has been examined and found to
  be correct; consequently it may signify: Pay the bearer; Let him *or* the merchandise pass.)
v/c, *vuelta de correo,* return mail
vda., *viuda,* widow
V.E., *Vuestra Excelencia,* Your Excellency
v.g. *or* v. gr., *verbi gracia,* for instance
V.M., *Vuestra Majestad,* Your Majesty
Vm. *or* Vmd., *Vuestra Merced,* Your Worship
VV., Vds., *ustedes,* you (*pl.*)

Ordinal numerals are expressed by using a superior *o* following the cardinal
numeral: 1°, *primero;* 2°, *segundo.*

SWEDISH

The alphabet. The Swedish language uses twenty-nine letters, the English alphabet with the addition of *å, ä,* and *ö.* These three letters are usually alphabetized in this order following *z.*

Capitalization. Proper nouns are capped, but not adjectives derived from them. Names of the months and of the days of the week are not capped: söndag, måndag.

Divisions. When a word is divided at the end of a line, at least three letters should be carried over unless the lines are very short. Compounds should be divided according to their construction, prefixes and suffixes kept intact. A Swedish word may be divided before a consonant standing between two vowels. The combinations *sch* and *sk* used for the sound *sj* are never separated and belong with a following vowel; *ng* always belongs to a preceding vowel unless *n* and *g* belong to different parts of a compound word.

mar-schera   manni-ska

When any other two or more consonants stand between vowels, a division may be made before the last.

PHONETICS

Our alphabet was taken from the Romans, who got it in turn from the Greeks. Of the six vowels used by the Greeks, however, the Romans took only five: *a* as in *bar, e* as in *obey, i* as in *machine, o* as in *note,* and *u* as in *rude.* As a result of growth and of changes in spoken English, each of these vowels came to represent many different sounds besides the original one. Note, for instance, the sound of *a* in the following words: *late, chaotic, bare, add, infant, was, talk, sofa.*

When a need arose for closer indication of sound, diacritical marks were adopted; and by means of them dictionary makers have for many years represented the various sounds of the vowels and of those consonants—like *c* and *g*—that have more than one pronunciation. (The various dictionaries have not chosen the same symbols to represent given sounds. Therefore, when pronunciation of a word is being looked up, it is advisable to refer to the key words at the foot of the dictionary page.)

To students of language, however, diacritical marks were not satisfactory as indicators of sound. They desired a new alphabet, internationally uniform, in which one sign or letter should represent only one sound, and each sound should be represented by only one symbol. In 1888 the International Phonetic Association devised and endorsed such an alphabet, the International Phonetic Alphabet, which is now in general use (with various modifications) for phonetic transcription of speech. Typical symbols used for English are as follows:

PHONETIC SYMBOLS

Vowels

Sound	Symbol	Example
a as in—		
father	[ɑ]	[fɑðə]
ask	[a]	[ask]
at	[æ]	[æt]
pale	[eɪ]	[peɪl]
air	[ɛə]	[ɛə]
sofa	[ə]	[sofə]
all, law	[ɔ]	[ɔl]
e as in—		
end	[ɛ]	[ɛnd]
eve	[i]	[iv]
fear	[ɪə]	[fɪə]
i as in—		
ice	[aɪ]	[aɪs]
it	[ɪ]	[ɪt]
o as in—		
own	[oʊ]	[oʊn]
obey	[o]	[obeɪ]
song	[ɒ]	[sɒŋ]
four	[ɔə]	[fɔə]
u as in—		
up	[ʌ]	[ʌp]
sure	[ʊə]	[ʃʊə]
pull	[ʊ]	[pʊl]
mute	[ɪu]	[mɪut]
use	[ju]	[juz] (v.)

Sound	Symbol	Example
o as in *word* *i* as in *fir* *u* as in *burn*	[ɜ]	[fɜ]
ou as in *house* *ow* as in *owl*	[aʊ]	[aʊl]
oi as in *oil*	[ɔɪ]	[ɔɪl]
oo as in *look* *u* as in *pull*	[ʊ]	[lʊk]
oo as in *do, ooze*	[u]	[du] [uz]
oo as in *poor* *u* as in *sure*	[ʊə]	[ʃʊə]
ea as in *bear* *e* as in *there*	[ɛə]	[bɛə]

Consonants

Sound	Symbol	Example
b as in be, rob	[b]	[bi]
c as in cost	[k]	[kɒst]
c as in cell	[s]	[sɛl]
ch as in church	[tʃ]	[tʃɜtʃ]
d as in do, nod	[d]	[du]
f as in fat, stuff	[f]	[fæt]
g as in girl	[g]	[gɜl]
g as in gem	[dʒ]	[dʒɛm]
h as in hue	[h]	[hju]
j as in judge	[dʒ]	[dʒʌdʒ]
k as in kit	[k]	[kɪt]
l as in lie	[l]	[laɪ]
m as in man	[m]	[mæn]
n as in net	[n]	[nɛt]
ng as in ring	[ŋ]	[ɹɪŋ]
p as in pair	[p]	[pɛə]
r as in red	[ɹ]	[ɹɛd]
s as in soil	[s]	[sɔɪl]
sh as in shawl	[ʃ]	[ʃɔl]
t as in tea	[t]	[ti]
th as in thin	[θ]	[θɪn]
th as in then	[ð]	[ðɛn]
v as in vise	[v]	[vaɪs]
w as in wit	[w]	[wɪt]
wh as in what	[ʍ]	[ʍɒt]
x as in box	[ks]	[bɔks]
y as in yes	[j]	[jɛs]
z as in zebra	[z]	[zibɹə]
z as in azure	[ʒ]	[æʒə]

In addition to the foregoing letter symbols the following signs are used:

A colon, preferably ː, after a symbol indicates that its sound is lengthened as compared with that of the unmarked symbol.

One dot, ·, is the sign of partial lengthening.

Stress is indicated by ˈ, secondary stress by ˌ, each of these accents being placed before the stressed syllable: buccaneer, ˌbʌkəˈnɪə.

A breve above a symbol indicates that the sound is unstressed: ĭ

A dot or a short vertical line under an l, m, n, or r indicates that the sound forms a syllable: bætl̩

Modifiers are signs indicating slight shifts in the position of the tongue.

Low modifier, ˔	Front modifier, ˟
High modifier, ˕	Back modifier, ˠ

The following are used under letters

₀ breath	ᵉ specially open vowel
˘ voice	⊓ dental articulation
. specially close vowel	ʷ labialization

Note the following:

High level pitch	ā	High falling	à
Low level pitch	a̱	Low falling	a̱
High rising	á	Rise-fall	â
Low rising	a̰	Fall-rise	ǎ

PART V.   GRAMMAR

Even the best of authors and editors occasionally have a problem with some aspect of grammar. The following chapter attempts to deal with common errors or problems of construction that may arise. The examples, culled from manuscripts and other written work, demonstrate situations that are likely to occur.

Where terms are not defined by context, a definition appears in the glossary of grammatical terms at the end of the chapter.

VERBS

Verbs show five changes of form: voice, mood, tense, person, and number. The following sections show common errors in the use of these forms.

VOICE

The active voice asserts that the person or thing represented by the subject of the sentence does something; the passive asserts that the subject is acted upon. In the active voice, "the mother calls." The passive voice is formed by adding the past participle of a verb to the proper form of the verb *to be*, for example: "The boy *is* called (*was called, has been called, may be called, will be called*)."

Although the passive has legitimate uses, it is a weak voice, more likely to be used effectively in exposition than in narration. Its main function in narration is when the acting agent is unimportant or obvious: for example, "the suspect was arrested." The ordinary use of the passive in exposition is seen in the following:

If new illustrations are prepared, they will temporarily be designated by fractions.

Where directness and vigor are desired, try the active voice first, even though the passive may have to be used sometimes to maintain parallelism.

Weak: His procrastination showed that will power was lacking in him.
Better: His procrastination showed that he lacked will power.

Weak: Everything that was done by us that day was a waste of time.
Better: Everything we did that day was a waste of time.

The copy editor should correct or query awkward passive expressions.

Weak: Useful wild animals known to the Indians were made known to the early explorers, and the "salt licks" that buffalo and other animals had sought out were also told about by the Indians.
Better: The Indians told the early explorers about useful wild animals and about the "salt licks" that buffalo and other animals had sought out.

Poor: This was not as suitable for many purposes as was the puddled iron, and the older methods have still continued to be used.
Better: ... the older methods are still in use.

A sudden shift from active to passive should be avoided.

Poor: We visited the museum, where many famous statues were seen.
Better: We visited the museum, where we saw many famous statues.

Poor: Such a book costs so little, and so much is learned from it.
Better: Such a book costs so little and teaches so much.

Poor: Stand on a book with the toes extending over its front edge. The toes are then bent up and down as far as they will go.
Better: ... Then bend the toes up and down as far as they will go.

MOOD

Mood refers to the form of the verb indicating the manner of doing or being. The moods commonly recognized are the indicative, the imperative, and the subjunctive. The infinitive is sometimes called a mood, as are certain verb phrases: those formed with *may, might, can, could* are called the potential; with *should, would,* the conditional; and with *must, ought,* the obligative.

Indicative. The indicative is a declarative or interrogative sentence:

The tree is tall.
Will he go home now?

Imperative. The *imperative* is a command:

Go home!
Add three eggs, beaten well.

The subject *you* is more often understood than written.

The subjunctive. Classically, the subjunctive expresses an improbable condition, one contrary to fact, or a wish, command, or desire. It is almost extinct in spoken English and is passing away even in written English. Fowler, in 1926, declared the usage moribund and said that when subjunctives were encountered they were often "antiquated survivals of pretentious journalism, infecting their context with dullness."

The following examples show the subjunctive in one of its few viable contexts. But observe the equally correct versions (in parentheses) that are more usual in speech and sound less pompous in writing.

He gave orders that the bills *be* paid.
(He gave orders for the bills to be paid.)

It is important that he *breathe* fresh air.
(He needs to breathe fresh air.)

How can one insure that the power of science *be* used for the benefit of man?
(How can one insure the use of the power of science for the benefit of man?)

The subjunctive is preserved in certain traditional phrases and idioms.

Far be it from me.   Be that as it may.
So be it.   Come what may,
If I were you.

The condition contrary to fact is the construction that gives the most trouble today in the use of the subjunctive. Following are correct examples of this use:

Janet wished bitterly that her brother were there to comfort her.
If such a device as this were not used, every time sex cells united the number of chromosomes would double.

However, many clauses introduced by *if* do not express a condition contrary to fact, but merely a condition or contingency. In such cases, the subjunctive is incorrect and betrays the kind of grammatical insecurity demonstrated by "between you and I." In the following examples, the correct examples use the indicative.

Wrong: If he were found guilty, he was probably outlawed.
Right: If he was found guilty, he was outlawed.

Wrong: Better yet, use its English equivalent if there be one.
Right: . . . English equivalent if there is one.

Wrong: Next I looked to see if the ground were clear.
Right: . . . was clear.

Clauses introduced by *as if* or *as though* usually express an unreal condition and require a past subjunctive.

Some patients feel as if they *were* falling.
Handle the baby as though it *were* your own.

Copy editors will often find that subjunctives need to be changed to indicatives in stilted writing, and that subjunctives seldom need to be inserted in vernacular writing.

Infinitives. English idiom is often violated by using an infinitive instead of a participle or a gerund.

Unidiomatic: These foods seem to have the property to prevent and cure rickets.
Correct idiom: . . . property of preventing . . .

Unidiomatic: They needed a simple method to keep their trading accounts.
Correct idiom: ...method of keeping...

Unidiomatic: Water transportation of heavy articles costs less than to send them
  by railroad or truck.
Correct idiom: ...costs less than sending...

Errors of this sort seem often to arise from confusion of constructions of synonymous words. For instance, the verb *help* may be followed by an infinitive, but *aid* may not.[1]

The discussions have helped to crystallize points of view.
The discussions have aided in crystallizing points of view.

Similarly: forbidden to do, prohibited from doing; right to deliver, privilege of
  delivering; eager to return, desirous of returning; obliges one to keep,
  necessitates one's keeping.

There are many other expressions similar to these. Some are listed in the section on the right preposition, pages 432ff.

Shift of mood. A common error in syntax is a shift in mood.

Imperative to indicative: Map the territory carefully, and you are also to note
  its topographical peculiarities.
Better: Map the territory carefully and also note its peculiarities.

Subjunctive to indicative: Whether their experience be unusual, or is only
  thought so by them, they are not excused from the ordinary sanctions of
  morality.
Better: Whether their experience is unusual, or is only thought so by them,...

TENSE

The present tense. The present tense is used to express action in the present, a state existing in the present, habitual or customary action, a general truth. It is also used to express future action when the time element is specified.

He is working in the garden.
He travels by bus.
Winters in this part of the country are very cold.
The ship sails next Tuesday.
We are going to Europe next year.

The present tense is sometimes used in relating an event that occurred in the past. Though this is legitimate and effective usage on occasion,

[1]See also Fowler's *Modern English Usage,* rev., Sir Ernest Gowers, 2nd ed. (London:
  Oxford University Press, 1965), article Analogy, 3.

writers in this form are often inconsistent. (In the following example all verbs after the first should be in the present tense.)

Inconsistent: Did you ever sit in on a ball game in the last half of the ninth — the score tied? The home team is up to bat. Two outs, three balls and the bases full! Oh, boy! You held your breath as the pitcher wound up. Then a vicious crack — and no, it can't be possible! Wait! You're grabbing on to the seat — then you let loose with a roar and the crowd went crazy. Man, it was a home run — four men in and it was all over.

The simple form *is* is sometimes incorrectly used for the progressive form *is being*.

Poor: Up to the period of depression in which this is written the balance has been favorable.
Better: Up to the period in which this is being written ...

The expressions *that is to say* and *the fact is* are idiomatically used with a sentence in the past tense. *Is reported, is said,* and the like are correct, rather than *was said*.

That is to say, he had now made up his mind.
The fact is that the Americans themselves had again and again supplied the corroborating information.
It was on this occasion that the General is reported to have said ...
Beyond was an empty shop — empty, that is, except for hats of all sorts and sizes.

The future and future perfect tenses. The future tense indicates action in the future. It is often misused for the future perfect tense, which indicates an action completed before a time in the future.

Soon a large body of knowledge *will have been* built up.   [Not *will be* built up.]
In March I *shall have been* out here five years.   [Not *shall be*.]

Past tenses. The simple past tense is used to express action that took place in the past or to express a condition that existed at a definite time in the past.

The modern fair had its origin back in those times.
The summer of 1949 was exceedingly hot.

They represent forms that lived and died long ago.   [Not *have lived*.]
We got our money's worth long ago.   [Not *have gotten*.]

The *present perfect tense* is used to express action beginning in the past and continuing to the present; or an action that has already come to pass at the time of speaking.

Wrong: From the time of Palestrina to the present day, Italian composers did not use so many folk songs in their works as the French and Germans did.
Right: ... Italian composers have not used ...

Wrong: Since that time the Roman arts became imitative.
Right: Since that time the Roman arts have become imitative.
Or: After that time the Roman arts became imitative.

The *past perfect tense* indicates an action completed before another action, also in the past.

The outflow of gold had been so heavy that the banks were forced to take action.

Sequence of tenses. Proper sequence of tenses requires that time relationships be accurately expressed.

The *present participle* represents action at the time expressed by the principal verb; the *past participle,* action distinctly before that of the principal verb.

It is an heirloom, *having been* first owned by my great-grandfather.   [Not *being* first owned.]
Having finished his dinner, he rose from the table.   [Not *finishing.*]

If the completion of the earlier action is virtually simultaneous with the second action, use of the past participle produces a pedantic sentence, and the present participle is better.

Stilted: The press secretary quoted the ambassador as *having said . . .*
Better: The press secretary quoted the ambassador as *saying . . .*

The infinitive is used for a statement true at the time of the principal verb. Violations of this rule are frequent with the verb *like.*

If Lincoln *had lived long enough to have* a part in the reconstruction of the South . . . [Not *lived long enough to have had.*]
It *might have been* to the interest of the Americans *to be* a little more ready themselves to forget recent disagreements.   [Not *might have been* to their interest *to have been.*]
How Ray *would have liked to range* alongside and wave a greeting to his comrades! [Not *would like to have ranged.*]

The general rule of sequence is to use past tense forms in a dependent clause when the governing proposition has a past tense form.

The witness said his name was Tom Jones and that he lived at 60 Summer Street. [Not *lives,* even if he does now live there.]
The control commission spokesman said that the teams would [not *will*] gather in the capital and that they had [not *have*] all their equipment ready.

This rule is not observed in expressing a statement of universal truth:

We were told in school that water boils at 212°F.
The speaker said that oak makes the best flooring.
Columbus proved that the earth is round.

The writers who used *is* in the following sentences were evidently dimly aware of this rule about universally true statements; but these are not examples of the rule. The verbs are not dependent and each sentence expresses a fact in the past.

Tell the story of the life of Captain John Smith. What was [not *is*] his chief literary work?
The printing press was [not *is*] perhaps the greatest invention of the latter part of the Middle Ages.
Palestine was [not *is*] the birthplace of Christianity.

Always consider the tense of a dependent verb form in relation to the time expressed in the verb upon which it depends.

Wrong: How would natural conditions be affected if water continued to contract until it freezes?
Right: ...until it froze?

Wrong: The affirmative would probably argue that wherever the flag goes it should carry with it citizenship and equal treatment, and that if people who have received United States citizenship should decide later that they preferred not to have it, that would be their responsibility.
Right: ...should decide that they prefer not to have it, that will be their responsibility.

Sometimes the subordinate verb fixes the time and the principal verb requires correction.

Wrong: If the unexpired subscriptions had amounted to $80,000, instead of $20,000, the adjusting entry would take the following form.
Right: . . . the adjusting entry would have taken the following form.

This rule does not operate when a past tense verb appears in a parenthetical construction.

Poor: The enforcement agency was ready, she said, and the work would go forward promptly.
Better: "The enforcement agency is ready," she said, "and the work will go forward promptly."
Or: She said that the enforcement agency was ready and the work would go forward promptly.

AUXILIARIES

An auxiliary verb is one used with another to indicate voice, mood, and tense. The auxiliaries are *be, can, have, may, must, ought, shall, will,* and their conjugational forms.

As is clear in some of the examples under Sequence of tenses, *can, shall, will,* and *may* follow the rule and become *could, should, would,* and *might* when governed by a verb in the past.

The witness further testified that it was his intention to continue to serve water
  to all consumers until such time as a mutual company *might be organized*.
  [Not *may be organized*.]
Plan how you would combine a wheel and axle, a pulley, and an inclined plane
  so that a man using a 150-pound force *could overcome* a resistance of 60,000
  pounds. [Not *can overcome*.]
He will be able to raise money only on his personal property and credit, which
  will be limited.   [Not *would be limited*.]

Could in an *if*-clause should be followed by *would*, not *will*.

Even if he could secure such men, the salaries he *would have to pay* would in
  many cases be so high as to eat up the profits.   [Not *will have to pay*.]

Can, could, will, and *would* should never follow *in order that*.

In order that the teacher *may be able to use* any textbook . . .   [Not *can use*.]
In order that she *may go*, he is hiring a substitute.   [Not *can go*.]

"Do." *Do* is a general-utility auxiliary in English, having uses hardly
paralleled in other languages. It is used with another verb to intensify the
meaning, and it is often used as a substitute for another verb. The latter
usage is carried very far.

You do snore (even though you say you do not).
She did do it.
He lied, though he says he didn't.

The use of *do* and *does* as auxiliaries for *have* and *has* has good sanction in American English where the sentence embraces the idea of "customarily" or "usually," and in circumstances where the British would
use *got*.

Do you have coffee with dinner or afterward?
The school does have openings for the fall in the ninth grade.
We do not have the key pages. (In England, "We *have not got* the key pages,"
  which jangles on many American ears.)

Misuse of *did* for *has* and *have* should not be allowed to pass in written
English.

Wrong: Did he come yet?
Right: Has he come yet?

Wrong: Did you *get* your acceptance yet?
Right: Have you *got* your acceptance yet?

"Shall" and "will." The auxiliaries *shall, will, should* and *would* are
classically used with verbs to express simple futurity and determination.
The distinctions formerly made between *shall* and *will* are breaking down

and are little observed in popular speech. Nevertheless the distinctions are observed by many careful writers.

1. In a declarative sentence simple futurity or mere expectation is expressed by *shall* (or *should*) in the first person, *will* (or *would*) in the second and third persons.

> I shall (he will, they will, and so on) go next week.
> I was afraid that I should be unable to come.
> The odds are that we shall never do it at all.
> We should hardly expect a tropical region to have much industry.
> The forests, as you would expect, are peopled sparsely.
> He found out how to design new auditoriums so that they would have just the right reverberation.

The word *one* is considered a third-person pronoun and is followed by *will* (or *would*) to express simple futurity.

> The winter temperatures are milder than one would expect.
> One would like to arrange the theaters of the world in exhibition.

2. In a declarative or an imperative sentence, determination, threat, command, willingness, and promise on the part of the speaker are expressed by *will* (or *would*) in the first person, *shall* (or *should*) in the second and third.

> *I will* (*you shall, he shall*) go next week.
> A final *vote shall* be taken to allocate power and choose *who shall* wield it.
> *He shall* make restitution.
> All *bills shall* be sent to the Committee on Revision and Printing.
> It is for them to determine when the *conversations shall* stop.

This distinction has been blurred for so long that experts debate whether General Douglas MacArthur's statement on leaving Corregidor in 1942, "I shall return," was an expression of determination or simple futurity. It is not even clear which would have been the stronger statement under the circumstances.

Should is also used in the sense of obligation and *would* in the sense of habitual behavior.

> A child should not be asked before the meal what he would like to eat.
> He would run away whenever my back was turned.

AGREEMENT

The rule of agreement is that in a sentence every part should agree logically with every other related part. A verb, therefore, should agree in number with its subject.

One of the serious results that come from the experience through which this country has been passing *is* loss of faith.
The *Stars and Stripes was* floating above us.
There *has been many* a benevolent dictator.
More than one book by this author *has* been cited.

The rule of course holds if the subject follows the verb.

The Normans, under which general term *are* comprehended the *Danes, Norwegians,* and *Swedes,* were accustomed to rapine and slaughter.
What *signify* good *opinions* unless they are attended by good conduct?
What type of veining *have most* of the long grasses?
Where *are* the *Khyber Pass* and *Mount Everest*?

This general rule is simple, but determining the number of the subject is sometimes complex, as a study of the following sections will show.
Some mistakes occur through failure to recognize a subject as a plural. (See Collective nouns, p. 354; also the glossary to this chapter.)

Existing *data prove* that . . .
The *media* in the United States *are* . . .

The number of the subject is sometimes obscured by words coming between it and the verb.

The *list* of broadcasters thus selected *is* arranged in alphabetical order.
The *extent* to which motor-driven vehicles and tractor-drawn farm machinery are being used on American farms *is* likely to lead . . .
A *number* of people go . . .[2]

An appositive does not change the number of the subject.

Our side, the Tigers, is ahead.
In this study the assumptions, the measuring stick, are placed clearly before the reader.

The number of the subject and verb is not affected by intervening words introduced by *with, together with, including, as well as, no less than, plus,* and similar expressions.

The farmer's happiness as well as his profits arises from being a part of a neighborhood instead of being merely a resident in it.
The load of ore, together with the consignment of pig iron, was delivered on time.
The rhyming no less than the meter of the poem is amateurish.
All living things, and therefore protoplasm, are composed of these substances.
Broth does not ferment or decay if the air, and consequently germs in the air, is excluded.

[2]*Number* preceded by *the* is singular, preceded by *a*, plural.

Writers sometimes use these phrases where, because the thought is plural, a simple *and* would be better. Copy editors should think of this if they come upon one of these phrases followed by a plural verb when the grammatical subject is singular.

> *Poor:* Heart disease, together with cancer, kidney diseases, and apoplexy are almost entirely diseases of middle life and older periods.
> *Better:* Heart disease, cancer, kidney diseases, and apoplexy are . . .
>
> *Poor:* This zygote, with the hyphae that develop from it, probably represent the diploid stage.
> *Better:* This zygote and the hyphae that develop from it represent . . .

The word *plus* is considered to behave like the preposition *with* rather than the conjunction *and*.

> The *Chamber of Commerce* plus its affiliates *awards* the medal.
> The *metal* plus the oxygen *weighs* more.
> *Erosion* Plus Dust *Ruins* Area Size of Kansas

Singular substantives joined by "and." A subject consisting of two or more substantives joined by *and* requires a plural verb.

> The *casting* of the line *and distribution* of the matrices *are* done automatically and do not interfere with the operation of the keyboard.
> To understand them and to enjoy them require a certain degree of talent.
> Whether the commission will set high standards and whether the courts will support the commission remain to be seen.

Sometimes the *and* is not expressed.

> His face, his eye, were hardening to the crisis.

The copy editor should not be misled by punctuation that makes a plural subject appear to be a singular subject followed by a parenthetical phrase.

> The great diversity of the risks covered, and the complex nature of the business, introduce production problems of an unusual character.
> How the minstrel and his people spy on the boy, and how he succeeds in finding a happy home after running away from them, are told in an interesting manner.

With two nouns connected by *and*, a singular verb may be used if the thought is definitely singular, as in the expression "name and address."

> Beating and counting the different kinds of measure is interesting and helpful practice for any chorus.
> As the hand and arm of the conductor pictures the singing of the phrase, . . .

Singular substantives joined by "or" or "nor." A subject consisting of two or more singular substantives joined by *or* or *nor* requires a singular verb.

> One or the other of the boys has been here.
> When sickness, infirmity, or reverse of fortune affects us, the sincerity of friendship is proved.
> Biting threads or cracking nuts with the teeth is an injurious practice.
> Neither tomb nor monument is needed.
> Salt or baking soda or a mixture of both possesses all the cleansing qualities of most toothpastes.

For a discussion of the use of *or* and *nor* in sentences without *either/neither*, see page 375.

When the copy editor comes upon a plural verb following *or* and a singular noun, he should consider whether the real error may not be in the use of *or* instead of *and*, rather than in the number of the verb. (*And* should be substituted for the italicized *or* in each of the following examples.)

> *Wrong:* Water, potato water, *or* milk are the liquids commonly used in making yeast breads.
> *Wrong:* Ether *or* acetone vaporize very readily, while mercury *or* glycerin, which boil at much higher temperatures, have . . .
> *Wrong:* When kerosene and a soap solution in water are violently shaken, the kerosene divides into colloidal-sized particles. The whipping of cream, beating of egg-whites, *or* mixing of mayonnaise are the same in principle.
> *Wrong:* Individuals *or* information opposed to time-honored beliefs were exceedingly unwelcome.

Plural and singular substantives joined by "or" or "nor." When a subject is composed of both plural and singular substantives joined by *or* or *nor*, the verb should agree with the nearer.

> Neither money nor men were lacking.
> Neither the national Constitution nor the national statutes undertake to prescribe any comprehensive regulations.
> Others are trapped by the fear that their interests or their property is being threatened.

Similarly, when the subject contains the correlatives *not only . . . but also*, the verb agrees with the nearer substantive.

> Not only the children but also the mother was ill.

Positive and negative subjects. When a negative subject is joined to a positive one, the verb should agree with the positive.

> It is Mary, and not her brothers, who drives the car.
> Not Mary but her brothers drive the car.
> Accuracy and not speed is the more important.

"Each," "every." *Each, every, either, neither, anyone, anybody, every one, everybody, someone, somebody, no one, nobody, one,* and *a person* call for a singular verb.

Each plant and animal has its peculiar character.
Is either of these candidates worthy of our support?
Every leaf, every twig, every drop of water, teems with life.
No one but schoolmaster and schoolboys knows...
The regions in which each of these is found in greatest abundance are shown on the map.

When a singular verb following *each* seems awkward, it may be that *each* is misplaced, as it is in the following.

Poor: Paul has divided his garden so that two plots each make 25 percent of the whole garden and five plots each makes 10 percent of the garden.
Better: Paul has divided his garden so that each of two plots makes 25 percent of the whole garden and each of five plots makes 10 percent of the garden.

When the subject is really plural, *each* before the verb often needs to be transposed to follow the verb.

Poor: John, Bob, and Harold each has 15 marbles.
Better: John, Bob, and Harold have 15 marbles each.

Poor: The upper and lower anteriors each has only one canal.
Better: The upper and lower anteriors have only one canal each.

In other cases the meaning would be more accurately conveyed without *each*.

Poor: Thorium and actinium each gives a series of similar disintegration products.
Better: Thorium and actinium give a series of similar products.

Poor: Each of the symbols C, H, and O stand for atoms of carbon, hydrogen, and oxygen.
Better: The symbols C, H, and O stand for atoms of carbon, hydrogen, and oxygen.

"None." *None* may be construed as either singular or plural, according to the thought to be conveyed: "no amount" (when the following noun is singular), or "no individuals" (when the following noun is plural).

None of the fruit was eaten.
None of the volcanoes in Chile are active.

When the meaning is "not one," it is better to use *not one* than *none* with a singular verb.

Not one of the guests has arrived.

Collective nouns. A collective noun takes a singular verb unless the individuals forming the group are to be emphasized.

His family has just moved in. Are your family well?
The committee adheres to its decision. The committee have signed their names to the report.
The audience struggle into their coats.
The infantry was dispatched.
The couple were married in 1952.
This people has become a great nation.
Practical cultivation of health is what the youth need in these critical years.
Black specks marked the spot where a herd of buffalo were lying.

Similarly, if the subject of a sentence is a group of words that conveys the idea of a number of individuals, the verb should be plural even though the governing noun is singular.

A racial majority of the population are . . .
Only a fraction of the total species of any given period are likely to leave recognizable fossil traces.
A whole galaxy of poets have celebrated the divine maiden snatched away to dwell in the underworld.
A wide range of bacterial phenomena are involved.
A considerable series of them are available.

Such groups of words are sometimes pitfalls, for in groups like the following, the noun is not always used in a collective sense.

What is the highest percentage? What percentage of speeches spread on the Congressional Record are actually made in Congress?
An average of fifteen for the month is about as good as one can expect. An average of fifteen men have been employed during the month.
A combination of numbers and letters at the left corresponds to the Dewey Decimal System. A combination of factors were responsible.
Another group of substances has been discovered. The group of words introduced by a preposition always act together as a unit to modify some other part of the sentence.
The number of relationships involved is enormous. An unreasonable number of boards have grown up in most states.

Subjects plural in form but singular in effect. The expression of a singular idea may look like a plural; a singular verb should follow.

Penal Acts was the legal name, Intolerable Acts what the colonists called them.
Caesar's *Commentaries on the Gallic War* is a model of historical writing.
"Cedar mountains" refers to the cedars of Lebanon in northern Syria, and "silver mountains" suggests the Taurus range.
Taps was blown.
Politics is a dangerous game. (See also nouns with only one form, under Spelling, pp. 476–477.)

The number of the verb in numerical statements depends upon the intention of the subject; that is, whether the number named is thought of as a whole or distributively.

Seven times nine is sixty-three.
Five-sixths of Spain is highlands.
Four years is too long a time to spend in college.
There was six feet of water in the hold.
Forty inches of air space filters out almost all the short waves.
Four-fifths of the words in common use are of Anglo-Saxon origin.
Millions of dollars were lost by the citizens of America.

Gerunds. A gerund used as the subject of a sentence should be followed by a singular verb.

Swimming is good exercise.
The whistling of the teakettle annoys the cat.

Relative clauses. The verb in a relative clause agrees with the antecedent of the relative pronoun, which is the nearest noun or pronoun and is often the object of a preposition, as in the phrase *one of those who, one of the things that.*

Recall here again one of the sayings that follow the title page of this book.
He is one of those men who talk much and think little.
An argument was one of the diversions that were always welcomed by a cowboy.

The following is not an example of the construction above.

Ask one of your classmates who knows the correct forms to listen as you read.
   (Ask one who knows.)

Expletives. After introductory *here* and *there* a singular verb is usually preferred when the logical subject consists of substantives joined by *and,* with the first one singular in number. To this there are frequent exceptions, such as the last two examples below.

There is always one or more who is not stupid.
There was one Jute, three Saxon, and four Angle kingdoms.
The sun had set three hours before, but there was no moon and no stars.
Here come Williamson and Friedberg.
There were, besides, the owner and his son.

"What." *What,* unless it clearly has a plural antecedent, as in the third example below, equals *that which* or *the thing that* and should be followed by a singular verb.

They journeyed through what is now the Dakotas.
What is perhaps of greater immediate interest to the chemist is variations of the thiazole structure.
My people were what were known as "God-fearing folk."

Predicate nouns. A verb should agree with its subject, not with its predicate noun, or complement.

One of the most important things in the North *is* the feet of dogs and men.
The night *sounds* of strange swamp creatures *were* the only thing that broke
the stillness.
The Vietnamese *are* an agricultural people of mixed race.
The *dagger* you lost *and* the missing *weapon are* one and the same.
All you told me about *was* the dangers.

In sentences like the last example above the complement may be felt so strongly to be the real subject that a plural verb may be used.

All you told me about were the dangers.

Sometimes grammatical correctness or euphony can be secured only by rewording the sentence.

Problem: The way a patient feels and looks are also important indications of
illness.
Better: A patient's feelings and appearance are also important indications of
illness.

Agreement in person. A verb must agree in person with its subject.

You have been with our dear friend.
Thou hast a wit of thine own.

When the subject consists of pronouns of different persons joined by *or* or *nor*, the verb agrees in person with the nearer.

Neither she nor they are involved.
You or he is the person who must undertake the business.

Since a relative pronoun agrees in person with its antecedent, the verb in the clause introduced by the pronoun should be of the person of the antecedent.

I, who am older, know better.
You, who are stronger, should be the one to go.

In sentences beginning with the expletive *it* and completed by a relative clause, the verb in the relative clause should be of the same person as the noun or pronoun that follows the principal verb. Do not confuse such sentences with those, like the third example below, that begin with a pronoun.

It is I that am going.
It is I who am hurt.
I am the one who is going.

NOUNS

English nouns do not alter their form when they are in the objective case, as do Latin nouns. Nonetheless, they cause considerable difficulty in other ways: number and possessive forms, among others.

NUMBER

Accord. The number of nouns or pronouns should be in accord throughout a sentence.

On the backs of the hands. [Not *back of the hands.*]
Read the questions one at a time, find the answers, and write them out briefly.
The bars tapered to a sharp point at one end and bore on their heavier end the name Eric Bernstein.

Singular with a plural possessive. To avoid ambiguity a singular noun is often used with a plural possessive when only one of the things possessed could belong to each individual.

Manufacturing helps many people in the smaller cities to earn their living.
Forbes knew most of them by their first name.
Some of them could not pay their rent.
They eyed each other furtively and cursed beneath their breath.
Four pilots crashed to their death.

Similarly:

Think of the last name of five pupils in the room.
The steam line ruptured, causing the death of seven longshoremen.
They doubled their efforts to discover the identity of two men who struck a man with their automobile and then fled.

Care must be taken not to apply the rule to the wrong noun.

Wrong: It is pretty clear that the smile on the *face* of the delegates, whenever they look at each other, is not a sincere one.
Right: . . . on the faces of the delegates . . . [*Smile* is the noun the rule applies to.]

"Kind," "sort." *Kind, sort, type, class, quality, brand, breed, species, size, variety,* and similar words are singular nouns with plurals regularly formed and used. The singular forms should therefore be modified by *this* and *that,* not *these* and *those.*

The singular forms of these words may be followed by a plural verb if individuals, rather than a class, are in mind.

This variety is hardy.
That make is most popular.
This variety are everbearing.
That breed are good watchdogs.

Expressions using *kind of, sort of,* and so on, should be in the singular, unless the plural idea is overriding.

> What type of cat appeals most to you?
> The variety of clam most used is called Little Neck.
> The kind of variation on which Darwin depended was the minute modifications everywhere evident.
> The variations employed in survival are a particular kind of variation that are of hereditary significance from the start.
> What kind of transactions decrease the proprietorship?
> What does the table show regarding concerns with very high profits? What kind of concerns are they?
> This is the magma of which the cone type of volcanic mountains are generally made.

The common error is failure to stick to the number selected.

> *Inconsistent:* The kind of apple we grow keeps well if they are separately wrapped.
> *Better:* The kind of apple we grow keeps well if each apple is separately wrapped.

> *Inconsistent:* This kind of cats are native to Egypt, but it is common in America.
> *Better:* This kind of cats are native to Egypt, but they are common in America.

The following would be better expressed entirely in the singular.

> *Inconsistent:* Waterless cookers are really a type of pressure cooker that depends . . .
> *Better:* A waterless cooker is really a type of pressure cooker that depends . . .

> *Inconsistent:* The owls are a sort of self-appointed night watchman who discover and eat great quantities of mice.
> *Better:* The owl is a sort of self-appointed night watchman who discovers and eats great quantities of mice.

While studying a sentence with *kind of* or *sort of* for grammatical correctness, the editor should not ignore the possibility that the phrase should be replaced or deleted. Such phrases are sometimes used to give the appearance of a casual writing style, or in a place where *virtual* or *near* is needed.

> *Poor:* The buildings were *kind of* a wreck. [Delete *kind of.*]
> *Poor:* Through his martial-law decrees, the President created a *sort of* dictatorship. [Delete *sort of,* or substitute *virtual,* depending upon the author's meaning.]

POSSESSIVE CASE

The following paragraphs discuss various uses of the possessive case. (For formation of the possessive case, see pp. 478–481.)

Inanimate objects. On rare occasions it may seem awkward to attribute possession to inanimate objects, e.g. "the house's roof." However, this awkwardness seems to have more to do with the sound than with the sense. Few editors today would cavil at "the farm's management," which was formerly considered less acceptable than "the management of the farm." Most writers today, however, would probably write it "the farm management" in the first place.

In some expressions, the idea of possession is so remote that the apostrophe is unnecessary and the phrase looks and sounds fine without it. The former possessive noun then becomes an adjective:

A two weeks waiting period is found in the laws of Alabama. (or, *a two-week . . .*)
The judge imposed sixty days sentence (or, *a sixty-day . . .*)

The modern tendency illustrated above to dispense with *of* and tighten the sentence by creating an adjective ("commissioner of parks" to "park commissioner") should be followed with care. "Critics of the administration" is not ambiguous; "administration critics" is. "Leaders of peasants" will translate easily into "peasants' leaders" but "peasant leaders" may not be correct.

The apostrophe is frequently omitted in names of organizations or buildings where the idea of possession seems obvious. In the absence of clear proof that the construction is correct or deliberate, the copy editor should query.

Farmers Loan and Trust Company   Peoples Savings Bank
teachers college   Consumers Union

But, according to the *Government Printing Office Style Manual*

Veterans' Administration

And the correct full form, not often needed, is

People's Republic of China

State prison, states' rights, state's attorney, state's evidence are preferred forms.

Possessive with an appositive. An awkward construction that editors should change is the use of an appositive with a noun in the possessive case. If the object possessed is named, the apostrophe should be used only after the appositive. If, however, the object is not mentioned and the appositive ends the sentence, being explanatory or emphatic in nature rather than restrictive, the sign of possession should be used on the first noun.

My teacher Mr. Smith's book.
Your estimable employer Mr. Tawney's private domain.
A voice spoke. It sounded like Mason's, the city editor.
He promised to have dinner at her aunt's, Susan Eldredge.

Double possessive. When the thing possessed is only one of a number belonging to the possessor, both the possessive case and *of* are used.

a friend of my brother's   a book of Ginn's
. . . who, as a devoted friend of Darwin's, employed . . .

Possessive with a gerund. In sentences like the following, in which a phrase containing a noun or pronoun and a participle is used as the object of a preposition, the participle may be construed as a noun (a gerund) and the preceding noun made a possessive, or it may be construed as a modifier of the noun. The idea of possession is much stronger in some sentences than in others, and sometimes failure to use a possessive might give the sentence a meaning different from that intended. The last example below is such a sentence: "the woman wearing pearls" conveys a different thought from "the woman's wearing pearls."

What do you think of my horse's running away today?
It does not follow from a word's being given as OF that it is obsolete.
An entire meal can be served without anyone's leaving the table.
I object to the woman's wearing pearls.

The failure to use the possessive with the gerund is discussed by Fowler at length under the entry "Fused Participle," and editors who have been brought up on this leap from the desk at the drop of "women having the vote." However, most sentences containing fused participles today are so heavy with superstructure that it is impossible to stuff 's in without creating pedantry. The copy editor should offer a smoother alternative to the author. Here are some simplified examples.

Wrong: I tried to prevent the boy falling off the ladder.
Correct: I tried to prevent the boy's falling off the ladder.
Idiomatic: I tried to keep the boy from falling off the ladder.

Theodore M. Bernstein, in *The Careful Writer,* offers a few fused participles that cannot be repaired, among them, "What are the odds against that happening again?"

Noun with an appositional phrase. In defining a noun with an appositional phrase, a writer can commit a solecism of the sort shown in the following examples.

Wrong: He was named executive vice president, the main paid post in the union.
Right: He was named executive vice president, the main paid official in the union.
Or: He was named to the executive vice presidency, the main paid post . . .

Wrong: They held a parade today, the first anniversary of the proclamation of the re publ ic.
Right: . . . a parade today, on the first anniversary . . .

Wrong: The bombing was said to have been carried out by E.T.A., the Basque initials for Euzkadi ta Azkatasuna, or Basque Nation and Liberty.
Right: . . . carried out by E.T.A., an organization whose initials stand for the Basque words . . .

This error, while common, causes editors much trouble. Allan M. Siegal, a teacher of editors at Columbia University and *The New York Times*, has devised a simple way of testing the correctness of such a sentence: mentally delete the object being defined or explained and see if the verb will work with the appositional phrase. In the first example above, no one can be named the main paid *post*, although one can be named the main paid *official*. And a parade cannot be *held* the first anniversary, although it can be *held on* the first anniversary.

PRONOUNS

CASE

The personal pronouns and the relative pronoun *who* have declensional forms to indicate case, and these in some constructions cause difficulty.

The subject of a verb should be in the nominative case.

In all that big theater there were only he and she.
It is not fit for such as we (are) to sit with the rulers of the land.
Is he as capable as she (is)?
Americans are as courageous as they (are).

An expression like *he says, you believe, she supposed, we pretend,* or *it was thought* between the relative pronoun *who* and its verb does not change the case of the pronoun. Such parenthetical phrases can be set off by commas, but the effect is the same if they are not.

Give the vocation of a person who you believe attained success.
Listen carefully to those who you have reason to believe know how to express themselves in well-chosen terms.

Whom is often misused for *who* because of failure to realize that the relative pronoun is the subject of the following verb, not the object of the preceding preposition or verb; the whole relative clause is the object.

I will exchange snapshots with *whoever writes.*
They should punish *whoever is found* to be guilty.
He was questioned as to *who* on his committee *was* to be given the position.

The object of a verb should be in the objective case.

Whom shall you appoint?
He is the free man whom the truth makes free.
I should consider him more capable than them (...than I should consider them.)
As a result, the company is unable to know whom it may be insuring.
Gentle reader, let you and me, in like manner, endeavor...
Ganowi contemptuously eyed him who had once been a swift killer.

Substantives connected by any form of the verb *to be* should agree in case. It should be remembered that the subject of an infinitive is in the objective case. Any doubt about the correct form can usually be cleared up by transposing the sentence.

What is he? Is it they? It is I [or *me,* colloquially].
If it had been we, how the tongues would have wagged!
Who did you think it was?
They declared the culprit to be him and no other. (They declared him to be the culprit.)
The man whom I thought to be my friend deceived me. (I thought him to be.)
Whom do you suppose him to be? (You suppose him to be whom?)

The object of a preposition should be in the objective case.

Between you and me...
Some of us fellows went fishing.
All went but me. None but me was able to come.
The man whom the committee agreed on was younger than any one of them.
Whom, by the way, do you think they have gone with?
I want a husband like him.

An appositive should be in the same case as the noun with which it is in apposition.

All are going—she, he, and we two.
He spoke to some of us—namely, her and me.
We all met—she, the officer, they you mentioned, and I.

The dative is not often used. An example is the familiar expression "Woe is me," literally, "Woe is unto me."

ANTECEDENTS

Since a pronoun is a word used in place of a noun to avoid repeating the noun, the word it replaces, its antecedent, must be expressed or clearly understood, and with this antecedent the pronoun should agree in person and number.

Doubtful reference. The antecedent of a pronoun should never be in doubt.

Vague: Dirt and bacteria cannot impregnate glass fibers, nor are *they* affected by mildew.
Clear: . . . cannot impregnate glass fibers, which are also unaffected by mildew.

Vague: "In all my years of experience with shoplifters," declares a certain well-known officer who has passed a dozen of *them* as chief detective . . .
Clear: . . . declares a certain well-known officer who has been for a dozen years chief detective . . .

Double reference. The same pronoun should not be used to refer to different antecedents in the same sentence or in the same paragraph unless the context makes the antecedent quite clear.

Tangled: When the baby is done drinking *it* must be unscrewed and laid under a faucet. If *it* does not thrive on fresh milk, *it* must be boiled.
Straight: When the baby is done drinking, the bottle cap must be unscrewed and laid under a faucet. If he does not thrive on fresh milk, it must be boiled.

Tangled: They became close friends and when Bigelow was released *he* asked for Ward's parole so that *he* might work in *his* company.
Straight: They became close friends, and when Bigelow was released he asked for his friend's parole so that Ward could work for the Bigelow company. (The use of *former* and *latter* would make this sentence worse.)

Missing reference. When an adjective looks like a noun, a writer may be encouraged to think he has an antecedent when technically he has none.

Wrong: The Wright hotel survived the earthquake, a testament to *his* confidence in its soundness.
Right: . . . a testament to the architect's confidence . . .

Relative pronouns. A relative pronoun normally refers to its nearest antecedent, and the reference should not be vague or faulty.

Vague: Vitamins are substances contained in various foods and are essential for growth, development, and other physiological processes.
Clear: Vitamins, contained in various foods, are substances that are . . .

Vague: Occasionally he would recall that Mrs. Durward was in reality a woman over forty, mother of a grown son, who according to all the usages of custom should be settling down.
Clear: Occasionally he would recall that Mrs. Durward was in reality over forty, the mother of a grown son, a woman who, according . . .

"It." More often than other pronouns, *it* is used without a clear antecedent.

Vague: How many people were living in South Dakota when it was organized as a state? How does *it* compare with the present population?
Clear: . . . How does *this* compare with the present population?

Vague: When the government workers who should be classed as administrators are enumerated, *it* reaches staggering proportions.
Clear: . . . are enumerated, the total is staggering.

Vague: It is well to mention in passing that the continuous creation of slang terms is part of the natural process of language making, that *it* is often forceful and picturesque, and that slang expressions frequently rise to be acceptable colloquialisms.
Clear: . . . language making, that slang is often forceful and picturesque, and that slang expressions frequently rise . . .

Use of *it* with indefinite reference is not accepted in written English.

Incorrect: It says right here in *The Telegraph* that . . .
Correct: The Telegraph says right here that . . .

As an expletive, *it* is sometimes added for smoothness or emphasis.

It will be best to go.

Much too often writers omit the pronoun *it* in places where correct interpretation of the sentence requires it.

Stir *it* until *it* is thick and then turn over. (. . . turn *it* over.)
If your ski underwear tickles you, turn inside out. (. . . turn *it* inside out.)

The use in one sentence of the pronoun *it* and the expletive *it* should be avoided.

Wrong: When the sun was out *it* was warm, but when *it* disappeared, *it* became chilly.
Better: When the sun was out, the air was warm, but when clouds passed, it was chilly.

The writer of the following did not understand the expletive use of *it*.

Wrong: Through the blackness moved a woman. She was Hilda, the housekeeper.
Right: . . . *It* was Hilda.

"They." The use of *they* as an indefinite pronoun is established colloquially and is not uncommon in writing. However, it can make a sentence heavy and can often be avoided.

Acceptable: In the Arctic, they do not worry about the common cold.

Poor: Why do they add copper to both silver and gold when these metals are prepared for commercial use?
Better: Why is copper added to both silver and gold . . .

Poor: If the design is one that they call a simple pattern, an attachment can be added to the regular loom.
Better: If the design is one that is called a simple pattern . . .

"This," "these." The antecedent of *this* or *these* should always be grammatically clear.

Antecedent in doubt: A thousand industries need such basic data on which to build progress. *This* is one purpose of the Bureau of the Census.
Better: A thousand industries need such basic data on which to build progress. *Providing these data* is one purpose of the Bureau of the Census.

Number. A pronoun should agree with its antecedent in number.

How can a tree (or any other green plant for that matter) develop into the great bulk that *it has?* [Not *they have.*]
Each of the calculations is introduced in a real setting, so that the pupil may be aware of its importance. [Not *their importance.*]
One should watch his step. [Not *their step.*]

What number *everyone/everybody* requires, however, is a matter of some dispute. Dr. Johnson, Jane Austen, and many modern writers have used *everyone . . . they* or *everybody . . . they* despite classic canons of grammar. A James Thurber alter ego in "The Interview" denounces *everybody . . . he* in a rage at the demolition of the idiom through such sentences as: "Has everybody brought his or her slate?" The sentence "*Everyone* was here but *they* all left" demonstrates the truth of a statement in the *Oxford English Dictionary:* "The pronoun referring to *every one* is often plural: the absence of a singular pronoun of common gender rendering this violation of grammatical concord sometimes necessary." However, "Everyone bought their own ticket" is clearly wrong and should be changed. An editor should examine each instance with care before acting.

A singular subject may look like a plural.

Read *Kent's Commentaries. It* will furnish you with a clear statement of the doctrine.

A pronoun referring to singular antecedents connected by *or* or *nor* should be singular.

Mary or Jane has lost *her* book.

A pronoun referring to a plural and a singular antecedent connected by *or* or *nor* should be plural.

Neither the members of the team nor the coach will give an inkling of *their* plans.

A pronoun referring to a collective noun should be singular unless the individuals forming the group are to be emphasized.

> The crowd surged aside. Then *it* gasped and held *its* breath.
> . . . advise the public how to protect *itself* against pickpockets.
> He pulled a tangle of rawhide thongs out of the tub where *they* had been soaking.
> A wise teacher never does for a class what *they* can do for themselves.

The error most often seen in this connection is a shift in number, as illustrated by the following sentences. In each, *their* should be *its*, unless the verb is changed to plural.

> The committee *adds* to *their* report.
> The race still *keeps* some traits of *their* barbarian forefathers.
> The school for chefs *is* giving *their* spring banquet at the Hotel Americana.
> Drake Hotel *pays their* bill September 6. Robinson Manufacturing *pays their* invoice September 9.

Gender. The gender of successive pronouns referring to the same antecedent should be the same for all.

> *Wrong:* After a time the butterfly comes out of *its* chrysalis shell. *It* clings to the shell, spreading and stretching *his* wings until they are dried and strengthened.
> *Right:* The butterfly comes out of its chrysalis shell. It clings to the shell, spreading its wings . . .

It is convenient to use *his* when both sexes are referred to, as in this example:

> Each student must give me *his* choice before tomorrow noon.

However, feminist writers and others have raised objections to this construction, and in manuscripts where such locutions are likely to be offensive, it is wise to use the more cumbersome form.

> In reading lyric poetry, each *boy* or *girl* must chart *his or her* own course.

Where possible, avoid the problem, as follows:

> In reading lyric poetry, each pupil must chart an individual course.

Person. Consistency should be maintained.

> Thou shalt be required to lie down in death, to go to the bar of God, and give up *thy account.* [Not *your account.*]
> If the members of the club will consult the legal department, *they* [not *you*] will be given complete legal advice on matters pertaining to ownership and operation of *their* car.

Reflexive pronouns. The reflexive pronouns *myself, yourself, himself,* and so on, should not be used needlessly in written English for *I, me, you,* and so on.

> My wife and I will go. [Not *My wife and myself.*]
> This is for you. [Not *for yourself.*]
> You and your family must be on board by nine o'clock. [Not *yourself and . . .*]
> The minister rebuked my brother and me. [Not *my brother and myself.*]

Reflexive pronouns should be so placed in the sentence that they refer to the proper noun.

> *Wrong:* Each of these cities is an interesting study in geography itself.
> *Right:* Each of these cities is itself an interesting study in geography.

Editorial forms. In speaking editorially, or in regal and formal style, the forms *we, our,* and *ourself* may be used for *I, my, myself,* but should not be used in combination with *I, my, myself.*

> . . . who sends this to delight *our* heart and stimulate *our* mind.
> We're old enough *ourself* to remember it.

"Each other," "one another." Many authorities insist that *each other* is properly used only of two; *one another,* of more than two. *Each other* has, however, been used in reference to more than two since Anglo-Saxon times. In current literary usage the distinction is nevertheless generally observed.

> John and Mary stared at each other in amazement.
> The members of the families exchanged gifts with one another.

"Either," "any one."

You may choose *either* of two or *any one* of three.

Authorities are not agreed about this distinction. In some contexts it tends to sound pedantic. In any case, for about three hundred years *either* has been used in reference to more than two.

"Anyone," "everyone," "someone." These indefinite pronouns must be carefully distinguished in use from *any one, every one, some one;* they should never be set as single words unless they are equivalent to *anybody, everybody, somebody.*

> She would not see anyone.
> You may have any one of these books you choose.
> You must decide upon some one course.
> Every one of these items should interest everyone.

ADJECTIVES AND ADVERBS

Adjectives and adverbs are grouped together in this discussion because both are modifiers, because both have positive, comparative, and superlative forms, and because errors in their use often arise from a failure to perceive that a noun is being qualified, not a verb, adjective, or adverb.

Many words can be used as either adjectives or adverbs and can be distinguished only by their function in the sentence; for instance:

best	fair	ill	low	short
better	fast	just	much	slow
cheap	first	late	near	soft
close	friendly	leisurely	pretty	straight
daily	full	likely	quick	very
deep	hard	little	right	well
direct	high	long	scholarly	wide
early	hourly	loud	sharp	wrong

Predicate adjectives. Verbs pertaining to the senses (*be, become, feel, seem, smell, sound, taste*) are followed by an adjective unless they refer to the operation of a sensory organ, as in "one feels badly with mittens on."

That is easy, seems rough, feels soft, smells sweet, sounds harsh, tastes sour.
I feel bad, fine, happy, sad.

Errors are more likely in the incorrect substitution of an adverb for an adjective than in the reverse. The copy editor should be on guard against carelessness or false elegance, as in the following examples:

He feels *poorly*. [Should be *ill*, or, if money is the problem, *poor*.]
It tastes *deliciously*. [Should be *delicious*.]

Other verbs may be followed by an adjective modifying the subject or by an adverb modifying the verb, depending upon the meaning.

He kept it *safe;* he kept it *safely*.
He held it *steady;* he held it *steadily*.
He appears *good;* he appears *well* in public.
Second-hand furniture sold *cheap*.
Millions of Americans listened *breathless* to the broadcasts.
Those whose names are written so *bright* on the pages of history.

Comparison. The comparative form of an adjective is used in comparing two persons or things, or in comparing a person or thing with a class.

The lower but more powerful branch...
If Pluto and Mars moved with the same velocity, which would have the greater momentum?
The prize-winning rose was bigger than a cabbage.

The word *other* or *else* is required with a comparative when a person or thing is compared with a class of which it is a part.

Dante was greater than all *other* poets of his day.
I like *The Iceman Cometh* better than anything *else* by Eugene O'Neill.

Similarly:

Wrong: Of all the heroes of the Celtic race, none were so greatly renowned as King Arthur. [*None* excludes King Arthur from the Celtic race.]
Right: ... no other was so greatly renowned as King Arthur.

Wrong: There is no jet crossing the ocean in as little time as ours.
Right: There is no other jet ...

The superlative is used to compare more than two persons or things. The word *all* (not *any*) is used with a superlative to include the subject of comparison within the class.

Shakespeare is considered the greatest of *all* English poets.
He is the most affectionate of *all* the puppies I have owned.

A more natural expression of the preceding sentence would be "most affectionate puppy I ever owned." Had the author written "most affectionate puppy of any I ever owned," the copy editor could simply have deleted "of any."

When the noun following a superlative denotes a thing in a general or absolute sense, it should be preceded by *of.*

A baby's velvety skin needs the tenderest *of* care.
The blackest *of* smoke rolled out of the chimney.
He protested that he had done it with the highest *of* motives.

Only things of the same class should be compared.

Wrong: Tom Spenser's explanations are the most sensible of all the boys.
Right: ... of all those given.

Wrong: Texas is larger than Sweden; and the Swedish population is about the same as New York City. [This sentence has a second fault. The subject under discussion was Sweden. *Swedish population* is ambiguous and might be interpreted the number of Swedish people in Texas.]
Right: Sweden is not so large as Texas, and its population is about the same as that of New York City.

Wrong: Why was the Secretary of State considered the most important position in the United States Cabinet?
Right: Why was the position of Secretary of State considered the most important one in the United States Cabinet?

Similarly, in comparisons expressed by use of *similar to* or *like:*

Wrong: The French government is a federation with districts having similar powers to our states.
Right: The French government is a federation with districts having powers similar to those of our states.

Wrong: Like Whittier, Longfellow's subjects are often drawn from New England legend.
Right: Like Whittier's, (*or* those of Whittier), Longfellow's subjects are often drawn from New England legend.

Comparisons are also expressed by using *as . . . as* or *so . . . as.* Many meticulous writers use *so . . . as* in negative statements, *as . . . as* in positive ones, but the distinction is of doubtful authority and may safely be ignored.

He is as rich as Jenkins but not so rich as Murdock.

Adjectives misused for adverbs. To qualify an adjective, an adverb must be used, since an adjective preceding another adjective automatically modifies the first following noun.

Wrong: It is amazing how helpless a normal capable man can feel.
Right: It is amazing how helpless a normally capable man can feel.

The opposite error is sometimes made:

Wrong: The British do not have a definitely written constitution.
Right: The British do not have a definite written constitution.

Nouns as adjectives. One of the commonest errors made by insecure writers and overlooked by inexperienced copy editors is the use of an adjectival form of a word where the sense calls for a noun form.

full-fashioned	*not* fully fashioned
drama class	*not* dramatic class
lead-colored	*not* leaden colored
ballistics expert	*not* ballistic expert
body odors	*not* bodily odors
north-bound bus	*not* northern bound bus

Wrong: They may be taken accidentally by children, who then develop poisonous symptoms.
Right: . . . develop symptoms of poisoning.

The correct form is not always so easily determined by the sense as in the examples above, and many contradictions in usage exist. For example, "music critic" and "dramatic critic" are both seen, although "drama critic" is more common. Grammarians cannot offer dependable rules for the use of nouns in this way. The writer and editor should be guided by logic, custom, and a firm sense of idiom. As in similar situations, the peril generally lies in any tendency toward false elegance. The writer of "... performed in the music shell" was justifiably outraged to find it "musical shell" in print; the burden of his angry letter to the editor was that he had been made to appear illiterate. Choosing the form depends upon a good ear. The editor who does not have one should ask someone who does.

"Above." Rules barring the use of *above* as an adjective have been modified. Sir Ernest Gowers's revision of Fowler says that there is no ground for "pedantic" criticism of the use, and Webster II offers *above* as an adjective without further notation. Nevertheless, many careful writers still prefer to say "the illustration given above" rather than "the above illustration."

"Barely," "hardly," "scarcely." These words should be completed by *when*, not by *than*.

> *Barely* had the captain appeared *when* the third mate fell unconscious.
> *Hardly* had we arrived *when* the storm began.
> *Scarcely* had the drawbridge been raised *when* Sir Bedevere came riding.

They should never be used with a negative, for they have inherently negative meanings.

> I could hardly believe it.
> I have scarcely any left.

"Different." The adjective *different* is usually followed by *from*, and some authorities consider any other phrasing improper. But *different to* and *different than* are common usage in England and have long literary usage to support them. *Different than* is being increasingly used in the United States when the object is a clause, probably because the construction required by *from* is often wordy.

> Conditions are now very different from what they were when the Constitution was drawn up and adopted.
> Cotton and linen are known as vegetable fibers and have different reactions than the animal fibers known as silk and wool (have).

"Due to." The rule is that *due*, an adjective, should be attached only to a noun or pronoun, and should not be used in place of *owing to, because of, on account of*, which are compound prepositions. To most minds the distinction between *due to* (adjective plus simple preposition) and *owing to* (participle plus simple preposition) seems artificial or forced; and that is no doubt the reason why it is so generally ignored, except by careful writers. The following are preferred forms.

> His failure was due to insufficient study.
> He failed owing to insufficient study.
> He failed because of insufficient study.

"Very," "very much." A traditional question is whether to use the form *very* or the forms *much* or *very much* before a participle used as an adjective—"I am very pleased" or "I am very much pleased." An absolute answer cannot be given. A simple adjective is modified by *very;* a participle that retains its verbal force strongly, by *much/very much.* But participles may slip over the stile and become adjectives without formal announcement. Moreover, several adjectives that have lost their verbal force still require *much/very much*—for example, "very much awake," "much aware."

very tired [not *much* tired]
very dignified [not *much* dignified]
very pleased [not *much* pleased]
(very) much inconvenienced [not *very* inconvenienced]
(very) much impressed [not *very* impressed]
(very) much concerned [not *very* concerned]

As a guide to the unsure, *much/very much* is what needs to be inserted in most cases; *very* is more often incorrectly used. When *much/very much* is not correct, it will sound awkward.

Sometimes either *very* or *very much* could be used correctly, conveying different meanings, however.

very worn [showing signs of long wear]
very much worn [worn a great deal or by many persons]

Sometimes neither is as good as some other adverb, perhaps *most.*

most uninspired [not *very uninspired* nor *much uninspired*]

ARTICLES

The indefinite articles *a* and *an,* and the definite article *the* (also classed as limiting adjectives) are often misused and often omitted without reason.

INDEFINITE ARTICLES

Use *an* before words beginning with a vowel sound; use *a* before words beginning with a consonant sound—including *e* as in *ewe, euphonic, o* as in *one, u* as in *union. An one, an unit, an historical, an hypothesis,* and the like are now little used.

Abbreviations. The same principle of sound governs the choice of *a* or *an* before an abbreviation when the name of the first letter begins with a vowel sound, as ms, NCO, E.E.C. The writer must decide whether the abbreviation is to be read as letters or as the words represented. If *ms* is to be read "manuscript," then "a ms version" is correct. If the writer expects *E.E.C. member* to be read "European Economic Community member," he will use *a* before it; otherwise he will use *an*. Similarly, "a $10,000,000 fire" (a ten-million-dollar fire), "an 11-ton truck" (an eleven-ton truck).

Series. When the article is used with two or more coordinate nouns, it should appear before each noun unless the things named comprise a single idea.

> He bought a horse and a cow. He owned a horse and buggy.
> A house and lot; a house and a barn.
> There was a good and a bad side to this.
> A man and a woman were sitting on the porch.
> In the group she saw a doctor, a lawyer, and a dentist.

Unnecessary articles. An object and its name should not be confused.

> *Wrong:* Look up the definition of a metal.
> *Right:* Look up the definition of *metal*.

> *Wrong:* A "cover" is a word used to designate the space for each guest's silver, dishes, napkin.
> *Right: Cover* is a word used to designate the space . . .
> *Or:* A "cover" is the space for . . .

> *Wrong:* Probably a "one-scene" play would be a better term than a one-act play.
> *Right:* Probably "one-scene play" would be a better term than "one-act play."

DEFINITE ARTICLE

Series. If the article *the* is used before two or more coordinate nouns denoting different objects, it should appear before each noun.

> Before 1760 it was necessary to depend upon the sun and the pole star in reckoning direction.
> The land between the Hudson and the Delaware . . .
> These coupons must be used between the first and the fifteenth of the month.

But note the effect of a following plural noun:

> The land between the Hudson and Delaware rivers . . .
> In the 16th and 17th centuries; in the first and second grades
> The first and the third Friday; the first and third Fridays

Use determined by meaning. Omission of *the* sometimes alters the meaning:

> We stand in awe before this great cosmic spectacle, in awe not tinged *with fear* of the ancients, but filling us with an unconquerable curiosity.
> We stand in awe before this great cosmic spectacle, in awe not tinged *with the fear* of the ancients, but filling us . . .

Elliptical style. Instructions are sometimes written in an elliptical style, omitting articles. Consistency should be observed, omitting all or none.

> With left palm toward *the* player, place point of knife against *the* thick of *the* hand near *the* little finger side, handle of knife toward *the* thumb side of hand. [Omit italicized *the's.*]

> *Idioms:* in place of *not* in the place of:
> out of the question *not* out of question
> by the thousand *or* by thousands

Missing articles. Some writers, particularly those who have worked for newspapers, will drop *the* at the beginning of a sentence or paragraph. It should be restored.

> Unexplained elements in the account are the disappearance of the wife and why the clock stopped at 6:05. (*The* unexplained elements . . .)

Similarly, it is *the* police or police*men*.

> His life was saved by the arrival of police (. . . of *the* police)

However, care should be taken that a *the* is not inserted before words that are meant to be indefinite in reference.

> Actresses may wait a lifetime for a role as meaty as this one. [Not *The* actresses . . .]

CONJUNCTIONS

Conjunctions are either coordinating (joining elements of equal rank: *and, but, or, nor, for, yet, so*) or subordinating (introducing a subordinate element: *when, where, since, though, so that*). Coordinating conjunctions used in pairs (*not only . . . but also*) are called correlative conjunctions.

COORDINATING CONJUNCTIONS

Faulty use of "and" before a "which" clause. Do not use *and* or *but* before *which* (or *who, whose, where*) when no relative clause precedes.

> *Wrong:* It was then proclaimed to the world in the solemn document known as the Declaration of Independence, and which has already been mentioned. [*And* should be deleted.]
> *Wrong:* One man may do a job of road rolling without any further assistance by making use of a new roller six feet high, and which is propelled by a gasoline engine. [*And* should be deleted.]

Faulty omission of "and." Use a conjunction, not a comma, between the parts of a compound predicate.

> *Poor:* The voter takes this ballot into a private booth, marks out the names of those running for Governor. (. . . booth *and* marks . . .)
> *Poor:* It hears appeals from local committees, fixes dates for primaries and for holding conventions. (. . . committees *and* fixes . . .)

Permissible omission. The conjunction is omitted between coordinate clauses if they are separated by a semicolon.

> Conrad has introduced us to the languors and the heats of the river mouths of the Malay Archipelago; Somerset Maugham has spun many a tale of cynical intrigue in the same steaming surroundings.
> A tool is for some ulterior purpose; a language exists as a world in itself.

In a series of three or more items, omission of *and* before the last may indicate that similar items could be added to those given. In other words, omission of *and* implies *etc.* or *and the like.*

> Use it in general cleaning—for floors, linoleums, refrigerators, tiling, marble.
> The idea is merely an outgrowth of the trend toward multiple-purpose rooms, which converts studies into music rooms, dining rooms into studies, studies into guest rooms.

"Or," "nor." As in the case of *shall* and *will*, firm rules governing use of *or* and *nor* have eroded, with *or* now taking precedence. *Nor* tends to reappear in sentences with long clauses, when the added negative feeling is required, and in material intended for oral delivery, when the consonant is more clearly heard.

There are, however, beyond the *either . . .or, neither . . . nor* situations, cases where *nor* must be used. In these cases, the negative aspect of the sentence is confined to the first part, and does not carry into the clause after *or/nor.* The presence of *any* in the clause often gives away the need for *nor.*

> The proposed amendment does not have the required approval of three-quarters of the states, *or* is any early change in this expected. [Should be: . . . *nor* is any early change . . .)

"So." As a coordinating conjunction *so* is permissible but is overused.

> *Permissible:* He arrived at last, *so* we sat down to dinner.
> *Permissible:* The simple partnership is a legal device of great antiquity; *so* there was little need of any essentially new rules of law.

So is too often used as a substitute for more exact connectives such as *therefore, accordingly, on that account.* It also sometimes weakens a sentence that should be rebuilt.

Permissible: The southwest was a region of small farms, *so* the general techniques of agricultural exploitation were closely comparable with French practice.
Better: The southwest was a region of small farms; *therefore* . . .
Or: The southwest being a region of small farms, the general techniques . . .

A sentence containing coordinate clauses connected by *so* would often be improved by subordinating one of the clauses.

Permissible: The mile was too small a unit for measuring astronomical distance, *so* the light year came into use.
Better: Since the mile was too small a unit for measuring astronomical distance, the light year came into use.

SUBORDINATING CONJUNCTIONS

"So that." *So that* means *in order that* or *with the result that.* Its use should be carefully distinguished from that of *so.*

I invested in bonds *so that* my money would be safe. (In order that it would be safe.)
I invested in bonds, *so* my money would be safe. (Consequently it would be safe.)

Some writers use *so* and *so that* when other forms of expression would be preferable. Sentences like the following are often seen.

Poor: The schedule is flexible enough so that a pupil can go to the shop for a longer period if his interest and need warrant it.
Better: The schedule is flexible enough to permit a pupil to go . . .

Poor: It was early enough so that we could hear the concerto.
Better: It was early enough for us to hear the concerto.

"But that." A study of examples is perhaps the easiest way to clarify the use and meaning of *but that.* Note first that *but that* is not interchangeable with *that.* These pairs of sentences express similar thoughts:

He was so strong that he lifted it easily.
He was not so weak but that he lifted it easily.

I am not so ill that I can't do it.
I am not so ill but that I can do it.

The two sentences below are almost opposite in meaning.

I cannot see that his chances are very good [= they do not look good to me].
I cannot see but that his chances are very good [= they do look good to me].

With no change of wording except *but that* for *that,* a different thought is conveyed:

> We are not sure that Father will be able to do it [= that he will have skill or time].
> We are not sure but that Father will be able to do it [= of those considered perhaps Father will do it].

A change from positive statement to negative statement changes the implication.

> It is impossible but that taxes will be higher. [Sure to be.]
> It is not impossible that taxes will be higher. [May be.]

Sentences like those above cause little trouble by comparison with sentences using *doubt,* verb or noun. The verb *doubt* requires an object— a noun, pronoun, or noun clause. Noun clauses are introduced by *that,* not *but that.*

> I do not doubt—
> that he lifted it easily.
> that they will win.
> that his chances are good.
> that Father will do it.

When the noun *doubt* is used after an expletive, think of *doubt* and its restrictive modifying clause as one unit.

> There is no—
> doubt that he lifted it easily.
> doubt that they will win.
> doubt that his chances are good.
> doubt that Father will do it.

But that is sometimes equivalent to *without.*

> It never turns but that it squeaks. (It never turns without squeaking.)

But that may be used instead of *except that.*

> We can arrive at no other conclusion but that organic evolution or changes in living things have taken place.

"But what." *But what* is used correctly when *except* could be substituted for *but.*

> He did nothing but what he pleased.
> The store had everything but what she was looking for.

But what used for *but that* is colloquial.

> I am not so ill but what I can do it.
> There is hardly a minute now but what our swift flight carries us over some interesting community.

"That." A common error is the use of two *that*'s to introduce a single clause. This is a mistake a copy editor should correct or query. The italicized *that*'s in the following sentences should be omitted.

> It is extremely important that when aiding with any test, or preparing patients for scientific tests, *that* directions be carefully followed.
> Insist that in securing material from an encyclopedia *that* the pupils take only what they can fully understand when they meet it.

When successive clauses introduced by *that* are numbered, the number and the conjunction should be in the same relative position before the clauses.

> *Wrong:* How does this topic show that (1) scientific knowledge advances slowly and by gradual steps and (2) that no great discovery was ever completely made by one man alone.
> *Better:* How does this show that (1) . . . that (2) . . .
> *Or:* How does this show (1) that . . . (2) that . . .

The conjunction *that* sometimes slips into a sentence incorrectly, as in the sentences below. The parenthetical phrases were evidently not recognized as such. A comma should be substituted for the italicized *that*.

> Near there, legend says *that* Norsemen trapped beavers long, long ago.
> This story was written by Homer, the blind poet, who begged for bread while he lived, but when he died, it is said *that* seven cities claimed the honor of being his birthplace.

"That," "which." People who handle *that* and *which* with perfect ease while speaking may go to pieces over these words when writing. The most common misconception, as it was in Fowler's time, is that *which* is more literary and therefore preferable. *That* is correct in restrictive clauses; *which* in nonrestrictive. Boiled down, the rule of thumb is that when a comma can be inserted, the word is *which*.

> *Wrong:* But the book *which* I picked up was purple.
> *Right:* But the book *that* I picked up was purple.
> *Or:* But the book I picked up was purple. [This clearly demonstrates the restrictive nature of the clause.]

> *Wrong:* The commander has given orders that all of the patrol boats in the navy *which* are sinking should be scrapped.
> *Right:* . . . boats in the navy that are sinking should be scrapped. [Clearly, all of the patrol boats in the navy are not sinking.]

The meaning of the sentence is the key when *that* and *which* are concerned; if context is insufficient, the copy editor should query.

"When," "where." Do not define a word by saying it "is when . . ." or "is where . . ."

> *Wrong:* Insomnia is when you can't sleep.
> *Right:* Insomnia is not being able to sleep.

> *Wrong:* Freezing is where water reaches 32°F.
> *Right:* Freezing is the solidification of water at 32°F.

Idiom requires that the word *example* or its synonyms should be followed by *of*, not *when* or *where*.

> *Wrong:* Can you give an example from your own experience *when* your own mood colored your feeling toward a particular place?
> *Right:* Can you give an example *of how* your mood colored your feeling?

> *Wrong:* Discuss examples in recent history *where* public opinion played an important part.
> *Right:* Discuss examples in recent history *of* public opinion's playing . . .

"Till," "until." *Till* and *until* mean *up to the time that* or *from this time to that time* and should not be used in place of *before* or *when*.

> I was out of work *until* June.
> It was not long *before* he called.

> *Wrong:* We had not been there long *until* the phone rang.
> *Right:* . . . *when* the phone rang.

"While." Formerly, many grammars insisted that *while* was an adverb of time which should not be used as a substitute for *although, whereas, but,* or *and* but which should be reserved for sentences like this one:

> You hold the tiller *while* I furl the sail.

While is used by many good writers, however, in a sense of *although* that does not defy the sense of *at the same time*. This usage, shown in the following examples, appears to be becoming acceptable.

> *While* she performs well, she does not appear at ease.
> *While* I disagree with you, I do not deny your right to speak.

The use of *while* that is demonstrated by the following is incorrect.

> *While* many women work the first shift, few are found on the second shift. [Should be: *Although* many women . . . or Many women work . . . but few are found . . .]

The most deplorable use of *while* is its substitution for *but* or *and* as an "elegant variation."

Wrong: While the council chairman was a Democrat that year, the post was held for twenty years thereafter by a Republican.
Right: The council chairman that year was a Democrat, but the post was held . . .
Or: Although the council chairman that year . . . the post . . .

Wrong: The car was wrecked and the groceries were scattered across the road, *while* the driver was bloody and bruised.
Right: The car was wrecked, the groceries were scattered across the road, and the driver was bloody and bruised.

CORRELATIVE CONJUNCTIONS

Correlatives. Correlative conjunctions are coordinating conjunctions used in pairs.

both . . . and	not only . . . but (also)
either . . . or	whether . . . (or)
neither . . . nor	though . . . yet

The constructions following correlatives should be parallel in form.

Both in England and in France; *or,* In both England and France.
Progress depends not only upon the proper method but also upon the proper motive.

Violations of this rule are often caused by misplacing the first conjunction.

Conjunction misplaced: Switzerland neither assumes obligations to send troops abroad nor to admit foreign troops to her soil.
Right: Switzerland assumes obligations neither to send . . . nor to admit . . .
Or: Switzerland does not assume obligations either to send . . . or to admit . . .

Ambiguous: We come in contact with many individuals who have names that are both difficult to spell and pronounce.
Probable meaning: . . . difficult both to spell and to pronounce.

Correction of a faulty correlative sometimes requires a change in the verb in one clause.

Faulty construction: It occurs not only in the blood of all vertebrates but is widely spread throughout the animal kingdom.
Better: Not only does it occur . . . but it is spread . . .

Faulty. You would have guessed that not only had she lived in the country all her life, but you would have known which country.
Better: Not only would you have guessed . . . but you would have known . . .

PREPOSITIONS

All but a few prepositions can be used as other parts of speech, most often as adverbs or subordinating conjunctions, to which prepositions are closely related. The more familiar prepositions are:

about	before	excepting	outside	to
above	behind	for	over	toward
across	below	from	past	under
after	beneath	in	pending	underneath
against	beside	inside	regarding	until
along	between	into	respecting	unto
amid	beyond	of	round	up
among	by	off	since	upon
around	concerning	on	through	with
at	during	onto	throughout	within
athwart	except	out	till	without

Compound prepositions:

according to	as to	aside from	instead of
apart from	on account of	because of	out of
owing to			

Terminal prepositions. The terminal preposition, as in "This is the book he told us about," "He is the man I spoke of," is frowned upon by those who do not know the history of English idiom, but has been well established for centuries. The probably apocryphal child who piled up five ("What did you bring the book that I didn't want to be read out of to up for?") is joined by no less a prose stylist than Churchill ("This is the kind of nonsense up with which I will not put") in defending the construction. "He is the man of whom I spoke" is wooden; "He is the man I spoke of" offends only the pedant. In any case, the copy editor should not create a strained sentence in order to avoid a terminal preposition.

Allowable omissions of prepositions. The *to* of an infinitive is often not expressed.

He doesn't dare tell his mother.
Use of it may help you understand the problems.

The *in* or *at* of an adverbial phrase may be dropped.

Bills of exchange are sold the same way.

Incorrect omissions. Do not omit a preposition needed to make clear the relation of a phrase to the rest of the sentence. The italicized *of* in the following sentence is needed.

His investigations are adding much to our accurate knowledge of the dawn of intelligence and *of* the ancestral sources of human behavior.

Series. The items of a series should be parallel in form. Omission of one needed preposition is a common error. (It is not to be assumed, of course, that every series introduced by a preposition should have a preposition in each part also; most of them do not.)

Wrong: Who were the first leaders in chemistry, physics, and in biology?
Right: . . . in chemistry, in physics, and in biology?

Wrong: They are equally disastrous to the farmer, to the consumer, and, in the final analysis, the railway engineer.
Right: . . . to the farmer, to the consumer, and to the railway engineer.

Part of a verb. Do not omit a preposition that is an inseparable part of a verb.

Wrong: The relationships between them are respectively orderly and peaceful without regard to what this order, peace, or equilibrium *is due.*
Right: . . . *is due to.*

Wrong: They quarreled long as to whom the bird rightly *belonged.*
Right: . . . whom the bird *belonged to.*

Poor: We never *tired watching* the tall mountaineers.
Better: We never *tired of watching* the tall mountaineers.

Faulty use of prepositions. Do not use a preposition before a noun clause used as an appositive.

Unnecessary preposition: It has the advantage over other alkalies *in that* its unused excess breaks up into water and ammonia gas. (. . . the advantage . . . that . . .)

Do not use a preposition before a restrictive appositive.

Unnecessary preposition: The name *of* artificial silk, first applied and later dropped, was incorrect. (The name "artificial silk" . . .)

Be watchful for the preposition that inadvertently slips into a sentence between a verb and its object.

Unnecessary preposition: The undermining of morals as a result of the war has been considered by some authorities *as* a greater loss than life and property. (. . . has been considered a greater loss.)
Unnecessary preposition: He was nicknamed *as* Thomas the Sudden. (He was nicknamed Thomas the Sudden.)
Unnecessary preposition: His term of office is *for* ten years. (. . . is ten years.)

Do not use two prepositions when one will convey the meaning: *for from, of between, for between, in behind, in between, to within,* are usually wrong.

Too many prepositions: Pasturage for a cow and a calf *for from* 42 to 68 days. (. . . for 42 to 68 days.)
Too many prepositions: Rights are selling *for between* $11 and $29. (. . . for $11 to $29.)

Do not use a preposition where the construction calls for a conjunction.

Faulty use of "to" after "between": The ratio *between* the width of the head *to* the length. (The ratio *between* the width *and* the length.)
Faulty use of "with" after "between": Comparisons drawn *between* the various tissues and the organs of this specimen *with* those of higher animals. (. . . between the tissues of this specimen and those . . .)

When ambiguity is created in reference to price rises, place the old price in the second position.

Ambiguous: The events increased the wholesale price from 73½ to 74 cents. [Unclear whether this is the price or the rise.]
Better: The events increased the wholesale price to 74 cents, from 73½.

"Among," "between." As a general rule, *among* refers to more than two and *between* to two; but *between* should be used of three or more items if each item is considered severally and individually.

We divided the money *between* the two.
We divided the money *among* the six.
Between morning, noon, and night . . .
Diplomatic talks *between* the members of the Common Market . . .
Between the rows of beans, plant lettuce.

The word or phrase used after *among* or *between* should be logical and not ambiguous.

Faulty: They divided the booty among one another.
Better: . . . among themselves.

Faulty: Place dunnage wood between each two layers.
Better: . . . between layers.

"In," "into," "in to." *In* denotes position; *into* implies motion from without to within. *Into* cannot be substituted for *in to* (adverb and preposition) without marring the sense of the sentence.

He was *in* the room.
He went *into* the room.
He went *in to* his family, who were already in the room.

"On," "onto," "on to." The distinctions between *on, onto,* and *on to* are similar to those between *in, into,* and *in to. On* denotes position upon something; *onto* indicates motion toward the upper surface of something. *On to* should be used when *on* belongs to the verb.

> The children played *on* the haymow.
> They slid down *onto* the hay below.
> The perpendicular line *onto* which we project the motion is . . .
> It can be done by sticking the butts *onto* wire hooks.
> They marched *on to* Concord.
> Hang *on to* the banister.

SENTENCE STRUCTURE

Errors in the construction of sentences arise from haste, from incomplete repairs, from lack of careful examination of the problem, and from half a dozen other sources. Sentences that appear to be gibberish in type will sometimes turn out to have resulted from incomplete or over-adequate deletion after a repair elsewhere in the sentence. In such cases, it is important to check the original sentence rather than to try to figure out what was intended.

The following examples, drawn from proofs, show the kinds of errors that writers, editors, and proofreaders must watch for.

Mixed constructions. Sentence constructions should be completed grammatically and logically.

> *Faulty:* Much of the direction of the institution is to be in the hands of a special
>   board, a majority of which should be composed of women.
> *Better:* . . . a majority of which should be women.

> *Faulty:* I had no illusions about the effects of books such as mine would have
>   been on public opinion.
> *Better:* I had no illusions about the effects of books such as mine on public
>   opinion.

Violations of parallelism. Parallelism is the principle that parts of a sentence that are parallel in meaning should be parallel in structure. Two or more sentence elements in the same relation to another element should be in the same form. For example, "Seeing is believing," or "To see is to believe." The following are examples of violations of the rule.

> *Noun and infinitive not parallel:* The duties assigned to the corps were the *oper-*
>   *ation* of tanks and *to recruit* and *train* the personnel. (. . . the operation of
>   tanks and the recruiting and training . . .)
> *Noun and gerund not parallel:* The solution lies not in *prohibition* or *censorship*
>   but in *developing* self-control. (. . . but in development of self-control.)
> *Gerund and noun not parallel:* The simplest treatment consists in *keeping* the
>   skin dry, the *avoidance* of overheating, and the *application* of talcum or corn-
>   starch. (. . . keeping . . . avoiding . . . applying.)

Noun and clause not parallel: All the grower was responsible for was attractive packing and that he should ship his goods promptly. (. . . attractive packing and prompt shipment of his goods.)

Gerund and infinitive not parallel: Our life in camp was more concerned with *cooking* and *washing up* than *to study* nature. (. . . cooking . . . washing up . . . studying nature.)

Infinitive and clause not parallel: Do you advise me *to go* to college now *or that* I wait till I am eighteen? (. . . to go now or to wait . . .)

Complete sentence and phrases not parallel: Some of the causes of failure are the following: Some important feature has been overlooked, lack of careful analysis before launching the enterprise, incorrect conclusions drawn from the analysis. (Some important feature has been overlooked, a careful analysis has not been made, incorrect conclusions have been drawn.)

Complete sentence and clause not parallel: The main pleas are, "You can't understand a people if you don't know their language" and that it "will promote the good neighbor policy." (. . . language" and "It will promote . . .")

Fraction and percentage not parallel: Poultry loses about *one third* of the live weight in dressing and drawing and another *12 percent* in bone and other inedible parts. (. . . about 33 percent . . . and another 12 percent . . .)

Items in a list not parallel: Other points made by the president at her news conference were the following:
    *Membership should be increased.
    *No rise in dues.
    *Making meetings more frequent. . . .
(*Membership should be increased. *Dues should be kept the same. *Meetings should be more frequent.)

Two questions, one parenthetical. A question that contains within itself a parenthetical question is often misconstructed, the parenthetical verb being construed as the main verb. So common is this error that every sentence containing the phrase *do you think* or *do you suppose* should be tested for correct form by reading a *that* after *think* to see if the sentence then conveys the meaning intended. Pattern sentences for correct construction:

Why do you think he did it?   [It is not certain that he did it.]
Why, do you think, did he do it?   [He did do it; what do you think were his reasons for doing it?]

Wrong: Why do you think each developed as it did?
Right: Why, do you think, did each develop as it did?

Wrong: Why do you suppose that so many of us do these things for nothing?
Right: Why, do you suppose, do so many of us do these things for nothing?

POSITION OF MODIFIERS

The meaning of a sentence often depends upon the position of a modifier. Compare the sentences in the following groups.

You also are guilty of this crime.
You are also guilty of this crime.
You are guilty of this crime also.

Even he is not frightened by your news.
He is not even frightened by your news.
He is not frightened even by your news.

Do not place a modifier where it can appear to modify the wrong word.

Wrong: He almost swam two hundred yards.
Right: He swam almost two hundred yards.

Wrong: All the men didn't go.
Right: Not all the men went.

Wrong: It scores the number of yards of the nearest line to which it strikes.
Right: . . . of the line nearest to which it strikes.

Adjectives. An adjectival modifier must be near its noun because a normal sentence contains several nouns; the opportunities for ambiguity are many.

Wrong: The people have a tendency to blame the party in power for panics and depressions, guilty or not guilty.
Right: . . . party in power, guilty or not guilty . . .

Faulty: At the age of eighty Aunt Sarah was a good hand with an ax. I watched her, as a little boy, go into the woodshed . . .
Better: . . . As a little boy, I watched her go into the woodshed.

Faulty: There is nothing more deadly than ridicule, and it is never deadlier than at the beginning of these movements. Scorched with laughter, now we can almost surely be confident that these "plots" will come to nothing.
Better: . . . We can be confident that these "plots," scorched with laughter, will almost surely come to nothing.

Adverbs within verbs. When an adverb is placed within a verb it should regularly follow the first auxiliary, not precede it—*may safely be used, will surely come.*

Faulty: The short *a*, for example, always must be modified.
Better: . . . must always be modified.

Faulty: There always have been circumstances . . .
Better: There have always been circumstances . . .

Split infinitives. An infinitive is split when an adverb is placed between *to* and the rest of the infinitive. Formerly completely proscribed, a split infinitive is now regarded as permissible if clearness, smoothness, or force would be lost in avoiding the split.

Poor: Reptiles too big to quickly adapt themselves to new conditions . . .
Better: Reptiles too big to adapt themselves quickly to new conditions . . .

The infinitive may have to be split to keep the adverb modifying it from being read as the modifier of some other word.

> *Split necessary:* To study current issues so that one may pass sound judgment on the problems and values of today is to equip oneself *to best serve* society.
> *Ambiguous:* To lose one's temper often signifies lack of self-control.
> *Clear: To often lose* one's temper signifies lack of self-control.

Do not place an adverb modifying an infinitive before the infinitive if in that position it might seem to modify a preceding verb.

> *Faulty:* It enables us better to understand ourselves.
> *Better:* It enables us to understand ourselves better.
>
> *Faulty:* Economic institutions enable mankind more efficiently to carry on the work of getting a living.
> *Better:* Economic institutions enable mankind to carry on more efficiently...
>
> *Faulty:* Teachers using these notes are urged carefully to study the many references to specific articles.
> *Better:* ... urged to study carefully...

There is no rule that a modifier of an infinitive must immediately follow the infinitive.

> *Faulty:* It is ordinarily not the best practice to invent deliberately a figure of speech.
> *Better:* ... to invent a figure of speech deliberately.

Ambiguous modifier. A modifier should not be placed so that it might modify either of two words.

> *Faulty:* Careful attention must be given them *both* in writing and in revising.
> *Better:* Careful attention must be given to *both* of them in...
> *Or:* ... to them in *both* writing and revising.   [Depending upon writer's intent.]

Phrases of time, place, or manner. A phrase of time, place, or manner that modifies the verb of a main clause may well stand first in the main clause; it might sometimes be ambiguous or awkward in any other position.

> *Faulty:* This may be done effectively with the hands *in loose soil.*
> *Better:* In loose soil this may be done...
>
> *Faulty:* The sea was up where this fossil was found *at one time.*
> *Better:* At one time the sea was up where this fossil was found.
>
> *Faulty:* The eggplant is very susceptible to serious injury *in all stages of growth* by a number of diseases.
> *Better:* In all stages of growth the eggplant is...

Do not place such a phrase first in the main clause if it does not modify the main verb. Test the correctness of arrangement by reading the phrase after the main verb.

Faulty: In very early times it is probable that a baited line was used without a hook.
Better: It is probable that in very early times a baited line was used . . .

The copy editor can sometimes correct sentences of this sort by making a parenthetical element of the subject and verb of the main clause.

Faulty: In the liquid state we think these molecules have a good deal of freedom.
Better: In the liquid state, we think, the molecules . . .

Faulty: During the early part of the twentieth century it is said that 40,000 artists were at work in Paris.
Better: During the early part of the twentieth century, it is said, 40,000 artists were at work in Paris.

Dangling modifiers. A phrase or clause that because of its position in a sentence seems to modify a word that it does not logically modify is called a dangling modifier. A dangler is most often at the beginning of a sentence.

Do not place a participial phrase at the beginning of a main clause unless it modifies the subject of the main clause. Test the correctness of a phrase so placed by reading it after the subject.

Faulty: Having recovered from his illness, his mother took him abroad.
Use a clause: When he had recovered from his illness, his mother . . .

Faulty: Destroying the vegetable cover by cultivation, the wind picks up the fine, loose soil and piles it in great heaps.
Recast: The vegetable cover having been destroyed by cultivation, with wind . . .

Faulty: Knowing this and hoping to find an easy route to the deposits by way of the Parana and Paraguay rivers, a Spanish settlement was founded on the site of Asuncion.
Change the subject: Knowing this and hoping to find an easy route to the deposits by way of the Parana and Paraguay rivers, the Spaniards founded a settlement on the site of Asuncion.

Do not begin a sentence with an *if* or *when* clause unless the clause modifies the main verb.

Faulty: If the atmosphere is very moist, explain why killing frosts are not likely.
Transpose the clause: Explain why killing frosts are not likely if the atmosphere is moist.

Faulty: When men had a desire to draw pictures, we can understand that they had made a great advance beyond the fist-hatchet stage.
Transpose the clause: We can understand that when men had a desire to draw pictures they had made a great advance . . .

Do not begin a sentence with an elliptical clause unless the omitted subject of the clause is the same as the subject of the main clause.

> *Faulty:* Though rather cold, I expected he would take his daily swim.
> *Better:* Though it was rather cold, I expected...

> *Faulty:* Though summer, I knew he would like to see a cheerful hearth.
> *Better:* Though it was summer, I knew he would like to see a cheerful hearth.

In using a nominative absolute or an absolute construction (*the day being cold,* we planned to go skating), be sure that the phrase either has a subject or is unable to collect one elsewhere in the sentence and, conversely, that it does not inflict its subject on the main part of the sentence.

> *Wrong:* It was his third pewter mug of the day, having won the doubles and the singles already.
> *Right:* It was his third pewter mug of the day, he having won the doubles and the singles.

Awkwardness. Do not place sentence elements in an awkward order.

> *Faulty:* Assembly Bill 241 allows with the consent of the refunding commission the district to issue additional bonds.
> *Better:* Assembly Bill 241 allows the district, with the consent of the refunding commission, to issue additional bonds.

> *Faulty:* The people in the same way are wiser politically by far than the most of our politicians.
> *Better:* In the same way, the people are far wiser politically than most of our politicians.

> *Faulty:* It called the purpose it had been created for to the attention of sculptors.
> *Better:* It called to the attention of sculptors the purpose for which it had been created.

OMISSIONS

Do not omit a word, phrase, or clause essential to a clear understanding of the structure and meaning of the sentence.

> *Verb omitted:* After what seemed a long time, *but probably* only a couple of minutes, she resumed her normal keel. (...but was probably...)
> *Noun omitted:* One level premium governs from *age 18 to 50.* (...age 18 to age 50.)
> *Part of noun omitted:* The force will number *6 to 700* men. (...number 600 to 700 men.)
> *Pronoun omitted:* The description of this fellow *and the man* wanted in Chicago tally exactly. (...and that of the man...)
> *Expletive omitted:* The total military forces numbered 644,540. In *addition were* thousands of civilian employees. (...there were thousands.)

Preposition omitted: Most protozoa divide much more rapidly at warm *than cold* temperatures. (. . . at cold . . .)

Preposition omitted: John always maintained that he was more grateful for that escape *than those* from dynamite and rifle fire. (. . . than for those . . .)

Participle omitted: Besides scholarship, a student in the university must have been active politically. (Besides possessing scholarship . . .)

Series. Series of words, phrases, or clauses may be defective because of the omission of prepositions or conjunctions. (See also p. 382.)

Faulty: It permits the use of different sizes and styles of type, spaces out the lines to an even right-hand margin, gives uniform impression and unusual legibility.

Better: . . . margin, and gives . . . [It does three things. Omission of *and* would indicate that it does several things, but only three are mentioned.]

Faulty: The discovery of the causes of tuberculosis by Koch, of diphtheria by Klebs and Loeffler, the invention of antitoxins by von Behring and Roux, control and prevention of diseases by Laveran, Ross, . . .

Better: The discovery of the causes of tuberculosis by Koch, and of diphtheria by Klebs and Loeffler, the invention of antitoxins by von Behring and Roux, control and prevention of diseases by Laveran, Ross, . . .

Faulty: Others involve safety as well—speed regulations, safety devices in factories, theaters, regulation of sale and use of firearms, explosives, and poisons.

Better: Others involve safety as well—speed regulations, safety devices in factories and theaters, and regulation of sale and use of firearms, explosives, and poisons.

Series are tricky forms to correct, and the copy editor needs to guard against haste. The following example at first reading seems to be faulty because of the omission of prepositions. On the contrary, a preposition should be omitted, because all four items of the series constitute one method of control. (The sentence is also faulty because the items of the series are not parallel in form.)

Faulty: Damping-off may be controlled by treating the seed before planting, proper thinning of seedlings, allowing the necessary ventilation, and by judicious watering.

Better: Damping-off may be controlled by treating the seed before planting, thinning the seedlings properly, allowing the necessary ventilation, and watering judiciously.

The omission of conjunctions sometimes makes a series of phrases out of two coordinate expressions.

Faulty: These laws require employers to install safety devices, maintain healthful working conditions and reasonable hours of labor.

Better: . . . to install safety devices and to maintain . . .

Faulty: They mark a beginning of, or effort at, self-regulation, elimination of child labor, long hours, and so on.

Better: . . . self-regulation and the elimination . . .

Words used in a double capacity. The omission of a word may result in some other word's being used in a double capacity, one of which is ungrammatical.

Faulty: This dedication will serve for almost any book that *has, is,* or shall be published.
Better: ... that has been, is being, or shall be published.

Faulty: Thus he did what many a man *has* and is *doing.*
Better: ... has done and is doing.

Faulty: Nickel steel came into use for the *French,* and afterward for all other European *navies.*
Better: ... for the French navy, and afterward for all other European navies.

Faulty: As a result of the fire, a child was killed, three nurses wounded, and the hospital destroyed.
Better: ... fire, a child was killed, three nurses were wounded, and the hospital was destroyed.

One word used for two phrases may be ungrammatical in one of them. Correction often requires a transposition of phrases.

Faulty: He was either forbidden or *discouraged to* play with other children.
Better: He was either forbidden to play with other children or discouraged from doing so.

Faulty: I must add that it is as *interesting* if not more so *than* any other part of the magazine.
Better: I must add that it is as interesting as any other part of the magazine, if not more so.

Faulty: It is one of the finest, if not *the finest,* theater *productions* of the year.
Better: It is one of the finest theater productions of the year, if not the finest.

Faulty: The fishing is as good, if not *better,* at this season *as* it is in the summer.
Better: The fishing is as good at this season as it is in the summer, if not better.

GLOSSARY OF GRAMMATICAL TERMS

(In many definitions the word *substantive* is used, for the sake of brevity, to denote a noun or pronoun or any other part of speech or group of words used as a noun.)

ablative In Latin and some other languages, the case that expresses chiefly separation, source, or instrumentality. Its place is taken in English by the prepositions *at, by, from, in, with.*
absolute construction *See* nominative absolute.
abstract noun *See* noun.
accusative In inflected languages (e.g., Latin), the case corresponding to the objective case in English.

adjective A word that modifies a substantive: *white* horse, *natural* piety, *young* child, *vegetable* matter, House *Beautiful*. Adjectives are classified syntactically as attributive and predicate. The former usually stand before the substantive; the latter, after the verb. The predicate adjective is a form of *complement* (q.v.).

Attributive: the *angry* man, the *yellow* leaf, make me *miserable*.

Predicate: The man is *angry*. The leaf is *yellow*. I am *miserable*.

Adjectives are classified by kind as follows:

Demonstrative or definitive: *this* (" this man"), *that, these, those*.

Descriptive: (1) common: *white, wise, heavy;* (2) proper: *French, Spencerian, April* (April showers).

Indefinite: *each, both, either, such, some,* and *so on,* when these modify a substantive — *each* man, *such* nonsense.

Interrogative: *which* ("which man?"), *whose, what.*

Limiting: *a, an, the.* (Also called articles.)

Numeral: *one, two, three, ten; first, tenth; quadruple, fourfold.*

Possessive: *John's, my.*

Pronominal: any possessive pronoun.

Relative: *whose, which, what,* when it both modifies a substantive and joins it to a qualifying clause ("We know which men are guilty").

Another part of speech or a phrase or a clause may be used as an adjective. The inflection of the adjective is called its *comparison* (q.v.). *See also* pages 368–370.

adjective clause   *See* clause.

adjective phrase   A phrase that qualifies a noun or pronoun in the sentence. Adjective phrases are classified, from the words introducing them, as *prepositional, participial,* and *infinitive.*

Prepositional: the top *of the table;* the girl *with the green eyes;* a cure *for flu.*

Participial: a man *selling bonds;* a dog *stunned by a blow.*

Infinitive: work *to be done;* sights *to be seen.*

adverb   A word that modifies a verb, adjective, or another adverb. The characteristic ending of adverbs is *-ly,* though some (e.g., *here, now, so*) do not have it. The inflection of the adverb, like that of the adjective, is called *comparison* (q.v.). The classification of adverbs as those of time, place, degree, quality, number, and manner is not important to grammar. *See also* conjunctive adverb.

Modifying a verb: sleep *peacefully;* put it *there;* do it *now.*

Modifying an adjective: *too* loud for comfort; *very* glad; *severely* ill.

Modifying an adverb: *exceedingly* poorly; *so* excruciatingly painful.

adverb clause   *See* clause.

adverb phrase   A phrase used as an adverb. The most common form is one introduced by a preposition — *"After knocking at the door,* we listened a long while" (preposition and gerund); *"After breakfast,* let's go for a walk" (preposition and noun). Another form is that introduced by an infinitive — "I seek only *to be agreeable."*

adverbial conjunction   A name, now little used, for *conjunctive adverb* (q.v.).

agreement   The proper relations of words, especially in case, number, and person, in a sentence. The rule of agreement is that in a sentence every part should agree grammatically (and logically) with every other related part. Exceptions are the *absolute construction* and *parenthetical expressions.* Three types of agreement are often violated: of adjective and noun, of pronoun and antecedent, and of verb and noun.

antecedent   A substantive to which a pronoun refers. *Examples:* "Mine be a cot beside a hill." "*To be or not to be,* that is the question." "The *man* who hath no music in his soul."

apposition A relation of one substantive to another, when the second, called an *appositive,* repeats the meaning of or identifies the first ("my sister *Cornelia;* Patsy, *a fox terrier").*

article The word *a, an,* or *the;* also called a *limiting adjective.*

attributive adjective *See* adjective.

attributive complement *See* complement.

auxiliary A verb used with another verb to indicate voice, mood, and tense. The auxiliaries are *be, can, do, have, may, must, shall, will,* and their conjugational forms.

case A form or position of a substantive indicating its relation to other words in the sentence. In English the cases are
> Nominative: that of the subject of a verb, of a predicate nominative (subjective complement), of a noun in apposition to another that is subject or complement, or of a substantive in certain nominative absolutes.
> Objective: that of the object of a verb, preposition, or verbal, and of the subject of an infinitive.
> Possessive: that denoting possession.
> (The classical names of cases, *genitive, dative, accusative,* are still widely used in English grammar. They are roughly equivalent to our *possessive* and *objective;* genitive for possessive, dative for objective of the indirect object, and accusative for objective of the direct object.)
> "The Athenians (*nominative*) defeated Darius's (*possessive*) fleet (*objective*)."
> "He (*nominative*) gave him (*objective*) a letter (*objective*)."

clause A group of words that contains a subject and a predicate and that is part of a sentence. Clauses are classified as
> Main, principal, or independent: This is the sentence itself in its simple form; the clause on which the rest of the sentence depends. A *simple sentence* is nothing but a main clause ("The son went into a far country").
> Subordinate or dependent: This is equivalent to a noun, an adjective, or an adverb, modifying a word in the main clause; i.e., it is subordinate to or depends on the main clause ("The son went into a far country, where he did eat husks like swine").
> Coordinate: This is a clause of the same rank as another; main as the other is main, subordinate as the other is subordinate ("There had been a heavy rain, and John said it was too wet to go on"). Coordinate clauses are connected by coordinating conjunctions: *and, but, or.*
> Adjective subordinate: "The witness *who proved the guilt of the prisoner* is his best friend."
> Adverb subordinate: "He will be condemned *when all the evidence is in."*
> Noun subordinate: *"That he is guilty* is evident to both judge and jury."

cognate object An object that enforces a verb by repetition.
> "I dreamed *a dream."*
> "He sleeps *the sleep* of the just."

collective noun The name of a group or aggregate—*congress, congregation, mob.* Such a noun takes a singular verb unless the individuals of the group are to be emphasized.
> "The crowd *shouts."* (Singular verb)
> "The crowd *ran* hither and thither." (Plural verb)
> *See also* page 354.

comparison Inflection of an adjective or an adverb to indicate degree. The degrees are called *positive, comparative,* and *superlative.* Monosyllables and some dissyllables are usually compared by adding *-r* or *-er* to form the comparative and *-st* or *-est* to form the superlative—*large, larger, largest; handsome, handsomer, handsomest; kindly, kindlier, kindliest.* The comparative and superlative of words longer than two syllables are formed by adding the adverbs *more* and *most* to the positive—*elegant, more elegant, most elegant;* shorter words may be compared this way, too: *kind, more kind, most kind.* Descending comparison is formed by adding *less* and *least* in the same way. Some adjectives, such as *unique* and *destroyed,* being absolutes, may not be compared. Also, some adverbs are not inflected. *See* adverb. Some adjectives have an irregular comparison:

bad, evil, ill	worse	worst
far	farther	farthest
	further	furthest
good, well	better	best
late	later	last, latest
little	less	least
	smaller	smallest
much, many	more	most
old	elder, older	eldest, oldest

complement   A substantive or an adjective that completes a predicate. A complement usually completes a verb of incomplete predication, such as *to be, to become, to appear, to look, to seem,* though it may complete a direct object or the object of an infinitive. Substantive complements of the first type are in the nominative case; those of the last two, in the objective case. All types of complement are illustrated in the following sentences:

"He is in danger." (*Danger* is a subjective complement—also called a predicate nominative—and is in the nominative case.)

"I call him a danger." (*Danger* is an objective complement, in apposition with him, which it completes, and is in the objective case.)

"He is likely to be a danger." (*Danger* is an objective complement, completing the infinitive *to be.* It is in the objective case.)

"He is dangerous." (*Dangerous* is an attributive complement—also called a predicate adjective—modifying the subject *he.*)

"I call him dangerous." (*Dangerous* is a predicate adjective, modifying the object *him.*)

"He is likely to become dangerous." (*Dangerous* is a predicate adjective, completing the adjective infinitive phrase *to become dangerous.*)

complex sentence   A sentence consisting of a principal clause and one or more subordinate clauses. *See also* clause.

"Who you are, I think I know."

"When I remember that day, I shudder."

"Do not forget that you are responsible."

compound preposition   A preposition consisting of more than one word: *in place of, from under, with regard to, out of.*

compound sentence   A sentence consisting of coordinate clauses.

"They look the same but they are different."

"I give the order; you follow directions."

compound-complex sentence   A sentence consisting of coordinate clauses, one or more of which contain subordinate clauses. Sometimes called *double* (or *multiple*) *complex.*

"The children looked for birds' nests while they were walking to Grandmother's, and they looked for tadpoles while they were coming back."

conditional sentence   A sentence containing a clause expressing condition or limitation. Such a clause is introduced by a subordinating conjunction: *if, as if, if not, unless, except, without, whether.*

conjugation   Inflection of a verb to denote voice, mood, tense, person, and number. The complete conjugation or paradigm of such will include all the forms of the following: voice—active and passive; mood—imperative, indicative, infinitive, subjunctive (and interrogative forms, if significant); tense—present, present perfect, past, past perfect, future, future perfect; person—

first, second, and third; and number—singular and plural. It might also include the tenses formed with *do, can, may, might, should, ought.*

conjunction A connective used to join sentences, clauses, phrases, or words. Conjunctions are classified as follows:

Coordinating: *and, but, or,* connecting elements of equal rank.

Correlative: *both...and, either...or, neither...nor,* coordinating conjunctions used in pairs.

Subordinating: *when, where, whither, whence, why, before, because, for, since, if, unless, though, although, than,* connecting clauses of unequal rank.

(Note that many of these are adverbs. They are often called *adverbial conjunctions.*)

conjunctive adverb An adverb used as a coordinating conjunction. It is not a true conjunction, but serves to carry the sense from one clause to another. The principal conjunctive adverbs are *however, so, then,* and *therefore;* others are *also, besides, consequently, furthermore, hence, nevertheless, otherwise, still, thus.* A conjunctive adverb is usually preceded by a semicolon and followed by a comma.

"I do not wish to go; besides, I am not ready."

"We have little money; nevertheless, we shall try to send you some."

connective A connecting word or particle, such as a preposition, conjunction, or relative pronoun.

coordinate clause *See* clause.

copulative verb A verb that links subject and complement. Sometimes *factitive verbs* (q.v.) are called copulative.

correlative *See* conjunction.

dangling modifier, dangling participle, unattached modifier, faulty reference A modifier with an unclear reference. It is commonly an adjective, a participle, or a participial phrase. The word "dangling" suggests that such a modifier hangs loose, unattached to any other member of the sentence, but the dangler is commonly attached to the wrong part.

"Being about to close our books, will you kindly remit." (The phrase seems to modify *you* but really modifies an unexpressed *we.*)

"A son of the governor, his name is often mentioned as a political prospect." (*His name* is not the son of the governor.)

"Having left my bed and board, I am no longer responsible for the debts of my husband." (The spouse has left, not the person speaking.)

dative In inflected languages, the case which expresses an indirect object. In English it is called *objective.*

declension The formation of the cases of nouns, pronouns, adjectives, by the addition of inflectional endings to the stem. In English the declension of nouns is very simple: *man, man's, man, men, men's, men;* that of the pronoun, more nearly complete: *I, my, me, we, our, us, mine, ours.*

defective verb A verb that lacks one or more of the usual parts or forms. For example, *must* has no inflection.

definite article The word *the.*

definitive adjective *See* demonstrative adjective.

degree *See* comparison.

demonstrative adjective *This, that, these, those,* when used to modify a substantive—"*This* book belongs to *that* boy." *This* and *these* are used for known, established, or nearer objects and persons; *that* and *those* for the more distant.

demonstrative pronoun *This, that, these, those,* and, idiomatically, *so* and *such.* Some grammarians call *he, she, it,* and *they* demonstrative. A demonstrative pronoun serves to indicate or point out the person or object referred to.

"I don't believe *that.*"

"This is the package the postman brought."

"Such is the difference between wishing and doing."

dependent clause　The same as subordinate clause, *see* clause.

descriptive adjective　*See* adjective.

direct address　The use of a noun or pronoun to name the person or thing spoken to.

"Fred, come here."

"Roll on, thou deep and dark blue ocean." (When used rhetorically, as here, direct address becomes the figure of speech *apostrophe.*)

direct discourse or **direct quotation**　The repetition without change of another's language. *Compare* indirect discourse.

"Have you finished?" she asked quietly.

direct object　The substantive in a sentence representing the person or thing acted upon. *See also* indirect object.

"Chickadees eat *sunflower seeds.*"

"Right makes *might.*"

"Please let us know *your decision.*"

distributive pronoun　*See* pronoun.

double comparative and superlative　Ways of forming the comparative and superlative—*more brighter, most unkindest.* It was once common in English, but is no longer correct.

double negative　The use of two (or more) negative particles to express a single negation—"I don't know nothing about it." Once in good use, it is now a solecism. It appears in more subtle forms in such expressions as "I couldn't hardly do it."

double possessive, double genitive　The use of both an apostrophe and *of* to form the possessive case. It appears most often when the thing possessed is one of a number belonging to the possessor—"Ben is a friend of John's." The usage is long established, as is clear when the pronoun is used; "Ben is a friend of *his*" is correct but "Ben is a friend of *him*" is not.

ellipsis　The omission of a word or words necessary to the complete grammatical construction of a sentence, but not necessary to the understanding of it. Faulty ellipsis, which is common, consists in omitting words needed for comprehension or correctness.

"The man I saw." (*Whom* elided; permissible.)

"I can and will see him." (*See him,* after *can,* elided; permissible.)

"It is my opinion he can do it." (*That* elided; faulty.)

"I am as old, if not older than you." (*As,* after *old,* elided; faulty.)

expletive　A word added, for smoothness or emphasis, to a sentence complete without it. The most familiar expletives are *it* and *there.* (In popular usage *expletive* may denote *interjection, exclamatory expression,* even *curse.*)

"There are thirty days in September."

"It will be best to go."

"Now what shall we tell him?"

factitive adjective　One that denotes a quality or state produced by the action of the verb.

"This medicine will make you *well.*"

factitive object　A substantive that both completes a predicate and describes or explains the object of the predicate. *See also* complement.

"The manager appointed him *chief.*"

factitive verb　A verb that has both a normal object and an additional object or a complementary adjective. *Make, consider, call* are the most common of such verbs. *See* factitive adjective *and* factitive object.

"We call her Penny."

"I consider him guilty."

"Shall we make a traitor senator?"

finite verb   One that with a subject alone can make a sentence; that is, one that does not require an object to make complete sense. The verb is *intransitive* (c.f.).
"I run."
"They rebel."
"He philosophizes."
future tense   *See* tense.

gender   There are two kinds of gender: natural and grammatical, or artificial, as it is sometimes known. Although the late linguist W. Cabell Greet said that grammatical gender probably predated natural gender in the development of language, English no longer has grammatical gender, with each substantive either masculine, feminine, or neuter. Gender in English nouns normally depends upon the sex of the object: the cow (she), the bull (he), and the house (it). Ships and countries are sometimes construed as feminine. English has what is sometimes called common gender, in which the noun may be either feminine or masculine—cousin, friend, employee. The only problem with gender is the type of pronoun to use. (*See* Gender, p. 366; *see also* feminine endings, p. 422.)
genitive   The case usually called *possessive* in English.
gerund   A nonfinite part of a verb having the form of the present participle (i.e., ending in -*ing*) and used as a substantive. It is sometimes called "the infinitive in -*ing*." It may take an object and be modified by an adverb. It can be distinguished from the present participle by the fact that the gerund may be preceded by *the* and followed by *of*—"the making of hay."
"Skating is a pleasant exercise."
"We like skating."
"She was surprised at its being I."
"Being in love is not an entirely happy condition."
gerund phrase   Either a phrase consisting of a gerund and its modifiers or a prepositional phrase in which a gerund is object of the preposition.
"We enjoyed painting the house."
"Doing the work hurriedly made me uncomfortable."
"In leafing over the album, I came upon a ten-dollar bill."
"After writing the essay, I went for a walk."

historical present   The present tense used in speaking of past events.
"While night hovers over the field of Waterloo, the Germans are marching."

idiom   A use of words peculiar to a particular language. An idiomatic phrase has a meaning as a whole that may not be suggested by its parts. One test of idiom is that it can be translated into another language only by an equivalent idiom. Examples of idiomatic phrases are: *to bring to pass, to call it a day, to come by, to go hard with, to put up with, to set about.* Such phrases are distinguished from figurative expressions, in which the words have their customary meanings but are used metaphorically: *to break the ice, to stand in one's own light, to bring to light.*
imperative   A mood of the verb expressing command.
indefinite adjective   *See* adjective.
indefinite article   The word *a* or *an*.
indefinite pronoun   *See* pronoun.
indicative   *See* mood.
indirect discourse or **indirect quotation**   The repetition without direct quotation of something said. It is expressed as a subordinate clause.
"They said that they saw it."
"He reported that the troops had left."
indirect object   A substantive designating the person to whom or for whom an action is done.

"Sing *us* a song."
"He gave *us* a book."
"We voted *James* a prize."

indirect question   A question expressed as a subordinate clause.
"She asked who had sent it."
"I asked whom he wanted."
"They asked what came next."

infinite verb   A participle, infinitive, or gerund.

infinitive   The fundamental form of the verb, commonly preceded by *to: to go, to work, to be, to wish.* The *to* is not a preposition but is historically a kind of prefix, and is therefore called the "sign" of the infinitive. The infinitive has two tenses: present, both active and passive—*to see, to be seen;* and perfect, both active and passive—*to have seen, to have been seen.* It has also a progressive form—*to be seeing, to have been seeing.* Functionally it is a nonfinite verb or a verbal noun (verbal) and may have modifiers, may take an object or complement, and may have a logical, though not a grammatical, subject. It may be used as an adjective, adverb, or noun. Without the *to* it is used to make up compound tenses with the auxiliaries *do, may, shall, will—I shall go, he may do it.*
As a noun: "*To name* him is *to praise.*"
As an adverb: "I come *to bury* Caesar, not *to praise* him."
As an adjective: "This house is *to let.*"

infinitive phrase   A phrase introduced by an infinitive. It may be used as an adjective, an adverb, or a noun.
"Do you dare to tell the truth?"
"To believe what you say, I should have to hear the evidence."
"To hear her talk, one would think she was musical."

inflection   Change in form of a word, especially in comparison, declension, or conjugation, to indicate voice, mood, tense, person, number, case, degree, and so on. The part of a word that remains when inflections are removed is known as the *stem.* Since English is a synthetic language, not an inflected one, inflections play a much smaller part in its grammar than they do in Greek, Latin, or German.

intensive pronoun   *Myself, yourself, himself,* and so on, when used in apposition to a noun or pronoun to increase its force: "I myself will do it." "I will do it myself." To be distinguished from the reflexive pronoun *myself, yourself,* and so on: "I burned myself." Intensive pronouns tend to be misused, perhaps because some think they sound more elegant. "My wife and myself want to ask you to dinner" represents this kind of error.

interjection   A word or words expressing sudden feeling: *O, oh, ah, pooh, rats, hurrah, amen.* The interjection is grammatically independent. Any part of speech may be used as an interjection—*Help! Well, well! What! The idea! Beautiful! Stuff! Bless me! Dear me!* Introducing words, such as *well, why,* may be parsed as *interjections* or *expletives* (q.v.), according to the amount of feeling they convey—"Well, look who's here!" "Well, after I had crossed the river . . ."

interrogative adjective   An interrogative pronoun used to modify a substantive —"Which man?" "Whose dog?"

interrogative pronoun   *Who, which, what* when used in questions: "Who is that?" "What is it?"

intransitive verb   One that does not require an object, that is, one that denotes an action, state, or feeling that terminates in the doer.
"The sun shines."
"Father and Mother walked, but Grandmother rode."
"I shall lie down."
"Let's sit down."
"The kite rises."

irregular verb   *See* verb.

limiting adjective   *A, an, the.* (Also called *article.*)
locution   A phrase; an idiom; an element in style; phraseology.

modifier   A word, phrase, or clause that affects (changes, restricts, enlarges) the meaning of another word or group of words.
mood or **mode**   A form of the verb indicating the manner of doing or being. The moods usually recognized in English grammar are the indicative, which includes interrogative and exclamatory forms, the subjunctive, and the imperative. The indicative indicates a fact; the subjunctive, a wish, condition contrary to fact, or demand; the imperative, a command.
Indicative: "That is a starling." "Is that a starling?" "That's a starling!"
Subjunctive: "If I were you, I'd wait." "Far be it from me to suggest such a thing."
Imperative: "Go!" "Come here!" "Let us start immediately."

nominative absolute   A substantive modified by a participle and not grammatically connected with the sentence. (The word *absolute* is used in the old sense —*absolve, absolution*—of *loose, untied,* because the absolute phrase does not modify.) The construction is to be used with caution.
"The vote taken, the committee proceeded with other business."
"The rain having washed out all traces, the clue proved useless."
nominative case   *See* case.
nominative of direct address   A substantive unrelated grammatically to any other word in a sentence and used as a salutation or form of address. *See* direct address.
"Boys, stop that noise."
"Sir, may I speak with you?"
"How are you feeling, Fred?"
nonfinite verb   *See* infinite verb.
nonrestrictive element   A word, phrase, or clause which, though it contributes to the general meaning of a sentence, does not limit or confine the meaning of the substantive it modifies. (The distinction between restrictive and nonrestrictive elements is important mainly in punctuation.) See pages 189–191.
noun   A word that is a name. A noun may be used as a subject, an object, a complement, an appositive, or a nominative absolute.
Nouns are classified as *common* and *proper.* A common noun is the name of one of a class or kind—*woman, deer, saw, wax;* a proper noun, the name of some particular person, place, or personified thing—*Italy, Chicago, Liberty, Mona Lisa, Mississippi, Vatican.*
Nouns are also classified as *abstract* and *concrete.* An abstract noun is the name of an idea, quality, action, or state, without regard to any person or thing. A concrete noun denotes a person or thing.
Abstract: "*Ripeness* is all." "*Beauty* is its own excuse for *being.*"
Concrete: "To hide a green *hill* in an April *shroud.*" "Fast-fading *violets* covered up in *leaves.*"
noun clause   A clause used as a substantive. Such a clause is usually introduced by a subordinate conjunction, an interrogative adjective or adverb, or a relative pronoun. It may be used as an appositive, an object, a predicate nominative, or a subject.
Subject: "*When he would arrive* was a mystery."
Object: "He asked me *when I should arrive.*"
Object of preposition: "Give the book to *whoever wants to read it.*"
Complement: "This map is *what you need.*"
Appositive: "The suggestion *that we should surrender* was greeted with jeers."
noun phrase   A phrase used as a substantive.
Gerund: "*Going to school* necessitated *walking five miles.*"
Prepositional: "*From here to there* is a long distance."
Infinitive: "*To see the parade* was Tom's ardent wish."

number   The singular or plural form of a noun, pronoun, or verb.

numeral adjective   A cardinal or ordinal number used to modify a substantive — *twelve* apples, *second* base, *fiftieth* year.

object   *Direct object:* a substantive that receives the action expressed by the verb and performed by the subject; *indirect object:* the person, animal, or thing to or for which the act is performed. *See also* cognate object.
"He sent me a message." (*Message* is the direct object and *me* the indirect. Note that the indirect object may be parsed as the object of a preposition — "He sent a message to me" — *to* being understood.)

parallelism   The principle that parts of a sentence that are parallel in meaning should be made parallel in structure. In its simplest form it is seen in "Seeing is believing" or "To see is to believe" (*not* "Seeing is to believe"). Parallelism has more important rhetorical than grammatical applications.

parenthesis   A word, phrase, or sentence inserted in a sentence by way of comment or explanation. The sentence is grammatically complete without it. It should always be set off by punctuation — commas, dashes, or parentheses.
"I was thinking — this was yesterday evening — that Charles (he is my brother) will be forty-five tomorrow."

participial adjective   A participle used to qualify a substantive — a *running* deer, a *defeated* candidate.

participial phrase   A participle with its modifiers.
"I saw the janitor *standing at the door.*"
"The vase, *broken by the fall,* disclosed the money."
"*Swinging his arms freely,* he walked rapidly away."

participle   A word that shares the properties of a verb and an adjective. It has, like a verb, tense and power to govern an object; and, like an adjective, power to qualify a substantive. The tenses of the participle are *present* and *past.* The present participle, ending in *-ing,* is used in forming the progressive conjugation. The past participle, ending in regular verbs in *-d* or *-ed,* is employed with auxiliaries in the perfect, past perfect, and future perfect. *See also* participial adjective, participial phrase, verb.

particle   A term sometimes used to denote the minor parts of speech, i.e., articles, conjunctions, interjections, and prepositions.

parts of speech   Adjectives, adverbs, conjunctions, nouns, prepositions, pronouns, verbs. In older grammars the article was designated a separate part of speech, but it is now considered a *limiting adjective.*

passive voice   *See* voice.

past and **past perfect**   *See* tense.

perfect   *See* tense.

person   Any one of the three relations in which a subject stands with respect to a verb, referring severally to the person speaking, the person spoken to, and the person spoken of. These forms are called the *first, second,* and *third* persons, singular or plural.

phrase   A group of words not containing a finite verb. Phrases are classified by the type of word that introduces them as prepositional, participial, and infinitive. They are further classified by their office as adjective, adverb, and noun. *See these headings.*

possessive adjective   A noun or pronoun in the possessive case used to modify a substantive — *my* hat, *Susan's* brother.

possessive case   *See* case.

predicate   The verb in a sentence, with or without its modifiers, as it relates to the subject. The verb alone is called the *simple* or *grammatical predicate;* with its modifiers and object or complement (if any), the *complete* or *logical predicate.* The term *predicate* is used correctly of the verb with relation to its

office in the sentence, as affirming or denying something about the subject.
Simple predicate: "I *have tried* to do it several times."
Complete predicate: "I *have tried to do it several times.*"
The predicate may be double or multiple.
"We *shall run* and not *be* weary, *walk* and not *faint.*"
"We laboriously *climbed, slipped, slithered, found* precarious toeholds and
handholds, and at last *pulled* ourselves over the edge of the cliff."

predicate adjective   See adjective *and* complement.

predicate noun, predicate nominative   See complement.

preposition   A word governing a noun or pronoun or its equivalent, the
preposition and the word it governs forming a *prepositional phrase*. The
preposition may be simple (e.g., *at, in, from, on*), compound (e.g., *in place
of, from under, with regard to, out of*), or verbal (e.g., *respecting, considering,
notwithstanding*). Several adjectives have prepositional force, *like* and *near*
being the most important.

prepositional idiom   A term applied to the use of prepositions (before a substantive or after a verb, adjective, or noun) natural to English. Since such use
often differs from that in other languages, it presents great difficulty to for-
eigners whose English is imperfect. The use of proper prepositions following
verbs, adjectives, or nouns is covered in a separate action (see pp. 432ff.).
Examples of idiomatic use of prepositions before nouns are: *in Boston, at the
seaside, out of season,* the girl *with the green eyes, below deck.*

prepositional phrase   A phrase consisting of a preposition and the substantive
it governs. It may be used as an adjective, an adverb, or a noun.
"*Over the hill* is the long way to go." (Noun)
"The animal *in the picture* is a panda." (Adjective)
"We went *to the theater* last night." (Adverb)

present perfect   See tense.

principal parts   The present, past, and past participle of a verb, as *sing, sang,
sung; herd, herded, herded; try, tried, tried.* The present participle—*singing,
herding, trying*—is also often called a principal part.

pronominal adjective   Any possessive pronoun used to modify a substantive—
his car, the car is *his.*

pronoun   A word used in place of a noun, to avoid repeating the noun. Pro-
nouns are classified as follows:
   Demonstrative: *this, that, these, those,* when used to point out, designate—
     "*That* is a marten and *these* are weasels."
   Distributive: *each, every, everyone, either, neither,* and so on.
   Indefinite: *any, few, many, none, one, some, such,* and so on.
   Intensive: *myself, himself, ourselves,* and so on—"I *myself* will do it."
   Interrogative: *who, which, what,* in questions—"*Who* is that?"
   Personal: *I, thou, he, she, it, we, you, they,* and their declensional forms—
     my, mine, me, for example.
   Reflexive: *myself, himself, ourselves,* and so on—"I have burned *myself.*"
   Relative: *who, which, what, that,* when introducing a subordinate clause.
The personal pronouns are declined in case, number, and person—*I, my* or
mine, me; we, our or *ours, us,* for example.
The indefinite pronouns are indeclinable, though some have a possessive
case; *one's, everybody's,* for example.
The relative pronoun *who* is declined—*who, whose, whom.*

proper noun   The name of a particular person, place, or personified thing—
John, Pittsburgh, Liberty.

reflexive pronoun   See pronoun.

reflexive verb   A verb whose object, expressed or implied, refers to the same
person or thing as the subject—*He dressed; He dressed himself.*

relative adjective   *Whose, which,* or *what,* when it combines the offices of adjective and relative pronoun. Also *whichever, whatever, whosoever.*
"He did not see which way to go."
"Take whichever book you please."

relative clause   A clause introduced by a relative pronoun or a subordinating conjunction. It serves as an adjective, qualifying a substantive in the main clause; or an adverb, modifying the verb in the main clause; or a substantive, used as subject, object, or complement of the main clause.
"*That he is a good man* is evident." (Subject)
"The trouble is *that I do not know him.*" (Complement)
"They killed *whatever was alive.*" (Object)
"Call me *whenever you are ready.*" (Adverb)
"The reason *that he did it* is mysterious." (Adjective)
"This is the place *where sandpipers love to come.*" (Adverb)
"It is a time *when great courage is needed.*" (Adverb)

relative pronoun   A pronoun introducing a subordinate clause. *See* pronoun.

restrictive modifier   One not merely descriptive or parenthetic. The test of whether a modifier is restrictive or not is to read the sentence without it and note whether the omission destroys the exact meaning of the sentence. *See also* nonrestrictive element.
"I shall send it *if you write me.*" (Restrictive)   .
"The man *I saw* is said to be a spy." (Restrictive)
"The man, *so open and plausible,* is said to be a spy." (Nonrestrictive)
"The books *that were burned* were on the third floor; the books *that were saved* were on the tenth floor." (Observe that the same grammatical sentence can have a different intent: "Where were the damaged books?" "The books, *which were burned,* were on the third floor." See section on *that* and *which*, p. 378.)

sentence   "A group of words so related as to convey a completed thought with the force of asserting something or of asking, commanding, exclaiming, or wishing." — Webster II.
"A grammatical unit comprising a word or a group of words that is separate from any other grammatical construction, and usually consists of at least one subject with its predicate and contains a finite verb or verb phrase; for example, *The door is open* and *Go!* are sentences." — American Heritage Dictionary. Sentences are classified by type of predication as declarative, interrogative, exclamatory, and imperative; by grammatical structure as simple, compound, and complex; and by rhetorical structure as loose, periodic, and balanced.

sequence of tenses   A phrase designating the idiomatic or logical order of tenses of verbs in a sentence. See pages 346–347.

singular and **plural**   Names of the grammatical *number* of substantives.

split infinitive   One in which the *to* and the infinitive proper are separated. See pages 386–387.

stem   *See* inflection.

strong verb   One forming the past tense by vowel change—*grow, grew*—and its past participle by *-n* or *-en*—*bite, bitten; give, given.*

subject   The substantive that in a sentence tells what the sentence is about; that which has a predicate. A distinction is made between the *simple subject* and the *complete subject,* the latter being the simple subject together with its modifiers. A subject may be compound or multiple. It may be a word, a phrase, or a clause. It is in the nominative case.
"All *birds* but a very few can fly." (Simple subject.)
"*All birds but a very few* can fly." (Complete subject.)
"*Certain birds, such as the auk, penguin, ostrich, and emu, which have lost*

the power of flight, are always in danger of extinction." (Complete subject.)
"*The auk, penguin, ostrich, and emu* have lost the power of flight." (Multiple subject.)

subjunctive mood   *See* mood.

subordinate clause   *See* clause.

substantive   A word, phrase, or clause used as a noun.

superlative   *See* comparison.

syntax   The section of grammar that deals with the relations of words in the sentence, their uses, agreements, and construction. More specifically, the syntax of a sentence-element is the part it plays in the sentence.

tense   The time of the action expressed by a verb. There are three times: past, present, and future—I loved, I love, I shall love; and three types: simple, progressive, and perfect—I love (*simple*), I am loving (*progressive*), I have loved (*perfect*).

ACTIVE	PASSIVE
Simple present: I see	I am seen
Present progressive: I am seeing	I am being seen
Present perfect: I have seen	I have been seen
Simple past: I saw	I was seen
Progressive past: I was seeing	I was being seen
Past perfect: I had seen	I had been seen
Simple future: I shall see	I shall be seen
Progressive future: I shall be seeing	I shall be seeing
Future perfect: I shall have seen	I shall have been seen

verb   A word or group of words that affirms (or denies) something about a subject. It expresses action, being, or state of being. In a sentence or a clause the verb is called the predicate when it is considered in relation to the subject. The inflection of a verb is called its *conjugation* (q.v.).

Verbs are classified as follows:

Finite: those that with a subject can make a sentence and that agree with the subject in person and number: *I go, he goes, they go; Jack swims; fish swim.*

Nonfinite or infinite: those that are employed as other parts of speech (though retaining some verbal force) and have no person or number: *to go, going.* Also called *verbal nouns, verbal adjectives,* or *verbals.*

Regular or weak: those that form their past tense and past participle in *-d, -ed,* or *-t: walk, walked, walked; bare, bared, bared; build, built, built.*

Irregular or strong: those that form their past tense by change of vowel and past participle in *-n* or *-en: be, was, been; throw, threw, thrown.* Some verbs are called irregular but not strong; those that form their past tense and past participle without change of vowel but show other irregularities: *cast, cast, cast; have, had, had.*

Transitive: in ordinary usage, those that have a direct object. There is disagreement concerning whether a verb that may have an object but has none is transitive. Compare: *She bakes; she bakes bread; the bread bakes.* (The first is potentially transitive; the second, transitive; the third, intransitive.) Many verbs are both transitive and intransitive but with a difference of meaning: *Birds fly. The boy flies a kite.*

Intransitive: those that (1) do not take an object or (2) in a given sentence do not have an object. *See* transitive. An intransitive verb may take a *cognate object* (q.v.).

Copulative or linking: those that take a complement: *be*—"I am he"; *feel*—"I feel tired"; *become*—"The child will become a man."

Auxiliary: those used with other verbs to form mood and tense: *be, can, may, might*—"I was sleeping"; "He can skate." Auxiliaries may be used as *notional verbs;* that is, have a sense of their own. "When Duty whispers low Thou *must,* the youth replies I *can.*"

Defective: those whose conjugation is not complete, such as *shall, may, can, ought.* Defective and irregular verbs are sometimes called *anomalous.*

verbal A word derived from a verb, combining something of the meaning and use of a verb with the use of a noun or adjective. Specifically, *infinitive, gerund, participle* (q.v.).

vocative The case of a noun of direct address or the noun itself.

voice The relation of a subject to its verb; that is, whether acting, acting upon or for itself, or acted upon. The voices are the active and passive; and these have two forms, the simple and the progressive.

Active: "The soldiers put up a tent."

Passive: "A tent was put up by the soldiers."

PART VI.   USE OF WORDS

Precision, freshness, appropriate use, sensitivity to idiom—these are qualities that writers must patiently seek. Those who blunder into wordiness, triteness, or flawed idiom not only offend their readers but undermine the substance of their writing. The responsibility for appropriate use of words lies with the writer, but the editor can assist in eliminating violations and lapses.

WORDINESS

The terms *redundancy, pleonasm, tautology, verbosity,* and *circumlocution* all refer to the same general fault—the use of superfluous words. But awareness of the distinctions between them helps in noting particular defects.

Redundancy and pleonasm. The term for use of more words than are necessary to express a meaning is *redundancy; pleonasm* refers specifically to the use of words whose omission would leave the meaning intact. Pleonasm may be used on occasion for rhetorical emphasis; pleonasm, however, usually occurs unintentionally. In the following group, the italicized adjectives duplicate part of the nouns they modify.

advance planning, warning
angry clash
heir *apparent*
awkward predicament
chief protagonist
close proximity
complete master, monopoly, ruin
definite decision
end product, result
essential condition
excess verbiage
fellow playmates, colleagues
final completion, settlement, outcome
first priority
fresh beginning
full satisfaction
funeral obsequies
general rule, public
good benefit
grateful thanks
habitual custom
hot-water heater
important essentials
innumerable numbers
integral part
invited guest
joint cooperation
body of *the late . . .* , widow of *the late . . .*
little sapling, duckling
local resident

lonely isolation
major breakthrough
more superior, preferable
mutual cooperation
necessary requisite
new beginning, creation, innovation, record, recruits
old adage
original source
passing phase, fad, fancy
past history
peculiar freak
knots *per hour*
pre-plan
present incumbent
proposed plan
prototype model
root cause
self-confessed
separate entities
serious danger
successful achievements
surrounding circumstances
temporary reprieve
total annihilation, extinction, reversal, destruction
true facts
universal panacea
usual customs
violent explosion
young infant, teenager

Sometimes the noun is the redundant part of the phrase.

anthracite *coal*   undergraduate *student*
barracks *buildings*   weather *conditions*
connective *word*   widow *woman*
doctorate *degree*

A preposition or adverb used as part of a verb may be redundant.

ascend, hoist, lift, zoom *up*   join *up*
attach, assemble, collaborate,   made *out* of
   cooperate, fuse, join, merge, unite   pare, swoop *down*
    together   penetrate *into*
christened *as*   persist *still*
connect *up/together*   protrude *out*
continue *on/still*   recall, recoil, return, revert,
eliminate *altogether*   remand *back*
feel *of*   separate *apart*
follow *after*   sink, swoop *down*
gather *up/together*   skirt *around*
hurry *up*   termed *as*

The italicized words in the following are redundant:

adequate *enough*   *every* now and then
and moreover   face *up to*
Angkor Wat *Temple*   frown, smile *on his face*
appear *to be*   in abeyance *for the time being*
appointed *to the post of*   intolerable *to be borne*
as never before *in the past*   last *of all*
as to whether   made *out* of
as yet   may, might *possibly*
atop *of,* inside *of*   mutual advantage *of both*
at some time *to come*   never *at any time*
but nevertheless   *on the occasion* when
Capitol *Building*   over *and done with*
commute to *and from*   over *with*
continue *to remain*   *quite* unique
dates *back* from   results so far *achieved*
descend *down*   shuttle *back and forth*
eliminate *altogether*   this *same*
entirely complete   throughout *the length and breadth*
equally *as*   topped *the . . . mark*

Both is often used redundantly; it should be omitted in sentences like
the following:

They are *both* alike.
They *both* go hand in hand.
Both the sons attend different universities.
Both of the other two are in the Marines.

The prepositional phrases in expressions like the following are redundant:

big *in size*	filled *to capacity*
biography *of his life*	graceful *in appearance*
bisect *into two parts*	lenticular *in character*
blue *in color*	prejudge *in advance*
classified *into groups*	smattering *of knowledge*
consensus *of opinion*	square, round *in shape*
contemporaneous *in age*	strangled *to death*
day dawned *in the east*	surgeon *by occupation*
depreciated, appreciated *in value*	2 p.m. *in the afternoon*
few *in number*	

Tautology. Another fault is tautology, which is defined in Webster II as "repetition of the same words or use of synonymous words in close succession." An example would be "an abandoned house left to rot." This can easily be edited to "a house abandoned to rot." Here are some examples frequently encountered:

and so as a result
any and all
cause of . . . is on account of
each and every
environment of rough surroundings
one and the same
pair of twins
separate and distinct
today's modern woman
when and if

Circumlocutions. Circumlocutions are not ungrammatical or repetitious, but are to be avoided as wordy.

ahead of schedule (early)
a large proportion of, percentage of (many)
am in possession of (have)
a percentage of (some)
are present in greater abundance (are more abundant)
at an early date (soon)
best of health (well, healthy)
by the name of (named)
call a halt (stop)
caused injuries to (injured)
destroyed by fire (burned)
draw the attention of . . . to (show, point out)
during the time that (while)
from the commercial standpoint (commercially)
gathered together (met)
give rise to (cause)
had occasion to be (was)
in advance of (before)
in the neighborhood, vicinity of (near, nearly, about)
in this day and age (today)

made an approach to (approached)
made a statement saying (stated, said)
made good an escape (escape)
placed under arrest (arrested)
put in an appearance (appear)
render assistance to (help)
retain position as (remain)
sparsely scattered (sparse)
suburban area (suburbs)
succumbed to injuries (died)
take action on the issue (act)
take into consideration (consider)
the house in question (this house)
was of the opinion that (believed, thought, said)
was witness to (saw)

WORDS AND PHRASES OFTEN USED SUPERFLUOUSLY

In the case of employees who are late, wages will be reduced. = Wages of
employees who are late will be reduced.
These books *in some instances* were . . . = Some of these books were . . .
The handwriting is *of* distinctive *character*. = The handwriting is distinctive.
the handwriting is *of* distinctive *character*. = The handwriting is distinctive.
The project is *of a* similar *nature*. = The project is similar.
The task is *of such difficulty* that . . . = The task is so difficult that . . .
He studied *for the purpose of* passing the examination. = He studied to pass
the examination.
The weather bureau predicts shower *conditions*. = The weather bureau predicts
showers.
She seldom loses money, *in spite of the fact* that she invests recklessly.[1] = She
seldom loses money, although she invests recklessly.
His work *along the line of* technical improvements has been outstanding. = His
work in technical improvements has been outstanding.
There *is a large number of* commercial varieties. = There are many commercial
varieties.
As long as we must go anyway, we may as well go cheerfully. = If we must go
anyway, we may as well . . .
He does not wish to review it *at the present time (at this juncture)*. = He does not
wish to review it now.
We sought advice *in connection with (in regard to, in relation to, with reference
to)* reservations. = We sought advice about reservations.
We are *currently* looking for a five-room house. = We are now looking for a
five-room house.
As far as I am concerned, foreign languages might as well not exist. = As for me,
foreign languages . . .
We expect results *in the not too distant future.* = We expect results soon.
The house is *situated on* Twelfth Avenue. = The house is on Twelfth Avenue.
I shall *give due consideration* to the matter. = I shall consider the matter.
He lives *in the vicinity of* the park. = He lives near the park.
He sought the funds *for the purpose of assisting* the victims. = He sought the
funds to help the victims.

---

[1]Strunk and White's *Elements of Style* says: *the fact that* "should be revised out of
every sentence in which it occurs."

TRITE EXPRESSIONS

Trite is a word derived from the Latin verb for *worn out;* cliché is the French word for a stereotype; *hackneyed* probably derives from a name applied to horses hired out for common use. All three terms have themselves become clichés applied, as their etymology indicates, to worn-out language—expressions once clever, pat, witty, now deprived of their original force through overuse. They come easily to the writer's pen, or typewriter, and can make a manuscript sound slick, tiresome, and false. Many such expressions have become trivial ways of describing something unpleasant or complicated, as in "cast a pall."

The writer must deal with clichés not by trying to avoid them altogether, but by being aware of their proper use. Tired as they may be, they remain part of the language, and sometimes their employment is more economical than the detours around them. At times, a cliché will serve better than a quasi-original phrase that fails. What are to be avoided are automatic phrases—*add insult to injury, as a matter of fact, each and every, it goes without saying, no sooner said than done*—that convey little information and can easily be omitted or replaced by a simpler, more precise expression.

The list of trite expressions below presents a selection of those that appear often in books, periodicals, and newspapers. Complete proverbs and quotations are omitted; overworked single words are noted in the section headed Appropriateness (see pp. 415ff.).

abject apology
abreast of the times
accidents will happen
acid test
active consideration
add insult to injury
adds a note of
after all has been said and done
age before beauty
agree to disagree
aired their grievances
all in a day's work
all over but the shouting
all things considered
all things to all men
all too soon
all walks of life
all work and no play
along these lines
among those present
ample opportunity
apple of the eye
armed to the teeth
as a matter of fact

as luck would have it
as the crow flies
at a loss for words
at a tender age
at long last
at one fell swoop
at first blush
auspicious event
auspicious moment
awaiting further orders

bag and baggage
bated breath
beat a hasty retreat
beggars description
beginning of the end
beneath contempt
benefit of a doubt
best-laid plans
better half
better late than never
better left unsaid
bewildering variety
bitter end

blanket of snow
blare forth
blissful ignorance
block out
bloody but unbowed
bloom of youth
bolt from the blue
bone of contention
breathless suspense
bright and shining faces
broad daylight
brook no delay
brute force
budding genius
built-in safeguards
burning question
burning the midnight oil
busy as a bee
by leaps and bounds
by the same token

calm before the storm
came in for their share of
capacity crowd
cast a pall
casual encounter
chain reaction
charged with emotion
checkered career
cherished belief
chief cook and bottle-washer
circumstances over which I have no
   control
city fathers
civic wrath
claimed him for its own
clear as crystal
colorful display
common or garden variety
conservative estimate
considered opinion
consigned to oblivion
conspicuous by its absence
controlling factor
crack troops
crying need
curiously enough
cut a long story short
cut down in his prime

daring daylight robbery
dark horse
dashed madly about
date with destiny
days are numbered
dazed condition
dead as a doornail

deafening crash
deadly earnest
deficits mount
deliberate falsehood
depths of despair
dialogue of the deaf
diamond in the rough
dig in one's heels
discreet silence
distinction with(out) a difference
dog in the manger
doom is sealed
doomed to disappointment
dotted the landscape
dramatic new move
drastic action
due consideration
dull, sickening thud
dynamic personality

each and every
easier said than done
eat, drink, and be merry
eloquent silence
eminently successful
engage in conversation
enjoyable occasion
entertaining high hopes of
epic struggle
equal to the occasion
ere long
errand of mercy
even tenor
exception that proves the rule
exercise in futility
existing conditions
express one's appreciation

failed to dampen spirits
fair sex
fall between two stools
fall on bad times
fall on deaf ears
far be it from me
far cry
fatal deed
fateful day
fateful scene
fate worse than death
favor with a selection
festive occasion
few and far between
few well-chosen words
fickle finger of fate
finishing touches
fit as a fiddle
floral tribute

food for thought
fools rush in
force of circumstances
foregone conclusion
foul play
from the sublime to the ridiculous

gala occasion
generation gap
generous to a fault
gild the lily
give the green light to
glowing cheeks
goes without saying
goodly number
grateful acknowledgment
grave concern
great open spaces
green with envy
greet the eye
grim reaper
grind to a halt

hale and hearty
hands across the sea
happy pair
hastily summoned
have the privilege
heartfelt thanks
heart of the matter
heart's desire
heated argument
heave a sigh of relief
height of absurdity
herculean efforts
high dudgeon
honored in the breach
hook, line, and sinker
hook or crook
hope for the future
hope springs eternal
hot pursuit
humble but proud
hurriedly retraced his steps

ignominious retreat
ignorance is bliss
ill-fated
immaculately attired
immeasurably superior
impenetrable mystery
implicitly trust
in close proximity
inextricably linked
infinite capacity
inflationary spiral
in no uncertain terms

in our midst
in short supply
internecine strife
in the limelight
in the nick of time
in the same boat with
in the twinkling of an eye
in this day and age
into full swing
iron out the difficulty
irreducible minimum
irreparable loss
it dawned on me
it goes without saying

just deserts

keep options open

labor of love
lashed out at
last analysis
last but not least
last-ditch effort
leaps and bounds
leave no stone unturned
leaves much to be desired
leave up in the air
let well enough alone
lend a helping hand
like a bolt from the blue
limped into port
line of least resistance
little woman
lit up like a Christmas tree
lock, stock, and barrel
logic of events
long arm of coincidence
long-felt need

make good one's escape
malignant fate
marked contrast
marshal support
masterpiece of understatement
matter of life and death
mecca for travelers
men were deceivers ever
method in his madness
milk of human kindness
miraculous escape
moment of truth
momentous decision
monumental traffic jam
more in sorrow than anger
more sinned against than sinning
more than meets the eye

more the merrier
Mother Nature
motley crew

narrow escape
nearest and dearest
needs no introduction
never a dull moment
never before in the history of
nipped in the bud
none the worse for wear
no sooner said than done

one and the same
on more than one occasion
on unimpeachable authority
open secret
order out of chaos
other things being equal
overwhelming odds
own worst enemy

pageant of history
paid the penalty for
pales into insignificance
paralyzed with fright
paramount importance
part and parcel
patience of Job
pay the piper
peer group
pet aversion
pet peeve
pick and choose
pinpoint the cause
place in the sun
play it by ear
point with pride
poor but honest
powder keg
powers that be
pretty kettle of fish
progressive and enlightened
pronounced success
pros and cons
proud but humble
proud heritage
proud possessor
psychological moment

quivered with excitement

ravishing beauty
red-letter day
regrettable incident
reigns supreme

reliable source
remedy the situation
riot of color
riot-torn area
ripe old age
round of applause
rude habitation

sadder but wiser
saw the light of day
scathing sarcasm
sea of faces
seat of learning
second to none
seething mass of humanity
select few
selling like hotcakes
shattering effect
shift into high gear
shot in the arm
sickening thud
sigh of relief
silence broken only by
silhouetted against the sky
simple life
skeleton in the closet
sleep the sleep of the just
snare and delusion
social amenities
specimen of humanity
spectacular event
speed the parting guest
spirited debate
steaming jungle
stick out like a sore thumb
stick to one's guns
strait and narrow path
straitened circumstances
such is life
sum and substance
sunny South
superhuman effort
supreme sacrifice
sweat of his brow
sweeping changes
sweet sixteen

take the bull by the horns
take up the cudgels
telling effect
tender mercies
terror-stricken
thanking you in advance
there's the rub
this day and age
this mortal coil

those present
throw a (monkey) wrench into the
  works
throw a party
throw caution to the winds
thunderous applause
tie that binds
time immemorial
time of my life
tired but happy
toil in the vineyard
tongue in cheek
too numerous to mention
tower of strength
trials and tribulations
tumultuous applause

unalloyed pleasure
uncharted seas
unprecedented situation
untimely end
untiring efforts

vale of tears
vanish into thin air

vest-pocket republic
viable alternative
view with alarm
viselike grip
voice the sentiments

watery grave
wax poetic
weaker sex
wear and tear
wee small hours
wend one's way
where ignorance is bliss
whirlwind tour
wide open spaces
words fail me
words fail to express
word to the wise
work one's wiles
worse for wear
wrapped in mystery
wreathed in smiles
wrought havoc

young hopeful

APPROPRIATENESS

The terms listed in this section often have old, conventional, and accepted meanings, but they have also acquired new senses and are not,
most of them, improper or incorrect when used in the proper context.
Thus the word *appropriateness* has been adopted to head this section,
rather than the more usual *propriety* or *correctness*, because the problem
raised by the words listed is one of determining the contexts in which
they are appropriate. The following section treats, in the main, the usages
described in the classifications below. A few doubtful usages and a few
illiteracies have been included because they appear from time to time in
written work.

The list has been designed to classify words according to their recent
standing in the several levels of English vocabulary. For the sake of clarity,
the list will use classifications as defined in *The American Heritage Dictionary of the English Language:*

Standard applies to words and expressions that conform to established
  educated usage in speech or writing.

Informal applies to words, expressions, and idioms acceptable in the
  natural spoken language but considered inappropriate as literary
  language and in the standard literary prose of ceremonial and official communications. (As the term is applied here, *informal* is for
  the most part the equivalent of *colloquial*.)

Dialect: "A regional variety of a language, distinguished from other varieties by pronunciation, grammar, or vocabulary; especially, a variety of speech differing from the standard literary language or speech pattern of the culture in which it exists."

Slang: "The nonstandard vocabulary of a given culture or subculture, consisting typically of arbitrary and often ephemeral coinages and figures of speech characterized by spontaneity and raciness."

Two other classifications have been included, in small numbers. One is *vulgar slang*, which comprises taboo words that may present problems to the copy editor. The other is *overworked*, applied mostly to words used so much that they have lost their impact.

Words move continually between classes. A slang term may shift to the informal level in months, and well-used informal expressions tend to become standard. Some words that may seem destined for a permanent place in the standard language either remain indefinitely at the slang level or fall into disuse. The list presented here is a reasonably good guide for its day. In case of doubt, consult a dictionary.

Because two dictionaries, the *American Heritage Dictionary* and the *Random House Dictionary*, have pioneered in annotating entries according to level of usage, they have been the chief sources for classifications in this list. A third source has been Theodore M. Bernstein's *The Careful Writer*. (All these works are listed in the bibliography starting on p. 487.) The following abbreviations are used: AHD = *American Heritage Dictionary;* RHD = *Random House Dictionary;* TCW = *The Careful Writer.*

addict   As a noun ("drug addict") — standard.

aggravate   For *irritate, annoy* ("That kind of music always aggravates me") — avoided in formal contexts; informal.

aim   For *intend* ("I aim to do it tomorrow") — chiefly dialect.

ain't   For *am not, is not, are not, has not, have not* — still condemned for standard or informal use, except in deliberately humorous contexts or for dialect.

alibi   As a noun, for *excuse* ("The pitcher had all sorts of alibis for losing the game") — accepted as informal, but in standard use retains something of its legal meaning of a claim of being elsewhere.

allergy   For *antipathy* ("He has an allergy to golf") — informal and overworked.

alright   For *all right* — not accepted.

and   For *to,* used between two finite verbs ("Try and stop me") — informal, not far from standard.

angel   For *financial backer,* especially in the theater — once slang, now informal approaching standard.

anxious   For *desirous, eager* ("I'm anxious to see the film") — gaining informal acceptance, but too loose for standard written work, where the word should suggest unease or apprehension.

anyplace   As one word — informal but gaining acceptance; *anywhere* is standard. As two words — standard ("I will follow you any place.") *Someplace* and *everyplace* are in the same category as *anyplace.*

aquacade   *See* motorcade.

around   For *about* ("around four dozen"); for *in this vicinity* ("Stick around"); *get around* for *circumvent* or for *have worldly experience* ("I can get around her"; "I get around") — all in wide informal use.

arty For *affectedly aesthetic*—informal.

author As a verb ("He authored a novel")—rare word returning to print; unnecessary and graceless in written work (AHD).

awhile Standard formation, meaning *for a brief time;* thus it should not be preceded by *for* ("Stay awhile").

awoken, woken For *awakened, waked*—though often called illiterate, better described as obsolescent. The confusion about the principal parts of these verbs is very old and still persists. The correct forms are as follows:

PRESENT	PAST	PAST PARTICIPLE
wake	woke *or less commonly* waked	waked *or rarely* woke *or* woken
awaken	awakened	awakened
waken	wakened	wakened
awake	awoke *or less commonly* awaked	awaked *or less commonly* awoke

Distinctions between the forms—their use in a transitive or intransitive sense and in active and passive constructions—are so indefinite as to be hardly useful.

awol (Pronounced *ā'wol*.) For A.W.O.L., *absent without (official) leave*—military slang.

aviatrix *See* feminine endings.

back of, in back of Both accepted in informal use and nearing standard use; in formal contexts, *behind* is still preferable ("behind the house" instead of "in back of the house").

bag For *something that someone is into, skilled in,* or *interested in* ("I guess photography just wasn't his bag")—slang. For *unattractive, disagreeable woman* ("Some old bag lives there")—derogatory slang. *Holding the bag* for *having total blame or responsibility devolve upon one*—informal. *In the bag* for *assured of success*—slang.

balance For *remainder* in a non-bookkeeping sense ("The balance of the party left for home")—gaining informal acceptance but still not in standard use.

ball *On the ball* for *alert, competent*—slang or informal. *Play ball* for cooperate—informal. *Balls* (as in testicles) for *courage*—vulgar slang but found in fiction and informal journalism.

balding Although vigorously condemned by its detractors, this adjective gains narrow approval from AHD.

barkeep For *barkeeper*—formerly slang, now standard.

bash Listed still as informal or slang, but meanwhile its old meaning, *strike a blow,* is fading before its use as a slang noun for *party.*

beam *On the beam* or *off the beam,* from a term for *radio guidance signal;* hence, *right* or *wrong, good* or *bad*—once slang, now informal.

beat Many slang uses, often verging on the informal: *beat up* for *thrash; beat out* for *defeat,* as in athletics; *beat down* for *force a seller to lower a price.* As a noun, *beat* is journalistic slang for either an exclusive piece of news or a responsibility regularly covered by a reporter. As an adjective, *beat* describes a literary movement of the 1950's, the *beat generation.*

bender *See* drunkenness.

better For *more than* ("He left here better than two hours ago")—informal, unacceptable in writing (AHD).

bilateral For *two-way, two-sided*—listed as standard, but overworked government jargon.

bit *Bit part* for a *small theatrical role*—standard; the related noun *bit* for a *kind of activity* or *behavior* ("do the intellectual bit")—a vogue informal expression. To *do one's bit* (do one's share)—a near-standard informal expression.

bitch As an intransitive verb for *complain,* and less frequently a transitive verb for *spoil, bungle* ("He bitched (up) the job") — slang and informal. *See also* woman.

black For *nonwhites,* especially those of dark-skinned African descent, has superseded the previously standard *Negro.* (*Black* is occasionally capitalized.) *Colored,* as a term for American blacks, still has scattered approval.

blimp For a *nonrigid airship* — formerly informal, now standard.

blow Has a variety of informal and slang uses, among them: for *depart* ("Let's blow!"); *blowhard* for *braggart; blow up* for *lose one's temper; blow in* for *arrive unexpectedly. Blow job,* vulgar slang for *fellatio,* is sometimes smuggled into written work.

blue For *despondent* — standard. For *risqué* — informal, despite years of use. *Blues* for *depression* or as a musical term — both standard.

blurb A coinage by Gelett Burgess, advertising and publishing jargon for a *brief, favorable publicity notice,* as on a book jacket or at the head of an article.

boggle Term of uncertain origin meaning *take alarm, be astounded* (as in the overworked phrase, "The mind boggles . . .").

bogus For *counterfeit* — standard. Also matter set by a union compositor, and discarded, that duplicates work that has been set by a nonunion shop — printing-publishing term.

bone up For *study intensely* — student slang, dated.

boner For *mistake* — slang, possibly informal.

boob All uses are slang: *stupid person; breast* (vulgar). *Booby* has the same meanings and standing.

booby hatch For *mental hospital* — slang.

booby trap For *concealed, triggered explosive* — once military slang, now standard for any situation catching a person by surprise.

boom For *rising prosperity* — standard.

boost, booster For *enthusiastic support* or *supporter* — once informal, now standard.

boss For a *workingman's superior* or for a *political leader* — all but standard. TCW cautions against formal use of *boss* in other contexts: "She's the boss in that family" or "electronics boss" for *executive* or *entrepreneur.*

bottleneck For *partial stoppage, obstruction* — World War II government jargon, then informal, now standard.

brain informal combinations: *brain child* for *original idea; brainstorm* for a *sudden idea;* and *brainstorming* for a *conference technique* designed to produce brainstorms (old Madison Avenue jargon). *Brain trust* for an *unofficial group of advisers* and *brain truster* for a member of such a troup — standard. *Brainwashing,* translated from a Chinese term for *indoctrination* — standard.

brass For *effrontery* — dying informal expression. For *military officers* or other types of officials ("corporate brass") — leaving slang for informal.

bread For *money* — slang.

break Most uses of the word, as verb or noun, are standard. Some exceptions: such sports jargon as *break* for the *path of a pitched curve ball* in baseball; and as an informal expression for *opportunity* ("What a break!").

broad *See* woman.

broke For *bankrupt* — informal. *Go for broke* for *stake everything on one effort* — slang.

brunch Once slang of British origin, now standard.

buck The following uses are informal: for *resist, fight* ("buck the odds"); *buck up* for *cheer up.* The following are slang: *buck* for *dollar* (old but still barred); for *lowest rank,* such as *buck private. Pass the buck* is sometimes listed as slang but is in good informal use. RHD lists *buck slip* for an *office routing slip* as standard.

bug For *insect*—standard. For a *disease-producing microorganism* ("flu bug")—informal. As a noun or verb for *secret listening device* or its use ("The agency installed a bug in the ambassador's heat lamp")—standard. As a verb, for *annoy, bother*—slang.

bulldozer As a *land-clearing tractor*—standard. *Bulldoze* for *intimidate*—slang.

bullshit Listed by AHD and RHD as vulgar slang, this term for *idle, exaggerated, uninformed* or *dissimulating talk* has been approaching informal status, especially among the young; it has been appearing as well in informal written contexts, and may ultimately displace the older, milder terms such as *bull* and *shoot the bull*. However, by that time it will be thoroughly overworked.

bum As a verb, for *beg* ("bum a cigarette")—informal. Used as an adjective for *inferior, malfunctioning* ("bum knee")—slang.

bunch For a *group of people*—informal.

bureaucrat Standard—and derogatory.

burgle Informal back-formation from *burglar*; *burglarize* is in standard or informal use.

bust Most uses of this vigorous term remain slang: for *break*, as a leg; for *burst*; for *go bankrupt*; for *collapse from effort* ("Pike's Peak or Bust!"); for *reduce in military rank*; for *tame* ("bust a bronco"); for *arrest*; and *bust up* for *separate*. The related *trust-buster*, however, though informal, is often used in standard written work.

butt in For *intrude*—informal. *Butt out* for *stop butting in*—slang.

buy As a noun, for *bargain*—substandard, except in commercial use.

cagey For *shrewd*—standard (AHD) or informal (RHD).

camp For "an affectation or appreciation of manners and tastes commonly thought to be outlandish, vulgar, and banal" (AHD)—standard. But slang when used as a verb, as in *camp it up* for *make an outlandish display* or for *flaunt homosexuality*.

can For *prison*, for *toilet*, for *buttocks*, or as a verb, for *dismiss* ("I got canned today!") or *record electronically*—slang.

careen For *move rapidly and out of control*—near standard, although a few authorities would still restrict *careen* to *lean* (as a ship) and use *career* for the other movement.

cat Slang—once used for *ill-tempered woman*, it is now more frequently used to mean a *man* or *fellow* ("That cat totaled his car").

catch on Informal—now less used to mean *understand* than for *become popular*.

cheapskate Slang.

chic For *stylish, style*—overworked.

Chicano For *person of Mexican descent in the United States* (short for *Mexicano*)—the term favored by those named by it.

Chinaman For *Chinese*—in poor taste.

chisel, chiseler For *cheat, cheater* ("He chisels on his income tax")—slang.

chortle A Lewis Carroll coinage, now standard for *chuckle loudly*.

cinch For *something certain* or *easy*—informal.

class For *merit, excellence*—standard. For *elegance* ("She has real class")—dated slang, as is *classy*.

clip For *rate of speed*—informal. *Clip joint* for a *place that overcharges customers*—slang.

close call For *narrow escape*—informal.

clout For *blow, hit, punch*—informal or slang. For *power*, especially government or political—jargon.

co-ed For *woman college student*—dated informal expression.

colored *See* black.

combine (Pronounced *com'bine.*) As a noun for *harvesting machine*—standard. As a *combination of persons* or *groups*—informal (RHD) and slightly derogatory.

come Slang and informal uses: *come across* for *do something demanded; come again* for *repeat; come down on* for *scold; come through* for *do as expected; come up with* for *produce;* and *come* for *experience orgasm* (vulgar).

commie, Commie For *Communist*—informal (AHD), derogatory.

-complected As in *light-complected*—regional, not in standard or good informal use. The correct form is *-complexioned.*

con For *swindle*—slang; similarly with *con* (confidence) *game.*

confab For *informal meeting*—informal, sometimes used in headlines.

confrontation For *facing of hostile forces*—overworked, after vogue use in the Cold War, in contexts where *debate, argument,* or *showdown* might serve better ("We had a confrontation with the dean this morning").

congressman For *member of the U.S. House of Representatives*—standard in all but the most formal contexts.

connection For *source of narcotics (The French Connection)*—slang, but also an overworked fad word.

consummate For *perform, accomplish*—standard, but pretentious and often misplaced ("to consummate a deal").

contact As a verb ("He'll contact me next week"), opposed by some authorities. Nonetheless, because it is less cumbersome than such equivalents as *get in touch with, contact* is rapidly becoming standard. The noun *contact,* for *source of assistance,* is already standard.

cool For *full, entire* ("a cool million")—informal. For *excellent,* often an exclamation ("Cool, man!"); for *composure* ("keep one's cool"); and *cool it* for *calm down*—slang.

cop For *policeman*—informal, and disliked by policemen. As a verb, *to steal*—slang.

cope Without *with* ("I am no longer able to cope")—semi-humorous informal expression rejected by the AHD usage panel for written work.

cop out As a verb, to *evade* or *avoid a commitment* ("The candidate copped out on the issue"); as a noun, *cop-out*—slang.

crack-up For *collision* or for *mental* or *physical breakdown*—near-standard.

crash For *intensive* ("crash program")—informal.

cuckoo For *silly, idiotic*—standard.

cup For *cupful* ("Take two cups of sugar")—informal. Also *baskets, tablespoons, teaspoons* in the same sense.

czar For *person in authority* ("liquor czar"; "czar of baseball")—semi-informal, dated, but still getting too much work.

dead For *tired*—informal, as is *deadhead* for *person riding on a free ticket* (dated). *Deadbeat* for *one who doesn't pay his bills,* and *deadpan,* adjective for *with no facial expression*—slang.

deal For a *degree, quantity* ("a great deal of money"); for a *transaction;* or for *treatment received* ("shabby deal")—all informal. *Make a big deal of*—slang. *New Deal, Fair Deal*—standard.

debut As a verb ("The pianist will debut tonight")—substandard.

debunk For *expose sham*—informal.

demise Often pretentious, should be used only for the death of a ruler or a person transmitting great wealth.

dialogue For *exchange of views*—overworked and stretched beyond its meaning of including only two parties ("The five nations carried on a dialogue").

do Slang and informal uses: *do time* (serve a prison term); *do out of* (cheat out of); *do in* (kill); *done in* (killed or tired); *do over* (redo).

doll *See* woman.

dope For *narcotic, marijuana*—in wide informal use, but sometimes acceptable in standard use, as in the verb *dope* for *administer narcotics* ("The horse was doped"). Slang uses and variations are numerous: *dope* for a *stupid person; dope* for *information; dopey; dope out; dope sheet* (for horse-racing); *dope fiend,* old term for narcotics user, revived as users' argot.

drag The old slang use—for *influence*—has almost vanished. More current slang uses include: for *something tiresome* ("What a drag it is to go to work every day"); for a *puff* (smoking); *in drag,* of a man, for *dressed as a woman.*

dream Adjective, as in *dream house, dream team*—overworked.

drunk As a noun for *drunken person* or *spree*—ranges from slang (TCW) to standard (AHD, RHD) in use. The terms *drunk driving* and *drunk driver* are now substandard, but may become standard, in the same way that *iced cream* became *ice cream.*

drunkenness The long list of slang terms associated with drunkenness are for the most part semi-jocular in tone and have long since lost much of their original force: *bombed, boozed up, crocked, fried, half-crocked, lit up, pickled, sloshed,* or *tanked up.*

dude Formerly, an *Easterner,* as on a *dude ranch,* or a *fop;* now any *man* or *fellow*—possibly ephemeral slang.

easy Can be used informally as an adverb in such a phrase as "Take it easy." But most adverbial uses ("I did it easy") are illiterate.

egghead For *intellectual* (from alleged hairlessness)—informal or standard; in use since Adlai E. Stevenson popularized the term in his 1952 presidential campaign.

emote Back-formation from *emotion*—informal, theatrical.

enthuse Back-formation from *enthusiasm*—AHD rejects it for standard use; RHD agrees but says that it is sometimes found in written work.

escalate Back-formation from *escalator,* for *intensify a war* (de-escalate is the reverse)—the AHD usage panel found *escalate* acceptable for written work, both as a transitive and an intransitive verb; overworked now, it may fall into disuse with the end of its applicability to the war in Indochina.

establishment, Establishment For *group with power to control a given field,* or for *the legitimate authority*—overworked.

everyplace *See* anyplace.

ex For *former* ("ex-president"; "ex-chairman")—in near-standard use; nonetheless, *former* remains preferable. *Ex* as a noun, shortened from *ex-wife* ("She's his ex")—slang, perhaps ephemeral.

exodus For *movement away, departure* ("holiday exodus")—overworked. Particularly jarring in such a phrase as "an exodus of Arabs from the West Bank."

extend For *send, give* ("Please extend my greetings to him")—falsely elegant.

factor For *component, element*—overworked. Even *part* would sometimes be preferable.

fall A versatile word in informal expressions: *fall down* for *prove unsuccessful; fall flat* for *fail;* and *fall for* for *become infatuated with* or for *be taken in by. Fall guy* for *scapegoat, victim*—slang.

fake Most uses have become standard, with the exception of a few slang uses such as *fake it* for *improvise,* from jazz.

famed Usage given impetus by *Time* magazine; now listed as standard in AHD and RHD.

fan For *devotee*—informal-to-standard, and rarely associated with its source, *fanatic.*

fault As a verb, for *criticize* ("It is hard to fault his persistence")—an old use, but one that has been only reluctantly accepted as standard.

faze For *disconcert*—informal, perhaps standard, less commonly used in its other forms—*feeze* or *feaze*.

feature As a verb, for *give prominence to*, as in a newspaper—standard in AHD, and RHD, but unacceptable to many editors.

feel *Feel like* for *be inclined to* ("I feel like leaving")—informal. *Feel out* for *ascertain a view indirectly*—also informal. *Feel up* for *caress*—vulgar slang.

feminine endings The *ette*'s and *ess*'s and *trix*'s that have long been attached to English nouns are falling into disuse. Even before feminine endings had started to disappear, editors and writers recognized the illogicality in calling Amy Lowell a *poetess* and James Russell Lowell a poet. *Sculptress, aviatrix, promptress, directress,* and *editress* are little used. *Suffragette* is a misnomer; the right title is *suffragist. Actress, waitress,* and *(drum)majorette* survive, but may not do so forever. A few usages are fixed: we will probably never call her Diana the Hunter.

fiend For one *addicted to a vice*, often jocular ("camera fiend")—informal. *See* dope.

figure For *conclude, predict*—informal, as in *it figures* ("It figures to be a rainy day").

finalize For *put in final form*—associated with bureaucratic jargon, and rejected by most authorities, especially in the light of such ready substitutes as *complete* or *conclude*.

fink For *strikebreaker* or, by extension, any *unworthy person;* also *fink out* for *abandon a commitment*—both slang. *Fink* as a verb for *inform* is in declining use.

fire For *discharge from a job*—informal; listed as standard in RHD, but rejected elsewhere for standard use.

fix Its standard use as a noun is for *predicament* ("in a fix"), but it is used also as a term for a *drug dosage*—slang. As a verb, its uses all appear to be close to standard, the closeness depending on whether one is fixing a clock, an engagement, a photographic negative, a cat, a jury, or blame.

flammable Synonymous with *inflammable*, and perhaps better because less subject to misunderstanding.

flip For *react strongly* ("You'll really flip!")—slang. *Flip* for *flippant*—slang or informal.

flunk, flunk out Durable student slang, nearly informal.

folks For *members of one's family* ("Meet the folks")—informal; TCW condemns all uses of *folks,* and calls the term a "casualism."

fraction For a *small part*—gaining standard use, but the writer should remember that a fraction can be as large as four-fifths or even ninety-nine-one-hundredths.

frazzled For *frayed* or for *worn out, nervous*—informal.

freak out For *react excitedly* or *lose touch with reality,* as with a hallucinogenic agent—slang. The noun form is *freak-out* or *freakout.*

freeze As a noun, for *economic control,* usually official ("price freeze")—standard. As a verb used in the same sense or for *render impossible of liquidation* ("freeze bank accounts")—also standard.

frisk For *search*—once slang, now standard.

function For *public* or *social ceremonial occasion*—standard, but too formal for most social events such as parties or dances.

funk For *fright*—informal. *Blue funk* for *depression* and *funky* for a *blues-like quality* in jazz or popular music—slang.

funny For *odd, curious* ("What a funny thing to say!")—standard in dictionaries, but TCW still regards it as informal.

gadget For *contrivance*—informal, becoming standard.

game For *vocation, business* ("the advertising game")—informal.

gangster Standard, after an informal start.

gay   For *homosexual*—slang in AHD and RHD, but in increasingly reputable informal and written use, because of its relatively neutral tone.

get   Informal-slang uses: for *retaliate against* ("I'll get you!"); for *annoy* ("Loud music gets me"); also *get around* for *convince with flattery* or for *be worldly;* and *get away with, get back at, get there, get with (it), get pregnant, get religion.*

go   For *ready, functioning* ("All systems go")—American space jargon.

goat   For *scapegoat, victim, villain* ("The shortstop was the goat in today's loss")—informal.

goner   For *person or thing dying*—informal.

good   As an adverb for *well* ("The car is running good today")—rejected for standard use; its closest approach to respectability is in the phrase *going good* ("Things are going good").

goof   For *stupid mistake* now more than for a *stupid person*—both still slang.

goon   For *hoodlum hired to commit violence,* especially in labor disputes— informal, as is *goon squad.*

gouge   For *swindle, extort from*—slang (AHD) or standard (RHD).

grab   For *seizure by unscrupulous means* especially "land grab"—not challenged in AHD or RHD. *Up for grabs*—informal.

graduate (v.)   "He *graduated* from college" and "He *was graduated* from college"—both forms standard.

graft   For *dishonest gain*—standard.

gripe   For *complaint* or for *complain*—informal.

groggy   For *dazed*—informal, near-standard, but no longer used in its original sense of *drunk.*

groom   For *bridegroom*—standard, now that there is little chance that the *groom* will be mistaken for someone who cares for horses; *bridegroom* is still preferable in formal contexts.

grouch   For *irritable person*—informal (RHD) or standard (AHD). Rarely used any more as a verb.

gumshoe   For *detective* (from silent rubber-soled shoes)—slang, in use since the 1920's, but fading.

gush   For *effusive language*—formerly informal, now standard.

guts   For *courage*—slang. *Gut* for *snap* course—Eastern U.S. college slang.

guy   For fellow, man ("He's a good guy")—informal, fading.

gyp   Noun or verb for *swindle*—informal. *Gyp joint*—slang.

half-breed   For *person with parents of differing ethnic types,* especially Caucasian and American Indian—standard, derogatory.

ham   For *actor given to over-acting*—long-standing slang; for *amateur radio operator*—informal.

hang-up   For an *inhibition* or for *compulsive behavior* ("School is one of Tommy's hang-ups")—informal.

hard-hat   For *required headgear in construction work* ("This is an all hard-hat job"—sign at a building project); by extension, a *construction or other blue-collar worker*—informal.

hardware   For *military weapons*—informal, adopted from Pentagon jargon. For *computer equipment*—standard.

hassle   As a noun, for *trouble, annoyance, fight, argument* ("Getting there is such a hassle"; "They gave me a real hassle about the car")—informal. As a verb, for *argue, annoy, aggravate* ("Look, don't hassle me about it")—informal.

heap   For *run-down car*—musty informal expression. *Heaps* for *a lot* ("Thanks heaps")—informal.

help but   *Cannot help but* for *unable to avoid* ("You cannot help but see it")— established informally, but condemned by AHD for written use.

highbrow   For *elite tastes*—informal, becoming standard with use in educated writing, as are *middlebrow* and *lowbrow.*

highfalutin, hifalutin, highfaluting For *pompous, pretentious*—informal, but in declining use.

hike All uses are listed in AHD and RHD as standard, but as a noun for *increase* ("pay hike") it is still informal in tone.

hijack For *steal in transit*—in use since Prohibition and becoming standard. *Skyjack*, sometimes used for *hijack* when referring to the *seizure of an airplane in flight*—remains slang.

hindsight Once a humorous informal expression, now standard but overused.

hit For *success* ("The play was a hit!")—standard. But *hit* for *beg money*, as well as *hit the road* (leave, depart), *hit the bottle* (drink), and *hit it off* (get along well)—informal.

hitchhike Standard.

hog As a verb, for *take more than one's share*—slang (RHD) or standard (AHD). Informal uses include *go the whole hog* and *live high off the hog*.

holdup For *armed robbery*—standard (AHD) or informal (RHD); by extension, an *overcharge*—informal.

homely For *unattractive*—standard.

homey For *comfortable, homelike*—informal (AHD) or standard (RHD).

honky, honkie For *white man*—derogatory slang, related to *hunky* and *bohunk*, earlier slang terms for *immigrant laborer*.

hoodlum For *thug*—standard. *Hood*, widespread shorter form, remains slang.

hooker For *prostitute*—slang, with some informal written use.

hootenanny Usually a *folk-music concert*—standard.

hopefully For *it is hoped* ("Hopefully, he arrived without incident")—still resisted by AHD, this use is becoming standard nonetheless because it is more economical than its equivalents.

ice Adjective, for *frozen* or for *with ice added*. The terminal "d" has been dropped from *iced cream* and *iced water* and is vanishing from *iced tea*.

image For *character projected to the public* ("The candidate sought a law-and-order image")—standard, but an overused vogue word.

implement As a verb, for *carry out*—standard but often used pompously.

individual For *person* ("an agreeable individual")—not unanimously accepted for standard use; most authorities prefer *individual* as a noun in contexts where the one is compared with the mass ("The individual in contemporary society").

input From data-processing technology, a fad word meaning *effort, contribution* ("My imput in that class was too low").

insane asylum The accepted term is *mental hospital*.

inside of *Of* should be omitted in formal writing, but such phrases as "inside of an hour" are in good informal use.

integrate For *eliminate racial segregation*—now standard in this use and obliterating earlier, more general meanings.

intrigue As a verb, for *arouse interest* ("That painting intrigues me")—in informal use since the 1920's but rejected by authorities for written work; recent opinion seems evenly divided.

invite As a noun, for *invitation*—informal, illiterate.

irregardless For *regardless*—weakly humorous or illiterate.

jag For *drinking bout* or other spree ("crying jag")—informal-to-slang.

jam For *predicament* ("in a jam")—informal; but *traffic jam* is standard.

jaywalk, jaywalker Standard.

jazz For *exaggeration, hyperbole, nonsense* ("all that jazz")—slang, as is *jazz up* for *embellish, enliven* ("They jazzed up the room with red paint"). But strictly musical uses of *jazz* are standard.

juke box For *coin-operated phonograph*—standard.

keel over   For *faint and fall*—standard, as is the original use, capsize.

kibitzer   For an *onlooker at a chess or card game who offers unwanted advice, meddler*—informal, as is the verb, *kibitz*.

kick   Informal and slang use as a verb: for *protest; kick around* for *treat roughly; kick in* for *contribute one's share; kick out* for *eject; kick the bucket* for *die; kick up, kick up a storm* for *make trouble; kick up one's heels* for *enjoy oneself freely. Kick* is also used as an informal and slang noun: for a *complaint;* for *power,* as in an engine; for an *intoxicant's impact,* as in a martini; for *feeling of excitement* ("I get a kick out of you").

kid   For *child*—informal, as is *kiddie. Kid* as a verb for *fool* ("Don't kid me") —informal.

kill   This word has many nonliteral meanings, but almost all of them are accepted as standard except, possibly, for *finish* ("kill a bottle of whiskey") or *killing* for a *successful deal* ("make a killing in the market").

kind   AHD condemns *kind* as a plural ("those kind of books") but approves the interrogatory form ("what kind of books are those?").

knock   Another word with many informal uses: for *disparage; knock off* for *stop work,* for *consume* (as a meal), for *deduct* (as from a bill), or for *kill. Knock out*—informal when used for *defeat* in general or for *exhaust oneself. Knockout,* a noun for *attractive person* or *thing* ("She's a knockout!")—dying slang.

kudos   Noun, construed as singular, for *prestige* (from Greek for glory)—originally British university slang, popularized in the U.S. by *Time* magazine. The form *kudo* does not exist in good usage.

lady   *See* woman.

lame duck   For *defeated officeholder serving out term*—standard.

lawman   Rejected in TCW as imprecise when specific officers such as sheriffs or policemen are meant.

layout   For *arrangement of elements on a printed page*—standard. For *residence, apartment, quarters* ("Quite a layout you have here")—informal.

learn   For *teach* ("The teacher learned him history")—an old use, now illiterate.

leave   For *let*—accepted as standard only in the case of a pronoun followed by *alone* ("Leave me alone"); not for "Leave me be" or "Leave us not misunderstand."

leery   For *suspicious*—informal.

let   Most uses are standard, and even those that are still listed as informal such as *let up* for *slacken, let on* for *reveal,* or *let up on* for *be lenient* are not far from standard. *Letdown* for *disappointment* or for *slowing down* also has become standard.

level   *On the level* for *truly* or *honestly,* and *level* as a verb ("Level with me")—informal.

like   For *as, as if*—rejected by a consensus of authorities (AHD). Many will accept "It looks like new" but not "It looks like it was new"; both uses are informal.

likes of   *The likes of* ("It was a sunset the likes of which is rarely seen in these parts")—condemned as a casualism, inappropriate for standard use.

lip   For *insolent talk* ("None of your lip")—slang; for *mouth* ("Button your lip")—slang.

liquidate   For *kill*—standard, but imprecise.

loan   For *lend*—accepted in U.S. business usage, but considered inferior to *lend* in such use as "I lent the money to a friend."

lobby   For *seek to influence legislators*—standard.

locate   For *situate* ("The house is located near town")—near standard, although it can be omitted in many instances. For *settle, establish oneself* ("He located in midtown")—informal.

loony For *odd, peculiar, demented*—informal.
louse For *contemptible person; lousy* for *contemptible*—slang.
lowbrow *See* highbrow.
lulu For *payment to legislators in lieu of itemized expenses* (lieu+lieu)—slang. For something *remarkable* or *extraordinary* ("His alibi was a lulu"; "it was a lulu of a performance")—slang.

mad For *angry*—informal, but often used in standard written work. For *enthusiastic* ("I'm mad about photography")—informal.
make Informal uses include: for *arrive in time* ("make a plane"); for *gain a position* ("make the team"); *make book* for *take bets; make it* for *achieve* a goal ("make it through school"); mildly vulgar for *have sexual intercourse; make up to* for *fawn on.*
marijuana A constantly recycling collection of slang words are used: *hemp, gage* or *gauge, muta, m.j., mary jane, weed, pot, grass, dope, stuff.* "Do you smoke?" does not necessarily refer to tobacco.
massive Vogue word, frequently used in contexts where *large* or *imposing* would do ("I have a massive pile of papers on my desk").
materialized For *happen* or *occur* ("What else materialized during the proceedings?")—substandard and pretentious. But accepted as standard (AHD) for the more specific *take form, take shape* ("The project materialized in a short time").
mean For *bad-tempered* ("I think you're mean") or for *in poor spirits* ("I'm feeling mean")—informal. *Mean for* for *intend that* ("He means for you to go")—substandard (AHD).
media For *means of mass communication,* singular ("Television is my favorite media")—in increasing use, but not accepted for standard or informal use by AHD. The accepted singular is *medium;* the plural is *media* or *mediums;* thus *mass media* refers to all the means of mass communication.
medic, medico For *doctor, medical student*—informal.
menial For *servant*—standard, but now considered derogatory.
middlebrow *See* highbrow.
mind For *obey* ("Mind your mother") or for *be cautious about* ("mind your step")—standard.
mini- Combining form for *something distinctively smaller than other members of its class*—standard but overused, not only employed in such coinages as *miniskirt* but stretched to include anything the writer considers less than full scale: *mini-concert, mini-holiday, mini-republic, mini-estate.*
minus For *without* ("He returned minus his hat")—informal, quasi-humorous.
miscegenation For *breeding between races.* A linguistic coincidence has caused this word to be misinterpreted, so that it is now often considered as derogatory: the prefix is not *mis* for *mistaken, wrong,* but *misce* from Latin for *mix.*
miss As a noun, form of address for *woman unmarried* or *of unknown status* —informal when not accompanied by a name ("Miss, may we have the check?"). Best avoided in written work ("a pretty young miss").
mixed blood For of *diverse ancestry*—scientifically inaccurate and somewhat derogatory. *Mixed marriage* for *interracial* or *interreligious marriage*—standard.
mixer For a *sociable person* or for a *party*—dated.
mob, the mob For *gang, crime, syndicate*—informal, as is *mobster* for a *member of a mob.*
moonshine For *illegally distilled liquor*—durable slang of Southern origin. Also *moonshiner.*
mortician Originally a euphemism for *undertaker.*
most For *almost* ("He had most everything")—substandard. For *very* ("They were most hospitable")—standard, but better in speech than written work.

mostly   For *for the most part* ("The campaign workers, mostly students, watched in silence"); should not be used to replace *most* (("You will be the one mostly helped").

motorcade   Adapted from *cavalcade*, a parade with horses, to describe a *procession of automobiles*—now standard, as is *aquacade* for a *swimming exhibition*.

movie, movies   Still somewhat informal, but RHD lists such derived terms as *moviegoer*, *movieland*, and *moviemaker* as standard. *Movieola*, a film-viewing device, is a trademarked term, capitalized.

nerve   For *effrontery*—informal.

news   Several coinages using *news* are becoming standard: *newscast, newsletter, newsmagazine* (sometimes still two words), *newsman, newsroom, newsprint* for wood-pulp paper. Newspaper officials object to *newsboy* and have tried unsuccessfully to substitute for it *newspaperboy*.

nice   Nearly all uses have become standard, but the word is still used more informally than in standard writing, partly because it is bland and imprecise— a *nice* coffee pot, a *nice* day, a *nice* man. It means something only when it is used as a synonym for *exact, discriminating*, as in "a nice distinction."

normalcy   Although Warren Harding was ridiculed for using the word, *normalcy* was, and is, a standard term, a possibly unnecessary alternative to *normality*.

nostalgia   Its use has expanded beyond the original meaning of a *longing for a former place or state* to cover any *wistfulness* or *melancholy* ("I felt a great nostalgia when I first visited the seashore"). The term is thus weakened.

nowhere   *Get nowhere* and *nowhere near* ("This is nowhere near as good a saw as that")—informal.

nut   For *eccentric person*—slang. *Nutty* for *eccentric*—informal.

off of   For *off* ("Get off of the curb!")—informal. For *from* ("I won five dollars off of him")—substandard.

OK or **okay**   AHD leans toward approving the term's use as a noun in business ("Please get his OK on this") but views with less favor the verb form ("Please OK this") and even less, the adjectival or adverbial ("Is everything working OK?"). Informal use remains widespread.

Okie   For Oklahoman—used disparagingly of poor migrants of the 1930's, but now accepted by Oklahomans, as in the song, "I'm proud to be an Okie from Muskogee."

operation   For *military, naval effort* ("Operation Overlord," for World War II invasion of Europe)—standard but trivialized by publicity use: "Operation Clean Streets"; "Operation Bargain Days."

orate   Back-formation from *oration*—close to standard.

orbit   As a verb, for *place in orbit* ("U.S. Orbits Sun Satellite")—a relatively new use, becoming standard alongside the older transitive use, as in "The moon orbits the earth."

out   Informal uses include: *out from under* for *freed from a burden or danger; on the outs* for *at odds with; out of sight* for *superior, excellent*, the last an old term enjoying a revival.

outside of   For *outside*—accepted in informal use. For *except* ("Nobody knew outside of him and me")—substandard.

over   For *more than* ("over five cents")—gaining standard use, but should not be associated with reduction: "A loss over last year's totals . . ."

overkill   As a noun, for *excessive action undertaken as a reaction to something* ("In the face of this criticism, his paintings began to show signs of technical overkill")—standard.

overly   Accepted as standard by AHD, but many writers object to it.

pad As a noun, for *one's apartment* or *one's own room* ("He had this pad like out of *2001*")—slang. As a verb, for *adding extraneous matter to make larger or longer* ("padding his report"; "padding her expense account")—standard.

parson For *clergyman*—provincial.

partial to For *fond of* ("I'm partial to cheese")—standard, but clashes somewhat with other uses of *partial*.

party For *person* ("That certain party of mine"—old song)—informal, stilted; for *person* as a legal term—standard. As a verb, for *attend parties* ("I'm going partying")—informal.

pass away For *die*—standard, but a euphemism. The similar Christian Science term *pass on* stems from religious doctrine.

patron For *customer*—commercialism.

paw (over) For *handle someone, something* ("Must you keep pawing over my books?")—informal.

payola For *under-the-table payments, especially for playing records on the air*—slang, becoming informal.

peeve For *annoyance*—back-formation from *peevish*, approaching standard use.

pep For *animation, energy*—informal and fading.

per An old rule declared that *per*, a Latin word, could precede only another Latin word, such as *cent* (for *centum*) or *capita*. Now *per* is accepted as standard for general use, as in "the price of eggs per dozen."

percent, per cent The term is becoming one word, and RHD lists it as one. *Percent* as a noun ("He offered me a percent of the profits") remains inferior to *percentage*.

-person Advocated by some as a replacement of the suffixes *-man* or *-woman*, when the sex of the functionary has not been previously specified (as *chairman, chairwoman, chairperson*).

phone As a noun or verb, for *telephone*—still listed as informal but actually standard.

phony For *fake, fraudulent* ("He spouted phony learning")—informal (RHD); standard (AHD).

photo For *photograph*—very close to standard.

piece For *distance* ("down the road a piece")—regional. For *woman* or for *intercourse*—vulgar slang.

pig For a *gross person*—informal. For *woman*—offensive slang. More recently, for *policeman, other authority*—derogatory slang.

pill *The pill*, for *oral contraceptive*—informal, with some standard use.

pitch For *sales talk*—slang, becoming informal.

plug For *promote, advertise covertly*—informal. Also, as a noun ("He got a plug on the late late show").

pretty For *somewhat, moderately* ("I feel pretty well now") or *very* ("It's raining pretty hard")—listed as standard, but informal in tone.

proceed For *go*—pretentious if used in simple contexts ("We proceeded to the stadium").

proportions For *size* ("The proportions of the ship were imposing")—accepted by a majority of the AHD usage panel, but resisted by those who wished to preserve the meaning of *proportion* as indicating a relationship between parts and the whole.

proposition For *enterprise, task* ("a challenging proposition")—informal. For a *dubious or immoral proposal* and as a verb ("Are you propositioning me?")—informal, and somewhat slangy.

prospect For *prospective customer*—once business jargon, now standard.

proven For *proved* (the preferred form)—widely used by many authorities as standard. It is favored slightly less as a participle than as an adjective ("A proven technique").

publicist For *press agent, public-relations person*—standard, and displacing the former meaning, *one who writes on public issues.*

pull For *influence*—slang. Informal uses as a verb include *pull* for *attract* ("pull a good crowd"), *soften* ("pull his punches"), *carry out* ("pull a stunt"; "pull off a deal"), and *draw out* ("pull a gun").

put Slang and informal uses: *put across*, for *carry out, execute*; *put down*, for *humiliate, criticize* (also *put-down*, a noun); *put-on*, as a noun, for *ploy, deception*, as an adjective, for *affected*; *put over on*, for *deceive, trick* ("He put one over on the manager"); *put up to* for *incite*; *stay put* for *remain in place.*

queer For *jeopardize* ("You'll queer the whole deal")—informal, near-standard. For *homosexual*—slang, derogatory.

quick For *quickly*—rejected for written work by AHD except in the form "Come quick."

quite As a synonym for *somewhat* or an *indefinite quantity* ("quite a few")—standard, as it was in its older sense of *totally* ("quite out of breath"). *Quite a* for *extraordinary* ("Quite a sight!")—informal.

quiz For *interrogate* ("The detective quizzed the prisoner")—narrowly accepted as standard. For *inquiry* ("Senate rackets quiz")—journalese.

quote For *quotation*—widely used, even in written work, but rejected for standard use by AHD.

racket For a *business*, humorously ("I'm in the dry-goods racket")—informal.

railroad For *force through quickly* or for *send to jail without fair trial*—both uses informal.

raise For *pay increase*—now standard.

rake-off For *share* (usually illicit)—slang, as persistent as the practice it refers to.

rap For *talk discursively*—slang of the 1960's, perhaps ephemeral. *Rap* for *blame*—*beat the rap, bum rap, take the rap*—also survives as slang.

rate For *deserve* ("He rates a medal") or for *have status* ("She rates high in scholarship")—both informal and both rejected for standard use by AHD.

rattle For *unnerve* ("Banks rattle me"—Stephen Leacock)—informal.

real For *very, really* ("It's a real nice day")—widespread regionalism.

reason why For *reason that* ("She would not tell me the reason why she did it")—standard. But not so the *reason why...because* construction ("The reason why I cannot go is because my mother is ill"). Substitute *that* for *because.*

relative, relation For *person akin to another*—dictionaries regard these terms as interchangeable, but the taste of authorities runs to *relative.*

reminisce Back-formation from *reminiscence*—now standard.

rendition For *performance*, as of music—standard, despite growth in meaning to cover not merely interpretation but the whole performance. Often the word is slightly pompous.

rile For *vex*—has moved from provincial to standard, according to AHD.

rip off As a verb, for *steal* or for *rob* ("First her guitar was ripped off, then they ripped off her apartment")—slang. As a noun, for *something that robs, cheats, or takes advantage of*, in terms of money, expectations, beliefs, and so on "That movie was a real rip off")—slang.

rooftop For *roof*—not strictly logical, but dictionaries becoming standard, especially in reference to flat roofs.

rookie For *recruit* or *first-year athlete*—slang or standard, depending on the use and authority.

row (Pronounced *rau*.) The standard use is *noisy brawl*; it should not be stretched, as it is occasionally by reporters, to cover any kind of disagreement, such as a political argument.

rubberneck   As a verb for *stare*, especially used in reference to people who slow down to see wreckage after an automobile accident — near-standard. As a noun for *sightseer* — slang.

said   For *aforementioned* ("Said tenant no longer resides at said address") — legal or business jargon.

same   For *aforementioned person, thing* ("I have written to the person you mentioned and have sent the package to same.") — legal-commercial jargon.

scab   For *strikebreaker* — long informal use; RHD lists as standard.

schoolteacher, schoolmistress, schoolmarm   Three similar terms, all listed as standard but now considered slightly derogatory. *Teacher* will suffice.

Scotch   As a noun, appropriate term for Scotch whiskey, the people of Scotland collectively, or the dialect spoken in Scotland. For individuals, Scotsmen prefer *Scotsman. Scottish* and *Scots* are the preferred adjectives.

scrounge   For *forage, borrow* ("Let's scrounge some supper") — slang (AHD) or standard (RHD).

sculpt   It sounds like a back-formation from *sculpture,* but has been in independent use for years. Although it is listed as standard, TCW prefers *sculpture* as the verb.

service   As a verb, for *serve* ("We are proud to service this community") — commercialized standard. The verb *service* is more accurately confined to senses involving actual maintenance.

shake-up   For *abrupt change in personnel* — in use since the 1880's in newspaper offices; now nearly standard.

ship   For *send by public transport* — informal, near-standard. For *send* ("Let's ship all the kids away for the summer") — slang.

shoplifter   Standard.

show   For *appear* ("He said he would come, but he didn't show") — informal, formerly *show up. No-show,* for a *booked passenger who does not appear* — airline jargon.

sideswipe   Formerly opposed by those who thought the syllable *-swipe* sounded informal; now considered standard.

sight, out of   See out.

skyjack   See hijack.

slick   For *popular magazine printed on glossy paper* — informal.

slow   For *slowly* — standard. Accepted uses include *drive slow, go slow,* my watch runs *slow* (an adjectival form).

slumlord   From *slum + landlord* — colloquial and in informal written use.

smog   From *smoke + fog* — standard.

sniffy, snippy   Both informal.

snooze   For *nap* — informal.

software   For *written or printed data,* especially for use in computers; contrasted with *hardware,* the machinery. Capable of misinterpretation when applied outside its own field.

some   For *remarkable* ("Some pig!" — E.B. White, *Charlotte's Web*) — informal. For *somewhat* ("The town has grown some") — substandard.

someplace   See anyplace.

sound out   For *test opinion* ("Let's sound out the board of directors") — standard, according to AHD.

spell out   For *state in detail* ("Can you spell out the requirements for admission?") — in good use, but now has nothing to do with spelling. *Spell out in detail* or *spell out the details* — redundant.

spook   For *ghost* — informal. For *frighten* ("The flashing lights spooked the horse") — slang. For *intelligence agent* — jargon.

square   For *conventional, unadventurous* ("He's so square he's cubed" — Walt Kelly, *Pogo*) — slang, almost informal.

stag For *all-male* ("stag dinner") — informal and probably dying.

stick Informal uses include *stuck on* for *reach an impasse* ("I'm stuck on fifteen across"); *stick around* for *stay, remain; stick it out* for *persist. Stick up for* for *support* — juvenile (TCW).

still and all For *still* — dialectical.

stoned For *under the influence of a drug* — slang.

straight For *conventional, law-abiding* — slang. For *not homosexual* — slang.

streamlined For *improved, modernized* — standard, but often stretched to inappropriate contexts ("A streamlined jury system").

suicide As a verb, for *commit suicide* — informal, not in wide use.

swap For *exchange* — informal, nearly standard, but seldom appropriate for formal contexts ("swapping prisoners of war").

take *Take it* for *endure; take it lying down* for *submit, give in; take it out on* for *relieve oneself of anger or frustration by venting it on another* — informal. *Take it on the chin* for *endure a specific blow* — slang.

teen Referring in general to the ages from 13 to 19. Used in *teens* and *teen-ager* — both standard; *teen* itself is turning up as a noun for *teen-ager.*

thusly For *thus* — non-standard.

tip-off For *warning* — informal. *Tip* for *advance information* — standard.

too When used in understatement ("Father was not too pleased with my grades," meaning he was not pleased at all) — standard. But *not too,* in the sense of *not very* ("I'm not too likely to be home by suppertime") is considered too casual for standard use by AHD. *Too* and *not too* can be dangerously ambiguous: "He was not too diligent for a doctor," for example.

total As a verb, for *destroy, demolish* ("Tony totaled the station wagon when he missed the turn") — slang.

transpire For *happen* — still substandard (AHD); the preferred use is for *come to light:* "It transpired that Sweden had a treaty with Tibet."

trigger For *set off* ("The bad news triggered a market slump") — the widened meaning is standard, but suffers from overwork.

trip As a noun, for an *extended hallucination produced by a hallucinogenic drug* or, by extension, for an *intense experience accompanied by a mental or emotional reaction* ("Seeing him again was a real trip") — slang. As a verb, for *experiencing such a hallucination or intense experience* ("They were tripping on LSD"; "He could really trip on that music") — slang.

truculent For *defiant, surly* — now standard.

try As a noun for *attempt* ("The old college try") — general but not unanimous approval as standard. *Try and* for *try to* — informal.

turn on For *give pleasure to* or for *use a drug* — slang, sometimes ambigious.

tycoon For *magnate, big-businessman* — informal, from the Japanese, approaching standard.

type For *sort.* AHD considers "That type of work..." acceptable, but not "That type work..." *Kind, style,* and *sort* are often preferable in such contexts. *Type* as a part of a class or group ("He was a professor type") — informal, overused. The combining form *-type* tends to be introduced informally and needlessly, as in a "ranch-type house" or a "musical-type production."

upcoming For *forthcoming, approaching* — rejected for standard use by AHD.

up tight For *tense, inhibited* — 1960's slang.

very Often used unnecessarily. For a discussion of *very* with participles, see page 372.

way For *far* ("Chinese surgery is way ahead of ours") — near-standard. But *a long ways* is substandard.

weed For *tobacco* — informal. *See also* marijuana.

where As a conjunction, for *that* ("I see by the papers where we're having a wet spring") — substandard. But sometimes *where* can be used for *when* "Trusting him was where we made our first mistake").

whitewash For *exonerate falsely* ("The report whitewashed the police department") — standard. For *keep from scoring* — sports expression.

whitey For *white person* — derogatory slang, contemptuous form of address.

wildcat strike For *unauthorized strike* — now in standard use.

win As a noun, for *victory* — accepted for sports stories but too informal for serious use elsewhere ("The South scored a win over the North at Bull Run").

wino For *wine addict* — slang.

-wise For *with reference to* ("Moneywise, I'm broke"; "Marketwise, it offered a real opportunity"). The formation is associated with commercial jargon and is substandard in other contexts ("Wisewise, how's he doing?" — an owl in a *New Yorker* cartoon).

women The use of *girl-woman-lady* by editors and writers has undergone a revolution in recent years. Blacks have long pointed out that using *girl* or *boy* to refer to adults betrays racism, but one still occasionally sees *girl* sticking out of a crowd of *women*. *Girl* can reveal also other biases: *salesgirl* has a belittling ring that *saleswoman* does not; *college girl* sounds more silly than *college student*. *Lady* is little used now except in a term such as *First Lady* or in the rare context where it should be paired with *gentleman*. Feminists abhor *doll, little woman, better half, weaker sex*, as well as all the grosser, more directly insulting terms in which the language abounds. The substitution of the suffixes *-woman, -women* for *-man, -men* in such words as *chairman, spokesman, councilman*, and *Congressman* is embraced by many feminists who hold positions with these titles. The usage is by no means established and editors should query writers before changing one way or another.

THE RIGHT PREPOSITION

The prepositional idiom — that is, a word and a particular preposition before a noun or after a verb or adjective — is the source of countless errors. Some are created through false analogy: although *forbid to* is correct, *forbid from* is often seen, because an analogy is made to *prohibit from*. Other errors arise through ignorance of nuance of meaning: *adapt for, adapt from, adapt to* have different uses. The idiom is subject to shifts, as a reading of Shakespeare shows. Correct current usages are simply idiomatic and must be learned or looked up.

abashed: *at, before, in.* He felt abashed *at* the other's actions. Pilate was abashed *before* Jesus. The culprit was abashed *in* the presence of the judge.

abate: *in, by* (or no preposition). The storm abated *in* fury. The nuisance was abated *by* law. The law abated the nuisance.

aberration: *from, of.* The action was an aberration *from* his usual course. It seems like an aberration *of* mind.

abhorrence: *of.* He expressed abhorrence *of* the action.

abhorrent: *to.* The suggestion was abhorrent *to* us.

ability: *at, with.* She shows ability *at* painting. She shows ability *with* paints.

abound: *in, with.* The Bible abounds *in* metaphors. The field abounds *with* crickets. (The difference is between inherent or essential qualities and unessential or accidental things or properties.)

abridge: *from*.   This version is abridged *from* the original book.

abridgment: *of*.   It is an abridgment *of* the book.

absolve: *from* (sometimes *of*).   He is absolved *from* blame.

abstain: *from*.

abstinence: *from*.

abstract (*v.*): *from*.   The dye is abstracted *from* the seed.

abut: *against, on*.   The cottage abuts *against* the cliff. The house abuts *on* the line he surveyed.

accede: *to*.

accessories: *in, of. In* a picture; *of* dress.

accessory: *after, before, to.   After* or *before* the fact; *to* a crime.

accident: *of, to*.   By accident *of* birth. An accident *to* a person.

accommodate: *to, with*.   We accommodated ourselves *to* the inconvenience. I accommodated him *with* a loan.

accompany: *by, with*.   They were accompanied *by* their dog. His gesture was accompanied *with* a smile.

accord (*n.*): *between, of, with*.   The accord *between* the two was obvious. An accord *of* interests would be advantageous. This passage is in accord *with* the rest.

accord (*v.*): *in, to*.   They accorded *in* their opinions. We must accord *to* him what he deserves.

accordance: *with*.

accordant: *with, to*.

according: *as, to*.   According *as* you may decide. According *to* the report, it was raining that day.

accountable: *for, to.   For* a trust; *to* an employer.

accuse: *of*.

acquaintance: *among, between, of, with*.   Many acquaintances *among* those present. Acquaintance *between* two. *Of* one *with* another; *with* a language.

acquiesce: *in*.   (Formerly *to* or *with*.)

acquit: *of, with*.   One is acquitted *of* a crime. One acquits oneself *with* credit.

adapt: *for, from, to*. The seats in the boat are adapted *for* seating ten persons. The model of the boat is adapted *from* that of a ketch. The boat is adapted *to* heavy weather.

addicted: *to*.

adept: *at* or *in*.

adequate: *for, to.   For* the purpose; *to* the need.

adhere: *to*.

adherence: *to*.

adhesion: *of, to.   Of* one *to* the other.

adhesive: *to*.

adjacent: *to*.

adjust: *for, to*.   I'll adjust your pillow *for* comfort. The dóg will adjust *to* its new home.

admit: *of, into* or *to*.   The quarrel admits *of* no compromise. I shall never admit her *to* (or *into*) my confidence.

admittance: *to*.

admonished: *by*.   (Biblical: *of*. Moses was admonished *of* God.)

advantage: *of, over*.   You have the advantage *of* me. You tried to gain an advantage *over* me.

adverse: *to*.

advert: *to*.

advise: *of* or *concerning*.   They advised us *of* (or *concerning*) his coming.

advocate: *of*.

affiliate: *with, to*.

affinity: *between, with*.   There was a strong affinity *between* them. They felt a strong affinity *with* their surroundings. *Affinity to* is also common, but is perhaps not as established: Water shows a strong affinity *to* salt.

aggression: *on* or *upon*.   (*Upon* preferred.)

aid (*v*.): *in*.

alien: *to, from*.

alienation: *from, of, between*.   He felt alienation *from* such ideas. She sued for alienation *of* affections. The alienation *between* the classes was notable.

allegiance: *from, to*.   A government exacts allegiance *from* the people. The people show allegiance *to* the government.

ally: *to, with*.   This species is allied *to* that. England allied herself *with* Greece.

alongside: (no preposition).   The boat lay alongside the wharf.

aloof: *from*.

alternate (*v*.): *with*.

amalgam: *of*.

amalgamate: *into, with*.

amazement: *at*.

ambition: *for*.

ambitious: *of*. (Formerly *for* or *after*.)

amenable: *to*.

amity: *between, with*.   *Between* nations; of one nation *with* another.

amorous: *of, by*.   *By* nature; *of* a desired person or thing.

amplify: *by, on* or *upon* (or no preposition).   *By* illustrative remarks; *on* (or *upon*) a statement; amplify a statement.

amused: *at* or *by, with*.   I was amused *at* (or *by*) his antics. He amused us *with* his antics.

analogous: *to*.   This rite is analogous *to* the other.

analogy: *between, with*.   The analogy *between* religion and the course of nature; *with* cellular growth.

anesthetize: *by*.

anger: *at, toward*.   *At* an insult or injustice; *toward* the insulter or offender.

angry: *at, with*.   *At* an action; *with* a person.

animadversion: *on* or *upon*.

animadvert: *on* or *upon*.

annoyed: *by, at* or *with*.   Be annoyed *by*; feel annoyed *at* (or *with*).

antipathetic: *to*.

antipathy: *against* or *to, between, for* or *to*.   *Against* (or *to*) a person; *for* (or *to*) a thing. Antipathy may exist *between* persons.

anxiety: *about* or *concerning, for*.   *About* (or *concerning*) the future; *for* another's safety.

anxious: *about, for*.   *About* a problem; *about* (or *for*) a person.

apart: *from*.

apathy: *of, toward*.   *Of* feeling; *toward* action.

append: to.

appetite: *for*.

apply: *for, to*.   We apply *for* a position *to* a person; and apply color *to* something.

apportion: *among, between, to*.   *Among* several; *between* two; *to* one.

appreciation: *of, for*.   The publisher gave an appreciation *of* Bigart's work. He expresses appreciation *for* the help.

apprehensive: *of, for*.   *Of* danger; *for* another's safety.

approximation: *of, to*.   *Of* one type *to* another.

apropos: *of* (or no preposition).

arrive: *at, in, upon*.   We arrive *in* a country or city; *at* (or *in*) a town. (*At* when we do not intend to stay; *in* when we do.) We arrive *in* port; *upon* the scene.

arrogate: *to, for*.   *To* oneself; *for* another.

aside: *from*.

aspiration: *after, toward*.   *After* righteousness; *toward* heaven.

aspire: *to, after, toward*.

assent: *to*.

assist: *at, in, with.* She assisted *at* the reception. The children assisted *in* painting the house. They assisted *with* the painting.

associate: *with.*

assuage: *by, with.* Assuage pain *with* hot cloths. Pain is assuaged *by* hot cloths.

assumption: *of.*

astonished: *at, by.* *At* that which is remarkable but deserving condemnation; *by* that which is remarkable, and worthy of praise.

attainable: *by* or *to.*

attempt (*n.*)**:** *at.* Smith's attempt *at* 18 feet failed at first.

attempt (*v.*)**:** *to.* I will not attempt *to* explain.

attend: *on, to,* (or no preposition). Attend *on* the Queen (but *on* is often omitted); *to* a duty.

attended: *with, by.* The sail to Napatree was attended *with* disaster. The star was attended *by* her maid and lawyer.

attest: *to* (or no preposition). Your remark attests your acumen (or *to* your acumen).

attribute (*v.*)**:** *to.*

attribute (*n.*)**:** *of.*

augmentation: *by, of.* The augmentation *of* our numbers *by* enlistment.

augmented: *by, with.*

augur: *from, of, for.* *From* signs; *of* success; *for* a cause.

authority: *of, on.* The right is invested in him by authority *of.* He is an authority *on.*

auxiliary: *of, in.* *Of* the army; *in* a good cause.

avaricious: *of.*

avenge: *on* or *upon.*

averse: *to.* (Rarely, *from.*)

aversion: *to* or *toward, for* or *toward.* *To* (or *toward*) a person; *for* (or *toward*) his acts.

bail (*n.*)**:** *on, in.* He was freed *on* (presentation of) bail; held *in* (default of) bail.

based: *on, upon, in.*

basis: *for, of.* *For* an argument; *of* facts.

becoming: *to, in.* Dress becoming *to* her; conduct becoming *in* him.

begin: *by, in, with.* He began *by* opening the book. The legislation began *with* (or *by* or *in*) the creation of . . .

beguile: *by, of, with.* *By* a sham; *of* our rights; *with* a dance or entertaining book.

benefactor: *of.*

bereaved, bereft: *of.*

bid (*v.*)**:** *for, on, in.* He bid *for* votes. He bid *on* the house. Because the price did not go high enough, the gallery bid *in* the vase. (This last usage is the only correct one for *bid in;* an item *bid in* reverts to the owner.)

blame: *for.*

blasé: *about.*

boast: *of, about.*

borne: *by, upon.* The truth was borne *in upon* me. His statement was borne *out* by the facts. We were borne *up by* our hopes. (In these sentences, *in, out,* and *up* are not strictly prepositions, but parts of the verbs preceding them.)

break (*v.*)**:** *with.* Break *with* precedent. (But break away *from.*)

break (*n.*)**:** *in.* A break *in* relations.

candid: *about* or *in regard to.*

capable: *of.*

capacity: *of, for, to.* *Of* a hundred gallons; *for* work; *to* sign a document.

capitalize: *at, on.*   *At* $25,000; *on* his mistake.
careless: *about, in, of.*   *About* dress; *in* one's work; *of* the feeling of others.
catastrophe: *of.*   The catastrophe *of* a play.
cause: *of, for.*   *Of* trouble; *for* warning.
caution: *against.*
center: *on.*   Rome centers *on* the Colosseum. (Not centers *about* or *around.*)
chagrin: *at* or *because of* or *on account of.*
characteristic: *of.*
characterized: *by.*
charge: *with.*   He is charged *with* espionage.
chide: *for.*
choose: *among, between.*   *Among* many; *between* two.
circumstances: *in, under.*   "Mere situation is expressed by '*in* the circumstances'; action is performed '*under* the circumstances.'" — *Oxford Dictionary.*
clear (*v.*): *from, of.*
coalesce: *in* or *into, with.*   *In* (or *into*) one; *with* one another.
coincident: *with.*
collide: *with.*   (Both objects in motion.)
common: *to.*
comparable: *to* or *with.*
compare: *to, with.*   "Shall I compare thee *to* a summer's day?" Compared *with* last year's crop, this year's is small. (*To* is a metaphoric or general comparison; *with* is comparison of size or other specific similarity or difference.)
compatible: *with.*
compete: *for, with.*   *For* a prize; *with* others.
complacent: *toward.*
complaint: *against, about.*
complaisant: *toward.*
complement: *of.*
complementary: *to.*
compliance: *with.*
compliment: *on.*
complimentary: *about* or *concerning.*
comply: *with.*
concentration: *of, on.*   *Of* attention; *on* a problem.
concerned: *about, in, to, with.*   *About* the welfare of a friend; *in* an affair; *to* be rightly understood; *with* business.
concur: *in, with.*   *In* a decision; *with* others.
conducive: *to.*
confide: *in, to.*   He confided *in* our discretion. He confided his savings *to* me.
confident: *of.*
conform: *to* or *with.*
conformable: *to* or *with.*
conformity: *to* or *with.*
congenial: *to, with.*
congratulate: *on* or *upon.*
connect: *by, with.*   Johnstown connects *by* good roads *with* Hicksville.
connive: *at.*   I shall not connive *at* any such deception.
conscious: *of.*
consent: *to.*
consequent: *to, on, upon.*   *To* the disagreement; *on* the investigation; *upon* his return.
consist: *in, of.*   Tolerance consists *in* respecting the opinions of others. Water consists *of* oxygen and hydrogen.
consistent: *in, with.*   He is not consistent *in* his statements. The statements are not consistent *with* her former ones.

consonant: *with* or *to.*   (*With* preferred.)
contact: *among, between, of, with.*   *Among* many; *between* two; *of* the mind *with* literature.
contemporaneous: *with.*
contemporary: *with.*
contemptuous: *of.*
contend: *with, against, about.*
contiguous: *to.*
contingent: *on* or *upon.* (*Upon* preferred.)
contrast (*v.*)**:** *with.*
contrast (*n.*)**:** *between, of, to.*   *Between* this and that. This presents a contrast *to* that. This is *in* contrast *with* that.
convenient: *for, to.*   *For* a purpose or use; *to* a place.
conversant: *with.*
convert (*v.*)**:** *from, to.*   *From* one purpose *to* another.
convict (*v.*)**:** *of.*
convince: *of.* (Never *to.*)
correlation: *between* or *of.*
correlative: *with.*
correspond: *to, with.*   It does not correspond *to* reality. Richard corresponds *with* me regularly.
culminate: *in.*
cure: *of.*

debar: *from.*
decide: *on, upon.*
defect (*n.*)**:** *in, of.*   *In* a machine; *of* judgment or character.
defend: *from, against.*   I defend my home *from* harm, *against* intruders.
deficiency: *in, of.*   *In* intelligence; *of* food.
defile (*v.*)**:** *by, with.*   *By* an act; *with* a substance.
definition: *of.*
demanding: *of.*
depend: *on, upon.*
deprive: *of.*
derive: *from.*
derogate: *from*
derogation: *of, from, to.*
desirous: *of.*   Desirous *of* learning.
desist: *from.*
despair: *of.*
destined: *to, for.*
destructive: *of, to.*   Poor food is destructive *of* health. Rabbits are destructive *to* young trees.
detract: *from.*
deviate: *from.*
devoid: *of.*
devolve: *from, on, to, upon.*   (The word means *pass* or *cause to pass from one to another.*) Authority devolves *from* the emperor *on* (or *upon* or *to*) the subjects. (*On* or *upon* has better authority than *to.*)
differ: *with, on, from.*   I differ *with* you *on* that. Hetty differs *from* Eugene in temperament.
different: *from.*
differentiate: *among, between, from.*   *Among* many; *between* two; this *from* that.
diminution: *of.*
disappointed: *in, with.*   *In* a person, plan, hope, result; *with* a thing.
disapprobation: *of.*

disapprove: *of.*
discourage: *from.*
disdain: *for.*
disengage: *from.*
disgusted: *at, by, with.* At an action; *by* a quality or habit of a person or animal; *with* a person, because of his general attitude or point of view.
dislike: *of.*
dispense: *with, from.* We dispense *with* formalities. I dispense you *from* your promise.
displace: *by.*
displeased: *at, with.* At a thing; *with* a person.
dispossess: *of, from.*
disqualify: *from, for.*
dissatisfied: *with.*
dissension: *among, between, with.* Among friends; *between* friends. He was in dissension *with* the world.
dissent (*v.*)**:** *from.*
dissimilar: *to.*
dissociate: *from.*
dissuade: *from.*
distaste: *for.*
distill: *from.*
distinguished: *by, for, from.* By talent; *for* honesty; *from* another person or thing.
distrustful: *of.*
divert: *from, to, by.* He diverted funds *from* the treasury *to* his own use. We were diverted *by* the child's playfulness.
divest: *of.*
divide: *by, into.* By cutting; *into* parts.
divorce: *from.*
dominant: *over, in.* Over others; *in* power or manner.
dominate: *over.*
dote: *on.*
drench: *with.*

eager: *for, to.* For success; *to* succeed.
educated: *about* or *concerning, for, in.* Concerning (or *about*) the needs of life; *for* living; *in* liberal arts.
effect: *of.*
eligible: *to* or *for.*
emanate: *from.*
embark: *on, upon.*
embellish: *with.*
emerge: *from.*
emigrate: *from.*
employ: *at, in.* At a suitable wage; *in* a gainful pursuit.
empty (*adj.*)**:** *of.* The house is empty *of* furniture.
enamored: *of, with.* Of a person; *with* a scene.
encouraged: *by, in.* Encouraged *by* success; encouraged another *in* his work.
encroach: *on, upon.*
endow: *with.*
engage: *in, upon.*
enjoin: *on* or *upon* (in law: *from*).
enraged: *against* or *with, at.* Against (or *with*) a person; *at* an action.
enter: *by, into, in.* By the window; *into* the spirit of it; items *in* a ledger.
entertained: *by, with.* By persons; *with* their doings.

enthralled: *by.*

entrust: *to, with.* The money was entrusted *to* me; I was entrusted *with* the money.

enveloped: *in.*

envious: *of.*

equal (*adj.*): *in, to.* *In* qualities; *to* a task.

equivalent: *in, to.* These two vessels are equivalent *in* volume. That remark is equivalent *to* saying No.

escape: *from, out of* (or no preposition: "He escaped justice").

essential: *in, of, to* or *for.* The first essential *in* study is concentration. He was well grounded in the essentials *of* mathematics. These things are essential *to* (or *for*) success.

estimated: *at.*

estrange: *from.*

estrangement: *of.*

evidence: *of.* He furnished evidence *of* the fact.

example: *of, from, to.* This is an example *of* the split infinitive. These examples *from* history are instructive. Let this be an example *to* you.

excerpt: *from.* (Never *of.*)

exclusive: *of.*

excuse: *for, from.* *For* an action; *from* an obligation.

exonerate: *from.*

expect: *from, of.* One expects profit *from* investment; honesty *of* a person.

expel: *from.*

experience: *for, in* or *of.* Experience *for* oneself, *in* (or *of*) travel.

expert: *in* or *at, with.* Expert *in* (or *at*) chess; *with* knitting needles.

expressive: *of.*

extract: *from.*

exude: *from.*

faced: *by, with.*

fail: *in, at, of.*

familiar: *to, with.* A scene is familiar *to* us; one person is familiar *with* another person.

fascinated: *by, with.*

favorable: *for, to.* Weather favorable *for* skating; I was favorable *to* his proposal.

fear: *of, for.* Fear *of* water; fear *for* another.

flinch: *at, from.*

fond: *of.*

fondness: *for.*

forbid: *to.* (Never *from.*)

foreign: *to.*

founded: *on, upon, in.*

free: *from, of* or *in.* *From* disease; *of* (or *in*) manner.

freedom: *from, of, to.*

friend: *of, to.*

frightened: *at, by.* *At* something sudden and threatening, as a gesture; *by* something alarming or inexplicable, as a lion or a ghost.

frugal; *of.*

fugitive: *from.*

grapple: *with.*

grateful: *for, to.* *For* your help; *to* you.

grieve: *at, for, after.*

guard: *against, from.* *Against* peril; *from* a person.

habitual: *with.*

hanker: *after.*

healed: *of, by.* She was healed *of* dermatitis *by* the doctor.

help (*v.*)**:** *with, to* (or no preposition). I will help *with* the laundry. I will help you *to* write it. I will help you write it. I will help you.

hinder: *from.*

hindrance: *to.*

hinge: *on.*

hint: *at.*

honor: *by, for, with.* I am honored *by* your invitation. We honored him *for* his honesty. He was honored *with* an invitation.

hope: *for, of.* Hope *for* better times; little hope *of* heaven.

identical: *with.*

identify: *by, to, with.* *By* credentials; *to* the police; *with* the man known to be innocent.

immanence: *of, in.* *Of* divine power *in* human life.

immerse: *in.*

immigrate: *to.*

impatient: *at, with.* *At* action; *with* persons.

impeach: *for* (or no preposition). Impeach *for* disloyalty; impeach one's motives.

impervious: *to.*

implicit: *in.*

impose: *on* or *upon.*

impress (*n.*)**:** *of, upon.* *Of* the design *upon* the coin.

impress (*v.*)**:** *into, upon, with.* A man *into* service; a duty *upon* a child; wax *with* a die.

impressed: *by, with.*

improve: *by, in, upon.* Trees may be improved *in* hardiness *by* grafting. I can improve *upon* that plan.

improvement: *in, of, upon.* *In* health *of* a patient. This machine is an improvement *upon* that.

inaccessible: *to.*

incentive: *to, for.*

incidental: *to.*

incongruous: *with.*

inconsistent: *with.*

incorporate: *in* or *into* or *with.* Incorporate laws *in* a constitution; incorporate this thing *in* (or *into* or *with*) the mass of others.

independent: *of.*

index: *of.* It is a rough index *of* the increase in knowledge.

indulge: *in, with.*

indulgent: *to, of.*

infer: *from.*

inferior: *to.*

infiltrate: *into.*

infiltration: *of, by.* The infiltration *of* the area *by* the guerrillas was complete.

influence (*v.*)**:** *by, for.* *By* actions *for* good.

influence (*n.*)**:** *of, over, upon.* *Of* a good man *over* others. One may exercise influence *upon* others.

infuse: *with.*

inimical: *to* or *toward.*

initiate: *into.*

innate: *in.*

inquire: *into, of.* *Into* causes; *of* a person.

inquiry: *about* or *concerning, of.* *About* (or *concerning*) any destination; *of a* bystander.

inroad: *into.*

inseparable: *from.*

insight: *into.*

inspire: *by, with.* *By* example; *with* courage.

instigate: *to.*

instill: *in* or *into.*

instruct: *in.*

intent: *on* or *upon.*

intention: *of, to.*

intercede: *for, with.* *For* a culprit; *with* a judge.

interest: *in.*

interfere: *in, with.* *In* an affair; *with* a person.

intermediary: *between, in.* *Between* persons; *in* a quarrel.

intervene: *in, between.* The referees intervened *in* the fight. The control commission intervened *between* the sides.

intimacy: *of, with.* *Of* association; *with* persons.

introduce: *to, into.*

intrude: *on, upon, into.*

inundate: *with.*

invest: *in, with.* We invest *in* stocks and bonds. We have invested the President *with* great power.

investigation: *of.*

involve: *in.*

isolate: *from.*

jealous: *of, for.* *Of* a person; *of* one's good name; *for* another.

jeer: *at.*

join: *to, with* Join this *to* that. Join *with* me in this venture.

justified: *in.*

label: *with.*

labor: *at, for, in, under, with.* *At* a task; *for* (or *in*) a cause; *under* a taskmaster; *with* tools.

lacking: *in.*

laden: *with.*

lag: *behind.*

lament: *for* or *over.*

laugh: *at* or *over.*

lend: *to.*

level: *at, to, with.* Level a gun *at.* The building was leveled *to* the ground. The line is level *with* the horizon.

liable: *for, to.* *For* illegal acts; *to* prosecution.

liken: *to.*

live: *at, by, in, on.* *At* a place; *in* a town; *by* peddling; *on* a street.

make: *of, from.* Make the boat *of* pine. My dress is made *from* an old one.

marred: *by.*

martyr (v.)**:** *for.* Martyred *for* his beliefs.

martyr (n.)**:** *to.* A martyr *to* rheumatism.

mastery: *of, over.* Mastery *of* a craft; *of* one man *over* another.

means: *of, to, for.*

meddle: *in, with.* I refuse to meddle *in* his affairs. Don't meddle *with* my things.

mediate: *between, among.*

meditate: *on, upon.*
militate: *against.*
minister: *to.*
mistrustful: *of.*
mix: *with, into.*
mock: *at, with.*   We may mock *at* a person; be mocked *with* vain desires.
monopoly: *of.*
mortified: *at.*
mortify: *by* or *with.*
motive: *for.*
muse: *on* or *upon.*

necessary: *for, to.*   A strong will is necessary *for* success. Rhythm is necessary *to* verse.
necessity: *for, of, to.*   He saw the necessity *for* quick action. I realize the necessity *of* consenting. A warm climate was a necessity *to* her.
need: *for, of.*   *For* improving conditions; *of* a new coat of paint.
neglectful: *of.*
negligent: *of, in.*

obedient: *to.*
object (*v.*)**:** *to.*
oblivious: *of.*
observant: *of.*
obtrude: *on* or *upon.*
occasion: *for, of.*   *For* thanksgiving; *of* distress.
occupied: *by, with.*   The park was occupied *by* joyous people. Are you occupied *with* work?
offend: *against.*
offended: *at, with.*   *At* an action; *with* a person.
opportunity: *of, to, for.*   *Of* enlisting; *to* enlist; *for* enlistment.
opposite: *of* (or no preposition).   Good is the opposite *of* bad. This house is opposite that.
opposition: *to.*
originate: *from, in, with.*   Baseball originated *from* the old game of rounders. The idea originated *in* his own mind. This plan originated *with* the board.
overlaid: *by* or *with.*
overwhelm: *by, with.*

parallel: *to* or *with.*
part: *from, with.*   *From* a person; *with* a thing.
partake: *of.*
partial: *to.*
participate: *in.*
patient: *in, with.*
peculiar: *to.*
permeate: *into, through.*
permeated: *by.*
permit: *of, to.*   The passage permits *of* different translations. I will not permit you *to* do it.
persevere: *against, in.*   *Against* opposition; *in* a pursuit.
persist: *against, in.*   *Against* objection; *in* an action.
persuade: *to.*
persuaded: *by, of, to.*
pertinent: *to.*

piqued: *at, by.*   *At* something done to us; *by* ridicule.

place (*v.*): *in.*   (Not *into.*)

plan (*v.*): *to.*

pleased: *at* or *by, with.*   He was pleased *at* (or *by*) the suggestion. The child was pleased *with* a rattle.

plunge: *in, into.*   *In* grief; *into* water.

possessed: *by* or *with, of.*   *By* (or *with*) a desire for money; *of* much property.

possibility: *of.*

practice: *of, in.*   *Of* a profession; *in* an art.

precedence: *of.*

precedent (*n.*): *of* or *for.*

precedent (*adj.*): *to.*

precluded: *from.*

predestined: *to.*

preface: *of* or *to.*

prefer: *to.*

pregnant: *with, by.*   A phrase pregnant *with* meaning. She was pregnant *by* the duke at the time she left.

prejudice (*v.*): *against, by.*

prejudice (*n.*): *against, for, in favor of.*

prejudicial: *to.*

preoccupied: *with.*

preparatory: *to.*

prerequisite (*n.*): *of.*

prerequisite (*adj.*): *to.*

prescient: *of.*

present (*v.*): *to, with.*   *To* a person; *with* a gift.

present (*adj.*): *to.*   *To* the senses.

preside: *at, over.*

presume: *on* or *upon.*

prevail: *against, on* or *upon, over, with.*   Patience will often prevail *against* force.   We prevailed *upon* (or *on*) him to accompany us.   David prevailed *over* Goliath.   He prevailed *with* youthful skill.

prevent: *from.*

preventive: *of.*

preview: *of.*

prior: *to.*

privilege: *of.*

prodigal: *of.*

productive: *of.*

proficient: *in, at.*

profit: *by, from.*

prohibit: *by, from.*   *By* law; *from* doing.

prone: *to.*

property: *of.*

propitious: *to.*

protest: *against.*

provide: *with, against, for.*   We provide you *with* food and clothes.   We provide *against* disaster.   We provide *for* your college.

punish: *with, by, for.*

punishable: *by.*

purchase: *of.*

purge: *of, from.*   *Of* the plague; *from* his mind.

pursuit: *of.*

put: *in* or *into, on, to.*   *In* (or *into*) use; *in* (or *into*) water; *on* the table; *to* work.

qualify: *for, as, to.*   *For* a competition; *as* a pilot; *to* act.

ranging: *between, from . . . to, within.*   *Between* boundaries; *from* 32 *to* 60 degrees; *within* a territory.

reason (*v.*): *about* or *on, with.*   *On* (or *about*) a subject; *with* a person.

reason (*n.*): *for.*

rebellious: *against* or *to.*

receptive: *to.*

recognition: *of.*

reconcile: *to, with.*   *To* a condition; *with* a person.

redolent: *of.*

regard: *for, to.*   *For* a person; *in* (or *with*) regard *to* a proposal.

regardless: *of.*   Regardless *of* whether she stays (*of* may not be omitted).

regret (*v.*): (no preposition).   We regret having done it.

regret (*n.*): *for, over* or *about* or *concerning.*   *For* one's misdoings; *over* (or *about* or *concerning*) misfortune.

relation: *of, to, with.*   *Of* this *to* that; *with* persons.

removal: *from, of, to.*   *From* a place; *from* normal; *of* snow; *of* difficulties; *to* another place.

renege: *on.*

repent: *of.*

replete: *with.*

reprisal: *for, against* or *upon.*   They took reprisal *against* (or *upon*) their neighbors *for* the attack.

repugnance: *between, against* or *for* or *of, to.*   *Between* versions of testimony; *of* a person *against* (or *for* or *of*) another; *to* a deed or duty.

repugnant: *to.*

request (*v.*): *to.*   I requested Ben *to* bring the letter from his teacher. (Not: I requested Ben *for* the letter from his teacher.)

request (*n.*): *of, for, to.*   I have a request *of* you.   A request *for* information.   A request *to* go home.

requirement: *for, of.*   *For* admission to college; *of* a tax.

requite: *with.*

resemblance: *among, between, of, to.*   *Among* members of a family; *between* two persons or things; *of* one thing *to* another.

resentment: *against, at* or *for.*   *Against* a person; *at* (or *for*) a wrong.

respect: *With* respect *to.*

responsibility: *for.*

restrain: *from.*

revel: *in.*

revenge: *for, upon.*   *For* a hurt; *upon* a person.

reward: *by, for, with.*   *By* conferring a prize; *for* herosim; *with* knighthood.

rich: *in.*

rid: *of.*

rob: *of.*   The family was robbed *of* its savings.

role: *of.*

sanction: *of, for.*   *Of* the law; *for* an act.

satiate: *with.*

satisfaction: *of, in, with.*   *Of* desire; *of* honor; *in* well-doing; *with* another's deeds.

saturate: *with.*

scared: *at, by.*

search (*v.*): *for*

search (*n.*): *for, of.*

seclusion: *from.*

secure (*adj.*): *in, of.* He felt secure *in* his position. We felt secure *of* their loyalty.

secure (*v.*): *by.*

sensible: *about, of.* Be sensible *about* this affair. We were not sensible *of* the cold.

sensitive: *to.*

separate: *from.*

serve: *for, on, with.* This will serve *for* (i.e., in lieu of) that. He will serve *on* the committee. He will serve *with* the army.

significant: *of.*

similar: *to.*

similarity: *to, of.* He shows some similarity *to* his brother. The similarity *of* the brothers was marked.

skillful: *at* or *in, with.* *At* (or *in*) art; *with* the hand.

slave: *of, to.*

solicitous: *of, for, about.*

solution: *of, to.*

sought: *for* or *after.*

sparing: *of.*

spy: *upon.*

stock: *in, of.* I do not take much stock *in* his plea. He took stock *of* his opportunities.

strive: *for, to, with, against.*

struck: *by.*

subject: *to, of.*

subscribe: *for, to.* *For* a purpose; *to* an occasion or action.

substitute (*n.* and *v.*): *for.*

suffer: *with, from.*

suitable: *for, to.* *For* a purpose; *to* an occasion.

superior: *to.*

supplant: *by.*

supplement: *by* or *with.*

surprised: *at, by.* *At* actions; *by* unexpected arrival or appearance.

surround: *by.*

suspect: *of.*

sympathetic: *with, to, toward.*

sympathize: *in, with.* *In* another's mood; *with* another person.

sympathy: *for, with, in.* *For* another; *in* his sorrow; *with* his desires.

synchronous: *to.*

tally: *up, with.* Tally *up* means "check up, reckon"; tally *with* means "agree."

tamper: *with.*

tantamount: *to.*

taste: *for, in, of.* A taste *for* simplicity. Good taste *in* house furnishings. A taste *of* onion.

taunt: *with.*

temporize: *with.*

tendency: *to, toward.*

theorize: *about.*

thoughtful: *of.*

thrill (*v.*): *at* or *to, with.* *At* (or *to*) the song of a thrush; *with* pleasure.

thrilled: *by.*

thronged: *with.* The square was thronged *with* people.

tinker: *at* or *with.*

tired: *of, from, with.* *From* boredom; *of* noise; *with* exercise.

tolerance: *for, of, toward.*

tormented: *by* or *with.*   Tormented *by* (or *with*) headaches.
transmute: *into* or *to.*
treat: *for, of, to, with.*   *For* peace; *of* a subject; *to* a soda; *with* care.
true: *to, with.*   *To* thine own self be true.   This cog is not true *with* the other.
trust: *in, to, with.*   I trust *in* you; *to* your judgment.   I trust you *with* my life.

umbrage: *at, to.*   Take umbrage *at;* give umbrage *to.*
unconscious: *of.*
unequal: *in, to.*   *In* qualities; *to* a task.
unfavorable: *for, to* or *toward.*   *For* action; *to* (or *toward*) an attitude or person.
unite: *by* or *in, with.*   *By* (or *in*) common motives; *with* another.
unmindful: *of.*
unpalatable: *to.*
unpopular: *with.*
use: *in* or *of.*
useful: *in, to.*   *In* (or *to*) a group; *in* an activity; *to* a person or cause.

variance: *with.*
vary: *from, with.*   This varies *from* that; each varied *with* the other.
vest (*v.*)**:** *in, with.*   Power is vested *in* a man; a man is vested *with* power.
vexed: *at, with.*   *At* a thing; *with* a person.
vie: *with.*
view: *of; in* view *of; with* a view *to.*
void: *of.*
vulnerable: *to.*

wait: *for, on.*   *For* something to happen or a person to come; *on* people at a
　　table: *on* another's convenience.
want (*n.*)**:** *of.*
wanting: *in.*
wary: *of.*
way: *of.*   What is the best way *of* calling the Souzas?
weary: *of.*
willing: *to.*
worthy: *of, to.*   *Of* note; *to* be called.
write: *on, off, out.*   *On* paper; *on* a subject; *off* a debt; *out* in full.

yearn: *after, for, over, toward, with.*   *After* solitude; *for* a loved one; *over* a
　　child; *toward* a nearby person; *with* compassion.
yield: *of, to.*

zeal: *for, in.*   *For* a cause; *in* work.

WORDS LIKELY TO BE MISUSED OR CONFUSED

The following list is a collection of lexicographic nightmares: homonyms; near synonyms with differing nuances of meaning; words whose meanings are frequently misconstrued altogether; trite, overworked, and stilted words. Some combinations are born to be typographical errors and are included as a warning: *casual, causal; marital, martial.*

abjure, adjure   *Abjure:* to renounce (an oath, for example); *adjure:* to beg or
　　request earnestly, to charge (a person under oath).

about, approximately, average *About* ("About a thousand people attended") is more vague than *approximately*, which suggests an attempt at calculation ("I am overdrawn approximately $125"); with *average*, neither *about* nor *approximately* is needed ("The classes average twenty-five pupils"). Sir Ernest Gowers *(The Complete Plain Words)* warns against the use of *very approximately* (as in *very roughly*).

absolutely, positively Unconditional terms, not appropriately used as substitutes for *very*, *indeed*, or *yes*.

absorb, adsorb *Absorb:* to take up or drink in; *adsorb:* to condense and hold (a gas) in a thin layer on the surface of a solid.

abysmal, abyssal The words mean the same, one being formed from *abysm* and the other from *abyss*; but *abyssal* is learned, used mostly in biology and geology, and is literal ("Abyssal depths of the ocean"), while *abysmal* may be figurative ("Abysmal depths of ignorance").

Acadia, Arcadia *Acadia:* Nova Scotia; *Arcadia:* a region in Greece, an imaginary abode of peace.

accept, except "We *accept* all the conditions, if we may *except* the clause that we found objectionable."

access, accession *Access:* opportunity for entrance ("We did not have access to the fort because we did not know the password"); a way of approach ("Access to the fort was barred by a gate"); a sudden seizure or its approach ("She felt an access of enthusiasm"). *Accession:* a coming to (something), arrival ("Accession to the throne is possible to a prince; access to it is possible to a petitioner"); a being joined with or by (something), an increase ("There has been no accession to—i.e., increase in—our army").

accident, mishap An *accident* can be good, bad, or neither; a *mishap* is always bad and also relatively unimportant when compared with a catastrophe or cataclysm.

acclivity, declivity One goes up an *acclivity*; down a *declivity*.

acetic, ascetic Vinegar is *acetic*; a monk is *ascetic*.

act, action When the two words are not interchangeable, *act* is simple and specific ("The Acts of the Apostles"); *action* complex and general ("The action of an acid"). The difference between "His actions are deplorable" and "His acts are deplorable" is that the first sentence refers to his habitual behavior and the second to certain doings.

adapt, adopt *Adapt:* to adjust, make suitable, remodel; *adopt:* to accept, receive as one's own.

addicted to, subject to A person is *addicted to* some act or thing through his own inclination, habit, or weakness; he is *subject*, by reason of his condition or position, *to* certain effects whose cause is outside himself and without his volition.

adhere, cohere Two or more separate things may *adhere* (stick together). Parts of the same thing *cohere* (hold together).

adherence, adhesion The words are often interchangeable, though *adhesion* is more likely to refer to the physical ("The adhesion of lime and brick"), *adherence* to the mental or social ("A man's adherence to a cause"). In such a sentence as "France's adhesion (adherence) to the pact is certain," either word is acceptable, though *adhesion* perhaps suggests a more definite and active sticking to than does *adherence*.

admission, confession In law-enforcement use, an *admission* (of guilt) is usually an oral expression; a *confession*, a written, signed statement.

adopted, adoptive An *adopted* child has *adoptive* parents.

advance, advancement *Advance:* progress; *advancement:* a specific promotion.

adversary, antagonist, enemy, foe, opponent All imply opposition, and *adversary*, *enemy*, and *foe* imply hostile opposition. In these three the idea of active hostility increases from the first to the third. An *antagonist* or an *oppo-*

nent may not be hostile at all. Lawyers who are antagonists in court may be good friends outside, and the same is true of opponents in politics or a game.

adverse, averse   *Adverse* (L. *advertare,* turn to): turned to or toward in opposition; *averse* (L. *avertare,* turn from): turned from in dislike or repugnance. We speak of adverse winds or circumstances and of a person's being averse to study. *Averse* implies feeling; *adverse* seldom does.

aerie, eerie   *Aerie:* the nest, on a crag, of a bird of prey; *eerie:* uncanny, weird, or frightened.

affect, effect   *Affect:* to influence, to pretend; *effect:* to accomplish, complete, bring about.

affective, effective   *Affective* relates to the emotions, *effective* means producing an effect of any sort. Confusion arises because some things may be both: Churchill's speeches were affective and effective.

aggravate, annoy   *Aggravate:* become worse ("The tension was aggravated by the failure of negotiations"); *annoy:* irritate ("Don't annoy the animals, dear").

aggregate, total   An *aggregate* is a collection of particulars in a mass or whole; a *total* is a whole or entirety without special reference to parts.

alibi, excuse   Strictly, *alibi* refers to the plea or fact of having been elsewhere at the time an act was committed, though colloquially it is used of any *excuse* designed to shift responsibility.

allege, assert   *Allege:* to say or affirm on insufficient grounds; *assert:* to say or affirm emphatically or with conviction.

alleged, accused, suspected   Theodore M. Bernstein's *The Careful Writer* points out the confusion created when these words are used indiscriminately to describe an *accused spy,* an *alleged assassin,* a *suspected thief.* An *accused lawyer* is an actual lawyer accused of something; an *accused spy* may not be an actual spy, but a person accused of being a spy. In addition, one *alleges* a crime, not a person, formally speaking, but use is establishing *alleged, accused,* and *suspected* as rough equivalents of *presumed.*

all ready, already   *Already* is an adverb meaning *before now* ("He has gone already"); *all ready* is an adjectival term meaning *completely prepared* ("I am all ready to go"), not to be confused with the usage "We are all ready to go" meaning that all of us are ready to go.

all together, altogether   *Altogether,* meaning *entirely, completely* ("They were altogether in accord"; "Apical growth may cease altogether"), is often misused for *all together,* meaning *collectively* ("Our state governments spend all together about two billion dollars a year"; "All together, the purges left hardly a single leader of the 1917 Revolution alive"; "They were all together in the room").

allude, elude, illude   These verbs are confused only because of similarity of sound or spelling. *Allude:* to refer indirectly; *elude:* to dodge, slip away from; *illude* (a rare word): to cheat, mock, play tricks upon (in the manner of a magician or prestidigitator). The nouns *allusion, elusion, illusion* are similarly distinguished.

allude, refer   We *refer* to a thing by a clear and direct statement; we *allude* by a passing, indirect, sometimes obscure remark or hint.

alternate, alternative   *Alternate:* first one and then the other ("The flag has alternate red and white stripes"); *alternative:* one without the other ("Alternative plans").

alternative, choice   Although *alternative* has come to include more than two possibilities, it implies a necessity to choose from a set of mutually exclusive conditions. *Choice* suggests the possibility of choosing or declining.

altitude, elevation   *Altitude* is absolute, *elevation* may be relative. The *altitude* of an object is its elevation above a norm, as sea level ("The altitude of the mountain top is 6,000 feet"). *Elevation* is any distance above something else ("We reached an elevation on the mountain halfway to the top"). *Elevation* has metaphorical uses that *altitude* has lost, as "elevation of soul or mind"; in astronomy, topography, and art it has technical uses.

amateur, novice, tyro *Tyro* (L. *tyro,* a recruit) or *novice* (L. *novus,* new, through the French) signifies a beginner, both words suggesting inexperience. *Amateur* (F. *amateur,* lover) refers to one who is a beginner only in the sense that he is not a professional: he may have been an amateur for years.

ambiguous, equivocal Both words mean *susceptible of more than one interpretation. Equivocal* describes a statement intentionally so worded that two or more meanings can be inferred; but the wording that makes a statement *ambiguous* may or may not be intentional. Actions, as well as words, may be described as equivocal, but not as ambiguous. Since the apparent meaning of an equivocal statement may not be the actual meaning, *equivocate* and *equivcator* have come to be euphemisms for *lie* and *liar.*

ambivalent, ambiguous *Ambivalent* refers to the presence in the mind of two conflicting wishes. *Ambiguous* can refer to a multiplicity of feelings.

among, amid *Among:* in the company of countable, disparate things ("Among my souvenirs . . ."); *amid:* with uncountable or inseparable things ("Amid the measureless grossness and the slag" — Whitman).

amoral, immoral, nonmoral, unmoral *Immoral:* in violation of morals; *amoral, nonmoral,* and *unmoral:* neither moral nor immoral. *Amoral* is a learned word growing steadily in popularity, perhaps because *nonmoral* and *unmoral* might be interpreted *immoral.*

amount, quantity *Amount* refers to an aggregate ("amount of money"), while *quantity* can refer to aggregate or number ("quantity of coins").

amusedly, amusingly "He told the story *amusingly.*" "We listened *amusedly.*"

anarchism, anarchy *Anarchism* is a philosophical or political term for the theory that all governments are wrong or unnecessary. *Anarchy* is the name of a state or condition of disorder during which government is absent or helpless. By extension it means disorder or lawlessness of any sort.

anachronism, anomaly An *anachronism* can pertain only to a displacement in time, such as a reference to automobiles in a play about the Civil War. An *anomaly* can refer to other types of incongruity.

ancient, antiquated, antique, antiquity *Ancient:* very old; *antiquated:* out of date or out of use or out of vogue. *Antique* as an adjective refers to style rather than to age, or to the old that is valued or prized; as a noun *antique* refers to an old and, often, valued object. The *antiquated* is not valued or prized. *Antiquity* refers to long ago, also to an object from such a time. By and large, *antiques* are in shops, *antiquities* in museums.

anhydrous, hygroscopic Words often confused even by scientists. *Anhydrous:* destitute of water, dried; *hygroscopic:* capable of absorbing moisture. *See also* hydroscope.

anile, senile *Anile:* like an old woman, feeble-minded; *senile:* like an old person of either sex. *Anile* is a rare word, except in crossword puzzles.

animal, beast, brute *Animal* is the mild and general word; there is not necessarily offense in saying that a man is an animal. *Beast* and *brute* we hesitate to use of any creature we are fond of. The *beast* is a creature of instinct and appetite; the *brute,* one that is dull of comprehension or that is of unreasoning strength and cruelty.

ante, anti *Ante:* before ("ante bellum"); *anti:* against ("antiwar rally").

anticipate, expect To *expect* an event is to think that it will happen; to *anticipate* it is to prepare for or act in advance on it.

apiary, aviary Bees stay in the first; birds in the second.

apperception, perception *Apperception:* self-consciousness, the act of mind by which it is conscious of perceiving; *perception:* knowledge derived from the senses, recognition of fact or truth.

apparently, seemingly, obviously *Apparently* and *seemingly* suggest that the senses or perception may be at fault and some doubt present, while *obviously* expresses certainty or complete clarity or recognition.

appear, seem There is often little need to differentiate the meanings of these words. *Appear,* however, by etymology refers to an effect upon the senses;

seem, to one upon the mind. The latter suggests more reflection than the former ("He appeared to be happy"; "He seemed to love her").

appraise, apprise *Appraise:* set a value on; *apprise:* inform.

appreciation, enjoyment *Enjoyment* is mainly emotional; *appreciation,* largely rational. We enjoy a picnic; appreciate a favor, music, a picture. In the arts enjoyment is necessary to appreciation, but appreciation not to enjoyment. By etymology *appreciate* suggests increase or enrichment, a meaning it still has ("The stocks appreciated in value").

apprehend, comprehend To *comprehend* is to grasp completely; to *apprehend* is to perceive the main drift or to look forward to with foreboding. *Apprehend* has also a specific use in the sense of *arrest, make a prisoner of.* This is very near the original Latin meaning, to *seize.*

apt, likely *Apt* has meanings which are not synonymous with those of *likely.* *Apt* suggests habitual tendency; *likely* emphasizes probability; but they are often interchanged ("He is apt to forget to wind his watch"; "We are likely to postpone the trip if it rains").

arbitrate, mediate To *arbitrate* is to hear evidence and make an award or decision. To *mediate* is to act as a go-between, to encourage contending sides to agree. Thus, in labor negotiations, the failure of *mediation* can sometimes lead to *arbitration.*

aroma, odor, scent An *aroma* is pleasant by definition; an *odor* may or may not be; a *scent* is the distinctive odor of a particular thing and is usually lighter and more subtle than an aroma or odor.

art, artifice Art refers to the product of imagination or creation; *artifice* to that of skill or cunning. The latter word often suggests imitation or fraud.

> Suppose you say your worst of Pope, declare
> His jewels paste, his nature a parterre,
> His art but artifice. — Austin Dobson.

ascension, ascent The words are as a rule interchangeable, but *ascension* is used mainly in dignified contexts—the ascension of Elijah or Christ—though the "ascension" of an aeronaut or a balloon is a stock phrase. We speak of the "ascent of man" and of the ascent, rather than ascension, of a mountain.

assay, essay As verbs, *assay:* analyze, test; *essay:* attempt. We assay a metal, essay a task. The words are etymologically the same, and *assay* was formerly used as *essay* is now.

assemblage, assembly An *assemblage* is diverse and unorganized; an *assembly* is organized and united.

assert, say, state *Say* is the least formal and most general of the three. *State* is more formal than *say* and means to *set forth in detail or completely.* To *assert* is to claim or state positively, sometimes aggressively.

astonish, surprise The distinction between these verbs is no longer as clear as it once was, but *surprise* carries more overtones of the unexpectedness of the thing, *astonish* is concerned with the reaction—being momentarily overwhelmed by the thing.

Attic, Greek, Hellenic, Hellenistic The first three words are often interchangeable. *Attic,* however, refers not to Hellas or Greece but to Athens or the dialect of Attica. It is therefore used of the highly civilized center of Greek culture ("Attic wit," "Attic purity"). *Hellenic* is sometimes used to distinguish Hellas or Greece of the great period, before 300 B.C., from the *Hellenistic* (or decadent), 300–150 B.C. Compare *classic, classicist.*

attorney, lawyer, counselor An *attorney* is any person designated to act for another; an attorney at law should be referred to as a *lawyer,* even though he may refer to himself as *attorney;* if he gives legal advice or acts as a trial lawyer, he is a *counselor.*

audience, spectators Technically, an *audience* hears, *spectators* see. However, *audience* is the more general term and may be used in reference to plays, concerts, movies, radio, and television; *spectator* is generally used in reference to sports and other such public events. *Audience* is also used for books in the sense of the general readership for a particular book.

auger, augur   *Auger:* a tool; *augur:* an omen.

aura, halo   An *aura* is an emanation or exhalation from a body and should as a rule be distinguished from a *halo* or other glow or flood of light surrounding a body but proceeding from an outside source.

avenge, revenge   Both mean to *punish as a payment for wrong done,* but *revenge* is egoistic and *avenge* not. We may speak of an avenging God, but hardly of a revenging God. In *avenging,* the punishment is righteous; in *revenging,* the question of right and wrong may not enter at all. The nouns *avenging* and *vengeance,* as distinguished from *revenge,* share the distinction.

avert, avoid   *Avert:* prevent; *avoid:* keep clear of ("If I had avoided that car, I might have averted the accident").

avocation, vocation   Although the words may be used in the same signification, careful writers use *vocation* of a person's calling by which he makes a living, and *avocation* of his way of occupying his leisure or of amusing himself.

bacchanal, bacchant   As nouns the two words are interchangeable and may be used of both males and females, the feminine of *bacchant* being *bacchante.* As adjectives, *bacchant* means *worshiping Bacchus, reveling; bacchanal* or *bacchanalian* also means this, but is more likely than *bacchant* to mean *drunken, licentious.*

baited, bated   The hook is *baited;* the breath is *bated.*

baleful, baneful   *Baleful:* malign, pernicious, or wretched, miserable; *baneful:* noxious, poisonous, dangerous. The little-used nouns *bale* and *bane* suggest the difference. *Bale:* evil or woe; *bane:* poison, or as dangerous as poison (as in *wolfbane, dogbane, henbane*). We speak of a baleful glance and baneful circumstances.

barbarian, barbaric, barbarous   A barbarian is not a civilized being, and therefore the adjective *barbarian* refers to a stage in human progress between the uncivilized and the civilized. *Barbaric* is used of certain qualities of barbarians, especially their crudeness or wildness ("barbaric splendor," "barbaric song"). *Barbarous* may be used in the sense of *barbarian* ("barbarous tribes") without implying anything except lack of civilization, but it usually denotes the cruelty which such tribes are likely to display ("barbarous treatment," "a barbarous crime").

bathos, pathos   *Bathos:* a ludicrous descent from the lofty to the commonplace, anticlimax; *pathos:* tender, sorrowful feeling or that which rouses it.

beau ideal   The phrase means *ideal beauty,* not *beautiful ideal.*

begin, commence, start   If any distinction exists between *begin* and *commence,* it is that the latter is more formal and bookish. *Start* is more definite than the other words. We speak of starting a race, a watch, a quarrel, but beginning the school year, a friendship, travel.

being, entity   The words are nearly synonymous. *Entity* is, however, used of inanimate objects and *being* is not—at least in popular usage.

below, under   A sailor says: "I'm going below," "Davy Jones's locker is under [*not* below] the ship," and "The keel is under the ship." These are correct uses of the words. *Below* is in contrast with *above; under,* with *over. Above* and *below* refer to differences of level; *over* and *under,* to levels with reference to something else that is between them.

beside, besides   *Beside:* at the side of; *besides:* in addition to, moreover. Although *beside* in the sense of *besides* is not incorrect, the distinction given is convenient and is generally observed by the careful writer.

bias, prejudice, bigotry   *Bias* is a predisposition for or against something; *prejudice* is usually against. *Bigotry* is unreasoning adherence to a set of beliefs. Thus a bigot, biased against other races, might make prejudiced statements about Indians.

bight, bite   The *Bight* of Biafra; a *bite* of toast.

bimonthly, semimonthly   *Bimonthly:* every two months; *semimonthly:* twice a month. Similarly, *biennial, semiannual; biweekly, semiweekly.*

blanch, blench   *Blanch:* to whiten; *blench:* to shrink back, flinch.

blatant, flagrant   *Blatant:* offensively noisy, resounding, echoing; *flagrant:* scandalous, notorious, heinous.

bloc, block   "The Communist *bloc* approved the move." "He sold a *block* of stock."

both, each, either   *Either* for *each* or *both* is becoming archaic. The current idiom is to say *on each side* or *on both sides* of the street rather than *on either side.*

bourgeois, bourgeoise, bourgeoisie   The first is a masculine noun or adjective, the second is a feminine noun, and the third is the collective noun for the class, used as a singular.

Breton, Briton, Britain   *Breton* is the name of the language, or a person who speaks the language, of Brittany in France. *Briton* is the name of a person who lives across the Channel, and *Britain* is where he lives. The similarity memorializes not 1066, but an emigration of Celts a century before.

business, busyness   "He owns his own *business,* that explains his *busyness.*"

cacao, cocoa, coca, coco   *Cacao:* the tree providing *cocoa* or chocolate; *coca:* a shrub providing a stimulant drug (cocaine); *coco:* the tree bearing coconuts.

callus, callous   *Callus* (noun): physical thickening of the skin; *callous* (adj.): used in the physical sense of toughened, but also in an emotional or moral sense.

can, may   *Can:* is able; *may:* is permitted. In spite of the efforts of generations of teachers, the two are popularly confused, especially in questions—"May [*not* can] I leave the room?" Note the difference between "Can we go to town? (Are we able? Do we have the time? the means of transportation?)" and "May we go to town? (Are we permitted?)."

canvas, canvass   *Canvas:* a heavy cloth; *canvass:* to solicit.

capital, capitol   *Capital:* the city that is the seat of government; *capitol:* the building in which a legislative body sits.

carat, karat, caret, carrot   *Carat:* a measure of weight for precious stones; *karat:* an expression of the purity of gold; *caret:* a punctuation mark; *carrot:* a vegetable.

carousal, carrousel (carousel)   "The party was no more than a drunken *carousal.*" "After a roller coaster ride, they decided to ride the *carrousel* (or *carousel*)."

causal, casual   Frequently, and disastrously, transposed. *Causal:* relating to cause and effect ("Excess of demand over supply has a causal relationship to inflation"). *Casual:* occurring by chance; informal ("Neville and Mildred have a casual relationship").

cement, concrete   Generally, *cement* consists of powdered rock and clay; *concrete* consists of cement, gravel, and water.

censure, criticize   To *censure* is to express disapproval or blame, to *criticize* may be this, but in a nobler sense it is to weigh the merits as well as the demerits of a person or work. Such phrases as *to criticize adversely, to criticize negatively,* suggest that the original meaning of *critic,* a judge, is still felt.

ceremonial, ceremonious   *Ceremonial:* connected with, consisting of, or suitable for a ceremony; *ceremonious:* having an air of or resulting from ceremony, formal. We speak of a ceremonious person as of one who is elaborately polite, formal (probably too formal).

certainty, certitude   Both signify an absolute conviction, but a *certainty* is based on reasoning and examination of the facts, *certitude* involves a belief or habit of the mind.

cheerful, cheery The words are usually interchangeable, though *cheery* suggests a more demonstrative gaiety of manner than does *cheerful*.

childish, childlike *Childish* refers as a rule to the less pleasant qualities of a child ("Don't be childish"); *childlike* to the more pleasant ("He had a childlike innocence").

Chile, chili The first represents a nation, the second is the name of a spicy dish, usually served in a bowl, or of the spice that flavors it.

chronic, acute A *chronic* illness goes on and on; an *acute* one is likely to come to a crisis.

cite, quote To *quote* is to repeat the words of the original; to *cite* is to refer to or to give the substance of the original.

classic, classical, classicist *Classic* and *classical* are interchangeable in most contexts. They are contrasted with *classicist*, in that they refer to the qualities, styles, productions of the ancients, while *classicist* refers to later imitations of these or to works inspired by Renaissance theory or practice derived from the ancients. The adjective *classicist* is giving way to *neoclassic*. As nouns, *classicist* may refer to either a follower of classicism or one devoted to the classics; while *classic* may refer to a work of art or literature of the highest rank or, colloquially, by extension, to any object or accomplishment adjudged to be first class. In the plural *the classics* refers to the works of the ancient Greeks and Romans or the greatest works of modern literature or music.

clench, clinch Although their uses can overlap, *clinch* is used more often to refer to the fastening of nails or bolts, and *clench* to a more general clamping-together, as in clenching one's teeth. However, only *clinch* may be used to refer to settling arguments or agreements.

clew, clue Two spellings of the same word, *clew* now being used as a term for *a corner of a sail* or *one of the cords by which a hammock is suspended*.

climactic, climatic *Climactic* refers to climax ("It was the climactic event of the political year"); *climatic*, to climate ("The heavy rains were an annual climatic occurrence").

climate, clime *Climate* refers to weather; *clime*, to a region, with or without regard to weather. The latter is mainly used in elevated contexts.

cognomen Often misused. Originally, in ancient Rome, a *cognomen* was a name added to a man's name to indicate the family or gens to which he belonged, as *Caesar* in *Caius Julius Caesar*; now a *surname*. Colloquially the word is now used of any name or nickname.

cohort Mistakenly used for *henchman* or *companion*. *Cohort* means a body of warriors or other group banded together.

coincidence A *coincidence* requires at least two concurrent incidents. Good weather is not a happy coincidence when one merely means a fortunate occurrence.

collision In past practice, *collision* referred to the impact of two bodies in motion, although recent dictionaries do not make this distinction. Many editors disapprove of: "The car collided with a brick wall."

comic, comical The *comical* is droll, laughable, funny; the *comic* is that which aims at or has its origin in comedy. The comic may not cause physical laughter at all—as in "serious comedy"—but only laughter of the mind. To call someone a great comic actor would be high praise, but to call him a comical actor might not be. As a noun *comic* has been debased in some meanings, though it still retains its original sense of *that which excites mirth* or *one who can cause laughter*.

commonplace, platitude, truism All three words are used of ideas or observations not worth making. The derogatory is strongest, however, in *platitude*, which by derivation suggests a flatness. *Commonplaces* and *truisms*, though trite, may still be valuable. We may say that poetry deals with the great commonplaces or truisms of life, but we should hardly say that it deals with platitudes.

compendious   Often used to mean encyclopedic, *compendious* in fact means concise or terse. A *compendium* is a brief, complete summary.

complacent, complaisant   *Complacent:* pleased with oneself; *complaisant:* desirous of pleasing, obliging. *Complaisant* is being supplanted by *complacent*, but the distinction is worth preserving ("His complacent manner was vaguely irritating"; "The king appeared complaisant toward his council").

complement, compliment   "My ability to edit *complements* your ability to write. If we are successful, we will receive *compliments*."

comprehensible, comprehensive   *Comprehensible:* capable of being understood; *comprehensive:* large in scope or content or of large mental power.

comprise, compose, constitute, consist of   *Comprise:* to contain, embrace ("The state comprises sixteen counties"); do not use *is comprised of*. *Compose* and *constitute* are used in an opposite manner, for the parts that make up the whole ("Together, the counties compose—constitute—a state"). *Consist of:* to be made up of, composed of ("The state's highway system consists of two expressways and a number of turnpikes and parkways").

compulsion, compunction   *Compulsion:* the state of being compelled or forced, an irresistible psychological impulse; *compunction:* uneasiness brought about by guilt, qualms.

concern, firm   *Concern:* a business or manufacturing organization; *firm:* a professional organization or partnership, such as a law firm.

conclude, decide   *Conclude* implies previous consideration or exercise of judgment. One may *decide* on the spur of the moment, but *conclude* after reflection.

condign   The word means *deserved, adequate,* or *fit,* not, as many writers seem to think, *severe.* Condign punishment is merited punishment.

condole, condone   *Condole:* to grieve with, sympathize with; *condone:* to set aside an offense as if it had not been committed. The nouns are *condolence* and *condonation.*

conducive, conductive   *Conducive:* helping, promoting a result; *conductive:* having the power to transmit or conduct. *Conductive* is used mainly in connection with electricity or fluids. The verbs *conduce* and *conduct* show a similar differentiation.

confidant, confident   In current usage *confidant* (fem., *confidante*) is a noun; *confident,* an adjective. A *confidant* is a person of whose discretion one may be *confident.*

congenital, congenial, convivial   *Congenital:* dating from birth; *congenial:* of similar sympathies; *convivial:* involving feasting and drinking.

conjurer, conjuror   Both spellings are used in the sense of *magician.*

connive, conspire   These words should not be used interchangeably. *Connive* has nothing to do with *conspire,* meaning to *plot, scheme,* but, with the preposition *at,* means to *tolerate* or *feign ignorance of* ("The mayor connived at the corruption among his appointees"). Despite the resistance of many editors, *connive* may ultimately become synonymous with *conspire;* two recent dictionaries list *conspire* as a secondary meaning.

connotation, denotation   The *connotation* of a word is what it suggests apart from its explicit and recognized meaning, its explicit and recognized meaning being its *denotation.* The *connotation* of a word may differ with every hearer or reader, because it is due mainly to associations of idea or feeling; the *denotation*—the meaning given in dictionaries—is relatively stable and fixed. Poetry is concerned largely with *connotation,* while science carefully avoids it.

consequent, consequential   *Consequent* (adj.): following as a natural result ("Sin and consequent remorse"); *consequential* (in its popular use): self-important ("A consequential manner"). *Consequential* also means *following as a sequel or logical consequence,* but *consequent* has nearly monopolized this meaning.

considerateness, consideration *Consideration* is taking thought for the comfort or feelings of others; *considerateness* is the general attitude or habit that shows consideration on particular occasions ("His temperamental considerateness made it natural for him to show consideration on this occasion").

contagion, infection There is much confusion concerning the medical use of the words. Perhaps the best usage is to call the transmission of disease directly or indirectly from person to person *contagion* and the communication of disease in any matter (by air, water, insects, as well as contact) *infection*.

contemporary, modern *Contemporary:* existing at the time, in the present if no other time is mentioned. *Modern* usually refers to an era of which one is a part, or to the recent past.

contemptible, contemptuous *Contemptible:* deserving contempt; *contemptuous:* exhibiting contempt.

contend, contest *Contend:* to maintain by argument ("I contend that he had no right to do it"); to strive in opposition or rivalry ("On the voyage the ship had to contend with a strong head wind"). In the first sense it is followed by *that*; in the second usually by *with*. *Contest:* to contend about earnestly, perhaps with physical force, or to strive to take, keep, or control ("The Italians contested the Austrians' passage of the Alps"); to challenge ("The mayor contested the validity of the election returns").

continual, continuous The *continual* lasts, but with pauses or breaks ("He continually annoyed her"); the *continuous* lasts without pauses or breaks ("The noise of the waterfall was continuous").

continuance, continuation, continuity *Continuance* refers primarily to lastingness; *continuation*, to prolongation. We speak, therefore, of the continuation (*not* continuance) of a story; but of the continuance of kindness. *Continuity* is the state or quality of being extended or prolonged or of being uninterrupted. The phrase "continuity of history" suggests unbrokenness primarily and lastingness only secondarily.

contumacy, contumely The words have no connection but are sometimes confused. *Contumacy:* stubbornness, perverse obstinacy ("The judge pronounced the prisoner's behavior mere contumacy"); *contumely:* haughtiness, scorn, insolence ("The proud man's contumely." — *Hamlet*).

convince, persuade *Convince:* to satisfy by argument or evidence; *persuade:* to induce. *Persuade* is followed by an infinitive; *convince* is not ("They persuaded me to go, but did not convince me that I had done the right thing").

corespondent, correspondent *Corespondent:* the person charged as the respondent's paramour in a suit for divorce; or a defendant in a suit in chancery or in an admiralty cause; *correspondent:* one party to an exchange of letters or a person hired to report news from a particular place.

corporal, corporeal *Corporal:* of the body itself ("Corporal punishment"); *corporeal:* like or of the nature of a body, material, tangible ("The specter appeared in corporeal form").

cost, price, value, worth The *cost* of a thing is whatever is paid for it; its *price*, what the seller asks for it; its *value*, the ratio it bears to a recognized standard; its worth, the ratio it bears to the buyer's desire or need.

councilor, counselor *Councilor:* a member of a council; *counselor:* one who gives advice, especially in law.

couple, two Popularly, *couple* is used in the sense of *two* or a *few* ("I'm going away for a couple of days"). Strictly, it refers to two joined in some way ("hounds leashed in couples," "a married couple").

covetous, envious, jealous A *covetous* person desires what another has; an *envious* one feels ill-will toward another because the latter has more than he; a *jealous* one resents another's intrusion upon what he possesses. Covetousness may involve envy and jealousy. Covetousness and envy are always bad; but jealousy may be good or bad, according to the occasion and the object.

One may, for example, be rightly jealous of one's good name; that is, resentful of slander. The Bible speaks of God as jealous—"I thy God am a jealous God."

credible, credulous   *Credible:* believable ("His story was credible"); *credulous:* too willing to believe ("His story was silly, but the credulous accepted it").

crevice, crevasse   A *crevice* may be in a wall; a *crevasse* is in a glacier or ice field.

currant, current   *Currant:* a kind of fruit. *Current:* a stream of water or the flow of electricity (*n.*); at the present time (*adj.*).

cynical   The word is so often used trivially that its traditional meaning should be kept in mind. It early meant *exhibiting an ignorant and insolent self-righteousness,* and, later, *a skeptical attitude toward virtue and idealism and towards lofty or noble motives.* Historically, then, it suggests *pessimistic* or *misanthropic.*

daemonic, demonic, demoniacal   *Demonic:* of or like a demon, as "demonic possession." *Demonic* and *demoniacal* are not clearly differentiated, though phrases like "demonic power" and "demoniacal leer" suggest that *demoniacal* connotes *wicked, devilish;* and *demonic, intense* (as a demon). *Daemonic* is sometimes used with a return to the Greek meaning of *possessing fascination* or *exercising hypnotic influence.*

damage, injury   *Injury* originally meant *that which is not right or just; damage* meant *loss.* As a rule we think of persons, feelings, rights, or reputation as being injured, our possessions, things, damaged. However, the distinction is not clearly drawn, for we also speak of injury to property. *Damage* involves loss; *injury,* hurt. *Damage* is impairment of value; *injury,* impairment of beauty, utility, integrity, dignity.

deadly, deathly   *Deadly:* lethal ("a deadly poison"); *deathly:* as in death ("a deathly stillness").

decadent, decedent   *Decadent:* in decay; *decedent:* a dead person.

deduce, deduct   *Deduce:* to derive by reasoning ("We deduced his meaning by his manner rather than by his words"); *deduct:* to take away or subtract ("You may deduct ten percent for cash").

deduct, subtract   *Deduct* applies to amounts or quantities; *subtract,* to numbers.

defective, deficient   *Defective:* having defects; *deficient:* lacking completeness. Think of *defect* and *deficit.*

definite, definitive   *Definite:* precise; *definitive:* conclusive, authoritative.

deism, theism   In ordinary use, *deism* is a historical term for a form of religious belief that flourished in the eighteenth century; *theism,* a general term for belief in God or gods. Theologians and historians of religion make more elaborate or subtle distinctions than this.

delay, postpone   *Delay:* to slow or hinder; *postpone,* to put off to another date ("Because the team was delayed on the road, the game was postponed").

delightedly, delightfully   *Delightedly:* manifesting delight; *delightfully:* affording delight.

delusion, illusion   *Delusion:* a false belief; *illusion:* a false perception. Illusions are more easily corrected than delusions.

demean   One use of this word is simply to note behavior or manner, as in *demeanor.* Some object to the other meaning, *debase* or *degrade,* as a corruption, but as Theodore M. Bernstein points out in *The Careful Writer,* the use has been standard since the dawn of the seventeenth century.

depot   Etymologically *depot* is connected with *deposit,* and refers to a place for deposit or storage, such as a warehouse. Its use for *railroad* or *bus station* is American.

deprecate, depreciate   *Deprecate:* to express disapproval or regret; *depreciate:* to lessen in value, to cry down.

deserving, worthy of *Deserving* and *worthy of* are synonymous and may be followed by a substantive expressing either praise or blame, reward or punishment. Present usage, however, tends to restrict *worthy of* to expressions of reward, though it does not hesitate to speak of "deserving punishment." To many minds there seems to be a contradiction in using *worthy,* meaning *meritorious, excellent, estimable,* in an expression of blame or punishment.

determinism, fatalism Popularly *fatalism* is the attitude of mind of those who believe that events, especially death, will come when they will and that it is useless to try to avoid them. Philosophically, *fatalism* is the theory that some external force controls our lives inexorably in every detail; *determinism,* the theory that the action of cause and effect makes our freedom of will only apparent or a delusion.

deterrent, detriment *Deterrent:* that which hinders, slows down ("Severe penalties may not act as deterrents to crime"); *detriment:* that which causes damage, injury, loss ("His lack of honesty proved a detriment to his candidacy").

detract, distract *Detract:* to reduce or take away from ("His unpleasant manner detracted from the credit his actions deserved"); *distract:* to divert or perplex ("A loud noise distracted our attention from the proceedings").

dialectal, dialectic A tendency is perceptible to limit *dialectal* to reference to dialect and *dialectic* to reference to dialectics; but the distinction, though useful, is not generally observed except among scholars.

dilemma Correctly, dilemma implies a choice between undesirable alternatives, but has been accepted as any choice, such as that between two candidates. It is not accepted yet as a synonym for *problem* or *quandary.*

discomfiture, discomfort Although their meanings are converging, according to the *American Heritage Dictionary,* some distinction remains. *Discomfiture* originally meant defeat, rout; *discomfort,* unease.

discover, invent We *discover* what is already in existence; *invent* something new.

discreet, discrete *Discreet:* wise, prudent, judicious; *discrete:* disconnected, separate, discontinuous.

disinterested, uninterested A person may be disinterested without being uninterested. A *disinterested* person is impartial; an *uninterested* one is indifferent.

distinctive, distinct *Distinctive* refers to things that set off a person or thing from others of the kind. *Distinct* means perceptible, unmistakable ("He had a distinctive gait, distinct at a great distance").

dock, pier, wharf A *dock* is the water beside or between *piers* or at a *wharf* for ships' use. A *pier* is a platform built out into the water; a *wharf* is any landing place. In popular usage, the three terms are used interchangeably in the sense of *pier.*

dominance, domination Though the words are often interchangeable, *dominance* suggests a condition or fact of authority ("The dominance of Rome in the first century does not suggest that its decline was to come so soon"), and *domination* an act or exercise of power, often arbitrary or insolent ("We rebelled against his domination").

dual, duel The first relates to *two* ("dual-control car"); the second describes a contest with swords or pistols.

durance, duress *Durance* is a term little used except in hackneyed phrases, such as "durance vile." It means *confinement. Duress* means *restraint,* not necessarily in confinement, and is used mostly in legal contexts.

eclectic *Eclectic* means chosen from diverse sources, not necessarily (as sometimes used) *discriminating.*

effete, effeminate By no means the same. *Effete:* sterile, worn out; *effeminate:* of a man, being soft, weak, womanish (derogatory in almost every sense).

egoism, egotism   The distinction is generally ignored, but is useful. *Egoism* suggests self-supremacy rather than self-conceit, and philosophically may be free from connotation of selfishness or vanity. *Egotism* is not philosophical but instinctive and usually expresses itself outwardly in speech and action.

elder, older   *Elder* can be used only to refer to persons ("elder brother," "elder statesmen"); *older*, to persons or things ("older workers," "older ideas," "older cars").

elemental, elementary   *Elemental* refers to the primal forces of nature or of feeling: fire, cold, passion, power, though it may also be used in the sense of *elementary*. *Elementary* usually refers to the simple or rudimentary parts, the first principles of anything: elementary grades in a school, elementary knowledge.

empty, vacant   Generally, *empty* is opposed to *full* ("an empty bucket"); *vacant,* to *occupied* ("a vacant building"). When *empty* is used to mean *not occupied,* it suggests less duration than does *vacant* ("an empty room," "a vacant room"). *Vacant* may be used figuratively to refer to a lack of intellectual content, knowledge, or intelligent expression ("the vacant mind," "a vacant stare").

endemic, epidemic   *Endemic:* prevalent in a given area or society; *epidemic:* spreading rapidly through a people or area, especially a contagious disease. The words share the Greek root, *demos* (people).

enervate, energize   *Enervate:* weaken; *energize:* instill with vigor.

English, British   *English:* (adj.) of England; (*n.*) the inhabitants or the language. *British:* (adj.) of Great Britain (England, Scotland, Wales) or of the United Kingdom of Great Britain and Northern Ireland; (*n.*) the inhabitants.

enormity   Should not be used to indicate size (*enormousness*), but *monstrous evil,* or *outrageousness.*

enough, sufficient   Despite much discussion no perceptible difference in meaning has been established. *Enough* is the homelier and blunter word.

entomb, entrap   *Entrapped* miners, for example, may be alive; *entombed* miners are dead.

envisage, envision   These words are sometimes used interchangeably, but *envisage* means to conceive of some future goal, while *envision* means simply to picture something in the mind.

epitome   The epitome is not the *acme* or *height of,* but is a *summary, abstract,* or *essence.*

equable, equitable   *Equable:* of even or balanced range or uniform condition, not varying ("an equable disposition"); *equitable:* fair, just, impartial ("an equitable decision").

eruption, irruption   *Eruption:* a bursting out; *irruption:* a bursting in ("After the eruption of the oil stored in the tanker, there was an irruption of water into the hold").

euphemism, euphuism   *Euphemism* means the substitution of an agreeable term for one considered over-bald, as in *Junoesque* for *large. Euphuism* indicates an elaborate, affected style, as in the poetry of Lyly.

exalt, exult   *Exalt:* raise in status, glorify; *exult:* rejoice.

exceedingly, excessively   *Excessively* is stronger than *exceedingly. Exceedingly* means *much; excessively, too much.*

exceptionable, exceptional   *Exceptionable:* objectionable; *exceptional:* out of the ordinary. Exceptional measures would be more than were expected; exceptionable measures, those that roused opposition. *Unexceptionable* means *not open to criticism or objection.*

excitation, excitement   *Excitation* is the rousing of feeling; *excitement* can be used for *excitation,* but also refers to feeling itself ("Excitation [*or* excitement] of the nerve centers by means of electricity"; "In the excitement of the moment, he forgot to look").

expedient, expeditious *Expedient:* advisable or advantageous; *expeditious:* quick, speedy. Expeditious action is fast action; expedient action is prudent or politic action.

extenuate Sometimes misused in the sense of *excuse. Extenuate* means, by derivation, *make thin;* it now means *represent as less blameworthy, mitigate, palliate.* When Othello says "Nothing extenuate..." he means that his deeds are to be told as they were, not softened for the sake of his reputation.

factitious, fictitious *Factitious:* contrived; *fictitious:* not real. Something *factitious* may nonetheless be genuine; something *fictitious* may not.

farther, further *Farther* refers to physical distance ("We can go no farther on this road"); *further* may be used in this sense but is most often used figuratively to refer to extent or degree ("We will proceed no further in this unpleasant business").

fatal, fateful *Fatal* has to do with death, destruction, failure. Originally *fateful* might be used of a happy fate, but now it usually signifies *fraught with malign fate or dangerous possibility.*

feasible, possible *Feasible* means *capable of being done or carried out.* In most contexts *feasible* and *possible* are interchangeable: "Transportation of troops by air is feasible" or "is possible." But often they are not. "A revolt is possible" and "a revolt is feasible" do not mean the same. *Feasible* should not be followed by *that,* as *possible* often is.

ferrule, ferule *Ferrule:* a ring or cap (on a cane, for example); *ferule:* a rod or ruler for punishing children.

fewer, less *Fewer* is used of numbers; *less,* of quantity ("Fewer men require less food").

flacon, flagon A *flacon* is a drinking vessel; a *flagon* has a spout and handle ("He downed the flacon of ale, then poured another round from the flagon").

flail, flay *Flail:* beat, thrash, strike; *flay:* strip the skin from. Should not be used interchangeably or loosely (as in "Senator Flays Administration Policies").

flair, flare *Flair:* instinctive power of discriminating, or taste combined with aptitude; *flare:* sudden or unsteady light or flame.

flaunt, flout Sometimes confused. *Flaunt:* make ostentatious display or vulgar show; *flout:* reject contemptuously, sneer at, jeer.

flee, fly *Fly* make take the place of *flee,* but *flew* cannot take the place of *fled* entirely—"The wicked flee when no man pursueth" might read "The wicked fly..." but "We fled (*not* flew) the country."

fleshly, fleshy *Fleshly:* having the qualities of flesh as compared with spirit ("The fleshly school of poetry"); *fleshy:* having flesh or like flesh ("A fleshy person or face or fruit").

flotsam, jetsam, ligan (lagan) *Flotsam:* wreckage found afloat; *jetsam:* a corruption of *jettison,* goods cast overboard and such goods washed ashore; *ligan:* goods thrown overboard with a marker buoy attached.

forceful, forcible *Forcible* implies the use of force; *forceful,* the predominance or domination of force. Compare "forcible enslavement" and "forceful leader."

foregone, forgone "That's a *foregone* conclusion." "We've *forgone* further discussion in order to have time to vote."

fortuitous, fortunate *Fortuitous:* unplanned, happening by change; *fortunate:* marked by good luck. Unexpectedly meeting an acquaintance on the street would be fortuitous; if it was someone you had been looking for, the meeting would also be fortunate.

founder, flounder *Flounder:* move clumsily; *founder:* to collapse utterly or (nautical) to sink. ("After floundering about in the storm for hours, the boat foundered").

fruition   *Fruition,* in its original meaning, is pride or enjoyment from use or possession ("We were stopped most suddenly and cruelly from the fruition of each other." — James Boswell). Except through abuse, *fruition* has nothing to do with fruit or ripening, but it has become accepted usage for the *condition of bearing fruit* and especially for the *realization of hopes.*

fulsome   *Fulsome* is not *plentiful* but *excessive, insincere, odious.*

gantlet, gauntlet   *To run the gantlet* (also spelled *gauntlet*): a punishment in which the victim is compelled to run between rows of men, who strike him. *Gauntlet:* a glove with a protective long wrist, a warrior's glove; hence *throw down the gauntlet:* challenge.

gibe, jibe   *Gibe:* a sneer, to scoff; *jibe:* (naut.) to swing from side to side; (colloq.) to agree.

glean   *Glean* means *gather leavings* or *gather laboriously,* little by little. It is incorrectly used to mean *acquire, get.*

graceful, gracious   *Graceful:* elegant, charming, or appropriate in carriage, movement, or manner; *gracious:* exhibiting or inclined to show favor, mercy, kindness, condescension. A graceful lady is not always a gracious lady.

grill, grille   *Grill:* a gridiron; something broiled; *grille:* a grating or screen, usually of wrought iron.

grisly, grizzly   *Grisly:* horrifying, ghastly; *grizzly:* grayish.

hanged, hung   A person condemned to death may be *hanged* by the neck until dead; but pictures and steaks are *hung.*

hardly, scarcely   These words are interchangeable in popular use, but in strict use *scarcely* refers to quantity and *hardly* to degree ("It is scarcely a mile to town"; "He will hardly reach town by noon").

healthy, healthful   *Healthful:* promoting health; *healthy:* in a condition of health. Although *healthy* is gradually driving *healthful* out of use, it is still preferable to say "healthful foods," not "healthy foods."

historic, historical   *Historic:* memorable or famous ("Historic shrine"); *historical:* having to do with history ("Historical evidence").

Hobson's choice   The phrase means *no choice at all,* not *dilemma.* The term comes from the name of a stablekeeper, who told customers to take the horse nearest the door or none.

homogeneous, homogenous   *Homogeneous:* of the same nature or constitution throughout; *homogenous:* a resemblance due to common origin (biological).

humane, humanitarian, humanistic   *Humane:* kind or merciful; *humanitarian:* benevolent, philanthropic; *humanistic:* pertaining to or characteristic of humanism.

hydroscope, hygroscope   *Hydroscope:* an apparatus for observing objects in the sea, or one for measuring time by the dropping of water; *hygroscope:* a device for measuring humidity.

hyperbola, hyperbole   *Hyperbola:* a geometrical figure; *hyperbole:* rhetorical exaggeration.

hypercritical, hypocritical   *Hypercritical:* overcritical, too exact; *hypocritical:* not practicing what one preaches.

hypothecate, hypothesize   *Hypothecate* is a legal or financial term meaning *to pledge, pawn, or mortgage; hypothesize* means *to make hypotheses.*

idle, idol, idyl (idyll)   "My broken leg forces me to be *idle.*" "The grave robbers stole the ancient *idol.*" "His eighteenth summer was an *idyl* (or *idyll*)."

ilk   The Scottish word *ilk* is often used incorrectly, as if it meant *name, family* or *kind.* It means *same* in Scottish, as it did in Anglo-Saxon. The phrase *of that ilk* means *of the same;* that is to say, the person's surname is the same as that of the estate; for example, Alexander Burnieside of that ilk, i.e., of Burnieside.

illegible, unreadable   That which is *illegible* cannot be read because of bad writing or printing; that which is *unreadable* is dull, dry, or irritating.

illude   *See* allude.

immanent, imminent   *Immanent:* indwelling, inherent ("God is immanent in nature"); *imminent:* impending ("The beginning of the battle was imminent").

immure, inure   *Immure:* wall in, imprison; *inure:* accustom, habituate ("Immured within the Bastille as he was, he became inured to pain").

impassable, impassible, impassive   *Impassable:* closed, offering complete obstruction to; *impassible:* unfeeling; *impassive:* unemotional, stoical, apathetic, calm, serene. *Impassible* and *impassive* are often interchangeable, but *impassible* is likely to be used in a nobler sense than *impassive,* the former suggesting the restraint of feeling, while the latter suggests the absence of feeling.

imply, infer   To use *infer* in the sense of *imply* or *suggest* ("I heard him infer that we had not told the truth") is a dubious colloquialism. Strictly, *infer* means *surmise, conclude* ("I inferred from his manner that he was disturbed in his mind"; "From these data we infer an inevitable conclusion").

impracticable, impractical   *Impracticable:* incapable of being done; *impractical:* unwise to do or inefficient in practical matters. A trip to Europe might be *impractical* because it would interfere with the conduct of one's affairs, or *impracticable* because one lacked the money for passage.

impress, impression   *Impress* is mainly used to mean *a mark made by pressure, an imprint* or *a characteristic trait, mark of distinction.* It shares the first meaning with *impression*—"the impress (or impression) made by a die"—but not the second—"The work bore the clear impress of genius." On the other hand, *impression* often means *a vague or indefinite remembrance or recognition* ("I have an impression that he is not entirely honest"). This meaning is not shared by *impress.*

inartistic, unartistic   *Inartistic:* not in accordance with the canons of art; *unartistic:* not gifted in or interested in art. A picture may be inartistic; a person unartistic.

incipient, insipient   *Incipient:* beginning to exist; *insipient:* stupid.

incubus, succubus   Originally an *incubus* was a male demon fabled to have intercourse with women in their sleep; hence it has come to mean *weight, load,* or anything harassing or oppressing. A *succubus* was a female demon fabled to have intercourse with men in their sleep.

indict, indite   *Indict:* of a grand jury, to bring a formal accusation against, as a means of bringing to trial; *indite:* write, write down, compose.

inflammable, inflammatory   *Inflammable:* readily set on fire. *Inflammatory:* tending to produce heat or excitement; also, causing inflammation. Dry shavings, passions, are inflammable; a seditious speech, a mustard plaster, inflammatory.

ingenious, ingenuous   *Ingenious:* skillful in contriving, inventive; *ingenuous:* artless, open, naïve, magnanimous, sincere, innocent in a good sense.

injury   *See* damage.

insoluble, unsolvable, insolvable   All three words are used in the sense of *not to be solved or explained,* but *unsolvable* and *insolvable* have only that meaning, while *insoluble* may mean *not to be dissolved.*

internecine   Although the word originally meant *mutually deadly,* now it can be used to describe *internal struggle,* not necessarily fatal. Thus the word has all but lost its original force.

intense, intensive   *Intense:* present to a high degree, extreme ("an intense light"). *Intensive:* concentrated, exhaustive ("an intensive review of tariff regulations").

join issue, take issue Both are correct forms and are used interchangeably in the sense of *disagree*. *Join issue* may, however, mean—and perhaps more properly—*to agree* on the grounds of a dispute. *Take issue* always means *disagree*.

judge, jurist A *jurist* is any person expert in law and not necessarily a *judge*, but a judge ought to be a jurist.

judicial, judicious *Judicial*: of or pertaining to judges or law courts; *judicious*: having to do with judgment, well-calculated, wise, prudent.

junction, juncture The words are interchangeable in the sense of *the act of joining or union*, but *juncture* is used in the special sense of *a point of time, a crisis or exigency*, while *junction* is used in the sense of *crossroads*.

kin *Kin* indicates relatives, collectively; it cannot be used to refer to an individual ("He's one of my kin," *not* "He's a kin of mine").

lack, absence *Lack*: deficiency of something needed ("The lack of fresh fruit brought on scurvy"); *absence*: non-presence of something that may or may not be necessary ("The absence of rain permitted the fair to open").

lament, mourn Both can mean *express grief*, but *lament* also may mean *complain, regret*. Thus, "The state mourns Senator Gerry" is preferable to "The state laments Senator Gerry."

last, latest *Last*: that which comes after all others, the end; *latest*: last in time. Idiomatically, however, *last* may mean *the next before the present* ("last Tuesday") and *the furthest from likelihood* ("the last person one would expect it of"). "The last news" means that there will be no more; "the latest news," that there may be.

levee, levy *Levee*: an embankment serving as a flood wall or as a landing-place; *levy*: a tax, or the act of imposing a tax.

liable, likely *Liable* suggests a legal obligation ("We are liable to lose our license if we drive so fast"). *Likely* expresses inclination or probability ("We are likely to arrive on time").

libel, slander In law, *libel* is written or printed; *slander*, spoken, but the distinction is ignored in popular usage and radio and television have blurred the legal line.

limpid Not *limp* or *frail*, but *clear*, like a brook.

literally, figuratively *Literally* is often used where *figuratively* is meant. "The horse literally stopped dead in its tracks." Unless the horse died, *figuratively* should be used. Better still, both terms can usually be omitted.

livid, lurid *Livid*: black or leaden (not red)—thus figurative when used to mean angry; *lurid*: from Latin for *ashy, pale*—thus ghastly, causing shock ("The actor became livid when he saw the lurid advertisements").

locate, situate *Locate* is used colloquially for *situate*, but in standard use *situate* refers to a site, and *locate* to finding or placing something ("Our house is situated on a hill; you can locate it by looking up"). Usually, *situated* can be left out entirely.

longshoreman, stevedore Often used interchangeably, but on the waterfront, a *longshoreman* is a worker, and a *stevedore* an employer of longshoremen.

luxuriant, luxurious *Luxuriant*: abounding, teeming; *luxurious*: given to indulgence of the senses, promoting bodily ease.

madam, madame *Madam*: a title of courtesy used before rank or title ("Madam Ambassador"); as a noun, *madam*, the keeper of a house of prostitution. *Madame*: roughly, the French equivalent of *Mrs.*; also, a title of courtesy.

mantel, mantle *Mantel*: relates to a fireplace; *mantle* to a cape or cloak.

marginal Should not be used in place of *minimal* or *minor*. A person with *marginal* mathematical skills does not necessarily have *small* skills, but rather skills barely within a given standard. In economics, *marginal* refers to enterprises that produce goods whose sale barely covers production costs.

marine, maritime *Marine:* of the ocean; *maritime:* bordering on or connected with the ocean ("marine animals," "the merchant marine"; "the Maritime Provinces," "The Maritime Commission").

marshal, Marshall, marital, martial *Marshal:* a military rank; *Marshall:* a Pacific island group; *marital:* having to do with marriage; *martial:* having to do with military affairs.

masterful, masterly *Masterful:* strong, domineering; *masterly:* showing superior skill, like a master. Thus, a musician gives a *masterly* (not *masterful*) performance. (*Cf.,* forceful, forcible.)

mean, median, average *Mean:* a middle point; usually the *arithmetic mean,* obtained by dividing the sum of quantities in a set by the number of terms in the set. *Median:* the middle value in a distribution—for example, in a series of numbers, the value that would be larger than half and smaller than half the series. *Average:* in ordinary use, the *arithmetic mean*; as an adjective, *usual, typical.*

meantime, meanwhile *Meantime* is usually a noun; *meanwhile,* an adverb ("In the meantime, he repaired his bicycle"; "Meanwhile, he repaired his bicycle"). Both words are overused.

minimize To *minimize* means to *reduce to the smallest possible degree* and and only secondarily to *belittle* or *depreciate* ("The treaty minimized the likelihood of war").

mitigate, militate Unrelated but often confused. *Mitigate:* moderate, alleviate ("Fond remembrance mitigated his sorrow"); *militate* (with *against*): have the force of evidence ("The weight of fact militates against that conclusion").

moot As an adjective, can mean either *arguable, under discussion,* or in legal use, *without legal significance, of solely academic interest.* Thus, a *moot court* at a law school offers students the chance to argue cases previously decided or imaginary.

morbid *Morbid* means *diseased, unhealthy,* in body or mind. We speak correctly of a morbid condition, mind, imagination. But popularly the word is often debased to mean *sad, unpleasant, unwholesome.*

mutual, reciprocal *Mutual* relates to both parties at the same time; *reciprocal,* to one with relation to the other, though there are contexts in which either is correct. A mutual agreement is, for example, the same as a reciprocal one. We cannot, however, speak of Jones's *mutual* dislike of Smith; here *reciprocal* would be correct.

mysterious, mystic *Mysterious:* full of mystery, difficult to explain. *Mystic:* primarily, pertaining to occult religious rites, inspiring a sense of wonder.

naval, navel, nautical *Naval:* having to do with the navy; *navel:* bellybutton. *Nautical:* pertaining to ships and seamen.

naked, nude Roughly synonymous, although *naked* is more forceful and also has more figurative uses. One would not write, for example, of *the nude truth,* or "*nude* to mine enemies."

nation, country *Nation:* the people of a single territory, or under a single government ("The nation has a republican form of government"); *country:* the geographical entity ("The country has two long rivers").

nauseous, nauseated *Nauseous:* causing nausea, sickening, repulsive; *nauseated:* experiencing nausea ("A nauseous sight"; "A nauseated airplane passenger").

noisome, noisy Unrelated but sometimes confused. *Noisome* means *offensive, foul, harmful* ("The noisome weeds, that without profit suck/The soil's fertility from the flowers"—Shakespeare, *Richard II*).

oblivious, ignorant   *Oblivious* should not be used for *ignorant,* which applies to things that you did not know in the first place. *Oblivious* implies forgetfulness or lack of awareness.

observance, observation   *Observance:* attending to or carrying out a duty, rule, or custom; *observation:* consciously seeing or taking notice. *Observance* is used of laws, rules, anniversaries; *observation,* of persons, things, events ("Reinforce imagination with observation"; "The observance of Sunday was strictly enforced").

obverse   The *obverse* of a coin is its face; antonym of *reverse.*

official, officious   *Official:* authoritative; *officious:* meddling.

Olympian, Olympic   *Olympian* has reference to Mount Olympus, home of the Greek gods in mythology. *Olympic* refers to Olympia, site of the ancient quadrennial games, and to their modern revival. By extension, *Olympian* means *majestic, superior.*

optimistic, hopeful   *Optimistic* is best used to describe a general attitude ("I am optimistic about the future"); *hopeful,* in regard to a specific occasion ("I am hopeful that the talks will succeed").

oral, verbal, vocal   *Oral:* by the mouth, spoken; *verbal:* of words, expressed in speech; *vocal:* of the voice, spoken. Clearly, these words overlap, but the standard preference is for *oral* to describe something not written down ("oral agreement") and *verbal* to contrast with physical ("a verbal exchange"). *Vocal* tends to refer to persistence and volume of speech.

partially, partly   The words are interchangeable in the sense of *in part, in some measure, not totally,* but only *partially* is used in the sense of *inclining to take one part or party, showing predilection or fondness.*

pedal, peddle   "Quang *pedals* his bicycle to the marketplace, unloads the baskets, and settles down to *peddle* his fruit."

pendant, pendent   *Pendant* is a noun; *pendent,* an adjective. *Pendant:* something that hangs ("She wore a jeweled pendant"); *pendent:* hanging ("Pendent nest and procreant cradle" — Macbeth).

piteous, pitiable, pitiful   *Piteous:* exciting pity; *pitiable:* exciting pity or contempt; *pitiful:* feeling or exciting pity or contempt. The words cannot be strictly differentiated.

place, put   *Put* is the more general term. *Place* denotes greater care and exactness. We put on our hat but place a crown on a king. Note that *put* may take *in* or *into,* but *place* takes only *in* ("I placed my model boat in the water"; "I put my hand in [*or* into] the water").

plebiscite, referendum   *Plebiscite:* a direct vote of a people on an issue, such as self-determination; it may or may not be binding as law. *Referendum:* the enactment of law by popular vote.

populace, populous   "The *populace* has been restive since the declaration of martial law." "Shanghai is the most *populous* city in the world."

practicable, practical   *Practicable:* feasible; *practical:* suited to use or action, useful.

precedence, precedents   *Precedence:* the condition of preceding or the act of going before; *precedents:* established usage or authority.

preceding, previous, prior   Dictionaries make the distinction that *preceding* refers to that which precedes without an interval and *previous* to that which precedes at any time. If, for instance, one were to refer on page 58 to something on page 57, the reference should be to "the preceding page," not to "the previous page." A reference to any other page should preferably be to "a previous page." Concerning acts or events occurring before a certain time or another act, the distinction is little observed. *Prior* is roughly synonymous with *preceding* and *previous* ("a prior commitment"), but *prior to* for *before* is to be avoided.

precipitous, precipitate *Precipitous:* steep; applied to physical characteristics ("a precipitous descent"). *Precipitate:* abrupt; applied to actions ("a precipitate reversal of policy").

predicate (*v.*) Does not mean *predict*, but *affirm, state* ("One cannot predicate virtue or vice of a creature which has no moral nature").

prescribe, proscribe "The doctor *prescribed* medicine for the heart patient. He *proscribed* any tension or exertion."

presumptive, presumptuous *Presumptuous:* more common, and rarely misused, for *presuming too much, taking excessive liberties. Presumptive:* founded on presumption (chiefly legal).

principal, principle Still confused, despite appearances on every list of misused words of this century and last. *Principle* (*n.*): a basic belief or truth ("Stick to your principles"); *principal* (*adj.*): most important ("His principal demand").

prone, supine *Prone:* face downward; *supine:* face upward.

prototype A *prototype* is a primal or original type ("The *Iliad* is the prototype of all classical epics"). The word is sometimes mistakenly used to mean the main or representative type or example.

progenitor A *progenitor* is an ancestor, not an inventor.

propaganda, publicity The difference between the terms is in persuasive intent. *Propaganda* tends to be material disseminated to sway people toward a given doctrine, ideology, or policy. *Publicity* is information disseminated to the public; it may or may not have persuasive content.

proportions, dimensions *Proportions* indicates relationship of one part to another or to a whole; *dimensions* refers to size.

protagonist Not the opposite of *antagonist* (adversary), but rather the principal character in a play (from the Greek *protos*, first), and thus the most prominent person in a given situation. A situation can have only one *protagonist*; hence, the plural is not used in reference to a single work or situation.

pupils, students Generally, "elementary-school *pupils*"; "high-school or college *students*."

qualitative *Qualitative* has nothing to do with merit or excellence, but with qualities or constituents of a subject, as in the chemistry term, *qualitative analysis.*

quota A *quota* is an *allotment*; a *share*, as of goods. It should not be used merely to indicate a number in general, as in "I have a big quota of magazine subscriptions."

rabbit, rarebit The original form was apparently *rabbit, Welsh rabbit* being a humorous name, like *Cape Cod turkey* for *codfish; rarebit* was substituted under a misapprehension of the joke.

rack, wrack, wreak *Rack:* (*v.*) spread out, torture, strain, as on a rack ("nerve-racking"). *Wrack:* (*v.*) wreck, destroy ("The hurricane wracked the city"); (*n.*) violent destruction ("wrack and ruin"). *Wreak:* (*v.*) vent (anger), inflict (vengence or punishment) ("They wreaked their vengence upon the children of their oppressors").

rare, scarce A thing may be *scarce* without being *rare*. Blackbirds are not *rare* birds, but they may be *scarce* in a place or season.

realtor The word was originally applied to a real estate agent or realty broker affiliated with the National Association of Real Estate Boards. The board still protests its use in other cases.

rebut, refute *Rebut:* answer, dispute, reject; *refute:* disprove.

recrudescence, resurgence Sometimes confused because of similar sounds. *Resurgence:* a revival, a sweeping back (as into power); *recrudescence:* a breaking-out after dormancy, as with a disease.

regretful, regrettable   "He is *regretful* about that *regrettable* action."

repel, repulse   Both mean to *drive off* or *resist*, but *repel* has the additional implication of *aversion* ("The idea repels me"). *Repulse* has nothing to do with *repulsive*; it can be used in the sense of *rebuff* ("They repulsed his efforts at peacemaking").

replace   There is no authority for the idea, often advanced, that *replace* must not be used in the sense of *substitute*. One meaning of *replace* is *put back in the same place,* but other meanings are *take the place of, supersede, find a substitute for.*

replica   Strictly, a *replica* is a duplicate (of a work of art) made by the original artist. Loosely, it is a duplicate or copy in general.

residue, residuum   Both words mean *that which is left over,* but *residuum* is a learned word or technical term in mathematics, physics, and chemistry. For popular use, *residue* serves.

respective   The word is really seldom needed, but such a sentence as the following is common: "Harvard and Yale men know far too little of their respective universities." The author meant "Yale and Harvard men know far too little about each other's university."

responsible for   Should not be used where the meaning would be better expressed by *produced, caused, resulted in.*

restrain, restrict   *Restrain:* to curb, check, repress; *restrict:* to restrain within bounds, limit, confine. *Restrain* is used in the sense of *restrict* in speaking of trade, power, or a title and in the sense of *confine, deprive of liberty* in speaking of insane or criminal persons.

restricted   Should not be used where the meaning would be better conveyed by *scant, scanty,* or *small*. Say "a man of scanty, or small, means," rather than "of restricted means." *Restricted* denotes *kept within limits or bounds, withheld from going too far.*

result, resultant   *Resultant* as a noun is a learned or technical term and should be reserved for scientific contexts—mechanics or algebra.

reverend   *Reverend* is an adjective, not a form of address.

reversal, reversion   *Reversal* should be thought of in connection with *reverse*; *reversion,* with *revert*. *Reversal* means *a turning round* ("His reply showed a complete reversal of his former opinion"); *reversion,* used mainly in science and law, means *a turning back* ("The actions of the maddened man suggested a reversion to savagery").

rightfully, rightly   The two words share the meaning of *justly, uprightly*—"The referee administered the claims rightly (or rightfully)." *Rightly* may also mean *exactly, correctly* ("You answered rightly"). *Rightfully* commonly means *properly* ("You rightfully should have received a share").

rob, steal   "A person or an institution is *robbed*; money or goods are *stolen*."

sadism   Often misused for *persecution* or *cruelty*. Its strict meaning is *sexual perversion exhibiting itself in cruelty to others.*

scrip, script   *Scrip:* a provisional document, a scrap of paper; *script:* a type of engraved matter imitating handwriting; handwriting; typescript in film and broadcasting.

seasonable, seasonal   *Seasonable:* appropriate to the season, as a hot spell in August; *seasonal:* according to the season, as a seasonal decline in employment.

sensual, sensuous   *Sensual* suggests the indulgence of the senses; *sensuous,* appeal to the senses. The former usually refers to the sexual, the latter seldom does. In older uses this distinction did not prevail. Keats speaks of music addressed to "the sensual ear." Milton describes poetry as "simple, sensuous, and passionate." (It is thought, in fact, that he invented the word.)

sewage, sewerage The best usage confines *sewage* to the waste matter run off in sewers; *sewerage*, to systematic drainage or a system of drains.

shambles *Shambles* originally meant a *scene of carnage*; usage has extended the meaning to a *scene of disorder* in general. It may not be used as a synonym for *ruins,* as in "a heap of shambles."

shamefacedly, shamefully The adverb *shamefully,* characterizing an act as indecent or disgraceful, should not be used for *shamefacedly* ("He later shamefacedly confessed that he had been stage-struck"; "He shamefully confessed his connection with the trade in narcotics"). Note that *shamefacedly* may not indicate real shame, but only embarrassment or modesty. In the first sentence, *ashamedly* would suggest real shame. In the second sentence *shamefully* is not entirely clear, for it may mean *with shame* or *disgracefully, indecently.* The group of words should be used with care.

signal, single The phrase is "single out" not "signal out," as one sometimes sees it. It is not correct to list more than one thing as *singled out.*

sleight, slight "*Sleight* of hand." "A *slight* to our honor."

speciality, specialty *Speciality:* the state or quality of being special, a peculiarity or distinguishing characteristic; *specialty:* an employment limited to one kind of work or an article or subject dealt in or with exclusively ("A speciality of function, by calling for a speciality of structure, produces..."). In British usage, *speciality* can be used for *specialty.*

specie, species *Specie:* coin or coins; *species:* kind or kinds ("Specie payments"; "One species, *Larix laricina,* is the common tamarack").

Spencerian, Spenserian *Spencerian:* of Herbert Spencer ("Spencerian philosophy"). *Spenserian:* of Edmund Spenser ("Spenserian stanza").

spiritism, spiritualism *Spiritualism* is the old word for the philosophy of Berkeley, for psychical research, and for a religious sect, but *spiritism* is preferred by some modern philosophers to distinguish their theory from older ones and by some scientific investigators of psychic phenomena.

stalactite, stalagmite *Stalactites* grow downward; *stalagmites,* upward.

stalemate, impasse A true *stalemate,* such as the one that ends a chess game, is final and irreversible. An *impasse* is a *dead end,* which at least allows the opportunity of turning around and going back.

stanza, verse The use of *verse* for *stanza* is popular but inaccurate. Strictly, a *verse* is a line of poetry or one of the numbered groups of lines in the Bible.

stimulant, stimulus *Stimulant:* anything that causes a rapid, temporary increase in activity ("Coffee is a stimulant"); *stimulus:* anything that stirs a response, long-term or short ("The stimulus of young minds improved her teaching").

stratagem, strategy *Strategy* is the general term for the art of planning and carrying out with skill and shrewdness; *stratagem,* the term for any particular device or trick for outwitting.

stultify *Stultify* is to be used with regard to its derivation from *stultus,* foolish. Its older meaning was *to make a fool of,* though it has been broadened to signify *to render useless* or *to disgrace, dishonor.*

subtile, subtle These words are constantly used interchangeably by good writers, though a tendency to distinguish them is observable. *Subtile* (sub-tile), the older spelling, is used mainly of physical matters, with the meaning *delicately or daintily made, rarefied, penetrating,* or *finely drawn* ("A spider's web is subtile"). *Subtle* (suttle) is mainly used of mental or spiritual matters, meaning *crafty, acute, refined, ingenious* ("Iago is a subtle villain").

suit, suite While both *suit* and *suite* may be used to refer to things related or grouped together because of a common function or form, *suit* is most often used to refer to clothing or playing cards, while *suite* refers to a retinue ("the king's suite"), a unit made up of connecting rooms ("a hotel suite"), or a set of furniture coordinated for use in one room ("a suite of living-room furniture").

synthetic   *Synthetic* means *concocted* rather than *false*. Avoid such terms as "synthetic hair," unless you mean specifically that it was manufactured from diverse parts.

tactile, tactual, tangible   Little or no difference is recognized between *tactile* and *tactual*. The art critic, however, speaks of *"tactile* values"—meaning the effect in a painting of tangibility or solidity. A biologist would speak of the horns of a snail as *tactual* organs. *Tangible* is the literary term and, unlike *tactual* and *tactile*, is used of things of the mind. It is closely synonymous with *palpable*.

tasteful, tasty   *Tasteful:* conforms to the principles of taste; displaying harmony, beauty, or other aesthetic qualities; or able to appreciate these qualities. *Tasty:* pleasing to physical taste, savory.

technics, techniques   *Technics:* the doctrine or study of an art or a technology; *techniques:* systematic procedures for performing complex tasks. Used interchangeably in the sense of *technical rules or methods*.

tend, trend   *Tend* and *trend* are both used in the sense of *have a drift or tendency*, but *tend* is used of both persons and things and *trend* only of things ("Man tends generally to optimism"; "The road tends to be rough"; "The path trends to the right").

terminal, terminus   Both words are used in the sense of *end*, as of a railroad, but *terminus* is also used for *boundary, limit, final goal*.

topee, toupee   *Topee:* a sun-helmet, a hat made of pith; *toupee:* a small wig.

tortuous, torturous   *Tortuous:* winding, twisting; *torturous:* having the quality of torture.

transcendent, transcendental   *Transcendent:* surpassing, of extreme excellence or greatness; *transcendental:* idealistic, beyond or above experience, visionary.

troche, trochee   *Troche:* a medicated lozenge; *trochee:* a type of metrical foot.

troop, troupe   *Troop:* an organized company, specifically of soldiers; *troupe:* a company, especially of actors.

troublesome, troublous   *Troublesome:* vexatious, causing worry. *Troublous:* uneasy, troubled. *Troublous* is most often a literary usage.

turbid, turgid   Unrelated words. *Turbid:* muddy, clouded ("Turbid waters"); *turgid:* inflated, bombastic ("Turgid oratory").

unquestioned, unquestionable   That which is *unquestioned* is not or has not been questioned; that which is *unquestionable* cannot be decently questioned. "His veracity is unquestioned" and "His veracity is unquestionable" have not precisely the same meaning.

venal, venial   *Venal* (L. *venalis,* sale): ready to sell that which should not be sold, as one's vote or influence ("A venal judge"); *venial* (L. *venia,* pardon): worthy of pardon, excusable ("A venial offense").

vernacular   Often misused in the sense of *dialect, patois*. The *vernacular* is the indigenous speech of a country. During the Middle Ages, for example, Latin was the literary or learned language of countries of which the vernacular was Italian, French, Spanish.

vice versa   Strictly, *vice versa* denotes a reversal of terms ("High wages mean high prices, and vice versa"; i.e., "High prices mean high wages"). It is often misused as if it meant *the opposite* or *the contrary* ("A small egg does not necessarily have a correspondingly small yolk, or vice versa." The real vice versa of this statement would be "A small yolk does not necessarily have a correspondingly small egg," though the author meant, of course, that a large egg does not necessarily have a large yolk).

virtu, virtue Both are from the Latin *virtus*, courage or virtue, but *virtu* is from the Italian and *virtue* from the French. *Virtu:* a love of or taste for objets d'art; the objects themselves — curios, antiques.

virtually In popular usage, *virtually* commonly means *very nearly* or *almost* ("I'm virtually famished,"; "It was virtually midnight before we got home"). More strictly the word means *in essence or effect, but not in fact* ("The bankers were virtually the rulers of the country").

virus A *virus* is a microorganism; the word *virus* should not be used as if it referred to a disease ("I'm having an attack of virus").

womanish, womanly Both mean *having the qualities of a woman,* but *womanish* is derogatory, usually meaning *effeminate* (q.v.) and generally only applied to males; *womanly* means in the manner of, befitting, or having qualities becoming a woman. The use is parallel to *childish* and *childlike.*

SPELLING

The writer who is a poor speller should work with a dictionary always at his side and should send out no manuscript, proposal, or outline without carefully checking doubtful words and proper names. For an editorial worker or proofreader, poor spelling amounts to an occupational disease that requires cure. Words often misspelled should be memorized or written on a list for future reference; there are special dictionaries and lists of words often misspelled that should be part of the working kit. In some cases, spelling is a matter of the publication's or publisher's own style (as in *theatre, theater*) and if no standard style is recommended, the first choice in Merriam-Webster is the usual preference. In addition, the reference books mentioned in the bibliography (pp. 487–488) may prove helpful.

English manuscripts being prepared for American publication require particular attention.

The following rules may be helpful in areas where most spelling errors occur.

THE "EI" OR "IE" DIFFICULTY

The spelling difficulty of *ei* or *ie* is mainly with those words in which the vowel sound is *ee*. "Write *i* before *e* when sounded as *ee*, except after *c*."

niece	believe	siege	grievous

Exceptions: either, neither, seize, weir, weird, sheik, leisure, inveigle, plebeian, obeisance.

"When the letter *c* you spy, place the *e* before the *i*."

ceiling   conceit   receipt

Exceptions: financier, specie.

In many words *ie* and *ei* have the sound of $\bar{a}$, $\breve{e}$, $\breve{i}$, or $\bar{i}$. Excepting *friend, sieve, mischief,* and *handerchief,* these are spelled with *ei*.

weigh   neighbor   sovereign   surfeit   height   foreign

There should be no spelling difficulty with words like *science, piety, deity,* in which the two letters have two vowel sounds, not a single sound as in the words noted above.

FORMING DERIVATIVES BY ADDING SUFFIXES

Words ending in "e." Words ending in silent *e* generally drop the *e* before a suffix beginning with a vowel.

age	aging	line	linage
blue	bluing, bluish	route	routing
cringe	cringing	sale	salable
force	forcible	true	truism

Exceptions: mileage, hoeing, shoeing, toeing, enforceable.

The *e* is retained also in *dyeing, singeing,* and *tingeing* (from *dye, singe,* and *tinge*) to distinguish them from *dying, singing,* and *tinging*.
Derivatives from proper names of persons retain the *e,* as *daguerreotype*.

Words ending in silent *e* generally retain the *e* before a suffix beginning with a consonant.

awe	awesome	move	movement
hate	hateful	polite	politeness

Exceptions: Many words ending in silent *e* immediately preceded by another vowel (except *e*) drop the *e* in forming derivatives, as *due, duly; argue, argument*.

Other exceptions are the words *awful, wholly, nursling, wisdom, abridgment, acknowledgment, judgment*.

Words ending in "ce" or "ge." Words ending in *ce* or *ge* retain the *e* before suffixes beginning with *a, o,* or *u,* so that the *c* and *g* will not be pronounced with the hard sound.

advantage   advantageous   change   changeable

enforce	enforceable	peace	peaceable

Exceptions: mortgagor, pledgor.

Words ending in "ie." Words ending in *ie* generally drop the *e* and change the *i* to *y* when adding *ing.*

die	dying		vie	vying

Words ending in "y." Words ending in *y* preceded by a vowel generally retain the *y* before any suffix.

annoy	annoyance	buy	buyer	enjoy	enjoyable

Exceptions: daily, gaily, gaiety, laid, paid, said, saith, slain.

Words ending in *y* preceded by a consonant usually change the *y* to *i* before any suffix except one beginning with *i.*

happy	happiness	fly	flier
country	countrified	body	bodiless

Exceptions: fryer, flyer (a crack train).

Derivatives of adjectives of one syllable ending in *y* generally retain the *y;* as *shy, shyness; wry, wryness.* The words *drier* and *driest,* from *dry,* are, however, commonly written with the *i.*

Before -*ship* and -*like,* as in *ladyship, citylike,* and in derivatives from *baby* and *lady* the *y* is retained.

Words ending in a consonant. Monosyllables and words of more than one syllable with the accent on the last syllable, ending in a single consonant preceded by a single vowel, double the final consonant before a suffix beginning with a vowel.

bag	baggage	corral	corralled
run	running	forget	forgettable
allot	allotted	occur	occurrence
control	controller	prefer	preferred

Exceptions: bus, buses; derivatives of the word *gas* (except *gassing* and *gassy*), *gaseous; chagrin, chagrined; transferable;* some derivatives in which the accent of the root word is thrown back upon another syllable, as *cabal', cab'alism; defer', def'erence; prefer', pref'erence; refer', ref'erence.*

Words accented on any syllable except the last, words ending in more than one consonant, and words ending in a single consonant preceded by more than one vowel, do not double the final consonant before an ending beginning with a vowel.

benefit	benefited		bias	biased

cancel	canceled (but cancellation)	tranquil	tranquilize (but tranquillity)
combat	combated	transfer	transferee
kidnap	kidnaper	repeal	repealed

Exceptions: outfitted, outfitter; cobweb, cobwebbed; handicap, handicapped; diagram, diagrammatic (but note diagramed); monogram, monogrammed; humbug and a few other words ending in *g* in which the *g* is doubled so that it will not be pronounced like *j*: humbugged, zigzagged. *Tranquillity, chancellor, crystallize,* and *metallurgy* get their double consonants from their Greek and Latin origins.

Words ending in "c." Words ending in *c* usually have a *k* inserted when adding a termination beginning with *e, i,* or *y,* so that *c* will not be pronounced like *s.*

arc	arced or arcked	picnic	picnicker
bivouac	bivouacked	panic	panicky
colic	colicky	——	politicking
frolic	frolicked	shellac	shellacked
havoc	havocking	traffic	trafficker
mimic	mimicking		

TROUBLESOME SUFFIXES

The terminations "-ceed," "-sede," and "-cede." Only three words of our language end in *-ceed;* one in *-sede.* Others with the same pronunciation end in *-cede.*

exceed	supersede	accede
proceed		antecede
succeed		concede

The terminations "-ise" and "-ize." Formerly the ending *-ize* was used if a word came from the Greek or Latin, and *-ise* if it came from the French; but in present-day American usage this rule is not followed. The following, together with their compounds and derivatives, use the *-ise.*

advertise	compromise	exercise	premise
advise	demise	exorcise	reprise
apprise	despise	franchise	revise
arise	devise	improvise	supervise
chastise	disguise	incise	surmise
circumcise	enterprise	merchandise	surprise
comprise	excise		

Words ending in "-able" or "-ible." The correct spelling of words with the terminations *-able* or *-ible* is often puzzling. The student of orthography can find rules for the use of these endings, but it is more practical to learn the spelling of each word and refer to the dictionary if memory fails for the moment.

combustible	digestible	repairable
comprehensible	reversible	definable
defensible	dependable	advisable

Words ending in "-ance" or "-ence." In all words in which the termination is preceded by c having the sound of s, or g having the sound of j, -ence, -ency, and -ent are used; where c sounds like k or g has its hard sound, the termination is -ance, -ancy, or -ant.

| beneficence | negligence | significance |
| coalescence | indigent | extravagant |

If the suffix is preceded by a letter other than c or g and the spelling is in doubt, the dictionary should be consulted.

FORMATION OF PLURALS

The plural of most nouns is formed by adding s or es to the singular, -es being added if the word ends in ch soft, s, sh, j, x, or z, which when pronounced would not unite with s alone.

| desks | chiefs | traces | churches | crucifixes | rushes |

Nouns ending in "f," "ff," or "fe." Most nouns ending in f, ff, or fe form their plurals regularly, but a few change the f, ff, or fe to ves. Staff, scarf, and wharf have two forms: staffs, staves; scarves, scarfs; wharves, wharfs.

beeves	knives	selves	thieves
calves	leaves	sheaves	wives
elves	lives	shelves	wolves
halves	loaves		

Nouns ending in "i." Nouns ending in i usually form their plurals by adding s.

| alibis | Hopis (or Hopi) | rabbis |

But alkalies is preferred.

Nouns ending in "o." Nouns ending in o preceded by a vowel form their plurals by adding s to the singular.

| cameos | tattoos | taboos | ratios |

No rule without exceptions can be given for the formation of the plural of nouns ending in o preceded by a consonant. The more familiar words add es to the singular, generally speaking, while words rather recently borrowed from other languages usually add s only.

heroes	mosquitoes	banjos	octavos
mottoes	commandos	halos	zeros
potatoes	infernos		

Nouns ending in "y." The plural of nouns ending in *y* preceded by a consonant or consonant sound is formed by changing the *y* to *i* and adding *es*.

| city, cities | fly, flies | colloquy, colloquies |

Change of vowel. In the following nouns, the singular and the plural differ in their vowels.

| foot, feet | louse, lice | mouse, mice |
| goose, geese | man, men | tooth, teeth |

Compounds ending with these words form their plurals in the same way.

| workman, workmen | dormouse, dormice | clubfoot, clubfeet |

Words ending in -*man* that are not compounds form their plurals regularly, by adding *s* only.

| caymans | humans | Germans | Ottomans |
| dolmans | talismans | Normans | Turkomans |

Similarly *mongoose*, ending in -*goose* but not a compound of *goose*, has the plural *mongooses*.

Nouns with two plurals. Some nouns have two plurals, with different meanings.

brothers (of the same parents)	brethren (of the same society)
dies (for stamping)	dice (for gaming)
geniuses (men of genius)	genii (imaginary spirits)
indexes (tables of contents)	indices (signs in algebra)
staves (poles, supports)	staffs (bodies of assistants)

Many nouns from foreign languages retain their original plurals, for instance:

alumnus, alumni	madame, mesdames
crisis, crises	monsieur, messieurs
datum, data	os, ossa (*bone*)
genus, genera	os, ora (*mouth*)
indicium, indicia	ovum, ova
insigne, insignia	paries, parietes
larva, larvae	phenomenon, phenomena
marchesa, marchese	planetarium, planetaria
matrix, matrices	phylum, phyla

Exception: Agenda, the plural form of a word in this group, is used with a singular verb: "The agenda *is* adopted."

Some nouns of this class also have a plural formed after English analogy. In the following, the form listed first is preferred.

adieu — adieus, adieux
appendix — appendixes, appendices
automaton — automatons, automata
bandit — bandits, banditti
beau — beaux, beaus
cherub — cherubim, cherubs
criterion — criteria, criterions
curriculum — curriculums, curricula
dogma — dogmas, dogmata

femur — femurs, femora
formula — formulas, formulae
fungus — fungi, funguses
gladiolus — gladioli, gladioluses
medium — media, mediums
memorandum — memorandums, memoranda
virtuoso — virtuosos, virtuosi
vortex — vortexes, vortices

The following Latin words have a plural formed by adding *es:*

apparatus, apparatuses
census, censuses
consensus, consensuses
hiatus, hiatuses

impetus, impetuses
nexus, nexuses
prospectus, prospectuses
sinus, sinuses

Singular forms used as plurals. Some nouns, mostly names of animals, use the singular form of spelling to denote more than one individual, as:

deer   sheep   swine   moose   fish

But these words, and some others that ordinarily have no plural, as *wheat, flour, coffee,* have a plural regularly formed that is used to denote more than one species or kind, or in some cases to emphasize the presence of the several component individuals.

fishes   trouts   wheats   coffees

Note: Names of shellfishes commonly use the regular plural forms

crabs   lobsters   scallops   shrimps

although *lobster* and *shrimp* are alternate plural forms.

Some nouns that have regularly formed plurals also can use the singular form as a plural.

dozen   head   score   youth
pair   couple   shot   heathen
brace   yoke

After the ceremony, the youths were given their diplomas.
All the youth of the city are ready for their tasks.

The word *people* is usually a plural, but it may be a singular, with a regularly formed plural, *peoples.*

Another people deserves brief mention.
. . . the benefits a people derives from foreign trade.
What peoples have had a part in the making of our civilization?

Nouns with only one form. Many names of tribes and races have the same form in the plural as in the singular.

Chinese   Portuguese   Norse

Such words as *assured, beloved, educated,* and so forth, used as nouns (instead of *assured person* or *persons*) are construed as singular or plural according to the context.

Forms in *-ics* are construed as singular when they denote a scientific treatise or its subject matter, as *mathematics, physics;* those denoting matters of practice, as *gymnastics, tactics,* are often construed as plurals.

Politics has been studied as a science since the days of Aristotle.
Politics in school affairs are always detrimental to educational progress.

Acoustics is a worthwhile study for a musician.
The acoustics in Philharmonic Hall were studied intensively.

Other nouns that have only one form and may be construed as singular or plural are the following:

aircraft	goods	rendezvous
alms	headquarters	remains
amends	mankind	samurai
bellows	means	scissors
bourgeoisie	offspring	series
chassis	pains (*care, effort*)	shambles
corps	pants (*clothing*)	species
counsel (*a person*)	précis	sperm
congeries	proceeds	sweepstakes
forceps	progeny	United States

Every means has been tried. All means have been tried.
It was a shambles.
The United States is; these United States are.

Plural in form, construed as singular:

aloes	measles	rickets
checkers (*a game*)	mews (*stables*)	shingles (*a disease*)
dominoes (*a game*)	mumps	whereabouts
heaves	news	works (*factory*)

There is a works of the Hempill Manufacturing Company in our town.

Note: The words *biceps, Cyclops, gallows,* and *summons* are not of this group; they have plural forms: *bicepses, Cyclopes, gallowses, summonses.*

Compound nouns. Compounds written as a single word form their plurals regularly, adding the sign of the plural at the end. Mistakes are infrequent except in the spelling of compounds ending in *-ful.*

cupfuls   handfuls   teaspoonfuls

Compound nouns written with a hyphen form their plurals by adding the sign of the plural to the word that essentially constitutes the noun.

sons-in-law   knights-errant   aides-de-camp
step-children   hangers-on   autos-da-fé
courts-martial

If the compound noun consists of two words, the sign of the plural may be used with both or with the noun portion.

knights bachelors   chargés d'affaires   coups d'état
knights templar   tables d'hôte

A substantive phrase containing a possessive — *master's degree,* for example — is changed to the plural by adding s to the second word.

master's degrees   debtor's prisons

Military and civilian titles consisting of two words form their plurals by adding the s to the significant word: *major generals, lieutenant colonels, surgeons general, sergeants major, attorneys general, consuls general, postmasters general, deputy chiefs of staff, general counsels.*

Proper names. The plural of proper names is formed regularly, by adding s or es to the singular.

George, Georges   Adonis, Adonises
Charles, Charleses   Hughes, Hugheses
the Kallikaks and the Edwardses   Ruckers, Ruckerses

Note: Many seem to have difficulty with the plurals of proper names ending in s, and errors like the following often appear:

Wrong: It can boast of fourteen Rubens and two Van Dycks.
Right: It can boast of fourteen Rubenses and two Van Dycks.
Wrong: There were several families of the name of Hanks. . . . The Hanks had come from Virginia.
Right: The Hankses had come from Virginia.

Proper names ending in *y* form their plurals regularly, and do not change the *y* to *i* as common nouns do.

Both Germanys
The Raffertys
The three Marys (*Three Maries* would mean three women named Marie.)
The two Kansas Citys (*Cities* would be misleading.)

Exceptions: Alleghenies, Rockies, Sicilies; Ptolemies.

Letters, figures, characters, signs. The plural of a letter, figure, character, or sign is expressed by adding to it an apostrophe and *s*.

Mind your *p*'s and *q*'s.	ABC's
There are two *m*'s in *accommodate*.	three R's
2300 r.p.m.'s	G.I.'s
five MIG's	Co.'s
during the 1850's	6's, +'s
Boeing 747's	

Exceptions: In stock and bond quotations, Govt. 4s, Bergen 8s, Treas. 3¾s. In golf scores, 3s, 4s.

In expressions like "twos and threes," "pros and cons," "ins and outs," "yeas and nays," a regularly formed plural is used.

Count by twos to 20; by fives to 50; by fours to 40.
He wanted to know the whys and wherefores.
The Number Twos become guards.
He voted with the Noes. The Yesses won.
It is usually instigated by the desire of the "have nots" to obtain some of the economic advantages possessed by the "haves."
Don'ts for Library Users.
There can be no ifs or ands or buts. (Meaning: There can be no conditions or objections.)

If the "ifs," "ands," and "buts" in the preceding example referred to the words as words, however, the correct form would be:

There can be no *if*'s or *and*'s or *but*'s.

That is, the plural of a word referred to as a word, without regard to its meaning, is indicated by apostrophe and *s*.

I used too many *and*'s.

FORMATION OF THE POSSESSIVE CASE

Common nouns. The possessive case of most singular nouns and of plural nouns not ending in *s* is formed by adding an apostrophe and *s*.

aunt's	son's	children's

Note: Some prefer *witness'*, *countess'*, and the like, but spoken English is more accurately reflected by a regularly formed possessive, as *countess's*, *boss's*.

The possessive case of plural nouns ending in s is formed by adding apostrophe alone.

aunts'   sons'   brothers'

Note: United States' even though the name is usually construed as a singular noun.

Various expressions ending in s or the sound of s form a possessive by adding the apostrophe alone.

for old times' sake   for appearance' sake
for goodness' sake   for convenience' sake
for conscience' sake

Proper names. The possessive form of almost all proper names is formed by adding apostrophe and s to a singular or apostrophe alone to a plural.

Jack's   James's   the Davises'
Burns's   Marx's   Schultz's
Dickens's   Adams's   Schultzes'

Wherever the apostrophe and s would make the word difficult to pronounce, as when a sibilant occurs before the last syllable, the apostrophe may be used alone.

Moses' laws   Isis' temple   Xerxes' army   Jesus' followers

But these forms are commonly replaced by

the laws of Moses   the temple of Isis   the army of Xerxes
                               the followers of Jesus

By convention, ancient classical names ending in s add only the apostrophe to form the possessive.

Mars' wrath   Achilles' heel   Hercules' labors

In forming the possessive of foreign names ending in a silent sibilant, as *Dumas, Vaux,* the usual practice is to use apostrophe and s whether the word is monosyllabic or not. Webster's Dictionary has the form *Rabelais's.*

Descartes's invention   Arkansas's problem   Des Moines's schools

Either is correct: *the Misses Topping's* or the *Miss Toppings'.*
Compound nouns. There are many compound nouns with first portions in the possessive case. In most cases, a singular possessive is preferred.

cow's milk   fool's gold   confectioner's sugar

Webster II prefers *rabbit foot* to *rabbit's foot; spider web* to *spider's web*. Although it also prefers *pigs' feet, pig feet* is often heard.

The spelling *oneself* is displacing *one's self,* the form that used to be the more common.

In modern usage the possessive of *somebody else* and similar expressions is formed by adding apostrophe and *s* to *else.*

someone else's   anyone else's   everybody else's

Phrases. The possessive case of a phrase or a combination of names is formed by adding the sign of the possessive to the last word only.

the Governor of Maine's   Simon the Pharisee's

In no case, however, should a sentence be permitted to take on the form identified by the editor and writer Harold Taylor as the "picnic's grandmother" construction: "I got it from the girl who gave the picnic's grandmother." The use of *of* is the customary remedy. Long titles often produce the same effect: "Minister of Posts and Telephone Ahmed Akbar's statement . . ." A double remedy is required here: "The statement of Ahmed Akbar, the Minister of . . ."

Joint ownership. If possession is common to two or more individuals, only the last name takes the possessive sign.

Teddy, Peggy, and Nancy's home
Lena and Bill's trip to the Fair
Tocqueville and Beaumont's Report
Burrows & Sanborn's store
Painters, Paper Hangers, and Decorators' Union
Drs. Sansum and Nuzum's theory

If possession is not joint, however, each modifying substantive should carry the apostrophe.

Teddy's, Peggy's, and Nancy's homes
Men's, women's, and children's shoes

A distinction should be made between true possessives like the examples above and expressions denoting something *for* two or more persons or groups.

Authors' and Printers' Dictionary
Editors' and Readers' Handbook
Boys' and Girls' Newspaper
Mothers' and Daughters' Banquet
Soldiers' and Sailors' Home

Stores, organizations, and books do not, however, always use the possessive form; careful checking of the phone book, letterheads, and title pages may be required. The old newspapermen's restaurant in New York, for instance, is *Artist and Writers* (according to folklore, only one artist was present when it was created as a speakeasy).

ORDINARY WORDS THAT CAUSE TROUBLE

Spelling errors can arise from confusion of homonyms—city council for city counsel, for instance. Inconsistency of usage gives rise to what look like spelling errors, but are actually variations of style: *ax*, then *axe*, then *ax*. Irregular spelling may also appear in material if there is a problem of British versus American usage; labour and labor, colour and color.

In addition to these sources of disorder, there are words that are constantly being misspelled. Following is a list of some words in frequent use whose spelling gives trouble.

accommodate	lightning
acknowledgment	liquefy
adviser	marshal
all right	minuscule
bouillon	nickel
canister	niece
diarrhea	paralleled
ecstasy	Philippines
embarrass	Portuguese
exhilarate	privilege
existence	resistance
fulfill	restaurateur
guerrilla	scurrilous
gypsy	sergeant
harass	siege
hemorrhage	skillful
inoculate	supersede
iridescent	tranquillity
irrelevant	vilify
judgment	weird
liaison	

FOREIGN WORDS AND PHRASES

Words are constantly being adopted from other languages—the terminologies of art, literature, dress, cookery, diplomacy, law, medicine, politics, military science, sport, abound in words of foreign origin, appropriated as the standard words for the things or ideas. These are ordinarily set roman. Many words and expressions are adopted because they are circumlocution savers; and these are italicized until usage accepts them as practically English. There are in addition many foreign phrases that are often used in an English text without any real excuse, since there is an equally short English equivalent that would be more readily understood. Use of them is often affected and pretentious.

It should be borne in mind that the number of words set in roman is constantly increasing; that phrases are less quickly adopted than single words; that newspapers and magazines italicize less than books; that the setting in which a word is used often determines whether it should be italicized. The following list, therefore, can be only suggestive.

abbé
ab extra
a cappella
accouchement
accoucheur
ad hoc
adieu
ad infinitum
ad interim
ad libitum
ad nauseam
ad valorem
affaire d'honneur
a fortiori
aiguille
à la carte
à la mode
alfresco
alma mater
alter ego
amende honorable
amour
amour-propre
ancien régime
Anglice
anno Domini
Anschluss
ante bellum
apache
aparejo
aperçu
apéritif
a posteriori
appliqué
appui
a priori
apropos
aqua
aqua fortis
aqua regia
aqua vitae
arête
argot
arrondissement
arroyo
artiste
assignat
atelier

attaché
auberge
au courant
au fait
au gratin
au jus
au naturel
au revoir
aurora australis
aurora borealis

barège
bateau
beau ideal
beau monde
beaux-arts
belles-lettres
berceuse
bête noire
bêtise
bezique
bibelot
bien entendu
bienséance
bijou
billet-doux
blasé
Boche
bona fide
bonhomie
bon mot
bonne
bon soir
bon ton
bon vivant
bon voyage
bouillabaisse
bourdon
bourgeois
bourgeoisie
bourse
boutonniere
brassard
bric-a-brac
bricole
briquette
burnoose

cachalot
cachet
cachou
cadre
café
caïque
caisson
camaraderie
canaille
canapé
canard
cancan
capapie
capias
caporal
capote
carcajou
carte blanche
cartouche
casus belli
catafalque
causerie
caveat
certiorari
cestui que trust
chaise longue
chamade
chanson
chanson de geste
chaparajos
chapeau
chaperon
charabanc
chargé d'affaires
charivari
charlotte russe
chasse
chassé
chasseur
château
chatoyant
chef-d'oeuvre
cheval-de-frise
chevelure
chiaroscuro
chose
cinquecento
claque
cloisonné
col
comme il faut
communiqué
con amore
concierge
conciergerie

concordat
confrere
congé
connoisseur
conservatoire
consommé
contra-
contretemps
cordon bleu
corrigendum
corveé
costumier
coup de grâce
coup de main
coup d'état
coup d'œil
coupé
couvert
crèche
crepe de Chine
crépon
cul-de-sac

daimio
danseuse
debacle
déclassé
décolleté
dedans
de facto
dégagé
de jure
delirium tremens
de luxe
démarche
demitasse
démodé
denouement
Deo volente
dernier
detente
de trop
deus ex machina
devoir
diablerie
dilettante
dinero
dishabille
distingué
distrait
dolce far niente
dossier
double-entendre
doyen
dramatis personae

eau de cologne
eau de vie
écarté
ecce homo
éclair
éclat
ecru
edition de luxe
élan
elenchus
embonpoint
émeute
émigré
en bloc
enceinte
en famille
enfant terrible
en masse
ennui
en route
entente
entourage
entr'acte
entree
entrepôt
entrepreneur
entresol
épée
epergne
ergo
erratum
esprit
esprit de corps
estaminet
et cetera
ex cathedra
exempli gratia
exequatur
ex libris
ex officio
ex parte
exposé
ex post facto
extempore

faïence
fainéant
fait accompli
farceur
fauteuil
faux pas
felo-de-se
feme
feme covert
feme sole

fete
feuilleton
fin de siècle
flânerie
flâneur
flèche
fleur-de-lis
force majeure
fortissimo
foudroyant
frappé
frater

gauche
gaucherie
gendarme
ghat
gratis
gringo
grisaille
gymkhana

habeas corpus
habile
habitué
hachure
hacienda
hara-kiri
hauteur
hegira
hors de combat
hors d'oeuvre

idée fixe
ignis fatuus
impasse
imprimis
infra dig
in propria persona
in re
in situ
intelligentsia
in toto
intrigant
ipse dixit
ipso facto

jalousie
jardiniere
jeu d'esprit
jongleur
jujitsu
julienne
jupe

kepi
kraal
kulak
Kulturkampf
Kyrie eleison

lacuna
laissez faire
lamé
lansquenet
lapis lazuli
lares and penates
lazzarone
leitmotiv
lese majesty
lingua franca
literatim
littérateur
locum tenens
loggia

maestro
magnum
magnum opus
malapropos
mal de mer
mañana
mandamus
manège
mardi gras
materia medica
matériel
mélange
melee
memorabilia
ménage
mésalliance
métier
métis
meum and tuum
mise en scène
mitrailleuse
modus operandi
modus vivendi
mondain
mot juste
muezzin
mutatis mutandis

naïve
naïveté
nee
ne plus ultra
névé
nil

nisi
nisi prius
noblesse oblige
nolle prosequi
nolo contendere
nol-pros
nom de guerre
nom de plume
non compos mentis
nonego
non est
non sequitur
nous
nouveau riche
nunc dimittis

obiter dictum
objets d'art
olla-podrida
omnium-gatherum
onus
opéra bouffe
opéra comique
ordonnance
outré

padre
paletot
panache
papier-mâché
par excellence
pari-mutuel
pari passu
parterre
parvenu
passé
pas seul
passim
pâté
paterfamilias
paternoster
patisserie
patois
pelerine
per annum
per capita
per centum
per contra
per diem
per se
persona grata
pièce de résistance
pied-à-terre
pince-nez
pis aller

planchette
plein-air
point-device
porte-cochere
portiere
poseur
poste restante
postmeridian
post meridiem
post-obit
pourboire
pourparler
pousse-café
pratique
précis
prie-dieu
prima facie
procès-verbal
prochein
pro forma
pro rata
protégé
pro tem
pro tempore
provenance
provocateur
puisne
purdah
purée

qua
quasi
quenelle
quidnunc
quid pro quo
qui vive
quondam
quo warranto

raconteur
raison d'être
rapprochement
rara avis
realpolitik
réchauffé
recherché
rencontre
repoussé
résumé
retroussé
revenant
riposte
risotto
risqué
rissole

ritardando
roué
roulade
rouleau

sahib
salmi
samurai
sanctum sanctorum
sangfroid
sans-culotte
sauté
savant
savoir-faire
séance
sec
semé
seriatim
sine
sine die
sine qua non
soi-disant
soiree
sotto voce
soubrette
soupçon
stemma
subpoena
sub rosa
sui generis
sui juris
summum bonum

table d'hôte
tabula rasa
tant mieux
tapis
terra firma
terra incognita
tertium quid
tête-à-tête
toilette
tour de force
tuyère

ultimo

vade mecum
vale
variorum
vers libre
versus
via
vice versa

vide ante	*volte-face*
vide infra	vox populi
videlicet	voyageur
vide post	
vide supra	Weltanschauung
virtu	
vis-à-vis	
viva!	*Zeitgeist*
vivandière	zenana
viva voce	Zollverein

BIBLIOGRAPHY FOR PARTS V AND VI

Dictionaries:

The American Heritage Dictionary of the English Language, edited by William Morris. Boston: American Heritage Publishing Company and Houghton Mifflin, 1969.

A Dictionary of American English on Historical Principles, edited by Sir William A. Craigie and James H. Hulbert. Four volumes. Chicago: University of Chicago Press, 1938-1944.

Dictionary of American Slang, edited by Harold Wentworth and Stuart Berg Flexner. New York: Thomas Y. Crowell, 1960.

A Dictionary of Clichés, by Eric Partridge. New York: Macmillan, 1940.

The New York Times Everyday Reader's Dictionary of Misunderstood, Misused, Mispronounced Words, edited by Laurence Urdang. New York: Quadrangle Books, 1972.

The Oxford English Dictionary. Thirteen volumes. London: Oxford University Press, 1928. Reissued as *The Compact Edition of the Oxford English Dictionary;* two volumes; 1971. See also *A Supplement to the Oxford English Dictionary,* Volume I (A-G), 1972.

The Random House Dictionary of the English Language, edited by Jess Stein. New York: Random House, 1966.

Webster's New International Dictionary of the English Language, 2nd ed. Springfield, Mass.: G. & C. Merriam Company, 1934–1961.

Webster's Third New International Dictionary of the English Language. Springfield, Mass.: G. & C. Merriam Company, 1961.

General Reference Works:

The Associated Press Stylebook, edited by G. P. Winkler. New York: The Associated Press, 1963.

BERNSTEIN, THEODORE M. *The Careful Writer: A Modern Guide to English Usage.* New York: Atheneum, 1965.

———. *Miss Thistlebottom's Hobgoblins: The Careful Writer's Guide to the Taboos, Bugbears and Outmoded Rules of English Usage.* New York: Farrar, Straus and Giroux, 1971.

CURME, GEORGE OLIVER. *Syntax.* Boston: D. C. Heath, 1931.

EVANS, HAROLD. *Newsman's English.* New York: Holt, Rinehart and Winston, 1972.

FOLLETT, WILSON, and others. *Modern American Usage: A Guide.* New York: Hill and Wang, 1966.

FOWLER, H. W. *A Dictionary of Modern English Usage*. London: Oxford University Press, 1926. Second edition, revised by Sir Ernest Gowers, 1965.

GOWERS, SIR ERNEST. *The Complete Plain Words*. London: Her Majesty's Stationery Office, 1954.

JESPERSEN, OTTO. *A Modern English Grammar* (in seven parts). London: George Allen & Unwin, 1954.

A Manual of Style. 12th ed. rev. Chicago: University of Chicago Press, 1969.

MARCKWARDT, ALBERT H., and FRED WALCOTT. *Facts About Current English Usage*. New York: D. Appleton-Century, 1938.

MENCKEN, H. L. *The American Language* and *Supplements* I and II. New York: Alfred A. Knopf, 1936, 1945–1948.

MORSBERGER, ROBERT E. *Commonsense Grammar and Style*. 2nd ed. rev. New York: Thomas Y. Crowell, 1972.

The MLA Style Sheet. 2nd ed. New York: Modern Language Association of America, 1970.

The New York Times Style Book for Writers and Editors, edited by Lewis Jordan. New York: McGraw-Hill, 1962.

NICHOLSON, MARGARET. *A Dictionary of American-English Usage*. New York: Oxford University Press, 1957.

PARTRIDGE, ERIC. *Usage and Abusage: A Guide to Good English*. Baltimore: Penguin Books, 1963.

SCHUR, NORMAN. *British Self-Taught: With Comments in American*. New York: Macmillan, 1973.

SELLERS, LESLIE. *Doing It in Style: A Manual for Journalists, P.R. Men and Copy-Writers*. New York: Pergamon Press, 1968.

STRUNK, WILLIAM, JR., and E. B. WHITE. *The Elements of Style*. 2nd ed. New York: Macmillan, 1972.

United States Government Printing Office Style Manual. Rev. ed., 1973.

BOOK PUBLISHERS

Current addresses of publishers may be obtained from the following sources: for American publishers, *Literary Market Place, Publishers' Trade List Annual,* and *Books in Print;* for Canadian publishers, *Canadian Books in Print,* from the Canadian Book Publishers' Council (Toronto, Canada), or for the French-language publications, the Conseil Supérieur du Livre (Montréal, Québec, Canada); for international publishers, *International Literary Market Place.*

PART VII. TYPOGRAPHY AND ILLUSTRATION

THE SIZE OF THE BOOK

The preceding sections — Parts I, II, III, IV, V, and VI — have dealt, fundamentally and in detail, with the preparation of manuscripts for the compositor. This work has been the responsibility of the editor and, even more, of the managing editor, if there is one, and the copy editor (sometimes called a manuscript editor). In most publishing houses the manuscript is next placed in the hands of the production department, which decides upon details of format and manufacture. In a few publishing houses some of these details may be left to the editor, but as a rule the editor's preparatory task is completed when the manuscript is editorially and factually approved. In many houses the managing editor is the link between the editorial department and the production department. It is his responsibility to make decisions in association with the production department and the editorial department, to see that they are carried out, and to handle the traffic as the manuscript advances through prescribed stages to a book.

The production department must have a thorough knowledge of all the processes involved in the manufacture of a book, from estimating copy and type to presswork and binding. The same may be said of its knowledge of materials.

In studying the pages that follow, the word *editor* should be interpreted loosely, as pointed out earlier (p. 57). The material in the following pages is intended to acquaint the editor or copy editor with the procedures followed in converting the words of the manuscript into type, and with the choices available in doing so — choices that affect the speed with which the finished book is delivered, its quality and appearance, and most importantly, the cost of producing it.

When a manuscript is accepted for publication, a number of basic decisions are made. These concern the trim size, type page, typeface and type size, number of pages, color, weight, and grade of paper, kind of binding, size of edition, and proposed price. These and associated decisions (endpapers, number and kind of illustrations, color plates, and so on) provide a blueprint for the creation of the book and enable the production department to make a cost estimate. Most of these decisions depend on the length of the copy. It is, therefore, *cast off*, or counted.

CASTING OFF

A castoff is secured by a count of words or of characters; a character count gives a more accurate estimate of the length of the copy.

Character counts. A character count is made from typewritten copy, as the number of characters to the lineal inch for most typewriters is invariable: pica typewriting has 10 characters to the inch and elite typing has 12. Punctuation marks and spaces between words are counted as characters. The length of a line in inches, multiplied by 10 or 12, depending on the style of typewriting, yields the number of characters in the line. To make a count of a paragraph or a page, a line of average length should be measured and the count of that line multiplied by the number of lines. A short line, as at the end of a paragraph or in dialogue, should be counted as a full line. To illustrate, suppose that on a given page of pica typewriting the lines average 6 inches in length. The characters in the line therefore number 6×10, or 60. If the page has 26 lines, there are 26×60, or 1560, characters on the page.

If the count is to be made of a book-length manuscript, each chapter should be counted separately, for accuracy, and the number of headings also noted. If different sizes of type are to be used, each size must be counted separately.

A simple gauge for making a character count of copy can be made from a standard $8\frac{1}{2} \times 11$ inch sheet of paper. Mark off one 11-inch side of the paper in 2-pica spaces, numbering each space consecutively 1, 2, 3, 4, and so on, from the top of the page. To find the line count for a double-spaced typewritten page, align the first 2-pica space mark of the line gauge with the base of the first line of type on the typewritten page; the number aligned with the base of the bottom line of type is the line count for the typewritten page. For the number of characters per line for elite type (12 to the inch), mark off the top of the line-gauge sheet in $\frac{1}{4}$ inches, numbering the marks consecutively 3, 6, 9, 12, 15, and so on, from the left. For the number of characters per line for pica type (10 to the inch), mark off the bottom of the line-gauge sheet in $\frac{1}{2}$ inches, numbering the marks consecutively 5, 10, 15, 20, and so on, from the left. To find the character count, use the character-count gauge for the size type (pica or elite) in which the manuscript page was typed. By placing this gauge on an average-length line, an average number of characters per line can be found. Multiply the average number of characters per line by the number of lines per page, and an average number of characters per typewritten manuscript page can be determined.

Word counts. Some editors prefer to make a word count, but a word count cannot be so accurate as a character count. An actual count of words with the idea that long words and short words will balance each other might give a fairly accurate estimate in some manuscripts but in others it would be wide of the mark. A writer's vocabulary (whether predominantly Anglo-Saxon or consisting of words from the Latin), the subject of his writing, its nature (technical or nontechnical), and the age group to which it is addressed all affect the count of actual words. If the method followed is to count two short words as one word and a very long word as two, the result depends on the estimator's judgment or accuracy in guessing. If a manuscript is partly typewritten, partly handwritten, and partly printed (*reprint*), estimating length may require a combination of character-count and word-count methods. It is not unusual for the contract for a book to specify the approximate number of words it will contain. To find out whether the author has fulfilled the contract in this respect, the number of characters divided by five will give the approximate number of words.

Having learned the length of the copy from a character count, as described above, the next decision concerns the physical size of the book. This will depend on the trim size (depth × width of the whole page, measured in inches), the size of the type page (length of text line and number of lines to a page, including any running heads or footnotes, measured in picas), and on the size and face of type in which the book is set.

TRIM SIZE

The trim size may be of almost any dimensions, but standard sizes are usually cheaper, and it is perhaps for this very reason that they have become standard. Often used standard sizes are:

5½ × 8¼
6⅛ × 9¼
7⅜ × 10

The first two are often used for fiction and ordinary nonfiction and the third for more important nonfiction, but any number of variations on these sizes are possible, and in practice there is great variation in the trim sizes of books. Art books frequently demand a larger trim size to do justice to the illustrations or plates; special books have special needs: the expression "coffee-table book" for oversize picture books is well known; reference books, by their nature, may dictate a size that varies markedly from any of those listed as standard; textbooks should not be so big and heavy that students cannot carry them around; ordinary books of fiction and nonfiction should not cause chaos in the bookshelf by their awkward size. The trim size, then, is determined by the needs of the book itself, the market for which it is intended, and experience with similar books.

At this stage, a complicated book, as a textbook carrying many halftones, graphs, tables, maps, and so on, or a scientific book with equations and unusual symbols or other hieroglyphs, may be turned over to a designer. The designer will consider the purpose of the book, perhaps the age of the reader for whom it is intended, and the similarity of its purpose and type to existing books. Page design, as an element of book design, is of necessity a variation on a theme—the arrangement of type in two-dimensional space; the arrangement is susceptible to infinite variation.

The next decision concerning the physical appearance of the book is to fix the size of the type page.

TYPE PAGE

The type page, or text page, is the area of the page that is occupied by type, and includes text, headings, footnotes, and folio. Its size, like that of the trim size, is determined to some extent by the kind of book under consideration, but generally speaking, the size of the type page is a matter of proportion — the amount of printed text on the page that is comfortable and easy to read and is pleasing to the eye.

The first thing to decide upon is the *measure,* or length of line, to be used. Much work has been done to discover the relationships of measure and type size that lead to accuracy and ease in reading. A study by Sir Cyril Burt, *A Psychological Study of Typography* (Cambridge University Press, 1959), using a standard 10-point Times Roman type, concluded that measures between 21 and 33 picas are most legible to the average reader. The optimum measures varied according to the subject matter of the book. For literary material, a measure of about 21 to 24 picas was found to be most readable, while in scientific and other material where the tendency is to skim rather than to read word-for-word, the optimum measure was found to be closer to 30 picas. The study also pointed out the relation between type size, leading, and measure. The longer measures in general called for a minimum 9-point type size, with 2-point leading being best for legibility.

Books of information, such as dictionaries, encyclopedias, handbooks, and other reference books, may be set in double columns in order to get more material on a page. This reduces the number of pages required and holds down the cost. In figuring the measure of the type page that is to be set double column, a wider total type page may be set than the 30 picas that is ordinarily the limit for the Linotype machine, because the column width can be 15. The arrangement of the type page is not limited to two columns; three, or even more, columns of various widths may be used. The number of columns and their widths are dependent on the trim size of the page and the need for ease of reading. Whether it is two, three, or more columns to a page, the material is set in one long column on the galleys and is arranged in the desired number of columns to a page in page makeup. For measurement of the type page, the space between columns, usually 1 to 1½ picas, is added to the line length. When a rule is used to separate columns the separation between them may be somewhat less. By the use of two or more columns, which breaks up the text block into shorter lines, a smaller typeface may be used, thereby further reducing the amount of paper required. Since in the kinds of books mentioned as being suitable for two or more columns, only short sections are consulted at one time, the smaller type size is appropriate and convenient.

When the width of the type page has been decided on, the length may be determined by some rule of good proportion. The skilled designer has a feeling for balance, and his good taste and judgment may be depended upon to guide him correctly.

For those of less experience, some rules of mathematical proportion have been laid down. These proportions vary and thus are apt to be a source of some confusion to the beginner. It may be said, however, that a type page of pleasing proportions is in approximately the ratio of 1 to $\sqrt{2}$, or 1 to 1.418. In other words, the length is 1.418 times the width. Variables of the proportion just set forth are $2:3$ and the still longer page in the proportion $3:5$. (In considering the page length, the running head and folio are included.)

Margins. Type page dimensions must also be considered in relation to margins, that is, the position of the printed page on the leaf. Type area and white space should be about equal, and the two facing pages of an open book should be treated as a unit in arranging the margins.

The margins around a printed (book) page should of course not be equal on all sides. The ideas of designers may differ about the exact position of the printed page on the leaf, since their objective is aesthetic as well as utilitarian. A rule of thumb is that the head margin (that at the top of the page, above the running head) is narrowest; the inner, or binding margin, which with the equivalent margin of a facing page is the *gutter* or *back margin,* is the next widest and should be wide enough so that no type is lost when the book is bound and opened to a double spread (books intended for library use are sometimes rebound—gutter margins should be wide enough to provide the extra paper that may be lost in rebinding); the outside margin is wider still; and the bottom, or foot, margin, excluding the folio if it is at the bottom of the page, is widest of all.

Sir Cyril Burt's *Psychological Study of Typography* studied the effects of various margin widths and concluded that as a general rule the width of the two side margins together should equal half the width of the type page (one-third of the width of the trim size). According to this study, extremely narrow margins can produce visual fatigue and wide leading requires wider margins; but most often the decision about the width of the margins is an aesthetic one.

The desirable width of margins depends on the nature of the book and the binding. Very wide margins would be used in a deluxe edition of a formal, literary work. Biographies and books of travel and of history generally have liberal margins. In novels and in textbooks the margins are often narrower, but they should not be narrowed to the extent that reading or study is difficult. Textbooks often require special consideration because schools are insistent on adequate inside margins. In textbooks, reference books, paperback books, and other books tending to have narrow margins, a greater inside margin increases ease of reading.

Specific requirements were noted in the Burt study for the margins in children's books. It found that narrow outside margins tend to lead the child's eyes off the page and concluded that wide side and bottom margins are especially important in books for young children. The study further noted that for the very young child, especially for those just learning to

read, the right-hand margins should be unjustified, having each line end
with the end of a phrase and using uniform spacing of about 1 em between
words.

THE MECHANICS OF COMPOSITION

The next question concerns the size and face of the type in which the
book is to be set. Each type character is produced from a *matrix*. A matrix
is any mold, model, or die that will produce a type character, and may be
made of metal or film or produced by electronic impulses. The difference
between different kinds of composition (composing words in type) is the
difference in the kind and use of the matrix. There are two fundamental
processes: hot metal composition and cold composition. *Hot metal* refers
to type that is produced from a cast mold, that is, a mold made from molten
metal. In its widest meaning, *cold type* refers to type that is produced by
processes that do not involve hot metal. More specifically, cold type refers
to composition on typewriter-like direct-impression, impact or strike-on
devices, such as the Varityper or IBM Selectric Composer. According to
Phototypography (Report No. 1, March 1972), the term *cold type* "should
not be used as a synonym for phototypography, photocomposition, or
phototypesetting." Composition that does not involve the use of hot metal
has developed rapidly since the 1950's. Over a hundred machines employing photographic means of typesetting were on the market by 1971. The
state of the craft is constantly evolving, as new and improved devices are
invented and marketed. Processes dependent on hot metal and those that
do not require it will be described later in this section.

Anyone dealing with compositors should have a sufficient knowledge
of type and its use to be able to give intelligent directions. Editors should
be informed in such matters, for their problems sometimes include the
selection of the kind of composition (hot metal, cold type composition, or
phototypesetting), style of face (old style or modern, serif or sans serif),
size of page. Necessary for intelligent decision-making is a knowledge of
the working materials of composition, how they are used, how they are
measured, and their adaptability.

HOT METAL COMPOSITION

Hot type may be set by hand or by machine. Two machine systems are
in use, the Monotype and the Linotype. The Monotype system requires
two machines—one a keyboard, by means of which a paper roll is perforated to correspond to type characters; the other a casting machine (commonly called a caster), which casts each character as a separate piece of
type. The operation of the caster is automatic, guided by the perforated

roll, casting each character in its proper order and each line to its proper measure (width), requiring only occasional attention from the machine attendant. Corrections of Monotype are made by hand, incorrect letters being replaced by letters taken from a type case containing the font, omitted words set by hand and inserted, and so forth.

The Linotype machine (and the Intertype, which resembles it) composes type molds (brass matrices) of letters into lines from which solid slugs or lines of type are cast, the whole operation taking place on one machine. Corrections are made by resetting and recasting the entire line in which the error occurs. No matter how slight the correction, the whole line must be reset.

The Monotype is especially adapted to tabular composition for which it is necessary to use column rules and boxheads. Involved formulas, equations, and the like are usually more efficiently composed by Monotype. (Some very skilled Linotype operators are able to set such matter quite well. A comparison of prices, with the factor of corrections taken into consideration, is often advisable before deciding which method of composition is most advantageous.)

The Linotype, largely by reason of its greater speed in turning out finished composition, is best adapted to newspaper and magazine work and any straight matter in which no great amount of resetting will be required. As a rule, foreign language matter is best composed by Monotype, although there are some Linotype operators able to do excellent work of this nature.

All composition, whether hand or machine, requires *justification*, which is the adjustment of the spacing within a line of type so that the line fits a specific, uniform measure on the type page. In hand composition this is accomplished by the insertion of spaces and quads. The Linotype (and Intertype) perform this operation automatically through the use of spacebands, which are wedge-shaped, and are dropped between the matrices. Justification of Monotype lines is provided for, also automatically, as it is being keyboarded, and the proper spaces are cast with each line.

TYPE

Movable types are made of an alloy of lead, tin, and antimony. Lead is used in the greatest proportion, but would be too soft used alone. Tin is added to give toughness and to form a union of all the metals; antimony gives hardness and has the peculiar property of expanding, rather than contracting, upon solidification.

Type that is to be used more than once is made of a harder alloy than is used by the Monotype or Linotype machines, which cast type in the composing room. After machine-set type has been used for printing or casting, it is remelted and prepared for re-use.

PARTS OF A PRINTING TYPE

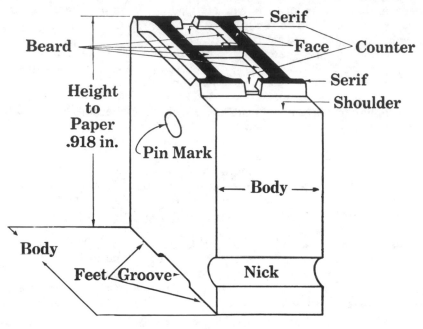

Figure 10

Parts of a type. (See Fig. 10.) The *face* of a type is the printing surface, which stands in relief on the upper end of the type.

The *body*, or *shank*, is all that part below the raised portion.

The *feet* of a type are the parts (the base) on which it stands, and they are formed by the *groove*, which runs across the lower end of the type.

The *shoulder* is the space on the upper end of the body above or below the outline of the face.

The *neck*, or *beard*, is the slope between the face and shoulder.

The *counter* is the depression between the lines of the face.

The *pin mark* is a small round depression in the side of the body, sometimes enclosing figures denoting the size.

The *nick* is a notch across the body of a type on the side which is uppermost when type is set. There may be one to four, variously grouped, no two fonts of the same size being marked alike. The nick serves the compositor as a guide in setting and enables him to distinguish between the different fonts when he is returning used type to the type cases ("distributing").

The *height to paper* of a type is the distance (.918 inch) from its face, or printing surface, to the base on which it stands. An object of this height is often referred to as being *type-high*.

Height of body refers to the size of the type body, or the size of type in points, and is often more than the *height of face,* which is the height of a letter or character as it appears when printed.

The face of a type may be further dissected. The thick strokes of a letter are the *heavy elements;* the lighter connecting lines, sometimes so fine that they are called *hairlines,* are the *lighter elements.* The main upright heavy element is the *stem,* sometimes called the *body mark* or *thick stroke.*

The *serifs* are the short cross lines placed at the ends of unconnected elements. Serifs may be thick or thin, long or short, straight or curved. Their formation largely determines the individuality of the face. Block-letter faces — Gothic, Sans-serif, Kabel, and many others — have none.

A *kern* is a part of a letter that projects beyond the body of a type, like the ends of the italic *f* and *j.*

An *ascender* is the part of a lowercase letter that extends above the body of the letter, as in *b, d, f, h.*

A *descender* is the part of a lowercase letter that extends below the base line of the letters, as in *q, p, y.*

A type font. A font of type is an assortment of types of one body-size and one style of face. Fonts vary in size from a job font, containing the smallest practicable assortment, to a font for intricate composition requiring accented letters, signs, and symbols. For machine composition printers require fonts of matrices (type molds), and such fonts include the companion italic. The less frequently used types are called side sorts, or peculiars.

An ordinary font of roman type for book work contains the following:

Capitals: A to Z
Small capitals: A to z
Lowercase: a to z
Accented or specially marked letters: á, à, â, ä, é, è, ê, ë, í, ì, î, ï, ó, ò, ô, ö, ú, ù, û, ü, ç, ñ
Numerals: 0 1 2 3 4 5 6 7 8 9
Solid fractions: ⅛, ¼, ⅜, ½, ⅝, ¾, ⅞
Superior numerals: 0 1 2 3 4 5 6 7 8 9
Punctuation marks: . , ; : ' ' " " ? ! ( ) [ ] - - —
Reference marks: * † ‡ § ‖ ¶
Signs: &, $, ¢, %, #, ℔, @, /
Spacing material: 5-em, 4-em (thin), 3-em[1] (thick), and en (nut) spaces; em, 2-em, 3-em, and 4-em quads.

Ligatures. Some fonts do not contain the ligatures fi, ff, fl, ffi, ffl, sometimes inaccurately called logotypes; others contain these and several others besides.

Spacing material. The quads are used to fill out a short line. The spaces are of five different thicknesses; alone or in combination, therefore,

---

[1] More accurate terms that printers find too cumbersome are 5-to-an-em, 4-to-an-em, 3-to-an-em.

they are adequate to space any line to the required width. (See Spacing, pp. 244–246. For spacing between lines, see Leading, pp. 243–244.)

Fractions. Fractions are of three kinds: solid, piece, and adaptable. A solid fraction is a single type, and fonts usually include the simple fractions listed above. A piece fraction, also called a split or built fraction, is in two parts:

1 2 3 4 5 6 7 8 9 0 $_1$ $_2$ $_3$ $_4$ $_5$ for the horizontal form; or

1/ 2/ 3/ 4/ 5/ 6/ 7/ 8/ 9/ 0/ /1 /2 /3 /4 /5 /6 /7 /8 /9 /0 for the diagonal form.

With these parts fractions of any size can be made: $\frac{1}{2}$ $\frac{49}{50}$ $\frac{449}{1000}$ $\frac{4}{5}$ (If a comma were inserted in the denominator of a horizontal fraction, the separating rule would be broken.) Large fractions can also be set with separate types for numerator, slant, and denominator: 25/53. Such fractions are the easiest to set because they require the least hand-work.

Superior numerals. Superior numerals are used for various purposes, but principally as references to footnotes and as exponents. As references, except in German composition, they follow all marks of punctuation except the dash. In other uses they precede punctuation.

> . . . per in.[2], or 49.
> See aery[2]. (A dictionary reference to the second meaning given for aery.)

Reference marks. The star, dagger, double dagger, section mark, parallels, and paragraph mark, in the order given, are often used as references to footnotes, appearing both in the text and before the footnote. (See p. 23.) They should, in English, follow all marks of punctuation except the dash.

The asterisk is sometimes used to indicate a particular trait or special circumstances. For example, in a list of items it may be placed before or after those most appropriate for a certain purpose, or those that are indispensable while others are merely desirable; in a list of books it may indicate those illustrated; in a list of names, those persons who have died (the dagger is often used for this purpose).

The "slant" or "slash." The solidus, called a slant or slash, has a variety of uses:

It separates the numerator from the denominator in fractions: a/b

It is the sign for shilling: 6/

Between *and* and *or* it means "or":

> Books of psychological and/or sociological interest.
> Lost or obliterated corners, and/or reference points.

In bibliographical matter it is a separatrix, indicating where one line ends and another begins:

> The Last Sheaf/Essay by/Edward Thomas/(device)/With a Foreword by/Thomas
> Seccombe-Jonathan Cape/Thirty Bedford Square/London

Short excerpts of poetry quoted in the text but not set as extract may be set with a thin-spaced solidus to indicate the end of one line and the beginning of the next.

> As we watch these characters building their frail word-shelters, we can't help agreeing with Mark Van Doren that "Wit is the only wall/Between us and the dark."

TYPE MEASUREMENT

The point system. Before 1878 type sizes were known by size names and varied from one another by irregular amounts. Since sizes cast by different foundries under the same names usually varied slightly and could not be used together, they caused printers much inconvenience and trouble. Standardization of type sizes was badly needed and was finally achieved about 1878 by the adoption of the point system devised by a French typographer named Pierre Simon Fournier. This is a system of measurement based on a unit called a point, practically 1/72 inch.[2]

1 point = 1/72 inch
12 points = 1 pica
6 picas = 1 inch

All sizes of type and materials cast according to this system are exact multiples of some other types or materials. Sizes are designated by points measured by the type bodies (see Fig. 10). There is a regular gradation of sizes differing from one another by half-points from 3-point to 7-point, by single points from 7-point to 12-point, by two points from 12-point to 24-point, and by even numbers of points beyond that size.

This line is set in 6-point
This line is set in 7-point
This line is set in 8-point
This line is set in 9-point
This line is set in 10-point
This line is set in 11-point
This line is set in 12-point
This line is set in 14-point
This line is set in 18-point

The em. Another unit of measurement in printing is the em, which is the square of the body size of the type. The name originated from the fact that the body of a letter M in a normal face is the same number of points

[2]Actually .013837 inch.

wide that it is high. The amount of type in a piece of composed matter is measured by ems of the size of type in which it is set; the work of compositors is measured by the number of ems they set.

Certain quite condensed faces, referred to as "lean" or "skinny" faces (Granjon and Caslon Old Face are examples) sometimes are measured and charged on the basis of a point size smaller in width (i.e., 11 point as 10 point), or a variation of such a basis.

For the measurement of column and page widths, printing space, and printing materials the pica em is used. This is twelve points high and twelve points wide and makes a convenient unit because it is approximately one-sixth of an inch. *Picas* is the shortened term to use. For example, a page (or a section) of type might be said to be set to a *measure* (width) of 21 picas, practically three and one-half inches.

The Monotype system uses also *unit* and *set-em* as terms of measurement. An em, regardless of size, is eighteen units wide; therefore M is an 18-unit character and other letters are varying units in width.

The set is the width of the 18-unit characters of a face expressed in points. Some faces are lean, condensed; others are fat, or extended. Therefore the width of an M might be greater or less than its height. The term *set-em* is used to designate an em which pointwise (in height) is the same as the point size of the face being measured and setwise is the width of the 18-unit characters of that face. Thus, the set-em of 10-point, 10½ set, would be 10 points high and 10½ points wide. For instance:

These lines are set in 12-point Cheltenham (No. 164), in which the M is 12 points high and 12 points wide.

These, for comparison, are set in 12-point Cheltenham, 10½ set, — a condensed face — in which the M is 12 points high but only 10½ wide.

Sizes of typefaces. As explained above, the sizes of type are designated by the number of points measured by their bodies, not by the height of their faces. It is extremely difficult for any but the experienced to distinguish sizes of type by their faces. Compare, for instance:

10-POINT CALEDONIA

The Monotype system of setting type was invented by Tolbert Lanston in 1888. It consists of two machines, a perforating machine with keyboard and a casting machine. The keyboard consists of 276 keys and is controlled and operated by com-

10-POINT BASKERVILLE

The Monotype system of setting type was invented by Tolbert Lanston in 1888. It consists of two machines, a perforating machine with keyboard and a casting machine. The keyboard consists of 276 keys and is controlled and operated by compressed air. The

10-POINT HELVETICA LIGHT

The Monotype system of setting type was invented by Tolbert Lanston in 1888. It consists of two machines, a perforating machine with keyboard and a casting machine. The keyboard consists of 276 keys and is controlled and operated by compressed air. The

10-POINT TIMES ROMAN

The Monotype system of setting type was invented by Tolbert Lanston in 1888. It consists of two machines, a perforating machine with keyboard and a casting machine. The keyboard consists of 276 keys and is controlled and operated by compressed air. The

10-POINT SCOTCH

The Monotype system of setting type was invented by Tolbert Lanston in 1888. It consists of two machines, a perforating machine with keyboard and a casting machine. The keyboard consists of 276 keys and is controlled and operated by compressed air. The

10-POINT OPTIMA

The Monotype system of setting type was invented by Tolbert Lanston in 1888. It consists of two machines, a perforating machine with keyboard and a casting machine. The keyboard consists of 276 keys and is controlled and operated by compressed air. The "lay" of

10-POINT BODONI BOOK

The Monotype system of setting type was invented by Tolbert Lanston in 1888. It consists of two machines, a perforating machine with keyboard and a casting machine. The keyboard consists of 276 keys and is controlled and operated by compressed air. The "lay" of the

10-POINT SANS-SERIF

The Monotype system of setting type was invented by Tolbert Lanston in 1888. It consists of two machines, a perforating machine with keyboard and a casting machine. The keyboard consists of 276 keys and is controlled and operated by compressed air. The "lay" of the keys

In the foregoing illustrations the designations Caledonia, Baskerville, Helvetica Light, Scotch, Bodoni Book, Times Roman, Optima, and Sans-serif are the names given to the respective typefaces, each of a different design. As they are all 10-point, the types are all of the same size (i.e., body size) from a printing standpoint. For "actual appearance" to the eye, however, their largeness or smallness varies — but the "size" (i.e., the face of the type) from this angle is not the point size. As indicated, 10-point type of any given typeface is the same as 10-point type for all other typefaces, but since the width of individual letters depends on the design of the face, 10-point alphabets in different fonts differ in length according to the design of the typefaces, and the number of characters per pica varies from typeface to typeface accordingly. Some very thin typefaces are known as "penalty" typefaces, since in composition costs they carry a penalty because they set considerably more than the normal number of characters per pica.

CLASSIFICATION OF TYPEFACES

All typefaces may be grouped in five general classes: roman, italic, script, gothic, and text. Roman, together with its companion italic, is the type used for books, magazines, newspapers, and all classes of ordinary reading matter. There are two styles of this face, differing from each other slightly in proportion, shape, and shading. The older form was cut in 1470 by Jenson at Venice, and is called Old Style (in England, Old Face). The principal identifying features of Old Style letters are the sloping serifs. Note their form in the following line.

Caslon — an Old Style Face
Note the sloping serifs in these letters: T R t r y c e
This is Caslon italic, designed to be used with Caslon roman.

The numerals for Old Style faces are termed *hanging numerals* because the 3, 4, 5, 7, and 9 "hang" below the base line of the numerals. The 6 and 8 extend above the other numerals. Many Old Style fonts now have so-called modernized (referred to as *lining*) numerals, which retain the characteristic features of the font but align at top and bottom.

Granjon hanging numerals: 1 2 3 4 5 6 7 8 9 0

Granjon modernized numerals: 1 2 3 4 5 6 7 8 9 0

The Old Style face was used until about 1800, when it was largely superseded by a new face called Modern, which was designed in 1783 by Bodoni. The serifs in this face are straight, the bottom of the t curves upward, y ends in a curve and a dot.

Bodoni — a Modern Face
Compare the letters T R t r y c e with the Old Style.
This is Bodoni italic.

The numerals in Modern fonts are lining, or ranging, numerals. All are the same height and none hang below the base line.

Bodoni numerals: 1 2 3 4 5 6 7 8 9 0

Both Modern and Old Style faces are in common use now, and numberless faces have been cut based upon these two styles. Each has italic forms. Commonly used Old Style faces are the Monotype Caslon, Baskerville, Century Old Style, and Garamont, and Linotype Granjon, Garamond, Elzevir, and Franklin. Scotch, Bodoni, Helvetica Light, Times Roman, and Optima are Modern faces.

Antique and boldface are heavy forms of Old Style and Modern faces.

This face is Old Style, Monotype No. 25

This face is Linotype Antique No. 3.

This is a Linotype face called Old Style Antique.

The older faces used before roman was designed are now used mainly in small jobs, as they are not suited for straight reading matter. The oldest style of typeface imitated the hand lettering that prevailed before movable types were invented, and in appearance is black and ecclesiastical. This style of face is called Text or Black-letter.

𝕷𝖎𝖓𝖔 𝕿𝖊𝖝𝖙   𝕮𝖑𝖔𝖎𝖘𝖙𝖊𝖗 𝕭𝖑𝖆𝖈𝖐

Script types are imitations of handwritings; their use is limited, as they are employed chiefly in announcements, invitations, display lines of checks, and similar matter.

Typo Script   *Trafton Script*

The Gothic types, especially some of the newer faces, have come into rather wide use for headings and captions. Gothic type has a perfectly plain face with lines of uniform thickness and without serifs. Its appearance, however, can be wholly changed by the addition of serifs, and each style of serif that is added to it seems to give it a new form. It is sometimes called Block-letter or Sans-serif. In England it is called Grotesque or Doric.

GOTHIC COPPERPLATE

Folio Bold

Compositors generally have fonts of a number of different typefaces, but few compositors have fonts of all available typefaces. Choice of typeface may, therefore, depend on the resources of the compositor chosen. Any number of type books are available to help in the selection of a face. Most production departments keep several on hand. The faces illustrated on the previous pages are frequently used. For general book work the typeface should be appropriate for the kind of material that is to be set, and so suited to its purpose that it is unobtrusive.

Display type. Type that is larger and heavier than ordinary text or body type is called display type. Display type may be set by hand or by semi-automated or automated equipment. It is used for headlines, on title pages, for part and chapter headings, on jackets, in advertisements, and so on. See examples in Figure 11.

Ornaments. In addition to fonts of typefaces and display type, compositors have a stock of ornaments that can be set. These are decorative devices that may be applied at chapter headings, at the end of chapters, on title pages, in advertisements, and elsewhere. See examples in Figure 12.

CHOICE OF SIZE AND FACE

The first consideration in deciding upon the size of type to be used is the age of the reader. Sir Cyril Burt made tests using a standard Times New Roman type, and reported *(A Psychological Study of Typography)* that 24-point type is best for preschool children and for those just learning to read, 18-point for children ages 7 to 8, 14- to 16-point for children ages 8 to 10, and 12-point type for children ages 10 to 12. The study found 10-point type optimum in a book designed for general reading by adults. (It should be noted that types of the same point size vary markedly in actual size, see pp. 501–503; for example, the 10-point Times New Roman used in Burt's study is equivalent to 11-point Baskerville.) College-level students found 9-point type as legible as 10-point, while older persons found 11- or 12-point type more legible. Nine-point type generally produced eyestrain in persons with visual difficulties. In the typefaces most commonly used in book printing, type sizes from 9-point to 12-point are all quite readable, provided optimum leading and line widths are employed for each type size.

Factors in choice of face. Factors to be considered in choosing a typeface are legibility, adaptability, and appropriateness. Choice may be affected by the printing process decided upon, the kind of paper used, the availability of harmonious secondary types, and the length of the book.

The Burt study confirmed that for adults there is no significant difference in legibility for the commoner book typefaces. In general, the typefaces and type sizes that people have become used to are most easily read

48 Point

Goudy Text

60 Point

Reiner Script

42 Point

48 Point

Dom Bold

30 Point

Century Nova Italic

28 Point

PROFIL

30 Point

Corvinus Skyline

24 Point

BODONI Shaded

24 Point

AUGUSTEA INLI

Figure 11
Display Type

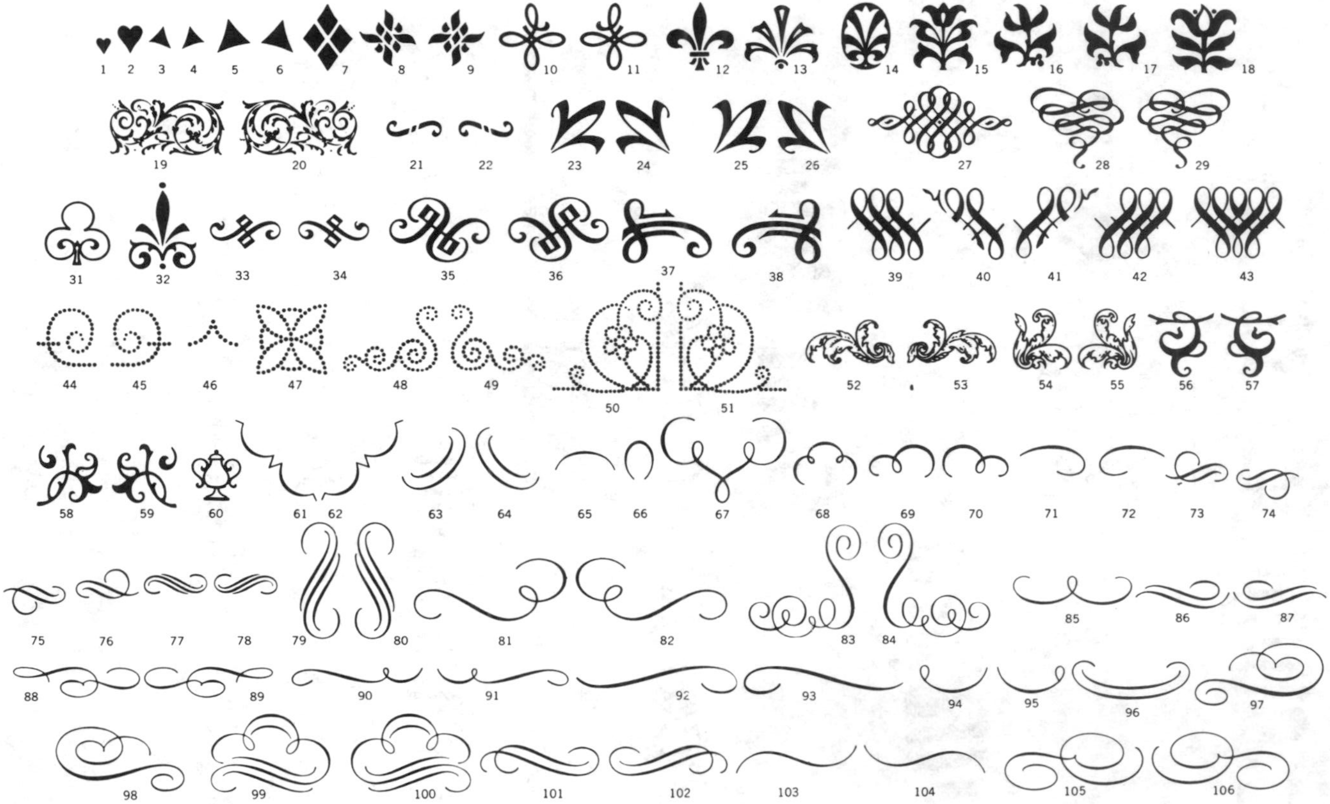

Figure 12
Ornaments

by them. With the children tested, the serif typefaces were found to be much more legible than were sans-serif, as serifs contribute to the uniting of letters into word-forms. The study also found that typefaces with longer ascenders and descenders and the older typefaces that tend to accentuate the differences between letters both made word-forms more legible to children. However, for numerals the more modern typefaces are preferred, as their lining, or ranging, numerals were found to be less confusing for children than were the hanging numerals of the older faces.

While most typefaces commonly used in book printing are about equally legible, they vary widely in adaptability. Some faces can be used effectively for many kinds of text, while others are more limited in their usefulness. Some are neat and businesslike, with an even and regular appearance; others have an irregular outline and variety in shapes of letters. Some are designed to convey ideas without making the reader conscious of type; others are designed to please the eye. Choosing a type appropriate to the text requires a judgment of the qualities of faces. The type book of the compositor who is to set for him is indispensable to the editor. A type book is made up to show the qualities of all the faces the compositor has and the effect of a page composed in them. *Kingsport Type Specimens*, for example, is a compilation in three volumes of display type and complete alphabets that are available at the Kingsport Press. (For other books on typography, see the Bibliography, pp. 532ff.)

COMPUTING THE LENGTH IN TYPE

Authors, editors, and printers often need to know how to estimate the length of copy and how to compute the space it will fill when it is set in type. The author's problem, often, is to know how much copy will be required to fill a given space, as when pages need adjustment, or an article or book must be of a certain length. The editor's problem may be how to get the copy within the space available for it, involving decisions of type size and face and amount of leading (space between lines of type). The compositor needs to have calculations of length of copy on which to base his cost estimates whenever he is required to submit a bid for composition. Casting off, or estimating the length of the manuscript, was described earlier (pp. 491ff.). Basic decisions concerning trim size, type page, typeface, and type size have also been discussed. When all of these steps have been taken, it is possible to project the number of pages of type that will be produced. Because of the many variables involved, as kind of type, size, style of face, and leading, computing the length in type is much more complicated than estimating copy length.

Computation. Printers' type books have simplified the editor's task by giving, along with their specimens of type, a notation of the number of words or characters to a given space. Henry D. Gold, *The Rapid Copy Fitter*

(20 Brick Drive, Merrick, New York, 1953), presents a list of about 300 typefaces and gives the number of characters that will be set per pica and per inch for type sizes from 6 to 14 points. A similar device for computing the length in type for various faces and sizes is the *Haberule*.

For example, this book is set in 9/11 Optima; that is, the body size of the type is 9-point and there is leading (see pp. 243–244) of 2 points between lines. By consulting *The Rapid Copy Fitter* we find that 9-point Optima sets 2.85 characters to the pica. The type page dimensions of the book are 26 × 44 picas, with 42 full lines to a page. The *Rapid Copy Fitter* has a table that shows the multiplication worked out for number of characters per pica times number of picas per line. By checking this table we find that 2.80–2.89 characters per pica times 26 picas equals 72 characters per line. Since our book is planned for 42 lines per page, 42 × 72 = 3024 characters per page. The total number of characters in the manuscript, which was determined by the castoff (see pp. 491–493), divided by 3024, will give the approximate length of the book in the type that has been selected. (For possible differences in set size of typefaces in hot metal and in photocomposition, see p. 515.)

Obviously, space that will be occupied by illustrations, tables, headings set off on the text page, and other interruptions in straight text, as well as front and back matter, must be added to the estimation of copy length. Such additions increase the size of the book, perhaps considerably, and affect the cost. Before additional material can be considered however, it is fundamental—in book work—to know the length in type that the text will run.

Adjustments. The computation described above applies to a specific trim size, type page, face, and size of type. When a computation of the length of a manuscript has been completed, it may become evident that the manuscript will not produce a book of the required length, and some adjustments must be made. Such adjustments affect the determination of the trim size, type page size, size and face of type, or the type arrangement within the book.

There are many ways of expanding a manuscript, such as using center heads and shoulder heads instead of run-in side heads, and presenting data in columns rather than run in. To gain length the book may be set in a typeface that has an alphabet length longer than average and looks best with wide leading, or the area of the type page may be reduced by enlarging the margins. If the manuscript is too long for a volume of the desired size, it may be edited for compactness; the size of the type page in proportion to the trim size may be increased; or a type may be chosen that requires only slight leading and sets more words, or characters, to a page. (One other factor in the appearance and physical size of a book is its bulk, that is, the thickness of the book without its cover. The bulk can be adjusted, within limits, by changing the weight and finish of the paper on which it is to be printed; see pp. 526–528.)

TYPESETTING

All of the foregoing regarding type applies to composition by Monotype or Linotype. With only slight modifications, most of the same faces and sizes exist for setting by cold type composition, photocomposition, and computerized composition.

In the typesetting process, the input of material to be set has historically been accomplished by an operator pressing the keys of a keyboard similar to that of a typewriter (Linotype and Monotype). The addition of a computer, fundamentally a high-speed counting device, to the input process, has greatly increased the speed at which material can be input for typesetting. The computer can be programmed to space, justify, hyphenate, and make other decisions automatically, decisions previously made by the operator. The keyboarder and the computer between them produce a tape that is used to drive the typesetter. All systems that are tape-driven are dependent on the computer.

The computer is used in conjunction with a variety of typesetting machines. Existing Monotype and Linotype machines (hot metal) have been adapted to make use of the computer; many other systems make use of it as well, including strike-on systems (cold type composition), Photon, Monophoto, Linofilm, Fototronic, and others (photocomposition). One great contribution of the computer is increased speed of input, because, as noted above, routine decisions are made automatically. Another is that one input can produce a variety of outputs. The same tape can be programmed to produce a different size and face, different measure and length of type page, and other differences in the design of the page. This is particularly advantageous when the probability of retrieval or revision of the material is high. However, there is a certain lack of flexibility. The computer can hyphenate at the ends of lines but prefers not to; it will, if at all possible, justify by spacing. This leads to uneven spacing of words within the line.

COLD TYPE COMPOSITION

The development of offset printing (photo-offset; see pp. 519–520) permitted totally photographic preparation of material for the press. This suggested methods of setting type that did not have to depend on cast metal to produce raised characters for the printing impression.

Keyboarding and setting. For material printed by photo-offset lithography, the office typewriter with a carbon film ribbon (producing sharper definition than provided by fabric ribbon) could be used to obtain "camera-ready" copy. The primary disadvantage to typewritten text was that it looked typewritten; but it was "type" nevertheless, and the typewriter did the "setting." Only time was needed to improve the mechanics involved so that a typewriter-like device could set in various faces and maintain the character alignment required for aesthetically pleasing printed matter.

Proportional-spacing typewriters (such as the IBM Executive) also permitted justification of set copy, although not without a double-typing process. Addressograph–Multigraph developed the Varityper, which combined proportional spacing with the ability to change type styles. Friden developed the Justowriter, a two-unit system employing punched paper tape produced by an *input* keyboard, which was "read" by an *output* device that automatically altered word spacing to provide justified copy. In this system no changes in typeface were possible.

Development of the IBM Selectric Composer, a proportional-spacing version of the Selectric typewriter, provided to the cold-type setter well-aligned characters in faces such as Baskerville, Bodoni Book, and Univers, with choices of roman, bold, italic, and bold italic complete with foreign language characters and piece accents. Camera-ready copy is produced by means of a single, sphere-shaped element, or printing font, about the size of a golf ball. The font bears all characters, numbers, and punctuation symbols. There are 150 type fonts available, and the single element permits rapid change from one type font to another. Type styles range in size from 6- to 12-point. There are 135 fonts of text typefaces, patterned after a broad range of modern and classical printing type styles; others are for printing Greek, mathematical and technical symbols; Spanish, French, British, German, and Nordic fonts are also made. Double-typing justification on the Selectric Composer involves only a simple color-coded setting that establishes the width of the space bar to expand set lines to the desired measure. Totally automatic justification on the Selectric Composer is accomplished by the addition of computerized magnetic tape units and keyboarding systems. Type set with the IBM Magnetic Tape/Selectric Composer may be automatically justified, centered, flushed right, or dot-leadered at a speed of 14 characters per second.

Input to strike-on equipment, such as the IBM Magnetic Tape/Selectric Composer, is generally in the form of punched paper or magnetic tape produced by a keyboarding unit. Coding that controls a computer component of the equipment (format, face changes, and so on) is keyboarded along with the text, so that totally automatic setting results.

Since the keyboarding step is essentially independent of the setting process in most computer-controlled cold type systems, the operator is free to concentrate more on accuracy than on format and fitting details. Such decisions as hyphenation are generally handled either totally or semiautomatically by the output equipment. The output equipment, also, can usually be programmed with details such as set measure, amount of paragraph indent, leading, and spacing variables (minimum and maximum interword spacing limits, for example). Thus a change in type specifications requires only a reprogramming to effect the change in set format; very little rekeyboarding is required.

Characteristics of cold type. The primary advantage of cold type composition is its speed. Keyboarding speed is increased by equipment that resembles a standard office typewriter. Results of keyboarding are immediately visible, and the operator may correct errors as they are made, *before type is set*. Actual setting need not wait for proofs to be pulled.

Composition of material that is difficult in hot type is easily accomplished with computer-controlled cold type equipment. Tabular material, for example, requires only a predefinition of table format (width of columns and placement of text within columns). Keyboarding need not be concerned with format; the arranging of columns is handled automatically during setting. The same is true of hanging indents, runarounds, outline format, centering, quadding to flush right, and so on.

Other advantages are the comparatively low cost of the equipment; the relatively low cost of operation—a good typist, as compared to an expensive highly skilled operator, can handle the composition (keyboarding); and the low cost of *standing type* (fully composed books may be kept for future changes without tying up the space and metal required in hot type). Also an advantage is the ability to store the paper or magnetic tape for future revision of catalogs and directory material. All of these advantages result in a significantly lower total cost for composition.

There are, however, some disadvantages associated with cold type production. Important among these is the problem of spacing. Typography as an art no longer exists when the spacing and other details are controlled by a machine rather than an experienced operator who instinctively sets and spaces for best effect. At the time this is written, cold type composition cannot produce the high quality obtainable with Linotype. The first product of the IBM equipment is a repro ready for page makeup. Because there is only one repro, cleanliness is of paramount importance; many hours of machine time can be destroyed by mere carelessness. This problem is particularly acute with carbon ribbon strike-on machines, since the type smears easily. To protect the one repro—the camera copy—copies are made of the repro and sent to the editorial department for corrections. On their return to the compositor, corrections or additions are set separately and pasted on the repro. Page makeup, following instructions, is made by the compositor in the form of a mechanical for each page.

Page makeup in cold type requires expertise and patience. Far from arranging hot metal slugs that can be held with ease, cold type requires the highest precision in cutting the paper and arranging it on a support (index stock or illustration board). Whereas in hot metal, the line slugs provide the proper leading, in making up cold type pages exact cutting, butting, and aligning are required. The mortising in of corrections is time-consuming and tedious. Too much handling during the makeup is sure to result in smeared, torn, or dirty results, requiring much resetting.

The advantages of low cost, speed, and versatility, however, far outweigh the disadvantages at makeup. Care taken early in the cold type process, during keyboarding, is repaid many fold by makeup not requiring correction mortising, leading out to fill allotted spaces, or resetting.

PHOTOCOMPOSITION OR PHOTOTYPESETTING

Another method of composition that does not involve the use of hot metal is photocomposition or phototypesetting. Phototypesetters set type by exposing a light-sensitive material (film or paper) to light projected through a negative of the character image. Relatively high speed phototypesetters employ a spinning disc containing the character images arranged in concentric circles; italic, roman, bold, and special characters may thus be selected automatically by controlling which circle of characters receives the projected light. Critical timing is provided by computerized control, which causes a high-speed flash of light when the correct character is in the proper place for exposure. Optics generally provide for movement across the set line. One master matrix of a character can be used to set many different sizes by optically changing the size of the individual character. Thus, a combination of discs can produce sizes from 5-point to 72-point, and can mix italic, bold, and roman within the same line. Since the characters are set by light passing through the negative of the character in the master matrix, the problem of type becoming worn or battered is eliminated.

Keyboarding. The copy to be set is first keyboarded. The keyboarding or inputting device is a machine (part of the photocomposition system) that enables the compositor to convert copy-edited manuscript into paper tape, magnetic tape, or scannable copy. In addition to transferring the manuscript to a tape or converting it to scannable copy, the inputting, or keyboarding, device is programmed to instruct the typesetting device how to set the copy, that is, to provide for variations of type—as roman, bold, italic—to set paragraph indents, to provide for runarounds, tabular material, and so on, to justify through a fixed measure, and to hyphenate when necessary.

No matter how sophisticated the equipment is, the conversion of the manuscript into machine-readable copy must be carried out by a fallible human being. This is the slowest and most expensive phase of the operation. An advantage of this kind of composition, as with cold type, is that the composition can be read out and corrected immediately.

Corrections. As the keyboarder presses the keys, on a machine that looks like a typewriter, the copy appears, as would typewriter paper from the roll, and can be read for correct coding and text. At the same time, the copy is being transferred, with all the coding, to a tape. Any errors can be spotted immediately and corrected on the tape. This is much faster than corrections can be made when hot metal is the method of composition. In the latter case, errors do not show up until proofs are pulled. A newer and more flexible method now being used in some of the more advanced typesetting firms is to make corrections or changes on a CRT (cathode ray tube) editing terminal. This is a device that looks like a TV screen with a typewriter attached to it. The operator can project a galley or page that is to be modified on the screen. Using the keyboard, words can be deleted and new ones entered; entire lines or phrases can be deleted, new ones added, and so on. It is possible to make any change to the text that is desired without having to create a new tape and merge it to the old one.

Type size and face. A wide variety of typefaces is available for phototypesetting. However, although many of these have the same names as hot metal typefaces, the set size of the typefaces in photocomposition varies somewhat from the set size of hot metal type of the same face. This factor should be taken into consideration when computing the length in type for a manuscript to be phototypeset. For many typefaces, the set size of type set on phototypesetters is slightly larger than the set size of the hot metal type and therefore will set fewer characters to a line. The difference might be significant in terms of the final number of pages. One way to check is to count the characters of several lines of sample pages and take an average.

Second-generation equipment. Phototypesetting systems have developed very rapidly in the last twenty-five years. The earlier systems, now called second-generation, were similar to Linotype and Monotype but were photomechanical, that is, characters on film were substituted for the metal matrices of the hot metal systems. Subsequent second-generation equipment eliminated some of the mechanical features of the earlier phototypesetters, many mechanical functions now being performed electronically. Second-generation equipment comprises "spinning discs or cylinders (font carriers) that contain several alphabets on their face in connective rows. On call, a letter is 'stopped in rotation' by a beam of high-intensity light recorded onto film."[3]

Harris–Intertype, Monotype, Friden, Mergenthaler, and Photon are among a number of manufacturers who produce machines that operate on variations of this principle. The Mergenthaler VIP System employs an input keyboard, a Fairchild Perforator, that punches an "idiot tape." This is coded for type size, roman, bold, italic, and so on, but is not coded for hyphenation or justification. The tape is put onto a VIP unit, which can hyphenate and justify. To proofread and correct the output of the VIP unit, the tape is put on a Mergenthaler correction terminal. The tape is run across a screen similar to a TV screen, and as corrections are necessary an operator at the keyboard connected to the screen makes them.

Third-generation equipment. The most advanced phototypesetting machines (CRT, third-generation phototypesetters) combine a cathode ray tube (CRT) capable of generating 6,000–10,000 characters per second and a computer programmed to control justification, hyphenation, and so on. (Devices capable of generating 60,000 characters per second for microfilm are possible.) The cathode ray tube employed here is an "electronic tube used by high-speed photocomposition machines to transmit the letter image onto film, photopaper, microfilm, or an offset plate. The image is actually comprised of a series of dots or line strokes, which vary in degree of resolution, depending on the system itself and the operation speed."[4] These machines can set at fantastic speeds. The equipment itself is expen-

[3]American Association of Advertising Agencies and Advertising Typographers Association of America, *Phototypography*, Report No. 1, March 1972.
[4]*Phototypography.*

sive, and the process is so fast that it is often a problem to supply it with enough copy to keep it busy. An enormous advantage of these systems is for the storage and retrieval of information. To cite an obvious example, an enormous dictionary can be put on tape, coded for the retrieval of a variety of different kinds of entries, stored in a small space, and revised at any time by merging a new tape with the original. The CRT system has been widely used for setting directories and long lists, such as telephone books, that must be amended frequently. Increasingly the system, as well as variations of it, is used for setting all kinds of copy—for books, newspapers, periodicals, and so on.

Combinations of the cathode ray tube and computer-controlled phototypesetting machines have been developed that are capable of reproducing, through the use of an electron beam, complex tabular material, multilevel formulas, line drawings and, more recently, halftone images.

Characteristics of photocomposition. As noted, the principal advantages of such systems are ease and immediacy of correction, speed, compact storage of composed material, retrieval of information, and ease and comparatively low cost of revising existing set copy. Another advantage is the flexibility the system gives to the designer. All elements: illustrations, tabular material, graphs, and runarounds can be easily positioned by the page makeup person using positive working film. The editor and compositor can immediately see what the page looks like. However, this process of handling and attaching bits of film is time-consuming.

Children's books, set in 20-point, 24-point, or larger, can now be set more efficiently. Before photocomposition, books set in these larger point sizes were set manually either by hand foundry (movable type) or on a Ludlow machine. Now they can be inputted on high-speed keyboards at reasonable cost.

Computerized composition performed on the latest third-generation systems (those that utilize the cathode ray tube to generate characters) has revolutionized typesetting. Many second-generation systems are currently widely used, and third-generation systems, in which great improvements in efficiency, materials, and equipment have been made, are rapidly taking over a share of the market. The foreseeable future of this method of typesetting is brilliant, and editors will do well to have some understanding of it.

SAMPLE PAGES

All of the methods of composition mentioned above offer a wide choice of faces; therefore, quality, cost, and possibly speed are usually the determining factors in deciding which method of typesetting is to be used.

Since few publishers have their own composing rooms, the production man will choose a compositor who can, in his judgment, best set the kind of work the publisher has in hand. An expert Linotype operator can set almost any kind of text, but texts such as chemistry and mathematics, with many equations and symbols, are more economically set in Monotype. Cold type, in its strictest sense, is not economical for a book requiring a number of different fonts and special characters, because, although the fonts are available, it is too time-consuming if they have to be changed often. Computer composition is very fast, but it may be more expensive. The advantages and disadvantages of the different kinds of composition must be weighed and a decision made based on cost, special qualifications of different compositors, need for speed, availability of desired faces, and, in some cases, accessibility to the publisher. Most important is the quality desired. As noted above, not all methods of typesetting produce the same quality.

PROOFS

Hot metal composition. If the book is set in Monotype or Linotype the first proof will be a set of *galleys*. These are long sheets of text, set full measure, sometimes including headings, except running heads, but with no space indicated for illustrations or tables. Footnotes, usually set in a smaller size type, may be set directly under the text reference or all together on separate galleys. (They are called galleys from the long and narrow metal pan, about half an inch deep and open at one end, in which lines of type are placed when they are first set. The first proof of composed type is, hence, a "galley" proof. For the handling of galleys in the publishing house, see pp. 70–74.) After the galleys have been corrected, they are returned to the compositor with instructions concerning the location and space requirements for illustrations and other non-textual matter noted on the galleys. If there are a number of tables in the book, they may be set separately by cold type composition, which is an economical way to set such matter. It is recommended that books containing a number of tables and halftones be dummied (see p. 524) in galleys.

Page proofs and repros. Proofs of the pages will be sent to the publishing house for checking and correction and returned to the compositor. Final proofs from the compositor are reproduction proofs, called *repros*. Two sets are usually sent, as insurance against loss or damage. Repros are also called *camera copy*. From them negatives are made by the printer for making plates used in the offset printing process. Repros are examined in the production department for the quality of the work—broken or weak type, dirt that will be reproduced on the negative, and errors. In each of the three stages from the compositor—galleys, page proofs, and repros— the compositor's readers have checked for type and errors, and the production department, or production editorial department, has also checked. Corrections of errors or changes in the material can be made at each stage, but changes are progressively more expensive in successive stages.

Cold type composition. The steps in the progress toward a book are slightly different when the book is set by cold type composition. In this case, camera copy, or reproduction proof, is produced in the first step. Electrostatic copies of this are sent to the production department, where they are checked for accuracy and then returned to the compositor. The compositor sets corrected lines and pastes them on the original camera copy. Following instructions supplied by the production department, the compositor uses camera copy that he has set or that is supplied by the production department and makes a mechanical of each page of the book. The difference between a mechanical and a dummy is that the former is camera copy, the latter illustrates how to arrange the pages of the book to make camera copy. With the completion of camera copy, the material (book) is ready for the printer.

Photocomposition. The proof sequence for material set by photocomposition is much the same as that for hot metal composition, and encompasses the following steps: (1) copy-edited manuscript to the compositor; (2) galleys to the publisher; (3) marked galleys and page dummy to the compositor, including all art work, figure legends, and so on; (4) first page proofs to the publisher; (5) marked page proofs to the compositor; (6) revised page proofs to the publisher; (7) okay, or okay with correction pages, to the compositor for final film; (8) film proofs (confirmation of final film) to the publisher for last okay and release of final film.

With CRT composition the galley stage may be eliminated. The computer can be programmed to go from machine-readable copy directly to pages, with running heads, folios, and footnotes. The pages have to be read carefully for typos and incorrect hyphenation (the computer can cause the reproduction of information that has been stored in its memory only, a new fact has to be cataloged before it will be reproduced automatically) and for evenness of page length. However, all the matter that would have been assembled in a series of galleys and page proofs by other means of composition appears together at once by CRT composition.

THE MECHANICS OF PRINTING

When the book is sent to the printer, a complete layout (see pp. 61–69) is sent with it.

There are two main methods by which books are printed: letterpress and offset lithography.

LETTERPRESS

Letterpress printing is essentially a matter of inking the surface of metal type set in forms and transferring the ink to paper by pressure. As carried out today, letterpress is simply a refinement of the earliest printing method used after the invention of movable type. To make sure that the pressure is transferred equally from the whole form or plate (which may contain irregularities) to the paper, certain adjustments, called *makeready,* must be made to the plate or to the bed or cylinder of the press. The quality of the impression on the paper depends on these adjustments. Once these adjustments have been made, however, letterpress can provide a consistently higher quality of impression than do other printing methods.

In letterpress printing, a separate plate must be made for each line cut or halftone appearing on a page. These separate plates must then be locked in with the type to complete the plate for the page. When a number of line cuts or halftones are to be used in a book, the added expense of making and integrating these separate plates can greatly increase the printing costs. However, when a heavily illustrated book requires the high degree of sharpness and clarity of impression possible with letterpress, and the final price of the book will cover the higher printing costs, letterpress printing may be used.

Letterpress printing was at one time the major method used in book printing, and it is still useful for certain purposes. Until recently, it was limited to printing from metal type, but new developments in the area of photomechanical platemaking are now being applied to letterpress, thus giving it some of the advantages of lithography. Other developments have made it possible for large printings of inexpensive books to be printed on a web-fed letterpress using rubber plates. These presses can produce at speeds that exceed those of any lithographic presses, but they do not yield a high quality of reproduction. Even with these technological advances, however, letterpress printing has so declined in use that by 1972 it accounted for only about 20 percent of all books produced.

OFFSET LITHOGRAPHY OR PHOTO-OFFSET

Offset lithography, or photo-offset, is surface printing (as opposed to printing produced by a die or mold) that makes use of a photomechanical process to transfer the image to a plate. The plate is then attached to a cylinder on the press. Offset lithography uses a rotary press in which the ink-wet plate, on a cylinder, transfers the ink impression to the paper traveling over an impression cylinder. Rotary presses can be sheet-fed, that is, individual sheets of paper travel through the press, or web-fed, in which the paper travels through the press from a roll. All offset presses are of the rotary type, which print at high speed and require little set-up time.

Most lithographic printing is done in a process utilizing surface plates. The image to be transferred to the plate is supplied by the compositor and reaches the lithographer in the form of camera copy (generally either reproduction proofs—repros—or mechanicals), which is then photographed by the lithographer, or in the form of photocomposition film, which needs no photographing. The negative (reversed) film of the copy photo or the negative photocomposition film acts as a stencil to create a photographic image on a metal plate with a photosensitized ink-receptive coating. Light, filtered through the negative film, hardens this coating into a printing surface that repels water but accepts the oil-base printing ink. The plate is placed on the cylinder of the press, where successive applications of water and ink charge the plate with an ink-wet image. The image on the plate is right-reading (can be read normally) and is offset from the plate to the rubber blanket and from there to the paper. Because the plate does not print directly onto the paper, there is less problem with the plate wearing down or filling up with paper fibers, and the resilience of the rubber blanket allows for greater press speed and provides a finer quality of reproduction on coarse paper.

Before making the plates from the film, the lithographer can provide the publishers with a set of blueprints (*blues*), a contact print of the film, for final proofreading. As changes in copy are prohibitively expensive at this point, blues should be checked only for spots and light or dark patches (they are not suitable for checking margins). After the plates are made, the lithographer provides press sheets, or press proofs, made as a first run of the press.

Two other types of plates, deep-etch and multimetal, are sometimes used in offset lithographic printing when the necessity for their particular qualities can justify the added cost. Deep-etch plates are processed with their printing area below the nonprinting surface and are thus less subject to wear than are surface plates. They are used for long runs (large printings of an edition) and for color-process printing that requires fine work. Multimetal plates, bimetal or trimetal, are electroplated after exposure to the film image, providing a long-lasting, high-quality plate. Because of their cost they are generally used only for long runs of over 100,000 copies.

One of the advantages of offset lithography is that it can print illustrations for about the same cost as printing type. Also, in offset lithography, images can be transferred to many types of paper, thus allowing for reproduction on inexpensive paper when necessary. However, the characteristics of the paper used have much effect on the quality of the lithographic printing, so care must be taken to select the correct type of paper to yield the desired quality of reproduction.

ILLUSTRATIONS

On pages 43–45, under the heading Illustrations, only so much instruction was given as seemed essential for authors whose work required illustration. More technical treatment of the subject is included here to familiarize the editor with terminology and methods, thus providing the groundwork upon which he can build by studying more comprehensive treatments of this very technical subject.

Preparation. When the book is ready for production and the illustrations have been selected, they should be numbered for clear identification and covered with tissue or tracing paper or stored in glassine envelopes to protect the surface from scratches and marking that would show up on a reproduction.

Sizing. Photographs are seldom the exact size needed to fit into the desired space on the page. Often the photographs must be reduced or, in some cases, enlarged to the appropriate proportions. Eight-by-ten glossy photographs are a convenient size for preparing for reproduction, but when they are not available other sizes may be used. Reduction in size results in sharper lines, within limits. It should be remembered, however, that in reducing a black-and-white illustration, the white area is reduced as much as the black. This can result in a filling in of black areas and consequent loss of detail. On the other hand, too great enlargement results in fuzzy, blurred lines and loss of tone.

The simplest method for sizing illustrations is the *proportional scale*. It is a device consisting of a small disk superimposed on a larger one. The circumference of each disk is marked off in graduated, numbered segments. When the size of the original illustration (either the length or the width, generally the longer measurement is used) located on the smaller disk is rotated to correspond to the desired reproduction size, a window in the smaller disk reveals the percentage of the original that the reproduction will be. For example, to fit an 8 × 10 illustration into a space 4 inches wide, line up the 10 on the small disk with the 4 on the larger disk. The percentage that appears in the window is 40 percent. The reproduction will then be 40 percent of the original. The other dimension will, of course, be reduced proportionately; in this case, 40 percent of 8 equals 3.20 inches. (There is another kind of proportional scale, resembling a slide rule, which also works on this same principle of matching present size with desired size and reveals reproduction percentage through a window in the sliding rule.)

Occasionally a photograph contains more background or foreground material than the editor wishes to use. The copy (photograph or drawing) can be cropped and the dimensions of the cropped copy used for sizing. Crop marks indicating the area to be reproduced should be noted on the cover tissue or in the margins of the illustration—not on the illustration itself.

After the percentage of reproduction has been read on the proportional scale, the cover tissue on the illustration should be marked (40%) and the width of reproduction, converted into picas (e.g., 4 inches × 6 picas per inch equals ←24→), with arrows marking the direction of measurement. (Before anything is marked on the cover or envelope, it should be removed from the illustration to avoid scratches or marks. Never write on the back of a print or other copy with a ballpoint pen or pencil, as the marks will show on the reproduction. If necessary to mark on the reverse of a photograph, use a felt pen or crayon marking pencil.) When all of the illustrations have been sized, a final list should be compiled of all illustrations to be used in the book. This list should include a notation of the reduced width, in picas, and the reproduction percentage for each illustration. The illustrations and the reduction directions can then be sent to the printer, along with a simple, clear set of general instructions and detailed notes on any special handling that is required.

Line cuts. Black-and-white illustrations can be reproduced by a process that produces either line cuts or halftones. The line cut, or line engraving, process reproduces black lines or dots on a white background. The illustration is photographed and its negative is used to make a plate. If the illustration is to be printed by letterpress, a photoengraving is made by exposing the negative on a zinc or copper plate, on which it is etched. This plate is then combined with the lead type for printing. For offset lithography the negative itself is laid out on a page (*stripped in*) with the negative of the type so that both negatives can be transferred together onto a zinc or aluminum plate for printing.

Halftones. Photographs, which require more variations in shading than is possible with line cuts, are reproduced by the halftone process. The original photograph must be rephotographed by the photoengraver, using a screen etched with a grid of black, opaque lines between the lens and the film in the camera. This *halftone screen* breaks the reflected light waves into a series of dots. The dots, which vary in size according to the amount of light reflected, follow the shading of the original and give the illusion of continuous tones. Halftone screens vary in size (number of lines per inch in the grid) from 50 to 175, with the finer screens giving greater illusion of continuous tone. When halftones are ordered, the size of screen to be used should be specified. When there is any doubt about the best size of screen to use, the coarser screen should be chosen; in book publishing, screens coarser than 120 are not generally used. The right screens for a given work should be selected by the photoengraver; for good results he needs to see the material to be reproduced and to know at least (1) the paper to be used and (2) the method of printing. He should know whether printing is to be from original halftones or from flat or curved electrotypes; if a cylinder press or a multicolor press is to be used. If he knows what kind of stock and what kind of inks are to be used for the color-process cuts, he can submit proof on the same stock and with the same inks. Screens are also available that produce halftones with straight, wavy, or circular lines instead of dots.

Halftones are usually rectangular, with square-cut (right angle) corners and are sometimes finished with a fine-line border. A variety of other shapes and appearances can be achieved through special effects. The background can be faded away for a *vignette*, broken abruptly with a very uneven outline for a *hard vignette,* or eliminated entirely for a *silhouette.* When pure white is needed to heighten contrast, the halftone dots can be cut away, or dropped out, in the lighter areas, giving a *highlight* or *dropout* halftone. Special effects can be done either by an artist or retoucher, before being sent to the printer, or by the printer, when he is provided with detailed instructions along with the illustrations.

Benday screens. For either line cuts or halftones, benday screens may be used to heighten or modify the value of the image of a line or halftone by reinforcing or lightening it. A *benday* is a piece of film with a screen pattern that creates an overall pattern of halftone dots or lines. Like halftone screens, bendays vary in size, but they also vary in strength (lightness or darkness, according to the size of the dots or lines) from 10 percent to 80 or 90 percent of black with increments of 10 percent.

Two-color illustrations. Color illustrations require more complicated printing processes involving more than one plate. Two-color illustrations can be produced by either a *duotone* or an *added color process*. Duotones consist of two halftones of the same subject, one in black and one in another color or in another tone of black, that are printed so that they are in register, that is, they correspond directly to each other. Line cut illustrations or black-and-white halftones may be enhanced by an added color process that uses another plate, a *tint block,* to print a flat surface of ink in register behind the illustration.

Four-color illustrations. Full-color or four-color illustrations can be made from either original copy (color photographs of paintings, drawings, or prints) or a transparency (color slide) and involve reproductions made from four different films of ink that, when superimposed on each other, simulate all the color value and tones of the original. Four negatives of the subject are shot, each through a different filter. The filters break down the subject into four colors, called *color separations:* magenta, yellow, cyan, and black that outlines and shades the illustration. The negatives shot for each color are checked and color-corrected if necessary. Each negative is then transferred onto positive film and this positive print is shot through a halftone screen to produce a halftone negative for platemaking. A separate plate is made for each color, and the colors are printed sequentially in register to produce the final full-color illustration.

Proofs. After the plates have been made, for either black-and-white or color illustrations, the printer provides proofs for checking. The proofs (sometimes called *browns* or *vandykes*) for black-and-white line cuts or halftones are checked by the editor for proper reproduction size. He also checks to see that the illustration has not been reversed and that there are no spots on it that were not on the original. Detailed notes and instructions concerning any of these problems are then sent to the printer, along with the proofs, for any necessary corrections.

Color proofs involve checking for all the problems discussed above along with specific points to be checked depending upon the kind of color printing involved. Two-color illustration proofs must be checked for register and for the quality of color. Checking four-color proofs is more complicated, as proofs (*progressives*) are provided in several stages. Since the colors are separated and printed sequentially (generally yellow, magenta, cyan, then black), a proof is made for each color and following each separate color proof is a proof of that color printed along with the colors preceding it in the sequence (e.g., yellow, magenta, yellow-magenta, cyan, yellow-magenta-cyan, black, final four-color proof). Each of these proofs is checked by the editor, generally with the help of a color expert, to determine how close the proof is to the original, and whether any color will need to be intensified or lightened. Four-color presses, which print all four colors in one operation, produce only a final four-color proof, which is then checked against the color of the original for any necessary color corrections.

Dummy. A dummy consists of large sheets or boards on which the actual trim size of the book has been drawn in blue. All the elements that will appear on each individual page of the finished book are indicated on the dummy pages. (A dummy is not necessary for a book that consists of straight text.)

When a dummy of a book is to be constructed, one set of text galleys is cut up and pasted to the dummy pages. Spaces are left for the illustrations and tables, and their captions and legends, which are set separately, are located by number on the dummy or pasted in to show position. Locations of running heads and folios are indicated, as well as any headings in the text that did not appear on the galleys. This dummy is then returned to the compositor with the corrected galleys to be set in page proofs.

The dummy is returned to the publisher, who holds it until the corrected proofs of the illustrations are returned by the printer as blues (or browns, or vandykes). These proofs are then pasted on the dummy to show the printer exactly where the illustrations are to appear. (Another kind of dummy is known as a *book dummy;* this is a set of unprinted pages of the number and kind of paper that will make up the final book, usually bound, whose purpose is to show the size, bulk, form, and general style of the finished product.)

Layout. Layout of a page is generally a matter of taste, although illustrations should not be placed too far down on the page or too near the fold margins. A book layout can be enhanced by special effects, such as by silhouetting or by bleeding an illustration. *Bleeding* is placing ⅛ inch of the illustration beyond the trim edge of the book so that the picture runs off the page. Maps and charts are often placed on the endpapers. Sometimes the illustrations are gathered in one section of the book and reproduced on a glossy coated paper. This is called an *insert,* a separate section that is either tipped in (pasted in by hand or machine) or bound into the book between two signatures. An insert should be at least four pages or in multiples of four, and a book may contain a number of inserts.

After the book dummy is completely pasted up, it is sent to the printer along with the specifications for gutter and fold margins, running heads, and folios for the final stages of production.

PAPER

The characteristics of paper depend upon both the raw materials from which it is made and its method of manufacture, and have a vital effect on the final product—the book. Paper may be chosen for its physical properties or appearance or for its adaptability to a particular printing process. It may be chosen with a particular effect in mind or for its cost, but most often paper selection is made on the basis of a combination of several factors.

General characteristics. The general nature of paper depends on the raw materials that go into it, of which pulp is the most important. The basic fibrous material of which paper is composed is cellulose, derived usually from wood pulp. When clean wood is ground up and the entire mass is used, including lignin and other impurities, the pulp obtained is called mechanical, or groundwood, pulp. Mechanical pulp is used largely for newsprint, but it may also be bleached and added to other grades of paper for increased opacity and lower cost. In time, high groundwood content papers (30 percent or more) tend to yellow and lose strength.

Chemical wood pulp, used in most book papers, is obtained by the chemical treatment, under heat and pressure, of wood chips. The resulting product is bleached and washed to remove lignin and other impurities, as well as residual chemicals. The resulting fiber yields are somewhat lower than in groundwood production, and costs are higher, despite recycling of 95 percent of the cooking chemicals in modern processes. Chemical pulps are cleaner and considerably longer lasting than mechanical pulps and are brighter in color as well.

Pulp may be made from waste paper, but the quality of pulp obtained varies widely depending on the raw material used. There is much interest in recycling as a better means of making use of fiber resources, but the cost of freight, handling, and sorting, as well as the difficulties involved in removing ink, dirt, plastic, metal, and other impurities, make the use of consumer waste an economic problem. Single large sources of waste, such as paper mill waste or tabulating card stock, are readily usable for paper-making fiber and, where available, contribute valuable material to the process. Technological improvements and economic pressures may be expected to increase the use of such secondary fibers.

"Rag" pulps, obtained from cotton fabrics or linters, are a less important source of book paper pulps than heretofore, and other fibrous raw materials, such as bagasse (sugar cane stalks) and esparto (a grass), are not in wide use in the United States.

Texture. The texture of paper depends on its raw material components, its bulk, and its finish. Paper may be *laid* or *wove,* the latter being more common in book use. Laid paper, made in imitation of the earliest European handmade papers, shows fine parallel lines (laid lines) watermarked in it, with other less frequent more pronounced lines (chain lines) crossing them at right angles. This pattern is produced by the pressure of a specially made screen-covered roll (dandy roll) during manufacture. Wove paper, which exhibits a more or less clear, even appearance when held up to a light, takes its name from the woven wires of the screen on which it is made.

Color. The color of white paper is its *shade* of white, and is produced by the addition of small amounts of pigments to the pulp or to the coating of coated paper. The distinction of shade is to a large extent subjective, although the eye is able to distinguish minute variations. Paper is made in standard shades, but mixtures of different grades of paper or of the products of different makers are to be avoided, for shade variation within a book can cause problems. The soft yellowish color called natural, or text shade, is often preferred in uncoated book paper, while offset papers used for juvenile books or coated papers used for illustration printing tend to be blue-white by comparison.

Brightness. The brightness of paper, as distinguished from its color, is its ability to reflect light of a specified wave length, and is not subjective but may be measured photoelectrically by standard test methods. Generally, chemical wood pulp papers of high quality tend to be brighter than ground-wood or waste-paper content grades. Sometimes fluorescent dyes or "optical brighteners" are added, and while these cannot fool a brightness meter, they can deceive the eye. Since they are likely to fade, they should not be used in book papers where lasting color is desired.

Opacity. With the increasing use of lightweight papers, opacity of paper has assumed greater importance than ever before. Opacity is adversely affected by high brightness and by light weight (because of less fiber), and may be improved by the addition of mineral fillers, such as clay or titanium dioxide, to the pulp mixture. Clay and other heavy fillers, if added in large quantity, may decrease the strength of the paper, and titanium dioxide, while much more effective than clay in small quantities, is a high-cost material. Both, however, have important uses as opacity improvers and may justify any loss of strength or any additional cost. Opacity may be measured optically by the use of a special meter, or may be expressed as printing opacity, which takes into consideration printing pressures and ink show-through.

Finish. Paper is classified as to finish both as an indication of surface texture and—for uncoated papers—as an approximate index to bulk. In order from rough to smooth, uncoated papers are termed Antique, Eggshell, Machine Finish, Smooth, English Finish, and Super.

Offset papers receive surface sizing to resist water penetration and allow for heavy ink coverage. They are less bulky, weight for weight, than other uncoated papers. Vellum Finish offset is between Eggshell and Machine Finish in bulk and surface smoothness, and Wove Finish offset is somewhat smoother than Machine Finish.

English Finish and Super, originally designed to provide reasonably good halftone printing without the gloss—or the higher cost—of coated papers, have been largely replaced by pigmented, matte, or blade-coated dull papers. In these grades, used extensively for textbooks requiring thin bulk, high opacity, and good illustrative printing, the surfaces have been impregnated, roll-coated (pigmented), or heavily coated (blade-coated dull), all on the paper machine. Their surfaces are smooth, mostly free from glare, and highly ink receptive. These new papers are made for letterpress, offset, or gravure printing.

Coated papers are those having the smoothest surfaces, the least bulk, and the greatest density. A book printed on coated paper will weigh more per inch of thickness than any other kind of book, but will provide optimum surface for the printing of halftone illustrations by any printing process. Coated papers are available in gloss or dull finishes and in various qualities according to brightness and refinement of surface.

Weight. Papers are classified by weight, both as a guide to their other properties and because book papers are priced by weight. The *basis weight* (or *substance weight*) of paper is the weight of a ream (500 sheets) of the basic size—in the case of book paper, 25 × 38 inches. For example, basis 50 (50 lb, sub. 50) book paper means that 500 sheets of size 25 × 38 weigh 50 pounds. Weight of 500 sheets of any other size of a particular paper may be computed by means of a simple proportion:

$$\frac{25 \times 38}{\text{basis weight}} = \frac{\text{size desired}}{x}$$

Commonly used basis weights of paper for books range from 50 lb to 80 lb, and are normally available within this weight range in 10-pound increments, although in-between weights may be made in large quantities. Weights ranging from 50 lb down to 30 lb, in 5-pound increments, are used where light weight is needed and usually carry an upcharge. Weights lighter than 30 lb are sometimes used for reference books and are of different construction because of much greater opacity requirements, while weights higher than 80 lb are sometimes used for special purposes but may present folding problems in the bindery. Weights per thousand sheets (M sheets) are often used instead of basis weights. The advent of a large volume of web, or roll, printing in the book field has placed a special need for emphasis on basis weight control. Roll paper is purchased on a *gross weight* basis, rather than by *nominal weight,* as in sheet paper. This means that a roll of a given number of pounds may contain less paper (mileage) than estimated if its basis weight is too high.

The metric system of paper weights is used in most countries of the world except the United States, and the recent changeover in Britain has altered British expression of paper weights and sizes accordingly. Weights are expressed as *grammage* (grams per square meter) and a whole new group of standard sizes ("A" sizes) have been adopted. Conversion tables are required to convert U.S. basis weight into grammage equivalents for British and European use.

Grain. The fibrous nature of paper causes it to have a grain, with the grain direction parallel to the long axes of the fibers as they are oriented on the paper machine. The grain direction is the machine direction: that is, *around* a roll of paper. Papers that are heavier than 50 lb, especially in bulky grades, must be specified with the grain parallel to the back of the book: that is, up and down the page. This prevents wrinkling in the gutter and at the edges of the book and permits it to open more easily.

When books are to be printed by offset lithography, especially in more than one color, the grain direction must be selected relative to the printing process first, rather than in relation to the book itself. In sheet-fed offset, the grain should be the long way of the sheet. In web-offset the grain will of course be determined by layout, roll width, and press cutoff.

Any questions about grain or other properties of paper should be resolved by consultation with paper supplier, printer or lithographer, and binder.

Bulk. Paper bulk is determined by measuring the number of leaves that measure an inch under standard pressure, and is generally expressed in pages per inch. It is related to *caliper,* which is the thickness in thousandths of an inch (points) of a single sheet. Bulk is important to know so as to visualize the finished book, as well as to design bindings, order dies, and so forth. Bulk varies with weight and finish and is related to surface smoothness. In book publishing usage, bulks are listed by each paper supplier for the grades he produces, and paper is generally ordered to bulk — recognizing, of course, that finish and/or surface impose limitations.

Permanence. The lasting qualities of paper are of concern to publishers, printers, and readers, and much has been done to improve permanence since the disastrous years of the late nineteenth and early twentieth centuries. Pulps and fillers, as well as sizing materials, have been improved, and papers have been made to withstand the test of time in both strength and color. Most researchers feel that an essential property of long-lived paper is that it be neutral or even slightly alkaline, but not acid. Changes in the papermaking process are making such papers available.

Computation of paper needed. When paper is ordered for a book, the size usually chosen is that which will print 32 or 64 pages on one side of a

STANDARD SIZES OF BOOK PAPERS, INCHES	PAGES TO A SHEET	SIZE OF UNTRIMMED LEAF, INCHES
25 × 38	32	6¼ x 9½
	64	4¾ × 6¼
28 × 42	32	7 × 10½
	64	5¼ × 7
38 × 44	32	7 × 11
	64	5½ × 7
30½ × 41	32	7⅝ × 10¼
	64	5⅛ × 7⅝
33 × 44	64	5½ × 8¼
35 × 45	64	5⅝ × 8¾
36 × 48	64	6 × 9
38 × 50	64	6¼ × 9½
41 × 61	64	7⅝ × 10¼
	128	5⅛ × 7⅝
42 × 56	64	7 × 10½
	128	5⅜ × 7⅞
43 × 63	64	7⅞ × 10¾
	128	5⅜ × 4⅞
44 × 64	64	8 × 11
	128	5½ × 8
46 × 69	128	5¾ × 8⅝

sheet (see table, p. 528). A sheet of this sort printed on both sides will contain 64 or 128 pages, which will be cut and folded in units of 16 or 32 pages, called signatures. Suppose a book is to be printed containing 384 pages, on a sheet that will print 32 pages on each side, a total of 64. Dividing 384 by 64 gives us 6, which is the number of 64-page *sheets* needed for one copy of the book. Hence for 5000 copies of the book, there would be needed 6 × 5000 sheets, i.e., 30,000 sheets or 60 reams. It is always necessary to add a percentage for waste and spoilage, usually 5 per cent for 5000 copies. So 63 reams should be ordered for this book.

If double size paper is to be used, printing 64 pages on each side (a total of 128 pages on a sheet), half this quantity, or 31½ reams should be ordered.

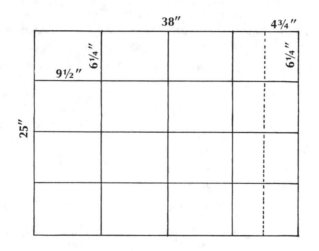

The untrimmed size of 32-page signatures can be seen at a glance, because the page dimensions are always one-fourth of the sheet dimensions. If 64 pages are printed on the same size sheet (32 pages on each side), the page length is then one-fourth the width of the sheet and the page width is one-eighth of the sheet length. When using a 64-page sheet, the pages must be so arranged in the form that when they are folded and cut they will come out in correct order. Arrangement of the pages in this manner is called *imposition*. Imposition applies equally to other sizes of sheet, as 48, 32, and 16.

In laying out a book, it is desirable to make it come out in even signatures—multiples of 32 pages—if possible. This can usually be accomplished by adjusting the front or back matter or both, by adding a flyleaf, or by leaving a few blank pages at the front or back of the book. If a number of pages are left over, they can be *canceled*, that is, cut off the sheet, but this is more costly, because of the time and labor involved.

BOOK SIZES

The common book-trade designations of sizes—quarto, octavo, twelvemo, and so on—were originally based on a sheet of paper measuring 19 × 24 inches. This sheet, folded once, formed a folio of two leaves (four pages), 12 × 19 inches in size, untrimmed. Folded twice, it formed a quarto of four leaves (eight pages), 9½ × 12 inches. Folded three times, it gave an octavo, eight leaves (16 pages), 6 × 9½ inches. Twelve pages printed on a sheet this size would be 4¾ × 8 inches. As paper-making developed, many different sizes of paper were manufactured and were designated by name rather than by size in inches. These names are in common use in England, but American paper manufacturers prefer to designate their book papers by size in inches, and most publishers use the size in inches of the trimmed page in describing their books.

If only approximate dimensions of book sizes are needed, the following designations of sizes may be of interest.

NAME	APPROXIMATE SIZE OF PAGE, INCHES	LIMIT OF OUTSIDE HEIGHT[5] CENTIMETERS	INCHES
Folio	16 × 25	30+	11.58+
Quarto	9½ × 12	30	11.58
Octavo	6 × 9	25	9.65
Crown Octavo	5⅝ × 8	25	9.65
Medium Octavo	6⅛ × 9¼	25	9.65
Royal Octavo	6½ × 10	25	9.65
Super Octavo	7 × 11	25	9.65
Imperial Octavo	8¼ × 11½	25	9.65
12mo	5 × 7⅜	20	7.72
16mo	4 × 6¾	17.5	6.75
18mo	4 × 6½	15	5.79
32mo	3½ × 5½	12.5	4.826
48mo	2½ × 4	10	3.86
64mo	2 × 3	7.5	2.89

BINDING

The binding of a book serves a number of functions: holding together and protecting the pages of the book, identifying and characterizing the book, and attracting and interesting the potential reader. When a dust jacket (a printed paper wrapper placed over the cover of the book) is also used, it generally assumes the reader-attraction function. The variety of bindings available for books fall into three main categories: case-bound, paper-bound, and mechanical.

[5]Adapted from the American Library Association listing, which gives the range of outside heights, in centimeters, of books to which the designations of book sizes may be applied. All books over 30 cm (11.58 in.) may be designated *folios*.

Case-bound books. Used for hardcover books, case binding involves a cover constructed of boards, generally cardboard, covered with cloth, plastic, paper, or a combination of these. Case-bound books often have paper jackets and sometimes a cardboard slipcase (an open-ended box into which a book, or set of books, fits so that only the spine of the book is visible). Occasionally glassine or clear plastic jackets will be used for protection on a case-bound book. Hardcover books vary in durability and cost of production according to the materials and production methods used. There are various grades of board and of cloth, plastic, and paper available. Cloth is generally more durable and more expensive than paper, but newly developed paper production methods now allow some high-grade paper to match cloth in performance. Plastic provides good protection and durability, but because of its qualities, it can present problems during various production processes. The cover design can be printed on the material before the cover is made up, or put on after the cover is made, either with silk screen or by hot or cold stamping.

Paper-bound books. Used for paperback or softcover books, paper binding consists of a thick grade of paper (resembling a light, flexible cardboard) that is usually coated on the side that receives the cover printing. Paper-bound books are seldom jacketed, but occasionally they are slipcased, usually in a set. Paper binding is less durable and less expensive than case binding, but new developments for both kinds of binding are narrowing the gap between the two. The several grades of paper covers available provide varying levels of protection and durability. Paper covers are printed with the cover design before being attached to the book.

Mechanical-bound books. Mechanical binding involves a mechanical device, such as a spiral, loose-leaf ring, post, or comb (a plastic device with a solid back and wide "teeth" that are bent to form the binding rings), that is used to join the covers and the pages of the book together. Books bound this way are rarely jacketed or slipcased. Mechanical binding is generally more expensive than case binding and is used only for books that require the special qualities of this kind of binding, which enables a book to lie flat, fold back flat, or stand as an easel, and often allows pages to be removed or rearranged. Use of this type of binding is limited usually to books no more than 1 inch thick. The cover materials used with the mechanical binding devices need only to resist soiling and abrasion, and can be like those used for either case binding or paper binding.

Binding processes. The binding process for all three types of binding begins with the folding of the printed signatures (see p. 529). At this point the endpapers, if any, and other extra pages are added, and for paperback and case-bound books, the signatures are then joined together by one of several processes. Signatures can be sewn together either by stitching through the gutter of each signature and joining the threads at the back (*Smyth sewn*) or by sewing signatures together through the side (*side sewn*). They are then *smashed* (compressed) and flexible glue is added to reinforce the back. When there is only one signature, it can be stapled or sewn through the gutter (*saddle-stitched*) or through the side. For *perfect binding,* the folds at the back are trimmed off, adhesive is applied to hold the pages together, and a cloth lining is glued to the back. For case-bound books, after any of these processes, the pages are then trimmed off and the back of the book is shaped and reinforced to take the binding.

While the signatures are being prepared, the cover is being made and printed in a separate process. In case-bound books, covers are then pasted to the endpapers and the reinforcing, or else to the first and last pages for a self-lining book. For paperbacks, most of which are perfect-bound, glue is applied to the back of the book, the paper cover is placed on the book, then the cover and pages are trimmed together. For certain mass-market paperbacks, the printing and binding are done in one process by special equipment. For mechanical binding, the pages of the folded signatures are trimmed, and holes are cut through the cover and the pages to receive the mechanical binding device.

BIBLIOGRAPHY

Dictionaries:

ALLEN, EDWARD MONINGTON. *Harper's Dictionary of the Graphic Arts.* New York: Harper & Row, 1963. *Compact glossary of technical and graphic arts terms.*
STEVENSON, GEORGE A. *Graphic Arts Encyclopedia.* New York: McGraw-Hill, 1968. *Illustrated dictionary of graphic arts and printing terms, covering recent technological advances as well as the basics.*

General:

BURT, SIR CYRIL. *A Psychological Study of Typography.* London: Cambridge University Press, 1959. *Report on elements of typography as they affect child and adult readers.*
JENNETT, SEAN. *The Making of Books.* New York: Frederick A. Praeger, 1967. *Well-illustrated survey of all phases of book production in the publishing house and of book manufacture, printing, and binding.*
TINKER, MILES A. *Legibility of Print.* Ames, Iowa: Iowa State University Press, 1963. *Report of psychological studies on effects of typography on speed and ease of reading; annotated bibliography.*

Graphic Arts:

ARNOLD, EDMUND C. *Ink on Paper: A Handbook of the Graphic Arts.* New York: Harper & Row, 1963. *History of graphic arts and survey of materials, mechanics, and techniques of printing; glossary of terms.*
The Penrose Annual: Graphic Arts International. Visual Communications Books. New York: Hastings House. *Well-illustrated annual publication, includes "The Penrose Survey" of recent printing and graphics developments and happenings as well as individual articles on technical and historical subjects.*

STONE, BERNARD and ARTHUR ECKSTEIN. *Preparing Art for Printing.* New York: Reinhold, 1956. *Well-illustrated handbook of tools, materials, and techniques needed in preparing artwork for production; short glossary.*

Typographic Design:

DAIR, CARL. *Design with Type.* 2nd ed. Toronto: University of Toronto Press, 1967. *Illustrated study of design possibilities of modern typography, including new trends and techniques; some type specimens.*

TSCHICHOLD, JAN. *Asymmetric Typography.* Translated by Ruari McLean. New York: Reinhold, 1967. *Translation of Tschichold's 1935 "Typographische Gestaltung," which laid the basis for modern typography in its emphasis on the integration of art and typography and on the use of asymmetric design.*

ZAPF, HERMANN. *About Alphabets: Some Marginal Notes on Type Design.* Rev. ed. Cambridge, Mass.: M.I.T. Press, 1970. *Statement of author's design principles, with examples of his typeface designs.*

Type and Typography:

BIEGELEISEN, J. I. *Art Directors' Work Book of Type Faces.* New York: Arco, 1963. *Type specimens, alphabetized by typeface name, with essays; also covers basics of typographical measurements and composition systems. Especially helpful for students.*

BIGGS, JOHN R. *An Approach to Type.* Rev. ed. New York: Pitman, 1962. *Surveys classification of typefaces, developments of letter designs, influence of paper on typography; for beginners and professionals.*

————. *Basic Typography.* New York: Watson-Guptill, 1969. *Study of basic design and production; good sections on color, layout, and paper.*

JASPERT, W. PINCUS and others. *The Encyclopedia of Type Faces.* 4th ed. New York: Barnes and Noble, 1971. *Useful collection of about 2000 typeface specimens, both display and text type; history and characteristics of each discussed.*

Phototypography. Prepared by American Association of Advertising Agencies in cooperation with Advertising Typographers Association of America. Report No. 1, March 1972.

ROSEN, BEN. *Type and Typography: The Designer's Type Book.* Rev. ed. New York: Reinhold, 1967. *Comprehensive type specimen book with full alphabets for 8 point to 72 point; arranged by type families, accompanied by essays.*

TSCHICHOLD, JAN. *Treasury of Alphabets and Lettering.* New York: Reinhold, 1966. *Collection of aesthetically pleasing types and letters, with notes; author's introduction analyzes letter forms and use of typographical elements.*

UPDIKE, DANIEL B. *Printing Types: Their History, Forms and Use: A Study in Survivals.* 3rd ed. 2 vols. Cambridge, Mass.: Belknap Press, 1962. *The definitive historical study of the development of typography; well-illustrated.*

ZAPF, HERMANN. *Manuale Typographicum: 100 Typographical Arrangements with Considerations about Types, Typography, and the Art of Printing.* New York: Museum Books, 1968. *Printed in eighteen languages; monumental study of complexities of type and typography.*

Composition and Printing:

KARCH, R. RANDOLPH and EDWARD J. BUBER. *Graphic Arts Procedures: The Offset Process.* Chicago: American Technical Society, 1967. *Introductory book, covering basic materials and techniques and recent technological advances; technical illustrations.*

LAWSON, L. E. *Offset Lithography*. Facts of Print Series. London: Vista Books, 1963. *Technical presentation of operating procedures and modern techniques.*

LEWIS, JOHN. *Anatomy of Printing: The Influence of Art and History on its Design.* New York: Watson-Guptill, 1970. *Historical survey of printing—pre-Gutenberg through computerized typesetting; presents design possibilities and aesthetics of new technology.*

SAYRE, I. H. *Photography and Platemaking for Photo-Lithography.* Chicago: Lithographic Textbook Publishing Co., 1969. *Technical survey of lithographic chemistry, layout, platemaking, photocomposition, and photographic techniques in printing.*

STEINBERG, S. H. *Five Hundred Years of Printing.* 2nd ed. rev. Baltimore: Penguin Books, 1966. *Historical study of the interaction between printing and culture.*

STRAUSS, VICTOR. *The Printing Industry.* New York: Printing Industries of America, 1967. *Comprehensive introduction to technology, techniques, and materials of book manufacturing.*

GLOSSARY OF PRINTING AND ALLIED TERMS

AA's Author's alterations, changes from copy made by the author after the manuscript is set in type. As printers charge for these corrections, a percentage (usually to 15%) of the original cost of composition is allowed the author for corrections; corrections in excess of this amount are charged against the author's royalties.

accents In the composing room *accent* means the mark indicating stress: heavy accent ('); light accent ('). In most fonts of type they differ from primes. The marks used over letters to indicate the quality of sound are named acute (é), grave (è), circumflex (ê), tilde (ñ), cedilla (ç). These are usually cast on the letters, but are sometimes separate pieces, called *piece accents*.

agate An old name for a size of type slightly smaller than 5½-point measuring fourteen lines to the inch; for years the standard for measuring advertisements. Called *ruby* in England.

airbrush A small, pencil-shaped gun, activated by compressed air, used to spray pigment on artwork to adjust or secure tonal effects.

alphabet length The number of points occupied by the 26 letters of the alphabet in lowercase. The relative condensation or expansion of typefaces is determined by comparison of alphabet lengths.

appendix Part of the end matter following the text; it may comprise notes, practice material, a glossary, tables, or other supplementary information regarding the text.

artwork Any material, such as drawings, paintings, photographs, and so on, prepared as illustrations for printed matter.

ascender The part of a lowercase letter that extends above the body of the letter.

aspects Signs used to denote certain positions of planets, such as ☌, ☍, □.

author's alterations *See* AA's.

author's proof Master proof, the galleys on which all changes and corrections must be marked before the proof is returned to the compositor for correction.

backbone, backstrip *See* spine.

back matter, end matter Material following the text, such as an index, appendix, glossary, and so on.

base line In type, the bottom alignment of capital letters.

battered type Marred or flattened type that prints imperfectly.

bed The surface of a press, against which type or plates are clamped and over which ink rollers and an impression cylinder pass in letterpress printing.

benday   A piece of film with a screen pattern that creates an overall pattern of halftone dots or lines. Benday screens may be used with either line cuts or halftones to change the value of tone or to alter the image of a line by reinforcing or lightening it.

bleed   Illustrations that extend to the edges of the page and are partially trimmed off when the page is cut for binding are said to *bleed*.

blind-blocking, blind-stamping, blind-tooling   Impressing a design on a book cover by hot tools only, using no ink or gold leaf.

blind folio   A page number that is counted during makeup but not printed.

block quotations   Excerpts in reduced type. Also called *extract*.

blueprints (blues)   Photographic prints made from negatives (or positives) of text or artwork for offset reproduction. In this sense, they are the last proofs before the book — or other material — goes to press.

boards   Stiff cardboard used for the sides of books, which may be covered with paper, cloth, leather, or other material.

boldface type (boldface)   Type that is heavier and thicker than the rest of the text type with which it appears.

book number   Standard book number. *See* SBN.

break up   (1) To separate solid matter into shorter paragraphs. (2) To dispose of pages of type that are of no further use, removing leads and foundry type, dumping machine metal to be remelted.

broadside   (1) A large sheet of paper printed on one side only, like a poster, or printed on both sides and folded in a special way. (2) A broadside table or illustration is one printed with the top at the left side of the page, requiring a quarter turn of the book to the right to be in position for reading.

brown, vandyke   A type of photographic print on special paper, used as a proof for halftones or line cuts.

bulk   (1) The thickness of a book, not including the cover. (2) The thickness of paper as measured by number of sheets per inch.

butted   Two short slugs joined to make a longer slug, with no space left between them, are said to be *butted*.

©   A copyright notice used for books, maps, works of art, models or designs for works of art, reproductions of a work of art, drawings or plastic works of a scientific or technical character, photographs, prints, and pictorial illustrations. It must be accompanied by the name, initials, monogram, mark, or symbol of the copyright owner.

calender   To make paper smooth and glossy by passing it through rollers.

camera copy   The final proofs from the compositor, often in the form of *reproduction proofs* (*repros*). From them negatives are made by the printer for making the plates used in offset printing.

cancel   (1) (*v.*) To cut out blank pages from a signature at the end of a book; also, to cut out printed pages. (2) (*n.*) A leaf reprinted to correct an error in a printed book and inserted by the binder in place of the incorrect page or pages.

caps   Capital letters.

caps and small caps, cap-and-small, c & sc   Two sizes of capital letters in one style of type. The small caps are the height of lowercase letters.

caption   The heading or title of a chapter, article, table, illustration, and so on, as distinct from a *legend,* which is explanatory or descriptive material accompanying an illustration, chart, table, and so on. The terms are often used interchangeably.

carding   Increasing vertical spaces between lines by less than a point (done to lengthen a type page). It is accomplished by inserting strips of cardboard between the slugs. In photocomposition, advancing the film to increase the space between lines and lengthen the type page.

caret   The proofreading symbol ∧, used to indicate that an insertion is to be made in a line.

case   The cover of a hardbound book. *See also* type cases.

cast off (1) To estimate the number of pages a given manuscript will make. (2) To estimate the number of pages of text from a set of galleys.

cast proof *See* foundry proof.

cast up To measure the amount of type set, in order to find the cost of composition.

cathode ray tube "(CRT) Electronic tube used by highspeed photocomposition machines to transmit the letter image onto film, photopaper, microfilm or an offset plate. The image is actually comprised of a series of dots or line strokes, which vary in degree of resolution, depending on the system itself and on the operation speed." (*Phototypography*)

center point, centered dot A period cast higher than the base line of the face. Used to show syllabication, to indicate multiplication, to separate words set in roman caps composed in the classic style of tablet inscriptions. Sometimes called a *space dot.*

character Numeral, letter of the alphabet, punctuation mark, or other symbol in a font of type.

character count A count of each letter, numeral, punctuation mark, space, and so on, in a piece of copy.

characters per pica The number of characters of a particular font that will fit into a 1-pica space.

chase A rectangular iron or steel frame in which pages of type, slugs, or plates are locked up to be printed or cast.

coded tape A tape produced by a keyboarding unit and bearing a code that activates typesetting equipment.

cold type composition "In a wider but erroneous sense, copy composed by means other than the use of metal type; more specifically, composition on typewriter-like direct-impression, impact or strike-on devices, e.g., Varityper, IBM Selectric Composer, etc. More properly called 'strike-on composition,' this term *should not* be used as a synonym for phototypography, photocomposition, or photo-typesetting." (*Phototypography*)

collate To arrange the folded sheets or signatures of a book or other printed matter in correct order.

collating marks A short rule (or letter, figure, or shortened book title) inserted in the form in position to print midway between the first and last pages of a signature. The mark on the first signature is placed near the top, that on the second somewhat lower, the third lower still, and so on. When the signatures are assembled, the marks show in a regular series, which would be broken if a signature were missing or duplicated.

colophon (1) A trade emblem used by a publisher on title pages of books, on his stationery, and stamped on book covers. (2) A statement placed at the end of a book, giving information about the book's production.

color separation The separation of the colors of a full-color transparency or print by photographing it with separate color filters.

comb binding A type of mechanical binding in which a plastic device forms a solid back, and from this back extend wide "teeth" that are bent back to form the binding rings.

compose To arrange letters, in the medium of type or film, for printing. A *compositor* is one who arranges the letters, manually, mechanically, or by computer, and *composition* denotes the act of arranging the letters or the product of this act; a *composing room* is the room in which composition is done; and a *composing stick* is a small metal frame in which the compositor can hand-set metal type.

computer composition "Unjustified or 'idiot' tape is produced on a keyboard and subsequently run through a computer which makes line-end, hyphenation and other typographic decisions on the basis of a program. The computer-produced second tape is then utilized as input for phototypesetting (or metal typesetting) equipment. Computer tapes can also be produced on the basis of punched cards." (*Phototypography*) (*See* idiot tape.)

condensed type A typeface that is narrow or slender in proportion to its height.

copy Material (manuscript, artwork, and so on) given to the compositor or printer to be set in type or prepared for printing.

copy-editing Correcting and preparing material for typesetting.

copyfitting Determining the amount of space required to set a given amount of copy in a specified type size and face, and a given measure. Also, the adjustments involved in making the copy fit the space in which it is to be printed.

copyright A protection granted by law to certain kinds of property, including literary property. The copyright assigns to the holder the exclusive right to make or have made copies of his work. *See* ©.

crop To eliminate portions of a photograph or illustration to make it a desired size or to eliminate unwanted details. Also, to add something to a photograph that was not in the original negative, reshoot it, and get a different picture.

cross-reference A reference made from one part of a book, article, and so on, to another part or page containing related information; also, a reference from one entry, as in an index, to another for additional information.

CRT *See* cathode ray tube.

cursive Running, flowing; applied to certain faces of type similar to italic but more decorative.

cut The printers' term for a halftone engraving or a zinc etching. Commonly used to mean any printed illustration.

cylinder press *See* printing press.

dead matter (1) Type, already set, no longer required for printing. (2) Material, as page proofs, blues, and so on, returned by the printer after a book has been printed.

deckle edge The untrimmed, rough edge of a sheet of paper, formed where the liquid pulp flowed under the frame (the deckle). The rough edges are often left untrimmed on handmade paper and can be artificially produced on machine-made paper.

delete (1) To mark for elimination a letter, word, or portion of a manuscript. (2) To instruct the compositor to take out type.

descender The part of a lowercase letter that extends below the base line of the letter.

dies Brass, copper, magnesium, or other alloy stamps used for impressing letters or designs on covers of books. Also called *brasses, binder's dies*.

digraph A combination of two letters to express a sound. *See* dipthongs.

diphthongs Vowel digraphs, as ae, oe. The word *diphthong* is often used to designate ae and oe set as ligatures: æ, œ.

display type Type, usually 18-point or larger and often in boldface and of a distinctive design, used for headings, in advertisements, or in text of a smaller size, to attract attention.

distribute To return types to their proper location in the cases after they have been used, or to melt down used type for re-use.

double, doublet Incorrect repetition of a word or words in the copy or the proof.

double spread Two facing pages.

drop folio A page number (folio) placed at the bottom of the page.

dummy (1) Large sheets or boards on which the actual trim size of the book has been drawn in blue. All the elements that will appear on each individual page of the finished book are indicated on the dummy pages—one set of galleys is cut up and pasted to the dummy pages, and spaces are left for illustrations, tables, captions, and so on. (2) A set of blank pages, of the size and number of the proposed book, bound to show its physical size.

duotone A process for producing two-color illustrations, in which two halftones of the same subject, one in black and one in another color or in another tone of black, are printed in register.

edge color or **stain**   Colored stain or gold leaf that is added to the edges of a book's pages.

edition   All the copies of a book, magazine, and so on, printed from the same type or plates (original or photographic copies). An *edition* refers to the format, as paperbound, clothbound, or to the text, as revised, expanded, and so on. It differs from *impression,* which is the reprinting of a work without change or with only slight changes. If extensive changes have been made in the text, the printing is a new edition.

electrostatic printing (xerography)   A dry photocopying or photographing process in which paper is given a positive charge. Where the printed image, projected through a lens, strikes the paper, the positive charge remains, the rest of the paper becoming negatively charged. Ink adheres only to the positively charged printing area.

electrotype, electro   A metal replica of a page of type or engravings. An impression is taken in a thin layer of softened wax, lead, or plastic. By an electrolytic process the impression so obtained is cast to form a copper, nickel, or steel shell, which is the electrotype. The shell is reinforced with a lead backing. Electros are used today only where high quality is necessary.

ellipsis, points of   Three points (periods or dots) used to indicate an omission.

em   In printing, a unit of measurement equal to the space occupied by letter *m* in the given font.

emage   The area of text measured in terms of ems of its type size.

embossing   Printing or stamping in relief to produce a raised image.

en   In printing, a unit of measurement equal to half of an em.

end matter   *See* back matter.

end paper   A folded sheet of paper different from the text paper, one-half of which is pasted to the inside of the front cover of a book; one-half of a similar sheet is pasted to the inside of the back cover.

engraver's proof   Proof of a line-cut or halftone engraving made by the photoengraver and sent to the editor for approval.

estimate   A calculation of the length of copy and the amount of space it will fill, accomplished by a word or character count, in order to determine the cost of the work.

even pages   The left-hand pages of a book, numbered 2, 4, 6, and so on. Often called *verso pages.*

expanded type, extended type   Type that is wider than usual in comparison with its height.

extract   Quoted material typographically set off from the rest of the text in a smaller type size, or shorter measure, or both. Also called *block quotations.*

face   A style of type; also, the printing surface of a type, which stands in relief on the upper end of the type body.

figures   (1) The Arabic numerals 0, 1, 2, . . . Type foundries supply scored figures ($\overline{1}, \overline{2}, \ldots$), pointed figures ($\dot{1}, \dot{2}, \ldots$), and canceled or scratched figures ($\not{1}, \not{2}, \ldots$). (2) Illustrations printed with the text.

flag   *See* masthead.

flat   A sheet on which negatives, or positives, are assembled in correct order, for stripping in, and from which an offset plate is made.

fluent   The sign of integration ($\int$).

flush   With no indention.

flyleaf   A blank leaf, or page, at the front or back of a book, not to be confused with endpapers.

folded and gathered   Sheets folded into signatures and collected (gathered) into the correct order for binding are said to be *folded and gathered.*

folio   A page number.

foldout (folding plate) An oversize leaf, often an illustration, that must be folded to fit within the trim size of a book or magazine.

font A complete assortment of type of one size and face, including caps, small caps, lowercase, punctuation marks, accents, and commonly used symbols.

footline The bottom line of a page. Only a drop folio or a signature may properly be set in this line.

form All the pages of a book either in repros or negatives, that are to be printed on one side of a sheet at one time, arranged in the correct order; also, type matter, engravings, or electrotypes locked up in a chase for casting or printing.

format Anything relating to the outward appearance, as the typeface, arrangement, makeup and binding, of a book or other printed piece.

foul galley or **proof** A galley or proof after the corrections indicated on it have been made and new proofs pulled.

foundry proof Proof taken of a form in chase immediately before casting.

four-color press A press that can print four colors simultaneously, on one pass of the paper, producing a vast range of colors.

front matter Printed matter preceding the text, such as the half-title, title page, copyright page, dedication, table of contents, preface, and so on.

full measure Extending across the entire width of the type page, without indention.

furniture Wooden or metal material, usually in multiples of 12 points, used for spacing in hot metal systems.

galley A long and narrow metal pan, about half an inch deep and open at one end, in which lines of type are placed when they are first set, whether by hand or by machine. The first proof of composed type is, hence, a *galley* proof.

grain The direction in which the paper fibers lie.

Greek ratio The law of proportion that "a line or measure is pleasingly divided when one part is more than a half and less than two-thirds the length of the other."

guidelines Lines drawn or printed in order to orient placement on dummies, layouts, and so on.

gutter The inner, or binding, margins of facing pages.

hair space Very narrow metal space used to separate words, for justifying lines, and for letterspacing.

half-title, bastard title The title of the book, standing alone on the page. It immediately precedes the text or the front matter, and may be in both places.

halftone A process in which a black-and-white photograph is rephotographed through a screen so that the gradations of light and dark in the original photograph are reproduced as a series of tiny dots that print as a continuous tone.

hanging indention Type set with the first line of the paragraph flush left, and the lines following it indented.

hanging numerals *See* Old Style numerals.

hard copy "A typewritten impression on ordinary paper produced by most keyboards for phototypesetting simultaneously with the paper or magnetic tape used to operate the photounit. This hard copy can help an operator spot an error before he finishes the line and it also is a convenience in marking instructions for the operator of the photounit." (*Phototypography*)

hardware "A term in computerized composition referring to the equipment as opposed to the software, or procedures and programming." (*Phototypography*)

head, heading Word or words identifying specific divisions, paragraphs, and so on, within the text, and differentiated in some manner from the straight text.

headband In binding, a decorative band used at the head and foot of the spine of a book.

head margin The blank area from the top edge of the page to the topmost printed element of the type page.

highlight (dropout) halftone A halftone in which the screen is dropped out to create areas of pure white.

idiot tape "Unhyphenated, unjustified tape, either paper or magnetic, intended to operate a phototypesetting machine (or metal typesetting machine) via a computer." (*Phototypography*)

imposition The arrangement of pages of type and/or illustration in a form so that the pages will be in proper order when the sheet is folded after printing.

imprint The name and address of a publisher, printed on the title page of a book, sometimes in conjunction with the colophon and the date of the printing.

indention The amount by which a line of type is less than full measure. Paragraph indention is space left blank at the left in the first line of a paragraph.

inferior letters or **numerals** Letters or numerals smaller than the body type, so cast on the type body that they print below the alignment of normal letters. Often called *subscripts*.

insert Illustrations, maps, or other material printed on different quality paper from the text and inserted in proper position before binding.

inset (1) A sheet or folded section of printed pages set within another in binding. (2) A small illustration, as a map, set in the text.

inside margin The blank area between the text and the binding edge of a page.

italic A sloping, slanted variation of a typeface: *italic type*.

joint The hinge where the sides of a casebound book are attached to the back.

justification The adjustment of the spacing within lines of type to fit the lines to a specific measure on the type page.

keep standing In hot metal systems, to hold composed type after printing or casting for a possible further printing.

keyboard An input mechanism that creates a type character on paper or magnetic tape, as in Monotype and phototypesetting, or releases a type character (mat) from the magazine.

keying Coding a manuscript heading or subheading with a symbol (letter or number) in order to identify it for the compositor.

kill (1) To delete copy or printed matter. (2) To indicate that composed type is to be melted down.

layout (1) The order of presentation of the various parts of a book. (2) The design of the book, including such elements as trim size, size of the type page, typeface, arrangement of textual matter, and so on.

leaders Periods or dashes used in tables of contents, programs, tables, and so on, to lead the eye across what would otherwise be open space.

leading (Pronounced *led'ing*.) (1) In metal typesetting, the space between the lines of type. (2) "In phototypesetting, better called line spacing and directly related to the film advance." (*Phototypography*)

leads Strips of metal 1, 2, or 3 points thick used to separate lines of type.

legend (1) The explanatory or descriptive material accompanying an illustration, chart, table, and so on, sometimes used interchangeably with the term *caption*, the heading or title of a chapter, article, table, and so on. (2) A key to the symbols on a chart or map.

letterpress A type of printing in which the surface of metal type set in forms is inked and the ink transferred to paper by pressure.

letterspacing Inserting thin spaces between the letters of a word to expand it.

Library of Congress card number (LC number) A number assigned to a book by the Library of Congress prior to publication. It is usually printed on the copyright page.

line copy Copy that can be reproduced without using a screen, as a pen-and-ink drawing.

line cut A photoengraving in which there are no gradations of tone. The printed effect is produced by lines, rather than dots as in halftones.

lining numerals Numerals of equal height, which, unlike Old Style numerals, do not descend below the line (as 3, 5) or ascend above the line (as 6, 8). Also called *lining figures*.

Linotype A typesetting machine that sets solid lines or slugs of metal type. Corrections are made by resetting and recasting the entire line in which the error occurs.

literals Alphabetic characters. In proofreading, to "read for literals" is to read for wrong fonts, defective letters, transpositions, spelling, and the like.

lock up In letterpress, to fasten pages of type in a chase so that they can be printed from or plated.

logotype (logo) (1) Two or more letters cast on a single body. (2) The product or company name set in a distinctive design and used as a trademark.

loose blues Blueprints of separate elements of the text, such as illustrations, before makeup.

lowercase Letters not capitalized.

Ludlow Typograph A line-casting machine, used particularly for advertising display lines, large-type headlines, and similar matter. Brass matrices are set by hand in a special stick. The assembled mats are then placed over a slot in a steel table and molten metal forced up from beneath, to form a solid slug, or line.

magazine On a Linotype machine the case in which the brass matrices (mats) of letters are stored.

majuscule A capital letter.

makeready In letterpress, the process of preparing type or plates for printing by making adjustments so that the printing surface receives and transfers ink evenly.

makeup The arrangement of composed type into pages, inserting folios, running heads, cuts, and so on.

margins The blank areas that border the printed type page.

master proof *See* author's proof.

masthead A statement of name, terms, ownership, platform, policies, and the like. In newspapers it is usually at the head of the editorial page. In magazines it is usually at the foot of the editorial page or table of contents. Sometimes called the *flag*.

matrix (pl. matrices) (1) In hot metal composition, the mold from which the face of a type is cast. Often called a *mat*. (2) In phototypography, type characters on film.

matte prints Photoprints with a dull, nonglossy finish.

measure The width of a full line of type on a type page.

mechanical A carefully prepared layout that shows the exact placement of each element on the sheet. It is used as camera copy in platemaking.

mechanical binding A mechanical device, usually plastic or metal, such as a comb or ring, that holds pages together.

minuscule A lowercase letter.

mold A negative (female) impression of type into which material is poured to produce a positive (male) duplicate of the type.

Monotype A typesetting machine. The material first is keyboarded and a paper tape is produced. This tape is run through a casting machine that casts each character as a separate piece of type. Corrections are made by hand.

montage A photograph in which several pieces of copy have been placed together to form a single unit.

mortising Cutting a space in a plate or film for the insertion of other type or film.

negative A photographic image in which the original image is reversed, the black areas white and the white areas black.

newsprint Inexpensive paper used mainly for newspapers.

nut In printing, another name for an en quad.

odd pages The right-hand, or *recto,* pages of a book: 1, 3, 5, and so on.

offcut (1) A portion of the printed sheet cut off and folded separately. (2) A part cut off a sheet of paper to scale it to press size.

off its feet Type is *off its feet* when it does not stand square upon its base.

offprint Separate reproduction of an article or book section that was originally part of a larger publication. Also called *reprint.*

offset (1) Surface printing that uses a photomechanical process to transfer the image to a plate that, when mounted on a cylinder, transfers the ink to the paper. (2) Transfer of ink which is not yet dry to the next sheet laid over it.

Old Style numerals (hanging numerals) A style of type in which certain numerals ascend (as 6, 8) or descend (as 3, 4, 5, 7, 9) from the x-height of the other characters of the face. Also called *Old Style* or *hanging figures.*

opaque In photoengraving and offset lithography, to paint out the areas on the negative that are not required on the plate.

optical center About one-eighth above the actual center of a page. A line that is to appear to be in the center of a page should be in this position.

paperback A book with a flexible paper cover, sold at a lower price than that of a comparable hardbound book. Paperbacks are divided into two types, quality or trade paperbacks, sold through book stores, and mass paperbacks, that sell through newstands, drugstores, and so on.

part title The number or title of a division of a book, more important than a chapter title, and usually printed alone on a separate page preceding the division to which it refers.

pasteup Preparation of type and illustrations in page layouts for photomechanical reproduction.

PE Printer's error, an abbreviation used in correcting and marking proofs; the charge for correcting pe's is borne by the compositor, not the author or the publisher.

peculiars Infrequently used characters of a font of type.

penalty copy Copy that is difficult to compose because it is faint, foreign, heavily corrected, and so on, and for which the typesetter charges an additional percentage above the regular typesetting rate.

perfect binding or **perfect-bound** Binding by which the pages of a book are held together with adhesive along the back edge after the backs of the signatures have been cut off and the edges roughened.

perfector press A press that prints on both sides of a sheet of paper in one pass through the press.

photocomposition *See* phototypesetting.

photoengraving A metal relief plate prepared by etching a photographically produced image on the metal with acid.

photomechanical plate A plate made as a result of photographing prepared material and used in mechanical printing processes.

photo-offset *See* offset.

phototext "Text matter composed or set onto film or paper by means of keyboard-operated or computer-tape-driven phototypesetting machines." (*Phototypography*)

phototypesetting "The composition of phototext and display letters onto film or paper for reproduction. Letters are projected from film negative grids, and are also stored in a binary form in computer core to be generated through a CRT system. *Photocomposition* is a synonym." (*Phototypography*)

phototypography "An encompassing term denoting the entire field of composing, makeup and processing phototypographically assembled letters (photodisplay and phototext, or type converted to film) for the production of image carriers by platemakers or printers." (*Phototypography*)

pied Type that has been mixed up or disarranged is said to be *pied.*

pica A printer's unit of measurement equal to 12 points, or about ⅙ of an inch, used especially in measuring the length of lines.

pigeonhole *See* river.

plant cost Factors in production that are not affected by the size of the edition, such as illustration cost, composition, plates.

plate (1) A sheet of metal, plastic, rubber, or other material converted, by one of various processes, into a printing surface and attached to the press for printing. (2) An illustration inserted into a book.

plate proof A proof of the finished plates, pulled on smooth stock so that imperfections in the type can be detected.

platen press *See* printing press.

point The basic unit of typographical measurement, equal to $1/72$ of an inch, or .01384 inch.

press run The number of copies that are to be printed. The run usually allows extra copies to be printed to allow for spoilage.

presswork The part of the printing process that involves the running of paper through the press, the actual printing of the work. The major steps in the printing of a book are composition, makeup, presswork, and binding.

printing press A machine in which the printing impression is transferred from inked plates or type onto the paper, either by direct impression or by offset. There are basically three kinds of presses—platen, flat-bed cylinder, and rotary. *Platen,* in which the paper is pressed by a flat surface onto a flat printing surface, and *flat-bed cylinder,* in which the paper is pressed by a cylinder onto a flat printing surface, are both used only in letterpress. Rotary presses, with various modifications, are used for letterpress, offset lithography, and gravure. In a *rotary press* the inked plates, attached to a cylinder, transfer the impression either directly onto the paper on a second cylinder or onto a rubber blanket cylinder that offsets the impression onto the paper on a third cylinder.

A *web perfecting press* prints both sides of a roll of paper, cuts the fold, and delivers in folded signatures. *See* perfector press, web-fed press.

A *multicolor press* prints several colors before the paper leaves the press. Also called a *chromatic press. See also* four-color press.

process printing A printing process in which a full color original is reproduced through the use of several (between two and four) halftone plates.

programming Providing an automatic machine, such as a computer, with a set of detailed instructions in order to produce the desired result when the machine is put into operation.

progressive proofs A series of proofs consisting of one plate in each of three or four colors (yellow, red, cyan, and sometimes black), and of combinations of two, three, and four colors. They are used to check color quality and as a printing guide.

proofreaders' marks Internationally known and understood symbols used (with some variations) to mark errors and changes on proofs.

proofs Printed examples from the compositor of the material he has set.

quad Metal blank spaces, less than type height, used to fill space in typesetting.

ragged right Text set with an unjustified right-hand margin.

ream 500 sheets of paper.

recto pages *See* odd pages.

reference mark A mark in the text used to refer the reader to a note pertaining to the material. A corresponding mark appears at the beginning of the notes. Reference marks may be asterisks, daggers, or more often, superior figures or letters.

register To print each impression in the correct position in relation to other impressions, as to superimpose color plates in process printing. If the impressions are not correctly aligned, they are said to be *out of register.*

reproduction proofs (repros) Final proofs intended for use as camera copy. *See* camera copy.

reverse Copy is said to be "reversed" when the colors are reversed, as when the white is printed as black, and the black as white.

river A streak of white space in printed matter caused by the spaces between words in several lines happening to fall one almost below another. Also called *river of white, pigeonhole, staircase.*

rotary press *See* printing press.

routing Cutting away unnecessary parts of a printing plate with an engraver's tool in order to prevent accidental printing.

rule A strip of brass or type metal, type high, used to print decorative or straight, thin or thick lines; in photocomposition, this is a decorative or straight line on the film.

runarounds Text set in shortened lines around an inserted illustration.

running head The author's name, book title, section, chapter, or other division title placed at the top of a page of text. It is separate from the text, and often combined with the folio. When it appears at the bottom of the page it is called the *running foot.*

runover, turnover (1) Lines other than the first line in a flush-and-hang paragraph. (2) The continuation of a heading onto a second line.

saddle-stitched Of a single signature, stitched through the back, the thread, silk, or wire showing on the back and in the middle fold.

sample pages Printed examples of selected pages made by the compositor according to the specifications of the production department and used to show the solutions to typographical problems, as tables, unusual symbols, equations.

SBN Standard Book Number, consisting of nine digits divided into three sections: the first part, the publisher prefix, identifies the publisher; the second, the title number, identifies the particular edition or title; and the third, one digit, is an arithmetic check against incorrect transcription of the entire number.

serif Short cross line placed at the end of the main strokes of a letter in certain typefaces.

set To compose or arrange type to be printed.

set solid Type set without leading additional to that provided by the shoulder of the type itself.

sheetwise A printing method in which the two sides of the sheet of paper are printed with different forms. Also known as *work-and-back.*

shelfback *See* spine.

shrink-wrap Cellophane put over a book, pad, and so on, in a heat process so that the edges are sealed; it is done to protect the book from warping, dust, damage, or other possible forms of injury.

side head A heading placed at the side of a page or column, set either as a separate line flush with the type page margin, or run in with the paragraph with which it belongs.

signature A sheet of paper folded so that, when cut, it will produce a certain number of pages (usually 32, but any multiple of 4 is acceptable).

silhouette halftone An illustration in which the background has been entirely cut away or masked.

sinkage White space, in addition to the top margin, left at the top of a page, as at the beginning of a new chapter. Sinkage throughout a book should be uniform.

skids Pallets used for delivering, storing, and moving paper, books, and so on.

slugs (1) 6- or 12-point thick leads, about ¾ inch high, used as spacing material between lines of type. (2) Lines of type cast as single metal bars on Linotype, Intertype, or Ludlow machines.

small capitals Capital letters smaller than the usual capitals of the given font, equal to the x-height of that font. Abbreviated as *s.c.* or *small caps.*

software "A term in computerized composition referring to procedures and programming as opposed to the hardware, or equipment." (*Phototypography*)

sorts　All the particular types in the boxes of a case, rather than a complete font. If few types remain in the boxes, the case is "low on sorts." When all the types in a box have been used, the box is "out of sorts."

spaceband　In a hot metal system the wedge-shaped device on a line-casting machine that automatically justifies a line of type.

spacebreaks　Spaces inserted in the text to indicate a change of subject, time, and so on. They are usually one or two lines deep.

space dot　*See* center point.

spacing　Insertion of quads of varying sizes to separate words, sentences, columns, and paragraph indentions, and to justify lines.

spine　The back of a book, connecting the front and back covers. Type set lengthwise on the spine commonly reads from the top down. Also called *backbone* or *shelfback*.

spoilage　Material lost in the production process through damage or imperfection.

spiral binding　A type of binding, common in notebooks, in which a cylindrical spiral of metal or plastic is wound through holes punched in the edges of the pages.

staircase　*See* river.

stereotype　A metal printing plate cast from a paper matrix made from a page of type.

stick　*See* compose.

stripping　"In phototypography, the assembly of film positive or negative elements for film mechanicals. Also, insertion of corrections in phototext or display." (*Phototypography*)

subheads　Headings that mark divisions of a chapter. They are subordinate to chapter heads.

subscript　A character or symbol printed partly below the base line of the text.

super　A strip of strong, thin cloth pasted over the back of the sections of a book and extending about an inch beyond the back at each side, added for reinforcement. Also called *crash*.

superior numeral　A small numeral used as a reference mark, printed above the x-height of the font. Also called *superior figure*.

superscript　A small numeral, fraction, or other symbol that prints above the x-height of the font and is used in mathematical notation.

swash letters　Ornamental italic letters used for headings and initials.

tailpiece　A small ornament or illustration at the end of a chapter.

teletypesetter　A machine that sets telegraphic news on either Linotype or Intertype machines as it comes off the wire.

text type　The size type customarily used for setting books or other large quantities. It is seldom larger than 14-point.

thirty or **30**　A term used in newspaper offices, signifying "the end."

tip in　To paste a leaf, or leaves, into printed sheets or bound books.

transparency　A black and white or full color positive on transparent film, viewed by means of transmitted light.

trim size　The final size of the whole page, including all margins, after trimming.

turned letters　A type placed feet up in composed matter to show that no type of the right letter is available. It shows in the proof as two black marks.

type cases　Shallow wooden trays divided into compartments of various sizes. There are about thirty styles of type cases, some holding a complete font, others only part of a font. The oldest, and the one most used before typesetting machines came into common use, required two cases for each font. When in use they were placed on sloping racks, one beyond the other and more steeply slanting. In the lower case were all the small letters, numerals, punctuation marks, and spaces and quads. In the upper were the capitals and small caps and a few extra characters. It was from their position in the type cases that capitals acquired the name "upper case letters," and small letters came to be called "lowercase." Matter in capitals is said to be "up," and in small letters it is "down."

type height The distance (.9186 inch in the United States) from the type's face, or printing surface, to the base on which it stands. Any object of this height is called type-high.

type (text) page That area of the page that contains all the printed matter, including footnotes, running heads, and folios; it is usually measured in picas.

type size The size of type, based on a unit of measurement called a point, in which 1 point equals $\frac{1}{72}$ of an inch, and 12 points equal 1 pica. All sizes of type and materials cast according to this system are exact multiples of this unit. Sizes are designated by points measured by the type bodies.

typographical error Commonly called a *typo,* it is an error made by the typesetter. Also known as *printer's error* or *p.e.*

upper case letters Capitals.

vandyke *See* brown.

verso pages *See* even pages.

vignette A type of finish on a halftone cut, in which the background is faded away. In a *hard vignette,* the background is broken abruptly with a very uneven outline.

web-fed press A press whose paper is fed from a continuous roll.

widow A short line at the top of a page; considered poor bookmaking.

wraparound plates Thin metal plates that are flexible enough to be wrapped around cylinders like offset plates, and are used for rotary or offset printing.

wrong font A type of a face different from accompanying letters.

wrong-reading Those prints or films that can be read correctly only with a mirror. The type reads from right to left on the emulsion side of the print.

x-height The height of the lowercase type characters such as the *x* in a font, exclusive of descenders or ascenders.

xerography *See* electrostatic printing.

INDEX

This book was composed by

WESTERN TYPESETTING and TYPOGRAPHICS

Kansas City, Missouri

Title: Words Into Type

Trim size: 6 x 9
Type page size: 26 x 44
Lines of text per page: 46
Head margin: 9/16"
Gutter: 3/4"

Text: 9/11 x 26 Optima Medium, paragraph indent 2 ems, lining figures throughout.
Extract: 8/9 x 26 Optima Medium, indent 2 ems at left, 12 pt. space above and below.
Footnote: 8 pt. x 26 Optima Medium solid, set flush, no rule, one line space above.

No. 1 head (Part title): Futura Medium—24 pt. caps centered—each Part title on right-hand page, 8 pica sinkage, 18 pt. below; two cols, balanced at top, of contents of part, set in 12 pt. Futura medium c/lc with 12 pt. Futura Bold condensed c/lc.
No. 2 head: Futura Medium—14 pt. caps—centered, 2 pi space above, 10 pt. space below.
No. 3 head: Futura Bold Condensed—10 pt. caps—centered, 1½ pi space above, 10 pt. space below.
No. 4 head: Futura Bold—9 pt., c/lc run in
No. 5 head: Optima Medium ital—9 pt., c/lc run in
No. 6 head: Futura light—8 pt. caps—centered, 1 pi space above, 10 pt. space below.

RH's: 10 pt. Optima Medium caps—fl. inside.
  Left hand page—WORDS INTO TYPE—Caps
  Right hand page—Part title—c/lc
  Folios—10 pt. fl. outside